Classifying Spaces of Degenerating Polarized Hodge Structures

Annals of Mathematics Studies

Number 169

Classifying Spaces of Degenerating Polarized Hodge Structures

Kazuya Kato and Sampei Usui

PRINCETON UNIVERSITY PRESS
PRINCETON AND OXFORD
2009

Published by Princeton University Press
41 William Street, Princeton, New Jersey 08540

In the United Kingdom: Princeton University Press
6 Oxford Street, Woodstock, Oxfordshire 0X20 1TW

Library of Congress Cataloging-in-Publication Data

Kato, K. (Kazuya)
Classifying spaces of degenerating polarized Hodge structures / Kazuya Kato and Sampei Usui.
p. cm. — (Annals of mathematics studies ; no. 169)
Includes bibliographical references and index.
ISBN 978-0-691-13821-3 (cloth : acid-free paper) — ISBN 978-0-691-13822-0 (pbk. : acid-free paper)
1. Hodge theory. 2. Logarithms. I. Usui, Sampei. II. Title.
QA564.K364 2009
514′.74—dc22 2008039091

British Library Cataloging-in-Publication Data is available

This book has been composed in LaTeX

Printed on acid-free paper. ∞

The publisher would like to thank the authors of this volume for providing the camera-ready copy from which this book was printed

press.princeton.edu

Printed in the United States of America

10 9 8 7 6 5 4 3 2 1

Suugaku wa mugen enten kou kokoro
Koute kogarete harukana tabiji

by Kazuya Kato and Sampei Usui,
which was translated by Luc Illusie as

L'impossible voyage aux points à l'in ni
N'a pas fait battre en vain le coeur du géomètre

Contents

Classifying Spaces of Degenerating Polarized Hodge Structures

Introduction

This book is the full detailed version of the paper [KU1].

In [G1], Griffiths defined and studied the classifying space D of polarized Hodge structures of fixed weight w and fixed Hodge numbers $(h^{p,q})$. In [G5], Griffiths presented a dream of adding points at infinity to D. This book is an attempt to realize his dream.

In the special case

$$w = 1, \quad h^{1,0} = h^{0,1} = g, \quad \text{other } h^{p,q} = 0, \tag{1}$$

the classifying space D coincides with Siegel's upper half space $\mathfrak{h}_g$ of degree g. If $g = 1$, $\mathfrak{h}_g$ is the Poincaré upper half plane $\mathfrak{h} = \{x + iy \mid x, y \in \mathbf{R}, y > 0\}$. For a congruence subgroup Γ of $\mathrm{SL}(2, \mathbf{Z})$, that is, for a subgroup of $\mathrm{SL}(2, \mathbf{Z})$ which contains the kernel of $\mathrm{SL}(2, \mathbf{Z}) \to \mathrm{SL}(2, \mathbf{Z}/n\mathbf{Z})$ for some integer $n \geq 1$, the quotient $\Gamma\backslash\mathfrak{h}$ is a modular curve without cusps. We obtain the compactification $\Gamma\backslash(\mathfrak{h} \cup \mathbf{P}^1(\mathbf{Q}))$ of the modular curve $\Gamma\backslash\mathfrak{h}$ by adding points at infinity called cusps (i.e., the elements of the finite set $\Gamma\backslash\mathbf{P}^1(\mathbf{Q})$). In the case (1) with g general, for a congruence subgroup Γ of $\mathrm{Sp}(g, \mathbf{Z})$, by adding points at infinity, we have toroidal compactifications of $\Gamma\backslash\mathfrak{h}_g$ [AMRT] and the Satake-Baily-Borel compactification of $\Gamma\backslash\mathfrak{h}_g$ [Sa1], [BB]. All these compactifications coincide when $g = 1$, but, when $g > 1$, there are many toroidal compactifications and the Satake-Baily-Borel compactification is different from them. The theory of these compactifications is included in a general theory of compactifications of quotients of symmetric Hermitian domains by the actions of discrete arithmetic groups. The points at infinity are often more important than the usual points. For example, the Taylor expansion of a modular form at the standard cusp (i.e., the class of $\infty \in \mathbf{P}^1(\mathbf{Q})$ modulo Γ) of the compactified modular curve $\Gamma\backslash(\mathfrak{h} \cup \mathbf{P}^1(\mathbf{Q}))$ is called the q-expansion and is very important in the theory of modular forms.

However, the classifying space D in general is rarely a symmetric Hermitian domain, and we cannot use the general theory of symmetric Hermitian domains when we try to add points at infinity to D. In this book, we overcome this difficulty. We discuss two subjects in this book.

SUBJECT I. TOROIDAL PARTIAL COMPACTIFICATIONS AND MODULI OF POLARIZED LOGARITHMIC HODGE STRUCTURES

A toroidal compactification of $\Gamma\backslash\mathfrak{h}_g$ is defined depending on the choice of a certain fan (cone decomposition). If the fan is not sufficiently big, we have a toroidal partial compactification of $\Gamma\backslash\mathfrak{h}_g$, which need not be compact and which is locally isomorphic to an open set of a toroidal compactification.

In this book, for general D, we construct a kind of toroidal partial compactification $\Gamma\backslash D_\Sigma$ of $\Gamma\backslash D$ associated with a fan Σ and a discrete subgroup Γ of $\mathrm{Aut}(D)$ satisfying a certain compatibility with Σ.

In the case (1), the classes of polarized Hodge structures in $\Gamma\backslash\mathfrak{h}_g$ converge to a point at infinity of $\Gamma\backslash\mathfrak{h}_g$ when the polarized Hodge structures become degenerate. As in [Sc], nilpotent orbits appear when polarized Hodge structures become degenerate. In our definition of D_Σ for general D, a nilpotent orbit itself is viewed as a point at infinity.

As is discussed in detail in this book, the theory of nilpotent orbits is regarded as a local aspect of the theory of polarized logarithmic Hodge structures. The "polarized logarithmic Hodge structure" (PLH) is formulated by using the theory of logarithmic structures introduced by Fontaine and Illusie and developed in [Kk1],[KkNc], and it is something like the logarithmic degeneration of the PH (polarized Hodge structure).

We give here a rough illustration of the idea of the PLH. Let X be a complex manifold endowed with a divisor Y with normal crossings, and let $U = X - Y$. Let H be a PLH on X with respect to the "logarithmic structure of X associated with Y." Then the restriction $H|_U$ of H to U is a family of usual PH parametrized by U. At $x \in U$, the fiber $H(x)$ of H is a usual PH. At Y, this family can become degenerate in the classical sense. At each point $x \in Y$, the fiber of H at x corresponds to a nilpotent orbit (in the classical theory) toward $x \in Y$. Schematically, we have the following:

$$
\begin{array}{ccc}
(H(x) : \text{a PH}) & = & (H(x) : \text{a PLH}) \\
{\scriptstyle \text{a fiber over a point } x \in U}\uparrow & & {\scriptstyle \text{a fiber over a point } x \in U}\uparrow \\
(H|_U : \text{a family of PH on } U) & \xrightarrow{\text{extension}} & (H : \text{PLH on } X) \\
{\scriptstyle \text{degeneration toward } x \in Y}\downarrow & & {\scriptstyle \text{a fiber over a point } x \in Y}\downarrow \\
(\text{a nilpotent orbit toward } x) & = & (H(x) : \text{a PLH}).
\end{array}
\tag{2}
$$

(See 0.2.20, 0.4.25.)

Our main theorem concerning Subject I is stated roughly as follows (for the precise statement, see Theorem 0.4.27 below).

Theorem. *$\Gamma\backslash D_\Sigma$ is the fine moduli space of "polarized logarithmic Hodge structures" with a "Γ-level structure" whose "local monodromies are in the directions in Σ."*

Roughly speaking,

$$
\begin{aligned}
&\Gamma\backslash D = (\text{polarized Hodge structures with a "}\Gamma\text{-level structure"}) \\
&\cap \\
&\Gamma\backslash D_\Sigma = \Gamma\backslash\{\sigma\text{-nilpotent orbit} \mid \sigma \in \Sigma\} \\
&\qquad = \begin{pmatrix} \text{"polarized logarithmic Hodge structures" with a "}\Gamma\text{-level structure"} \\ \text{whose "local monodromies are in the directions in } \Sigma.\text{"} \end{pmatrix}
\end{aligned}
$$

Here, a σ-nilpotent orbit is a nilpotent orbit in the direction of the cone σ. For $\sigma = \{0\}$, a σ-nilpotent orbit is nothing but a point of D; hence we can regard $D \subset D_\Sigma$.

In the classical case (1), $\Gamma\backslash D_\Sigma$ for a congruence subgroup Γ and for a sufficiently big Σ is a toroidal compactification of $\Gamma\backslash D$. Already, in this classical case, this theorem gives moduli-theoretic interpretations of the toroidal compactifications of $\Gamma\backslash \mathfrak{h}_g$.

The space $\Gamma\backslash D_\Sigma$ has a kind of complex structure, but a delicate point is that, in general, this space can have locally the shape of a "complex analytic space with a slit" (for example, $\mathbf{C}^2$ minus $\{(0, z) \mid z \in \mathbf{C}, z \neq 0\}$), and hence it is often not locally compact. However, it is very close to a complex analytic manifold. $\Gamma\backslash D_\Sigma$ is a logarithmic manifold in the sense of 0.4.17 below. Infinitesimal calculus can be performed nicely on $\Gamma\backslash D_\Sigma$. These phenomena were first examined in the easiest nontrivial case in [U2].

One motivation of Griffiths for adding points at infinity to D was the hope that the period map $\Delta^* \to \Gamma\backslash D$ ($\Delta^* = \{q \in \mathbf{C} \mid 0 < |q| < 1\}$) associated with a variation of polarized Hodge structure on Δ^* could be extended to $\Delta \to \Gamma\backslash(D \cup$ (points at infinity)) ($\Delta = \{q \in \mathbf{C} \mid |q| < 1\}$). By using the above main theorem and the nilpotent orbit theorem of Schmid, we can actually extend the period map to $\Delta \to \Gamma\backslash D_\Sigma$ for some suitable Σ (see 0.4.30 and 4.3.1, where a more general result is given).

SUBJECT II. THE EIGHT ENLARGEMENTS OF D AND THE FUNDAMENTAL DIAGRAM

In the classical case (1) above, there is another compactification $\Gamma\backslash D_{\mathrm{BS}}$ of $\Gamma\backslash D$ (Γ is a congruence subgroup of $\mathrm{Sp}(g, \mathbf{Z})$) called the Borel-Serre compactification, where D_{BS} is the Borel-Serre space denoted by $\bar{D}$ in [BS], which is a real manifold with corners containing D as a dense open set.

For general D, by adding to D points at infinity of different kinds, we obtain eight enlargements of D with maps among them which form the following fundamental diagram (3) (see 5.0.1).

Fundamental Diagram

$$\begin{array}{ccccccc}
 & & & & D_{\mathrm{SL}(2),\mathrm{val}} & \hookrightarrow & D_{\mathrm{BS},\mathrm{val}} \\
 & & & & \downarrow & & \downarrow \\
D_{\Sigma,\mathrm{val}} & \leftarrow & D^{\sharp}_{\Sigma,\mathrm{val}} & \rightarrow & D_{\mathrm{SL}(2)} & & D_{\mathrm{BS}} \\
\downarrow & & \downarrow & & & & \\
D_\Sigma & \leftarrow & D^{\sharp}_\Sigma & & & &
\end{array} \tag{3}$$

Note that the space D_Σ that appeared in Subject I sits at the left lower end of this diagram. The left-hand side of this diagram has Hodge-theoretic nature, and the right-hand side has the nature of the theory of algebraic groups. These are related by the middle map $D^\sharp_{\Sigma,\mathrm{val}} \to D_{\mathrm{SL}(2)}$, which is a geometric interpretation of the SL(2)-orbit theorem of Cattani-Kaplan-Schmid [CKS].

In the case (1) with $g = 1$, the largest Σ exists. For this Σ, $D_\Sigma = \mathfrak{h} \cup \mathbf{P}^1(\mathbf{Q})$, and the above diagram becomes

$$\begin{array}{ccccccc} & & & & \mathfrak{h}_{\mathrm{BS}} & = & \mathfrak{h}_{\mathrm{BS}} \\ & & & & \| & & \| \\ \mathfrak{h} \cup \mathbf{P}^1(\mathbf{Q}) & \leftarrow & \mathfrak{h}_{\mathrm{BS}} & = & \mathfrak{h}_{\mathrm{BS}} & = & \mathfrak{h}_{\mathrm{BS}} \\ \| & & \| & & & & \\ \mathfrak{h} \cup \mathbf{P}^1(\mathbf{Q}) & \leftarrow & \mathfrak{h}_{\mathrm{BS}} & & & & \end{array} \tag{4}$$

The space $\mathfrak{h}_{\mathrm{BS}}$ is described as follows. It is the union of open subsets $\mathfrak{h}_{\mathrm{BS}}(a)$ for $a \in \mathbf{P}^1(\mathbf{Q})$. $\mathfrak{h}_{\mathrm{BS}}(\infty) = \{x + iy \mid x \in \mathbf{R}, 0 < y \le \infty\} \supset \mathfrak{h} = \{x + iy \mid x \in \mathbf{R}, y > 0\}$. The action of $\mathrm{SL}(2,\mathbf{Q})$ on $\mathfrak{h}$ extends to a continuous action of $\mathrm{SL}(2,\mathbf{Q})$ on $\mathfrak{h}_{\mathrm{BS}}$, and we have $g(\mathfrak{h}_{\mathrm{BS}}(a)) = \mathfrak{h}_{\mathrm{BS}}(ga)$ for $g \in \mathrm{SL}(2,\mathbf{Q})$ and $a \in \mathbf{P}^1(\mathbf{Q})$. In particular, all $\mathfrak{h}_{\mathrm{BS}}(a)$ are homeomorphic to each other. The map $\mathfrak{h}_{\mathrm{BS}} \to \mathfrak{h}_\Sigma = \mathfrak{h} \cup \mathbf{P}^1(\mathbf{Q})$ for the biggest Σ is the identity map on $\mathfrak{h}$ and sends elements of $\mathfrak{h}_{\mathrm{BS}}(a) - \mathfrak{h}$ to a for $a \in \mathbf{P}^1(\mathbf{Q})$.

In the case (1) for general g, the fundamental diagram becomes

$$\begin{array}{ccccccc} & & & & D_{\mathrm{SL}(2),\mathrm{val}} & = & D_{\mathrm{BS},\mathrm{val}} \\ & & & & \downarrow & & \downarrow \\ D_{\Sigma,\mathrm{val}} & \leftarrow & D^\sharp_{\Sigma,\mathrm{val}} & \rightarrow & D_{\mathrm{SL}(2)} & = & D_{\mathrm{BS}} \\ \downarrow & & \downarrow & & & & \\ D_\Sigma & \leftarrow & D^\sharp_\Sigma & & & & \end{array} \tag{5}$$

In this case, for a subgroup Γ of $\mathrm{Sp}(g,\mathbf{Z})$ of finite index and for a suitable Σ, $\Gamma\backslash D_\Sigma$ is a toroidal compactification [AMRT] of $\Gamma\backslash D$, $\Gamma\backslash D_{\Sigma,\mathrm{val}}$ is obtained from $\Gamma\backslash D$ by blow-ups, and the maps $\Gamma\backslash D^\sharp_\Sigma \to \Gamma\backslash D_\Sigma$ and $\Gamma\backslash D^\sharp_{\Sigma,\mathrm{val}} \to \Gamma\backslash D_{\Sigma,\mathrm{val}}$ are proper surjective maps whose fibers are products of finite copies of $\mathbf{S}^1$. On the other hand, $\Gamma\backslash D_{\mathrm{BS}}$ is the Borel-Serre compactification [BS] of $\Gamma\backslash D = \Gamma\backslash\mathfrak{h}_g$. The spaces D_{BS} and $D^\sharp_\Sigma$ are real manifolds with corners (they are like $\mathbf{R}^m \times \mathbf{R}^n_{\ge 0}$ locally). Already, in this classical case, the fundamental diagram (5) gives a relation between toroidal compactifications of $\Gamma\backslash\mathfrak{h}_g$ and the Borel-Serre compactification of $\Gamma\backslash\mathfrak{h}_g$, which were not known before (see 0.5.28 below).

For general D, these eight spaces are defined as

D_Σ = (the space of nilpotent orbits) (1.3.8),
$D_\Sigma^\sharp$ = (the space of nilpotent i-orbits) (1.3.8),
$D_{\mathrm{SL}(2)}$ = (the space of SL(2)-orbits) (5.2.6),
D_{BS} = (the space of Borel-Serre orbits) (5.1.5),
$D_{\Sigma,\mathrm{val}}$ = (the space of valuative nilpotent orbits) (5.3.5),
$D_{\Sigma,\mathrm{val}}^\sharp$ = (the space of valuative nilpotent i-orbits) (5.3.5),
$D_{\mathrm{SL}(2),\mathrm{val}}$ = (the space of valuative SL(2)-orbits) (5.2.7),
$D_{\mathrm{BS},\mathrm{val}}$ = (the space of valuative Borel-Serre orbits) (5.1.6).

The space D_{BS} was constructed in [KU2] and [BJ] independently, by using the work [BS] of Borel-Serre on Borel-Serre compacifications. The spaces $D_{\mathrm{SL}(2)}$, $D_{\mathrm{SL}(2),\mathrm{val}}$, and $D_{\mathrm{BS},\mathrm{val}}$ are defined in [KU2].

Roughly speaking, these eight spaces appear as follows:

$\Gamma\backslash D_\Sigma$ is like an analytic manifold with slits,
$D_\Sigma^\sharp$ and $D_{\mathrm{SL}(2)}$ are like real manifolds with corners and slits,
D_{BS} is a real manifold with corners,
$\Gamma\backslash D_{\Sigma,\mathrm{val}}$ and $D_{\Sigma,\mathrm{val}}^\sharp$ are the projective limits of "blow-ups" of $\Gamma\backslash D_\Sigma$ and $D_\Sigma^\sharp$, respectively, associated with rational subdivisions of Σ,
$D_{\mathrm{SL}(2),\mathrm{val}}$ and $D_{\mathrm{BS},\mathrm{val}}$ are the projective limits of certain "blow-ups" of $D_{\mathrm{SL}(2)}$ and D_{BS}, respectively.

The maps $\Gamma\backslash D_\Sigma^\sharp \to \Gamma\backslash D_\Sigma$ and $\Gamma\backslash D_{\Sigma,\mathrm{val}}^\sharp \to \Gamma\backslash D_{\Sigma,\mathrm{val}}$ are proper surjective maps whose fibers are products of a finite number of copies of $\mathbf{S}^1$, where this number is varying.

Like nilpotent orbits, SL(2)-orbits also appear in the theory of degenerations of polarized Hodge structures [Sc], [CKS]. The fundamental diagram (3) shows how nilpotent orbits, SL(2)-orbits, and the theory of Borel and Serre are related.

In this book, we study all these eight spaces. To prove the main theorem in Subject I and to prove that $\Gamma\backslash D_\Sigma$ has good properties such as the Hausdorff property, nice infinitesimal calculus, etc., we need to consider all other spaces in the diagram (3); we discuss the spaces from the right to the left in the fundamental diagram (3) to deduce the nice properties of $\Gamma\backslash D_\Sigma$, starting from the properties of the Borel-Serre compactifications (which were proved in [BS] by using arithmetic theory of algebraic groups).

The organization of this book is as follows. In Chapter 0, we give an overview of the book. In Chapters 1–4, we formulate the main theorem in Subject I. In Chapters 5–8, we prove the main theorem, considering all eight enlargements of D in the fundamental diagram (3), and we also prove various properties of the eight enlargements. In Chapters 9–12, we give complementary results.

The authors are grateful to Professor Chikara Nakayama for ongoing discussions and constant encouragement. He carefully read various versions of the manuscript and sent us lists of useful comments and advice. In particular, we owe him very heavily for the proof of Theorem 4.3.1. The authors are also grateful to Professor

Kazuhiro Fujiwara for stimulating discussions and advice and to Professor Akira Ohbuchi for inputting the figures in the electronic file.

Parts of this work were done when the first author was a visitor at Institut Henri Poincaré and when the second author was a visitor at Institute for Advanced Study; the hospitality of each is gratefully appreciated.

The first line of the Japanese poem (5-7-5 syllables) in the book's epigraph was composed by Kato and then, following a Japanese tradition of collaboration, the second line (7-7 syllables) was composed by Usui. We are very grateful to Professor Luc Illusie for his beautiful translation.

This work was partly supported by Grants-in-Aid for Scientific Research (B) (2) No. 11440003, (B) No. 14340010, (B) No. 16340005; (A) (1) No. 11304001, (B) No. 19340008, (B) No. 15340009 from Japan Society for the Promotion of Science.

Chapter Zero

Overview

In this chapter, we introduce the main ideas and results of this book.

In Section 0.1, we review the basic idea of Hodge theory. In Section 0.2, we introduce the basic idea of logarithmic Hodge theory. In Section 0.3, we review classifying spaces D of Griffiths (i.e., Griffiths domains) as the moduli spaces of polarized Hodge structures. In Section 0.4, we describe our toroidal partial compactifications of the classifying spaces of Griffiths and our result that they are the fine moduli spaces of polarized logarithmic Hodge structures. In Section 0.5, we describe the other seven enlargements of D in the fundamental diagram (3) in Introduction and state our results on these spaces.

In this chapter, we explain the above subjects by presenting examples. Hodge theory (Section 0.1) and logarithmic Hodge theory (Section 0.2) are explained by using the example of the Hodge structure on $H^1(E, \mathbf{Z})$ of an elliptic curve E and its degeneration arising from the degeneration of E. This example appears first in 0.1.3 and then continues to appear as an example of each subject. The classifying space D (Section 0.3) and its various enlargements (Sections 0.4 and 0.5) are explained by using the following three examples: (i) $D = \mathfrak{h}$, the upper half plane; (ii) $D = \mathfrak{h}_g$, Siegel's upper half space ((i) is a special case of (ii). In the case (ii), we mainly consider the case $g = 2$.); (iii) an example of weight 2 for which D is not a symmetric Hermitian domain. These examples appear first in 0.3.2 and then continue to appear as examples of each subject.

In this chapter, we do not generally give proofs.

0.1 HODGE THEORY

0.1.1

First we recall the basic idea of Hodge theory.

For a topological space X, the homology groups $H_m(X, \mathbf{Z})$ and the cohomology groups $H^m(X, \mathbf{Z})$ are important invariants of X.

If X is a projective complex analytic manifold, the cohomology groups $H^m(X, \mathbf{Z})$ have finer structures: $\mathbf{C} \otimes_{\mathbf{Z}} H^m(X, \mathbf{Z}) = H^m(X, \mathbf{C})$ is endowed with a decreasing filtration $F = (F^p)_{p \in \mathbf{Z}}$, called Hodge filtration.

The cohomology group $H^m(X, \mathbf{Z})$ remembers X merely as a topological space, but, with this Hodge filtration, the pair $(H^m(X, \mathbf{Z}), F)$ becomes a finer invariant of X which remembers the analytic structure of X (not just the topological structure of X) often very well.

0.1.2

For example, in the case $m = 1$, the Hodge filtration F on $H^1(X, \mathbf{C})$ is given by $F^p = H^1(X, \mathbf{C})$ for $p \leq 0$, $F^p = 0$ for $p \geq 2$, and F^1 is the image of the injective map

$$H^0(X, \Omega^1_X) \to H^1(X, \mathbf{C}). \qquad (1)$$

Here $H^0(X, \Omega^1_X)$ is the space of holomorphic differential forms on X, and (1) is the map that sends a differential form $\omega \in H^0(X, \Omega^1_X)$ to its cohomology class in $H^1(X, \mathbf{C})$: Under the identification $H^1(X, \mathbf{C}) = \mathrm{Hom}(H_1(X, \mathbf{Z}), \mathbf{C})$, the cohomology class of ω is given by $\gamma \mapsto \int_\gamma \omega \quad (\gamma \in H_1(X, \mathbf{Z}))$.

For the definition of F^p of $H^m(X, \mathbf{C})$ for general p, m, see 0.1.7.

0.1.3

Elliptic curves. An elliptic curve X over $\mathbf{C}$ is isomorphic to $\mathbf{C}/(\mathbf{Z}\tau + \mathbf{Z})$ for some $\tau \in \mathfrak{h}$, where $\mathfrak{h}$ is the upper half plane. For $X = \mathbf{C}/(\mathbf{Z}\tau + \mathbf{Z})$ with $\tau \in \mathfrak{h}$, $H_1(X, \mathbf{Z})$ is identified with $\mathbf{Z}\tau + \mathbf{Z}$, $H^1(X, \mathbf{Z})$ is identified with $\mathrm{Hom}(\mathbf{Z}\tau + \mathbf{Z}, \mathbf{Z})$, and the Hodge filtration on $H^1(X, \mathbf{C}) = \mathrm{Hom}(\mathbf{Z}\tau + \mathbf{Z}, \mathbf{C})$ is described as follows. The space $H^0(X, \Omega^1_X)$ is a one-dimensional $\mathbf{C}$-vector space with the basis dz, where z is the coordinate function of $\mathbf{C}$, and where we regard dz as a differential form on the quotient space $X = \mathbf{C}/(\mathbf{Z}\tau + \mathbf{Z})$ of $\mathbf{C}$. Let $(\gamma_j)_{j=1,2}$ be the $\mathbf{Z}$-basis of $H_1(X, \mathbf{Z})$ that is identified with the $\mathbf{Z}$-basis $(\tau, 1)$ of $\mathbf{Z}\tau + \mathbf{Z}$, and let $(e_j)_{j=1,2}$ be the dual $\mathbf{Z}$-basis of $H^1(X, \mathbf{Z})$. Since $\int_{\gamma_1} dz = \tau$ and $\int_{\gamma_2} dz = 1$, the cohomology class of dz coincides with $\tau e_1 + e_2$, and hence $F^1 H^1(X, \mathbf{C})$ is the $\mathbf{C}$-subspace of $H^1(X, \mathbf{C})$ generated by $\tau e_1 + e_2$.

0.1.4

Elliptic curves (continued). If X is an elliptic curve, we cannot recover X merely from $H^1(X, \mathbf{Z})$. In fact, $H^1(X, \mathbf{Z}) \simeq \mathbf{Z}^2$ for any elliptic curve X over $\mathbf{C}$, and we cannot distinguish different elliptic curves from this information. However, if we consider the Hodge filtration, we can recover X from $(H^1(X, \mathbf{Z}), F)$ as

$$X \simeq \mathrm{Hom}_{\mathbf{C}}(F^1, \mathbf{C})/H_1(X, \mathbf{Z}) = \mathrm{Hom}_{\mathbf{C}}(F^1, \mathbf{C})/\mathrm{Hom}(H^1(X, \mathbf{Z}), \mathbf{Z}). \qquad (1)$$

Here $H_1(X, \mathbf{Z})$ is embedded in $\mathrm{Hom}_{\mathbf{C}}(F^1, \mathbf{C})$ via the map $\gamma \mapsto (\omega \mapsto \int_\gamma \omega)$ ($\gamma \in H_1(X, \mathbf{Z})$, $\omega \in \Gamma(X, \Omega^1_X)$), and the isomorphism $X \simeq \mathrm{Hom}_{\mathbf{C}}(F^1, \mathbf{C})/H_1(X, \mathbf{Z})$ sends $x \in X$ to the class of the homomorphism $F^1 \to \mathbf{C}$, $\omega \mapsto \int_\gamma \omega$, where γ is a path in X from the origin 0 of X to x (the choice of γ is not unique, but the class of the map $\omega \mapsto \int_\gamma \omega$ modulo $H_1(X, \mathbf{Z})$ is independent of the choice of γ). In (1), the middle group is identified with the right one in which $\mathrm{Hom}(H^1(X, \mathbf{Z}), \mathbf{Z})$ is embedded in $\mathrm{Hom}_{\mathbf{C}}(F^1, \mathbf{C})$ via the composition $\mathrm{Hom}(H^1(X, \mathbf{Z}), \mathbf{Z}) \to \mathrm{Hom}_{\mathbf{C}}(H^1(X, \mathbf{C}), \mathbf{C}) \to \mathrm{Hom}_{\mathbf{C}}(F^1, \mathbf{C})$, which is injective.

If $X = \mathbf{C}/(\mathbf{Z}\tau + \mathbf{Z})$ with $\tau \in \mathfrak{h}$, this isomorphism $X \simeq \mathrm{Hom}_{\mathbf{C}}(F^1, \mathbf{C})/H_1(X, \mathbf{Z})$ is nothing but the original presentation $X = \mathbf{C}/(\mathbf{Z}\tau + \mathbf{Z})$ where $\mathrm{Hom}_{\mathbf{C}}(F^1, \mathbf{C})$ is identified with $\mathbf{C}$ by the evaluation at dz.

0.1.5

Now we discuss Hodge structures.

A *Hodge structure* of weight w is a pair $(H_{\mathbf{Z}}, F)$ consisting of a free $\mathbf{Z}$-module $H_{\mathbf{Z}}$ of finite rank and of a decreasing filtration F on $H_{\mathbf{C}} := \mathbf{C} \otimes_{\mathbf{Z}} H_{\mathbf{Z}}$ (that is, a family $(F^p)_{p\in\mathbf{Z}}$ of $\mathbf{C}$-subspaces of $H_{\mathbf{C}}$ such that $F^p \supset F^{p+1}$ for all p), which satisfies the following condition (1):

$$H_{\mathbf{C}} = \bigoplus_{p+q=w} H^{p,q}, \quad \text{where } H^{p,q} = F^p \cap \bar{F}^q. \tag{1}$$

Here $\bar{F}^q$ denotes the image of F^q under the complex conjugation $H_{\mathbf{C}} \to H_{\mathbf{C}}$, $a \otimes x \mapsto \bar{a} \otimes x$ ($a \in \mathbf{C}$, $x \in H_{\mathbf{Z}}$).

We have

$$F^p = \bigoplus_{p'\geq p} H^{p',w-p'}, \quad F^p/F^{p+1} \simeq H^{p,w-p}. \tag{2}$$

We say $(H_{\mathbf{Z}}, F)$ is of Hodge type $(h^{p,q})_{p,q\in\mathbf{Z}}$, where $h^{p,q} = \dim_{\mathbf{C}} H^{p,q}$ if $p + q = w$, and $h^{p,q} = 0$ otherwise (these numbers $h^{p,q}$ are called the *Hodge numbers*).

0.1.6

For a projective analytic manifold X and for $m \in \mathbf{Z}$, the pair $(H_{\mathbf{Z}}, F)$ with $H_{\mathbf{Z}} = H^m(X, \mathbf{Z})/(\text{torsion})$ and F the Hodge filtration becomes a Hodge structure of weight m.

For example, if X is the elliptic curve $\mathbf{C}/(\mathbf{Z}\tau + \mathbf{Z})$ with $\tau \in \mathfrak{h}$, then $H^{1,0} = \mathbf{C}(\tau e_1 + e_2)$, $H^{0,1} = \mathbf{C}(\bar{\tau} e_1 + e_2)$, and $H^1(X, \mathbf{C}) = H^{1,0} \oplus H^{0,1}$ since $\tau \neq \bar{\tau}$.

The theory of homology groups and cohomology groups is important in the study of topological spaces. Similarly, Hodge theory (the theory of Hodge structures) is important for the study of analytic spaces.

0.1.7

For a projective analytic manifold X, the Hodge filtration F^p on $H^m(X, \mathbf{C})$ is defined as follows. Let

$$\Omega_X^\bullet = (\mathcal{O}_X \xrightarrow{d} \Omega_X^1 \xrightarrow{d} \Omega_X^2 \xrightarrow{d} \cdots)$$

be the de Rham complex of X where $\Omega_X^p = \bigwedge_{\mathcal{O}_X}^p \Omega_X^1$ is the sheaf of holomorphic p-forms on X ($\mathcal{O}_X$ is set in degree 0). Let $\Omega_X^{\geq p}$ be the degree $\geq p$ part of $\Omega_X^\bullet$. Then F^p is defined as

$$F^p := H^m(X, \Omega_X^{\geq p}) \hookrightarrow H^m(X, \Omega_X^\bullet) \simeq H^m(X, \mathbf{C}).$$

Here $H^m(X, \Omega_X^{\geq p})$ and $H^m(X, \Omega_X^\bullet)$ denote the mth hypercohomology groups of complexes of sheaves. The canonical homomorphism $H^m(X, \Omega_X^{\geq p}) \to H^m(X, \Omega_X^\bullet)$ is known to be injective. The isomorphism $H^m(X, \Omega_X^\bullet) \simeq H^m(X, \mathbf{C})$ comes from the exact sequence of Dolbeault

$$0 \to \mathbf{C} \to \mathcal{O}_X \xrightarrow{d} \Omega_X^1 \xrightarrow{d} \Omega_X^2 \xrightarrow{d} \cdots.$$

We have isomorphisms

$$H^{p,m-p} \simeq F^p/F^{p+1} \simeq H^{m-p}(X, \Omega_X^p),$$

where the second isomorphism is obtained by applying $H^m(X, \ )$ to the exact sequence of complexes of sheaves $0 \to \Omega_X^{\geq p+1} \to \Omega_X^{\geq p} \to \Omega_X^p[-p] \to 0$.

0.1.8

It is often important to consider a *polarized Hodge structure*, that is, a Hodge structure endowed with a polarization.

A *polarization on a Hodge structure* $(H_{\mathbf{Z}}, F)$ *of weight* w is a nondegenerate bilinear form $\langle \ , \ \rangle : H_{\mathbf{Q}} \times H_{\mathbf{Q}} \to \mathbf{Q}$ ($H_{\mathbf{Q}} := \mathbf{Q} \otimes_{\mathbf{Z}} H_{\mathbf{Z}}$) which is symmetric if w is even and is antisymmetric if w is odd, satisfying the following conditions (1) and (2).

(1) $\langle F^p, F^q \rangle = 0$ for $p+q > w$.

(2) Let $C_F : H_{\mathbf{C}} \to H_{\mathbf{C}}$ be the $\mathbf{C}$-linear map defined by $C_F(x) = i^{p-q}x$ for $x \in H^{p,q}$. Then the Hermitian form $(\ ,\ )_F : H_{\mathbf{C}} \times H_{\mathbf{C}} \to \mathbf{C}$, defined by $(x, y)_F = \langle C_F(x), \bar{y} \rangle$, is positive definite.

Here, in (2), $\langle \ , \ \rangle$ is regarded as the natural extension to the $\mathbf{C}$-bilinear form. The Hermitian form $(\ ,\ )_F$ in (2) is called the *Hodge metric* associated with F. The condition (1) (resp. (2)) is called the *Riemann-Hodge first* (resp. *second*) *bilinear relation*.

0.1.9

For a projective analytic manifold X, we have the intersection form

$$\langle \ , \ \rangle \colon H^m(X, \mathbf{Q}) \times H^m(X, \mathbf{Q}) \to \mathbf{Q}$$

induced by an ample line bundle on X (see [G1], [GH]). The triple $(H^m(X, \mathbf{Z}), \langle \ , \ \rangle, F)$ becomes a polarized Hodge structure.

Elliptic curves (continued)

For the elliptic curve $X = \mathbf{C}/(\mathbf{Z}\tau + \mathbf{Z})$ ($\tau \in \mathfrak{h}$), the standard polarization of X gives the antisymmetric pairing $\langle \ , \ \rangle : H^1(X, \mathbf{Q}) \times H^1(X, \mathbf{Q}) \to \mathbf{Q}$ characterized by $\langle e_2, e_1 \rangle = 1$. This pairing is nothing but the cup product $H^1(X, \mathbf{Q}) \times H^1(X, \mathbf{Q}) \to H^2(X, \mathbf{Q}) \simeq \mathbf{Q}$. It satisfies

$$\langle \tau e_1 + e_2, \tau e_1 + e_2 \rangle = 0,$$

$$(\tau e_1 + e_2, \tau e_1 + e_2)_F = i^{1-0} \langle \tau e_1 + e_2, \bar{\tau} e_1 + e_2 \rangle = i(\bar{\tau} - \tau) = 2\,\mathrm{Im}(\tau) > 0.$$

Hence $(H^1(X, \mathbf{Z}), \langle \ , \ \rangle, F)$ is indeed a polarized Hodge structure.

0.1.10

It is often very useful to consider analytic families of Hodge structures and of polarized Hodge structures.

Let X be an analytic manifold.

After Griffiths ([G3], also [D2], [Sc]), a *variation of Hodge structure* (VH) *on* X *of weight* w is a pair $(H_{\mathbf{Z}}, F)$ consisting of a locally constant sheaf $H_{\mathbf{Z}}$ of free $\mathbf{Z}$-modules of finite rank on X and of a decreasing filtration F of $H_{\mathcal{O}} := \mathcal{O}_X \otimes_{\mathbf{Z}} H_{\mathbf{Z}}$ by $\mathcal{O}_X$-submodules which satisfy the following three conditions:

(1) $F^p = H_{\mathcal{O}}$ for $p \ll 0$, $F^p = 0$ for $p \gg 0$, and $\mathrm{gr}_F^p = F^p/F^{p+1}$ is a locally free $\mathcal{O}_X$-module for any p.
(2) For any $x \in X$, the fiber $(H_{\mathbf{Z},x}, F(x))$ is a Hodge structure of weight w.
(3) $(d \otimes 1_{H_{\mathbf{Z}}})(F^p) \subset \Omega_X^1 \otimes_{\mathcal{O}_X} F^{p-1}$ for all p.

Here (3) is called the *Griffiths transversality*.

A *polarization of a variation of Hodge structure* $(H_{\mathbf{Z}}, F)$ *of weight* w on X is a bilinear form $\langle\ ,\ \rangle : H_{\mathbf{Q}} \times H_{\mathbf{Q}} \to \mathbf{Q}$ which yields for each $x \in X$ a polarization $\langle\ ,\ \rangle_x$ on the fiber $(H_{\mathbf{Z},x}, F(x))$. In this case, the triple $(H_{\mathbf{Z}}, \langle\ ,\ \rangle, F)$ is called a *variation of polarized Hodge structure* (VPH).

0.1.11

Let X and Y be analytic manifolds and $f : Y \to X$ be a projective, smooth morphism. Then for each $m \in \mathbf{Z}$, we obtain a VH (variation of Hodge structure) of weight m on X:

$$H_{\mathbf{Z}} = R^m f_* \mathbf{Z}/(\text{torsion}),$$
$$F^p = R^m f_*(\Omega_{Y/X}^{\geq p}) \hookrightarrow R^m f_*(\Omega_{Y/X}^{\bullet}) \simeq \mathcal{O}_X \otimes_{\mathbf{Z}} H_{\mathbf{Z}}.$$

Indeed, on each fiber $Y_x := f^{-1}(x)$ $(x \in X)$, $H_{\mathbf{Z},x} = H^m(Y_x, \mathbf{Z})/(\text{torsion})$ and $F^p(x) = H^m(Y_x, \Omega_{Y_x}^{\geq p})$ form the Hodge structure.

If we fix a polarization of Y over X, this VH becomes a VPH (variation of polarized Hodge structure) on X ([G3]; cf. also [Sc], [GH]).

0.2 LOGARITHMIC HODGE THEORY

From now on, we discuss degeneration of Hodge structures by the method of logarithmic Hodge theory. The logarithmic Hodge theory uses the magic of the theory of logarithmic structures introduced by Fontaine-Illusie. It has a strong connection with the theory of nilpotent orbits as discussed in Section 0.4.

In the story of Beauty and the Beast, the Beast becomes a nice man because of the love of the heroine. Similarly, a degenerate object becomes a nice object because of the magic of LOG.

$$\begin{array}{ccc} \text{(Beast)} & \xrightarrow{\text{Love Of Girl}} & \text{(a nice man)}, \\ \text{(degenerate object)} & \xrightarrow{\text{LOG}} & \text{(a nice object)}. \end{array}$$

The authors learned this mysterious coincidence of letters from Takeshi Saito.

0.2.1

Elliptic curves (continued). We observe what happens for the Hodge structure on H^1 of an elliptic curve when the elliptic curve degenerates. In this section, the idea of logarithmic Hodge theory is explained by use of this example.

Let $\Delta = \{q \in \mathbf{C} \mid |q| < 1\}$ be the unit disc. Then we have a standard family of degenerating elliptic curve

$$f : E \to \Delta,$$

which is a morphism of analytic manifolds, having the following property.

(1) For $q \in \Delta$ with $q \neq 0$, $f^{-1}(q) = \mathbf{C}^\times / q^{\mathbf{Z}}$. This is an elliptic curve. In fact, taking $\tau \in \mathbf{C}$ with $q = \exp(2\pi i \tau)$, we have $\tau \in \mathfrak{h}$ (since $|q| < 1$), and

$$\mathbf{C}/(\mathbf{Z}\tau + \mathbf{Z}) \xrightarrow{\sim} \mathbf{C}^\times / q^{\mathbf{Z}}, \quad (z \bmod (\mathbf{Z}\tau + \mathbf{Z})) \mapsto (\exp(2\pi i z) \bmod q^{\mathbf{Z}}).$$

(2) $f^{-1}(0) = \mathbf{P}^1(\mathbf{C})/(0 \sim \infty)$.

The definition of E will be given in 0.2.10 below. E is a two-dimensional analytic manifold and $f^{-1}(0)$ is a divisor with normal crossings on E. These look like the left-hand side of Figure 1. This family degenerates at $q = 0$ as is described on the left-hand side of Figure 1. All the fibers $f^{-1}(q)$ for $q \in \Delta^*$ are homeomorphic to the surface of a doughnut, whereas the central fiber $f^{-1}(0)$ has a degenerate shape.

However, as we will see below, the central fiber recovers its lost body as in the right-hand side of Figure 1 by the magic of its logarithmic structure. We will explain this magical process in the following.

$$(z_1, z_2) = (r_1 u_1, r_2 u_2) \; \leftmapsto \; (r_1, r_2, u_1, u_2)$$
$$\updownarrow \qquad\qquad\qquad \updownarrow$$
$$q = z_1 z_2 = sv \; \leftmapsto \; (s, v) = (r_1 r_2, u_1 u_2)$$

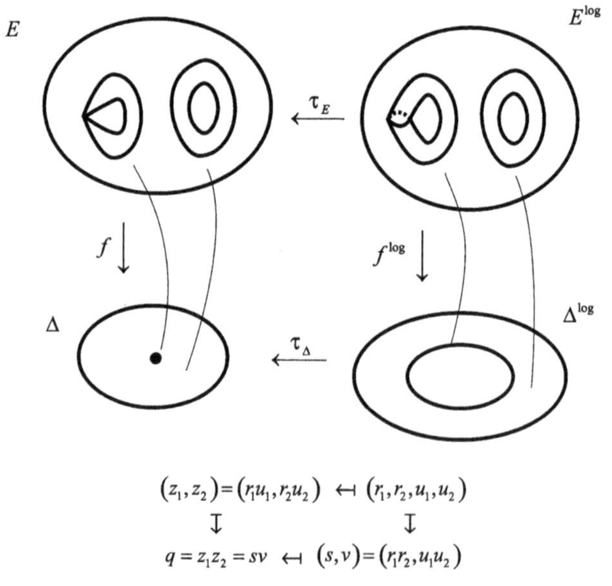

Figure 1

0.2.2

Elliptic curves (continued). Let $\Delta^* = \Delta - \{0\}$, let $E^* = f^{-1}(\Delta^*) \subset E$, and let $f' : E^* \to \Delta^*$ be the restriction of f to E^*. Since $f' : E^* \to \Delta^*$ is projective and smooth, the polarized Hodge structures on H^1 of the elliptic curves $\mathbf{C}^\times/q^{\mathbf{Z}}$ $(q \in \Delta^*)$ form a variation of polarized Hodge structure $(H'_{\mathbf{Z}}, F')$. Here, $H'_{\mathbf{Z}} = R^1 f'_* \mathbf{Z}$ is a locally constant sheaf on Δ^* of $\mathbf{Z}$-modules of rank 2, and the filtration F' on $\mathcal{O}_{\Delta^*} \otimes_{\mathbf{Z}} H'_{\mathbf{Z}} = R^1 f'_*(\Omega^\bullet_{E^*/\Delta^*}) = R^1 f'_*(\mathcal{O}_{E^*} \xrightarrow{d} \Omega^1_{E^*/\Delta^*})$ is given by $(F')^p = \mathcal{O}_{\Delta^*} \otimes_{\mathbf{Z}} H'_{\mathbf{Z}}$ for $p \leq 0$, $(F')^p = 0$ for $p \geq 2$, and $(F')^1 = f'_*(\Omega^1_{E^*/\Delta^*}) \subset \mathcal{O}_{\Delta^*} \otimes_{\mathbf{Z}} H'_{\mathbf{Z}}$.

0.2.3

Elliptic curves (continued). This variation of Hodge structure on Δ^* does not extend to a VH on Δ. First of all, the local system $H'_{\mathbf{Z}}$ on Δ^* does not extend to a local system on Δ. $H'_{\mathbf{Z}}$ extends to the sheaf $R^1 f_* \mathbf{Z}$ on Δ, but this sheaf is not locally constant. The stalk $(R^1 f_* \mathbf{Z})_0 = H^1(f^{-1}(0), \mathbf{Z})$ of this sheaf at $0 \in \Delta$ is of rank 1, not 2. In fact, $e_1 \in H^1(\mathbf{C}/(\mathbf{Z}\tau + \mathbf{Z}), \mathbf{Z}) \simeq H^1(\mathbf{C}^\times/q^{\mathbf{Z}}, \mathbf{Z})$ for $q \in \Delta^*$ (0.1.3) extends to a global section of $R^1 f_* \mathbf{Z}$ on Δ, but e_2 is defined only locally on Δ^*, depending on the choice of τ with $q = \exp(2\pi i \tau)$. There is no element of $(R^1 f_* \mathbf{Z})_0$ that gives e_2 in $H^1(\mathbf{C}/(\mathbf{Z}\tau + \mathbf{Z}), \mathbf{Z}) \simeq H^1(\mathbf{C}^\times/q^{\mathbf{Z}}, \mathbf{Z})$ for $q \in \Delta^*$ near to 0.

We show that by a magic of the theory of logarithmic structure, $H'_{\mathbf{Z}}$ does extend over the origin as a local system in the logarithmic world (0.2.4–0.2.10), and the variation of polarized Hodge structure $(H'_{\mathbf{Z}}, F')$ also extends over the origin as a logarithmic variation of polarized Hodge structure (0.2.15–0.2.20).

0.2.4

By a monoid, we mean a commutative semigroup with a neutral element 1. A homomorphism of monoids is assumed to preserve 1.

A *logarithmic structure* on a local ringed space $(X, \mathcal{O}_X)$ is a sheaf of monoids M_X on X endowed with a homomorphism $\alpha : M_X \to \mathcal{O}_X$, where $\mathcal{O}_X$ is regarded as a sheaf of monoids with respect to the multiplication, such that $\alpha : \alpha^{-1}(\mathcal{O}_X^\times) \to \mathcal{O}_X^\times$ is an isomorphism.

We regard $\mathcal{O}_X^\times$ as a subsheaf of M_X via α^{-1}.

0.2.5

Example. A standard example of a logarithmic structure is given as follows. Let X be an analytic manifold, let D be a divisor on X with normal crossings, and let $U = X - D$. (That is, locally, $X = \Delta^n$ a polydisc with coordinates $q_1, \ldots, q_n$, $D = \{q_1 \cdots q_r = 0\}$ $(0 \leq r \leq n)$, and $U = (\Delta^*)^r \times \Delta^{n-r}$.) Then

$$M_X = \{f \in \mathcal{O}_X \mid f \text{ is invertible on } U\} \overset{\alpha}{\hookrightarrow} \mathcal{O}_X$$

is a logarithmic structure on X. This is called the logarithmic structure on X associated with D.

0.2.6

Elliptic curves (continued). Let $f : E \to \Delta$ be as in 0.2.1. Define the logarithmic structure M_Δ on Δ by taking $X = \Delta$ and $D = \{0\}$ in 0.2.5, and define a logarithmic structure M_E on E by taking $X = E$ and $D = f^{-1}(0)$.

Then the stalk of M_Δ is given by $M_{\Delta,q} = \mathcal{O}^\times_{\Delta,q}$ if $q \in \Delta^*$, and $M_{\Delta,0} = \bigsqcup_{n\geq 0} \mathcal{O}^\times_{\Delta,0} \cdot q^n$ where the last q denotes the coordinate function of Δ.

0.2.7

For a complex analytic space X endowed with a logarithmic structure M_X, let

$$X^{\log} = \{(x, h) \mid x \in X; h \text{ is a homomorphism } M_{X,x} \to \mathbf{S}^1 \text{ satisfying (1) below}\}.$$

Here

$$\mathbf{S}^1 = \{z \in \mathbf{C}^\times \mid |z| = 1\}$$

is regarded as a multiplicative group.

$$h(u) = \frac{u(x)}{|u(x)|} \text{ for any } u \in \mathcal{O}^\times_{X,x}. \tag{1}$$

We have a canonical map

$$\tau : X^{\log} \to X, \quad (x, h) \mapsto x.$$

The space $X^{\log}$ has a natural topology, the weakest topology for which the map τ and the maps $\tau^{-1}(U) \to \mathbf{S}^1$, $(x, h) \mapsto h(f)$, given for each open set U of X and for each $f \in \Gamma(U, M_X)$, are continuous.

0.2.8

Elliptic curves (continued). For Δ with the logarithmic structure M_Δ, the shape of $\Delta^{\log}$ is as in Figure 2. Here, in $\Delta^{\log}$, $\{0\} \subset \Delta$ is replaced by $\mathbf{S}^1$ since $h : M_{\Delta,0} \to \mathbf{S}^1$ can send $q \in M_{\Delta,0}$ to any element of $\mathbf{S}^1$. Thus, roughly speaking, $\Delta^{\log}$ has a shape like Δ^* ($\Delta^{\log}$ is an extension of Δ^* over the origin without a change of shape).

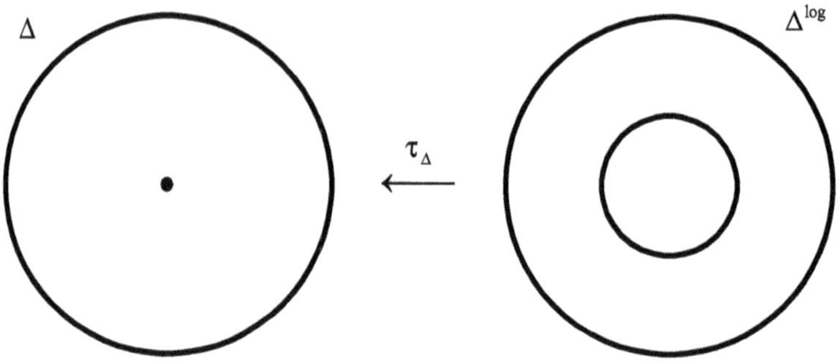

Figure 2

Precisely speaking, the inclusion map $\Delta^* \to \Delta^{\log}$ is a homotopy equivalence. From this, we see that $H'_{\mathbf{Z}}$ on Δ^* extends to a local system on $\Delta^{\log}$ of rank 2.

We will see in 0.2.10 below, for the continuous map $f^{\log} : E^{\log} \to \Delta^{\log}$ induced by $f : E \to \Delta$, all the fibers $(f^{\log})^{-1}(p)$ of any $p \in \Delta^{\log}$ are homeomorphic to the surface of a doughnut even if p lies over $0 \in \Delta$, as described on the right-hand side of Figure 1. The above local system of rank 2 on $\Delta^{\log}$ is in fact the sheaf $R^1 f_*^{\log}(\mathbf{Z})$.

0.2.9

Example. In 0.2.5, consider the case $X = \Delta^n$, $D = \{q_1 \cdots q_r = 0\}$. Then $X^{\log} = (|\Delta| \times \mathbf{S}^1)^r \times \Delta^{n-r}$, where $|\Delta| = \{t \in \mathbf{R} \mid 0 \le t < 1\}$, which has the natural topology. The map $\tau : X^{\log} \to X$ is given by $((r_j, u_j)_{1\le j\le r}, (x_j)_{r<j\le n}) \mapsto (x_j)_{1\le j\le n}$ where $x_j = r_j u_j$ for $1 \le j \le r$. For $x \in X$, the inverse image $\tau^{-1}(x)$ is isomorphic to $(\mathbf{S}^1)^m$ where m is the number of j such that $1 \le j \le r$ and $x_j = 0$.

0.2.10

Elliptic curves (continued). Let $f : E \to \Delta$ be the family of degenerating elliptic curves in 0.2.1. Explicitly, this E is defined as $X/\sim$, where $X = \{(t_1, t_2) \in \mathbf{C}^2 \mid |t_1 t_2| < 1\}$ and $\sim$ is the following equivalence relation. Let $g : X \to \Delta$, $(t_1, t_2) \mapsto t_1 t_2$. For $a, b \in X$, if $a \sim b$, then $g(a) = g(b)$. The restriction of $\sim$ to $g^{-1}(q)$ for each $q \in \Delta$ is defined as follows. Assume first $q \neq 0$. Consider the map $g^{-1}(q) \simeq \mathbf{C}^\times \to \mathbf{C}^\times/q^{\mathbf{Z}}$ where the first isomorphism is $(t_1, t_2) \mapsto t_1$. For $a, b \in g^{-1}(q)$, $a \sim b$ if and only if the images of a, b in $\mathbf{C}^\times/q^{\mathbf{Z}}$ coincide, i.e., $(b_1, b_2) = (q^n a_1, q^{-n} a_2)$ for some $n \in \mathbf{Z}$. Next assume $q = 0$. Consider $g^{-1}(0) = \{(t_1, t_2) \mid t_1 t_2 = 0\} \to \mathbf{P}^1(\mathbf{C})/(0 \sim \infty)$, where the arrow sends $(t_1, 0)$ to t_1 and $(0, t_2)$ to t_2^{-1}. Then, for $a, b \in g^{-1}(0)$, $a \sim b$ if and only if the images of a, b in $\mathbf{P}^1(\mathbf{C})/(0 \sim \infty)$ coincide, i.e., $a = b$ or $\{a, b\} = \{(t, 0), (0, t^{-1})\}$ for some $t \in \mathbf{C}^\times$.

The projection $X \to E$ is a local homeomorphism. The analytic structure and the logarithmic structure of E are the unique ones for which this projection is locally an isomorphism of analytic spaces with logarithmic structures.

We have

$$X^{\log} = \{(r_1, r_2, u_1, u_2) \in (\mathbf{R}_{\ge 0}) \times (\mathbf{R}_{\ge 0}) \times \mathbf{S}^1 \times \mathbf{S}^1 \mid r_1 r_2 < 1\},$$
$$\Delta^{\log} = |\Delta| \times \mathbf{S}^1$$

($|\Delta|$ is as in 0.2.9) and the projection $X^{\log} \to \Delta^{\log}$ is $(r_1, r_2, u_1, u_2) \mapsto (r_1 r_2, u_1 u_2)$. The projection $X^{\log} \to E^{\log}$ is a local homeomorphism. For $a = (r_1, r_2, u_1, u_2)$, $a' = (r'_1, r'_2, u'_1, u'_2) \in X^{\log}$, in the case $r_1 r_2 \neq 0$, the images of a and a' in $E^{\log}$ coincide if and only if there exists $n \in \mathbf{Z}$ such that

$$r'_1 = r_1(r_1 r_2)^n, \quad r'_2 = r_2(r_1 r_2)^{-n}, \quad u'_1 = u_1(u_1 u_2)^n, \quad u'_2 = u_2(u_1 u_2)^{-n}.$$

In the case $r_1 r_2 = 0$, the images of a and a' in $E^{\log}$ coincide if and only if either $a = a'$ or

$$\{a, a'\} = \{(c, 0, u_1, u_2), (0, c^{-1}, u_1(u_1 u_2), u_2(u_1 u_2)^{-1})\} \quad \text{for some } c \in \mathbf{R}_{>0}.$$

For example, if $p = (0, 1) \in |\Delta| \times \mathbf{S}^1 = \Delta^{\log}$, we have a homeomorphism $(f^{\log})^{-1}(p) \simeq [0, \infty]/(0 \sim \infty) \times \mathbf{S}^1$ which sends the image of $(c, 0, u_1, u_2) \in X^{\log}$ in $E^{\log}$ to (c, u_1), and the image of $(0, c, u_1, u_2)$ to (c^{-1}, u_1). Hence we have a homeomorphism $(f^{\log})^{-1}(p) \simeq \mathbf{S}^1 \times \mathbf{S}^1$.

We show that, for each $p \in \Delta^{\log}$, there are an open neighborhood V of p and a homeomorphism

$$(f^{\log})^{-1}(V) \simeq V \times \mathbf{S}^1 \times \mathbf{S}^1$$

over V. This shows that any fiber of $f^{\log} : E^{\log} \to \Delta^{\log}$ is homeomorphic to $\mathbf{S}^1 \times \mathbf{S}^1$.

Let $B :=$ (the complement of $(1, 1)$ in $[0, 1] \times [0, 1]) \subset (\mathbf{R}_{\geq 0}) \times (\mathbf{R}_{\geq 0})$, and let $A := B \times \mathbf{S}^1 \times \mathbf{S}^1 \subset X^{\log}$. Then the projection $A \to E^{\log}$ is surjective, and we have a homeomorphism

$$A/\sim \xrightarrow{\sim} E^{\log},$$

where for $a, a' \in A$, $a \sim a'$ if and only if either $a = a'$ or

$$\{a,\ a'\} = \{(1, r, u_1, u_2),\ (r, 1, u_1(u_1u_2), u_2(u_1u_2)^{-1})\}$$

for some $r \in |\Delta|$ and for some $u_1, u_2 \in \mathbf{S}^1$. Note that this equivalence relation $\sim$ in A comes from the equivalence relation associated with the local homeomorphism $X^{\log} \to E^{\log}$.

Take a continuous map $s : B \to [0, 1]$ such that we have a homeomorphism

$$B \xrightarrow{\sim} |\Delta| \times [0, 1], \quad (r_1, r_2) \mapsto (r_1r_2, s(r_1, r_2)),$$

and such that

$$s(r_1, r_2) = 0 \Leftrightarrow r_2 = 1, \quad s(r_1, r_2) = 1 \Leftrightarrow r_1 = 1.$$

For example, the function $s(r_1, r_2) = (1 - r_2)/((1 - r_1) + (1 - r_2))$ has this property.

Let $p = (r_0, u_0) \in \Delta^{\log}$ ($r_0 \in |\Delta|$, $u_0 \in \mathbf{S}^1$) and take an open neighborhood V' of u_0 in $\mathbf{S}^1$ and a continuous map $k : V' \times [0, 1] \to \mathbf{S}^1$ such that $k(u, 1) = k(u, 0)u$ for any $u \in V'$. For example, if $u_0 = e^{i\theta}$ with $\theta \in \mathbf{R}$, we can take $V' = \{e^{i\lambda} \mid a < \lambda < b\}$ for any fixed $a, b \in \mathbf{R}$ such that $a < \theta < b$ and $b - a < 2\pi$ and $k(e^{i\lambda}, t) = e^{i\lambda t}$ $(a < \lambda < b, 0 \leq t \leq 1)$. Let

$$V = [0, 1) \times V' \subset \Delta^{\log}, \quad V'' = \{(u_1, u_2) \mid u_1u_2 \in V'\} \subset \mathbf{S}^1 \times \mathbf{S}^1.$$

Then we have a homeomorphism

$$B \times V'' \xrightarrow{\sim} V \times [0, 1] \times \mathbf{S}^1,$$
$$(r_1, r_2, u_1, u_2) \mapsto (r_1r_2, u_1u_2, s(r_1, r_2), u_1k(u_1u_2, s(r_1, r_2))).$$

The subset $B \times V''$ of A is stable for the relation $\sim$, and this homeomorphism induces a homeomorphism

$$(B \times V'')/\sim \xrightarrow{\sim} V \times ([0, 1]/(0 \sim 1)) \times \mathbf{S}^1.$$

Since $(B \times V'')/\sim \xrightarrow{\sim} (f^{\log})^{-1}(V)$, we have a homeomorphism

$$(f^{\log})^{-1}(V) \xrightarrow{\sim} V \times ([0,1]/(0 \sim 1)) \times \mathbf{S}^1$$

over V. Note that $[0,1]/(0 \sim 1) \simeq \mathbf{S}^1$.

See [U3] for a generalization of this local topological triviality of the family $E^{\log} \to \Delta^{\log}$ over the base.

Thus Figure 1 is completely explained.

0.2.11

The above magic for $f : E \to \Delta$ is generalized in logarithmic complex analytic geometry as follows. In logarithmic complex analytic geometry, we consider mainly logarithmic structures called fs logarithmic structures.

We say a monoid $\mathcal{S}$ is *integral* if $ab = ac$ implies $b = c$ in $\mathcal{S}$. An integral monoid $\mathcal{S}$ is embedded in the group $\mathcal{S}^{\mathrm{gp}} = \{ab^{-1} \mid a, b \in \mathcal{S}\}$. We say a monoid $\mathcal{S}$ is *saturated* if it is integral and if $a \in \mathcal{S}^{\mathrm{gp}}$ and $a^n \in \mathcal{S}$ for some integer $n \geq 1$ imply $a \in \mathcal{S}$. We say a monoid is *fs* if it is finitely generated and saturated.

For an fs monoid $\mathcal{S}$, $\mathcal{S}^{\mathrm{gp}}$ is a finitely generated abelian group, and $\mathcal{S}^{\mathrm{gp}}$ is torsion free if $\mathcal{S}$ is torsion free.

Let X be a local ringed space.

Let $\mathcal{S}$ be an fs monoid which is considered as a constant sheaf on X, and let $h : \mathcal{S} \to \mathcal{O}_X$ be a homomorphism of sheaves of monoids. The *associated logarithmic structure* on X is defined as the push-out $\tilde{\mathcal{S}}$ of the diagram

$$\begin{array}{ccc} h^{-1}(\mathcal{O}_X^\times) & \longrightarrow & \mathcal{S} \\ \downarrow & & \\ \mathcal{O}_X^\times & & \end{array}$$

in the category of sheaves of monoids, which is endowed with the induced homomorphism $\alpha : \tilde{\mathcal{S}} \to \mathcal{O}_X$ (for an explicit description of the push-out, see 2.1.1).

A logarithmic structure on X is *fs* if it is locally isomorphic to the one above.

The logarithmic structure in 0.2.5 associated with a divisor with normal crossings on an analytic manifold is an fs logarithmic structure. This can be checked locally by the fact that the logarithmic structure on $X = \Delta^n$ associated with the divisor $q_1 \cdots q_r = 0$ $(0 \leq r \leq n)$ is induced from the homomorphism $\mathbf{N}^r \to \mathcal{O}_X$, $(n_j)_{1 \leq j \leq r} \mapsto \prod_{j=1}^r q_j^{n_j}$. Note that $\mathbf{N}^r$ (the semigroup law is the addition) is an fs monoid.

An fs logarithmic structure M_X on X is integral, and hence M_X is embedded in the sheaf of commutative groups M_X^{gp}. For an fs logarithmic structure M_X on X, the stalk $(M_X/\mathcal{O}_X^\times)_x$ at $x \in X$ is a sharp fs monoid, and, in particular, torsion-free. Here we say that a monoid $\mathcal{S}$ is sharp if $\mathcal{S}^\times = \{1\}$, where $\mathcal{S}^\times$ denotes the set of all invertible elements of $\mathcal{S}$. Hence $((M_X/\mathcal{O}_X^\times)_x)^{\mathrm{gp}} = (M_X^{\mathrm{gp}}/\mathcal{O}_X^\times)_x$ is a free $\mathbf{Z}$-module of finite rank.

For instance, in the above example $X = \Delta^n$, if $x = (x_j)_{1 \leq j \leq n} \in X$ and if the number of those j satisfying $1 \leq j \leq r$ and $x_j = 0$ is m, then $(M_X/\mathcal{O}_X^\times)_x \simeq \mathbf{N}^m$

and this monoid is generated by (q_j mod $\mathcal{O}_{X,x}^\times$) for such j. (Note that $q_j \in \mathcal{O}_{X,x}^\times$ for the other j's.) We have $(M_X^{\mathrm{gp}}/\mathcal{O}_X^\times)_x \simeq \mathbf{Z}^m$.

An analytic space with an fs logarithmic structure is called an *fs logarithmic analytic space*.

For an fs logarithmic analytic space X, the canonical map $\tau : X^{\log} \to X$ is proper, and $\tau^{-1}(x) \simeq (\mathbf{S}^1)^m$ for $x \in X$, where m is the rank of $(M_X^{\mathrm{gp}}/\mathcal{O}_X^\times)_x$.

Here "proper" means "proper in the sense of Bourbaki [Bn] and separated" (see 0.7.5). We keep this terminology throughout this book.

0.2.12

Example: *Toric varieties.* Let $\mathcal{S}$ be an fs monoid, and $X := \mathrm{Spec}(\mathbf{C}[\mathcal{S}])_{\mathrm{an}} = \mathrm{Hom}(\mathcal{S}, \mathbf{C}^{\mathrm{mult}})$ (here $\mathbf{C}^{\mathrm{mult}}$ denotes the set $\mathbf{C}$ regarded as a multiplicative monoid) be the analytic toric variety. Then $\mathcal{S} \subset \mathbf{C}[\mathcal{S}] \to \mathcal{O}_X$ induces a canonical fs logarithmic structure. We have

$$X^{\log} = \mathrm{Hom}(\mathcal{S}, \mathbf{R}_{\geq 0}^{\mathrm{mult}} \times \mathbf{S}^1)$$

(here $\mathbf{R}_{\geq 0}^{\mathrm{mult}}$ denotes the set $\mathbf{R}_{\geq 0}$ regarded as a multiplicative monoid).

Using this, we have a local presentation of $X^{\log}$ for any fs logarithmic analytic space X. Let X be an analytic space, let $\mathcal{S}$ be an fs monoid, let $\mathcal{S} \to \mathcal{O}_X$ be a homomorphism, and endow X with the induced logarithmic structure. Then

$$X^{\log} = X \times_{\mathrm{Hom}(\mathcal{S}, \mathbf{C}^{\mathrm{mult}})} \mathrm{Hom}(\mathcal{S}, \mathbf{R}_{\geq 0}^{\mathrm{mult}} \times \mathbf{S}^1) \tag{1}$$

(the fiber product as a topological space).

For a morphism $f : Y \to X$ of local ringed spaces and for a logarithmic structure M on X, the inverse image f^*M of M, which is a logarithmic structure on Y, is defined as in 2.1.3. If M is an fs logarithmic structure associated with a homomorphism $\mathcal{S} \to \mathcal{O}_X$ with $\mathcal{S}$ an fs monoid, the inverse image f^*M is the fs logarithmic structure associated with the homomorphism $\mathcal{S} \to \mathcal{O}_Y$ induced by f. Hence the inverse image of an fs logarithmic structure is an fs logarithmic structure. If $f : Y \to X$ is a morphism of analytic spaces, for an fs logarithmic structure M on X and for f^*M on Y, we have $Y^{\log} = Y \times_X X^{\log}$ as a topological space. The description of $X^{\log}$ in (1) is explained by this since, in that case, the logarithmic structure of X is the inverse image of the canonical logarithmic structure of $\mathrm{Spec}(\mathbf{C}[\mathcal{S}])_{\mathrm{an}}$.

Example. Let $x = 0 \in \Delta$, and define the logarithmic structure M_x of x as the inverse image of M_Δ. This logarithmic structure is induced from the homomorphism $\mathcal{S} = \mathbf{N} \to \mathcal{O}_x = \mathbf{C}$, $n \mapsto$ (the image of q^n in $\mathcal{O}_x$) $= 0^n$ (note $0^0 = 1$). Hence $M_x = \bigsqcup_{n \geq 0}(\mathbf{C}^\times \cdot q^n) \simeq \mathbf{C}^\times \times \mathbf{N}$ with $\alpha : M_x \to \mathcal{O}_x = \mathbf{C}$, $c \cdot q^n \mapsto c \cdot 0^n$ ($c \in \mathbf{C}^\times, n \in \mathbf{N}$). Thus a one-point set can have a nontrivial logarithmic structure. We have $x^{\log} = \mathbf{S}^1$ for this logarithmic structure M_x.

0.2.13

The morphism $f : E \to \Delta$ in 0.2.1 is an easiest nontrivial example of logarithmically smooth morphisms (see 2.1.11 for the definition) of fs logarithmic analytic spaces. A logarithmically smooth morphism can have degeneration in the sense of classical complex analytic geometry, but, with the magic of the logarithmic structure, it can behave in logarithmic complex analytic geometry like a smooth morphism in classical complex analytic geometry.

A wider example of a logarithmically smooth morphism is a morphism with semistable degeneration $f : Y \to \Delta$ (Δ is endowed with M_Δ), that is, a morphism which, locally on Y, has the form $\Delta^n \to \Delta$, $(q_j)_{1\le j\le n} \mapsto \prod_{j=1}^r q_j (1 \le r \le n)$ with the logarithmic structure of Δ^n in 0.2.9. Indeed, it is shown in [U3] that, if f is proper (0.7.5), the associated continuous map $f^{\log} : Y^{\log} \to \Delta^{\log}$ is topologically trivial locally over the base.

Kajiwara and Nakayama [KjNc] proved the following:

Let $f : Y \to X$ be a proper logarithmically smooth morphism of fs logarithmic analytic spaces. Then, for any $m \ge 0$, the higher direct image functor $R^m f_*^{\log}$ sends locally constant sheaves of abelian groups on $Y^{\log}$ to locally constant sheaves on $X^{\log}$.

An fs logarithmic analytic space X is said to be logarithmically smooth if the structural morphism $X \to \operatorname{Spec}(\mathbf{C})$ is logarithmically smooth. Here $\operatorname{Spec}(\mathbf{C})$ is endowed with the trivial logarithmic structure $\mathbf{C}^\times$. An fs logarithmic analytic space X is logarithmically smooth if and only if, locally on X, there is an open immersion of analytic spaces $i : X \hookrightarrow Z = \operatorname{Spec}(\mathbf{C}[\mathcal{S}])_{\mathrm{an}}$ with $\mathcal{S}$ an fs monoid such that the logarithmic structure of X is the inverse image of the canonical logarithmic structure of Z (i.e., the logarithmic structure of X induced by the homomorphism $\mathcal{S} \to \mathcal{O}_X$ defined by i).

0.2.14

Logarithmic differential forms. The name "logarithmic structure" comes from its relation to differential forms with logarithmic poles.

For an fs logarithmic analytic space X, the sheaves of logarithmic differential q-forms ω_X^q on X ($q \in \mathbf{N}$) are defined as in 2.1.7. If X is an analytic manifold and D is a divisor on X with normal crossings, and if X is endowed with the logarithmic structure associated with D, ω_X^q coincides with the sheaf $\Omega_X^q(\log(D))$, the sheaf of differential q-forms on X which may have logarithmic poles along D. In general, ω_X^q is an $\mathcal{O}_X$-module, there is a canonical homomorphism of $\mathcal{O}_X$-modules $\Omega_X^q \to \omega_X^q$, ω_X^q is the qth exterior power of ω_X^1, and ω_X^1 is generated over $\mathcal{O}_X$ by Ω_X^1 and the image of a homomorphism $d\log : M_X^{\mathrm{gp}} \to \omega_X^1$. For a morphism $Y \to X$ of fs logarithmic analytic spaces, the logarithmic version $\omega_{Y/X}^q$ of $\Omega_{Y/X}^q$ is also defined (2.1.7).

Example. Let $X = \Delta^n$ with the logarithmic structure as in 0.2.9. Then for $x = (x_j)_{1\le j\le n} \in X$, the stalk of ω_X^1 at x is a free $\mathcal{O}_{X,x}$-module with basis $(\omega_j)_{1\le j\le n}$ where $\omega_j = d\log(q_j)$ if $1 \le j \le r$ and $x_j = 0$, and $\omega_j = dq_j$ otherwise.

For a logarithmically smooth morphism $Y \to X$ of fs logarithmic analytic spaces, the sheaf $\omega_{Y/X}^1$ is locally free although $\Omega_{Y/X}^1$ may not be locally free if degeneration occurs in $Y \to X$. Consider the case $n = r = 2$ of the last example, and consider the logarithmically smooth morphism $X \to \Delta$, $(x_1, x_2) \mapsto x_1x_2$. The sheaf $\Omega_{X/\Delta}^1$ is generated by dq_1 and dq_2 which satisfy the relation $q_1dq_2 + q_2dq_1 = d(q_1q_2) = 0$.

This indicates that $\Omega^1_{X/\Delta}$ is not locally free. However, $\omega^1_{X/\Delta}$ is generated by $d\log(q_1)$ and $d\log(q_2)$, which satisfy $d\log(q_1)+d\log(q_2) = d\log(q_1q_2) = 0$. This shows that $\omega^1_{X/\Delta}$ is a free $\mathcal{O}_X$-module of rank 1 generated by $d\log(q_1)$.

Example. Let $\mathcal{S}$ be an fs monoid and let $X = \operatorname{Spec}(\mathbf{C}[\mathcal{S}])_{\text{an}}$ endowed with the canonical logarithmic structure. Then we have an isomorphism

$$\mathcal{O}_X \otimes_{\mathbf{Z}} \mathcal{S}^{\text{gp}} \xrightarrow{\sim} \omega^1_X,\ f \otimes g \mapsto fd\log(g).$$

Example. Let $x = 0 \in \Delta$ and endow x with the inverse image of M_Δ. Then ω^1_x is the one-dimensional $\mathbf{C}$-vector space generated by the image of $d\log(q) \in \Gamma(\Delta, \omega^1_\Delta)$. Thus a one point set can have a nontrivial logarithmic differential form.

Now we talk about Hodge filtration and logarithmic Hodge theory. The key point is that, for an fs logarithmic analytic space X, we define a sheaf of rings $\mathcal{O}^{\log}_X$ on $X^{\log}$. Roughly speaking, $\mathcal{O}^{\log}_X$ is the ring generated over $\mathcal{O}_X$ by $\log(M^{\text{gp}})$. First, by considering the example $f : E \to \Delta$ of 0.2.1, we will see why such a ring is necessary.

0.2.15

Elliptic curves (continued). Let $f' : E^* \to \Delta^*$ be as in 0.2.2. Let $H'_{\mathbf{Z}} = R^1 f'_*(\mathbf{Z})$ and let $\mathcal{M}' = R^1 f'_*(\Omega^\bullet_{E^*/\Delta^*})$.

We have seen that $H'_{\mathbf{Z}}$ extends to the local system $H_{\mathbf{Z}} = R^1 f^{\log}_*(\mathbf{Z})$ on $\Delta^{\log}$ (0.2.8).

On the other hand, the $\mathcal{O}_{\Delta^*}$-module $\mathcal{M}'$ with a decreasing filtration F' (0.2.2) extends to a locally free $\mathcal{O}_\Delta$-module $\mathcal{M} := R^1 f_*(\omega^\bullet_{E/\Delta}) = R^1 f_*(\mathcal{O}_E \xrightarrow{d} \omega^1_{E/\Delta})$ of rank 2 with the decreasing filtration defined by $\mathcal{M}^0 = \mathcal{M}, \mathcal{M}^1 = f_*(\omega^1_{E/\Delta}) \hookrightarrow \mathcal{M}$, $\mathcal{M}^2 = 0$. We have

$$\mathcal{M} = \mathcal{O}_\Delta e_1 \oplus \mathcal{O}_\Delta \omega \supset \mathcal{M}^1 = \mathcal{O}_\Delta \omega, \tag{1}$$

where we denote by the same letter e_1 the image of the global section e_1 of $R^1 f_*\mathbf{Z}$ under $R^1 f_*\mathbf{Z} \to R^1 f_*(\omega^\bullet_{E/\Delta})$, and we denote by ω the differential form $(2\pi i)^{-1} dt_1/t_1$ on E with logarithmic poles along $f^{-1}(0)$ (where t_1 is the coordinate function of X in 0.2.10).

Although the relation between $H'_{\mathbf{Z}}$ and $\mathcal{M}'$ is simply $\mathcal{M}' = \mathcal{O}_{\Delta^*} \otimes_{\mathbf{Z}} H'_{\mathbf{Z}}$, the relation between $H_{\mathbf{Z}}$ and $\mathcal{M}$ is not so direct. They are related only after being tensored by a sheaf of rings $\mathcal{O}^{\log}_\Delta$ defined below.

Let q be the coordinate function of Δ. Then on Δ^*, via the identification $\mathcal{M}' = \mathcal{O}_{\Delta^*} \otimes_{\mathbf{Z}} H'_{\mathbf{Z}}$, we have

$$\omega = (2\pi i)^{-1} \log(q) e_1 + e_2 \tag{2}$$

where e_2 is taken by fixing a branch of the multivalued function $(2\pi i)^{-1}\log(q)$, and the same branch of this function is used in the formula (2). This formula (2) follows from the fact that the restriction of ω to each fiber $\mathbf{C}^\times/q^{\mathbf{Z}} \simeq \mathbf{C}/(\mathbf{Z}\tau + \mathbf{Z})$ ($q \in \Delta^*$, $\tau = (2\pi i)^{-1}\log(q)$) is $dz = \tau e_1 + e_2$ on $\mathbf{C}/(\mathbf{Z}\tau + \mathbf{Z})$.

Locally on $\Delta^{\log}$, e_2 extends to a local section of $H_{\mathbf{Z}}$, and we have

$$H_{\mathbf{Z}} = \mathbf{Z}e_1 \oplus \mathbf{Z}e_2. \tag{3}$$

When we compare (1), (2), and (3), since $\log(q)$ in (2) does not extend over the origin in the classical sense, we do not find a simple relation between $H_{\mathbf{Z}}$ and $\mathcal{M}$. However, $\log(q)$ exists on $\Delta^{\log}$ locally as a local section of $j_*^{\log}(\mathcal{O}_{\Delta^*})$ where $j^{\log}$ is the inclusion map $\Delta^* \hookrightarrow \Delta^{\log}$. In fact, if $y = (0, e^{i\theta}) \in \Delta^{\log} = |\Delta| \times \mathbf{S}^1$ ($\theta \in \mathbf{R}$), where $|\Delta|$ is as in 0.2.9, if we take $a, b \in \mathbf{R}$ such that $a < \theta < b$ and $b - a < 2\pi$, and if we denote by U the open neighborhood $\{(r, e^{ix}) \mid r \in |\Delta|, a < x < b\}$ of y in $\Delta^{\log}$, the holomorphic map $re^{i\theta} \mapsto \log(r) + i\theta$ defined on $\Delta^* \cap U = \{(r, e^{i\theta}) \mid 0 < r < 1, a < \theta < b\}$ is an element of $\Gamma(\Delta^* \cap U, \mathcal{O}_{\Delta^*}) = \Gamma(U, j_*^{\log}(\mathcal{O}_{\Delta^*}))$, which is a branch of $\log(q)$. All branches of $\log(q)$ in $j_*^{\log}(\mathcal{O}_{\Delta^*})$ are congruent modulo $2\pi i\mathbf{Z}$. Let $\mathcal{O}_{\Delta}^{\log} \subset j_*^{\log}(\mathcal{O}_{\Delta^*})$ be the sheaf of subrings of $j_*^{\log}(\mathcal{O}_{\Delta^*})$ on $\Delta^{\log}$ generated over $\tau^{-1}(\mathcal{O}_{\Delta})$ by $\log(q)$. Here $\tau^{-1}(\ )$ is the inverse image of a sheaf. Then from (1), (2), and (3), we have

$$\mathcal{O}_{\Delta}^{\log} \otimes_{\tau^{-1}(\mathcal{O}_{\Delta})} \tau^{-1}(\mathcal{M}) = \mathcal{O}_{\Delta}^{\log} \otimes_{\mathbf{Z}} H_{\mathbf{Z}}.$$

0.2.16

The sheaf of rings $\mathcal{O}_X^{\log}$. For an fs logarithmic analytic space X, we have a sheaf of rings $\mathcal{O}_X^{\log}$, which generalizes the above $\mathcal{O}_{\Delta}^{\log}$.

First we consider the case $X = \mathrm{Spec}(\mathbf{C}[\mathcal{S}])_{\mathrm{an}}$ for an fs monoid $\mathcal{S}$. Let U be the open subspace $\mathrm{Spec}(\mathbf{C}[\mathcal{S}^{\mathrm{gp}}])_{\mathrm{an}}$ of X, and let $j^{\log} : U \to X^{\log}$ be the canonical map. Then we define $\mathcal{O}_X^{\log}$ as a sheaf of subrings of $j_*^{\log}(\mathcal{O}_U)$ generated over $\tau^{-1}(\mathcal{O}_X)$ by the logarithms of local sections of M_X^{gp} (these logarithms exist in $j_*^{\log}(\mathcal{O}_U)$ and are determined mod $2\pi i\mathbf{Z}$ just as in the case of Δ).

The definition of $\mathcal{O}_X^{\log}$ for a general fs logarithmic analytic space X is given in 2.2.4. If the logarithmic structure of X is induced from a homomorphism $\mathcal{S} \to \mathcal{O}_X$ with $\mathcal{S}$ an fs monoid, we have

$$\mathcal{O}_X^{\log} = \mathcal{O}_X \otimes_{\mathcal{O}_Z} \mathcal{O}_Z^{\log} \quad \text{with} \quad Z = \mathrm{Spec}(\mathbf{C}[\mathcal{S}])_{\mathrm{an}}$$

(note that $\mathcal{O}_Z^{\log}$ is explained just above), where we denote the inverse images on $X^{\log}$ of the sheaves $\mathcal{O}_X$, $\mathcal{O}_Z$, and $\mathcal{O}_Z^{\log}$, by $\mathcal{O}_X$, $\mathcal{O}_Z$, and $\mathcal{O}_Z^{\log}$, respectively, for simplicity. We have a homomorphism $\log : M_X^{\mathrm{gp}} \to \mathcal{O}_X^{\log}/2\pi i\mathbf{Z}$, and $\mathcal{O}_X^{\log}$ is generated over $\tau^{-1}(\mathcal{O}_X)$ by $\log(M_X^{\mathrm{gp}})$. For an fs logarithmic analytic space X and $x \in X$, if the free $\mathbf{Z}$-module $(M_X^{\mathrm{gp}}/\mathcal{O}_X^{\times})_x$ is of rank r with basis $(f_j \bmod \mathcal{O}_{X,x}^{\times})_{1 \le j \le r}$ ($f_j \in M_{X,x}^{\mathrm{gp}}$), then for any point y of $X^{\log}$ lying over x, the stalk $\mathcal{O}_{X,y}^{\log}$ of $\mathcal{O}_X^{\log}$ is isomorphic to the polynomial ring in r variables over $\mathcal{O}_{X,x}$ by

$$\mathcal{O}_{X,x}[T_1, \ldots, T_r] \xrightarrow{\sim} \mathcal{O}_{X,y}^{\log}, \quad T_j \mapsto \log(f_j)$$

($\log(f_j)$ is defined only modulo $2\pi i\mathbf{Z}$ but we choose a branch (a representative) for each j). Let

$$\omega_X^{q,\log} = \mathcal{O}_X^{\log} \otimes_{\tau^{-1}(\mathcal{O}_X)} \tau^{-1}(\omega_X^q).$$

Then we have the de Rham complex $\omega_X^{\bullet,\log}$ on $X^{\log}$ with the differential $d : \omega_X^{q,\log} \to \omega_X^{q+1,\log}$ defined as in 2.2.6.

Example. Let $X = \Delta^n$ endowed with the logarithmic structure in 0.2.9. Let $x \in \Delta^n$ and let y be a point of $X^{\log}$ lying over x. Let m be the number of j such that $1 \le j \le r$ and $x_j = 0$. Then the stalk $\mathcal{O}^{\log}_{X,y}$ is a polynomial ring over $\mathcal{O}_{X,x}$ in m variables $\log(q_j)$ for such j.

Example. Let $x = \mathrm{Spec}(\mathbf{C})$ endowed with an fs logarithmic structure (we will call such x an *fs logarithmic point*). Then $M_x = \mathbf{C}^\times \times \mathcal{S}$ for some fs monoid $\mathcal{S}$ with no invertible element other than 1, where $\alpha : M_x \to \mathbf{C}$ sends $(c, t) \in M_x$ ($c \in \mathbf{C}^\times$, $t \in \mathcal{S}$) to 0 if $t \neq 1$, and to c if $t = 1$. Let r be the rank of $\mathcal{S}^{\mathrm{gp}}$ which is a free $\mathbf{Z}$-module of finite rank. We have

$$x^{\log} \simeq (\mathbf{S}^1)^r, \quad \omega^1_x \simeq \mathbf{C}^r, \quad \mathcal{O}^{\log}_{x,y} \simeq \mathbf{C}[T_1, \ldots, T_r] \quad (y \in x^{\log}).$$

0.2.17

Let X be an fs logarithmic analytic space, let $x \in X$, and let y be a point of $X^{\log}$ lying over x. The stalk $\mathcal{O}^{\log}_{X,y}$ is not necessarily a local ring, and has a global ring-theoretic nature. Let $\mathrm{sp}(y)$ be the set of all ring homomorphisms $s : \mathcal{O}^{\log}_{X,y} \to \mathbf{C}$ such that $s(f) = f(x)$ for any $f \in \mathcal{O}_{X,x}$. If we fix $s_0 \in \mathrm{sp}(y)$, we have a bijection

$$\mathrm{sp}(y) \xrightarrow{\sim} \mathrm{Hom}((M^{\mathrm{gp}}_X/\mathcal{O}^\times_X)_x, \mathbf{C}^{\mathrm{add}}), \tag{1}$$
$$s \mapsto (f \mapsto s(\log(f)) - s_0(\log(f))) \quad \text{for} \quad f \in M^{\mathrm{gp}}_{X,x}.$$

Here $\mathbf{C}^{\mathrm{add}}$ is $\mathbf{C}$ regarded as an additive group.

Let $(H_{\mathbf{Z}}, F)$ be a pair of a local system $H_{\mathbf{Z}}$ of free $\mathbf{Z}$-modules of finite rank on $X^{\log}$ and of a decreasing filtration F on the $\mathcal{O}^{\log}_X$-module $\mathcal{O}^{\log}_X \otimes_{\mathbf{Z}} H_{\mathbf{Z}}$ such that F^p and $(\mathcal{O}^{\log}_X \otimes_{\mathbf{Z}} H_{\mathbf{Z}})/F^p$ are locally free as $\mathcal{O}^{\log}_X$-modules for all p. Then, for each $s \in \mathrm{sp}(y)$, we have a decreasing filtration $F(s)$ on $H_{\mathbf{C},y} = \mathbf{C} \otimes_{\mathbf{Z}} H_{\mathbf{Z},y}$ (called the *specialization of F at s*) defined by $F^p(s) = \mathbf{C} \otimes_{\mathcal{O}^{\log}_{X,y}} F^p_y$. Here $\mathcal{O}^{\log}_{X,y} \to \mathbf{C}$ is s. We will see later that a nilpotent orbit can be regarded as the family $(F(s))_{s \in \mathrm{sp}(y)}$ associated with such $(H_{\mathbf{Z}}, F)$. The reason why an orbit of Hodge filtrations, called a nilpotent orbit (not a single Hodge filtration), appears in the degeneration of Hodge structures is, from the point of view of logarithmic Hodge theory, that the stalk $\mathcal{O}^{\log}_{X,y}$ is still a global ring and we have many specializations at a point y.

0.2.18

Elliptic curves (continued). Let $f : E \to \Delta$ be as in 0.2.1, and consider $H_{\mathbf{Z}} = R^1 f^{\log}_*(\mathbf{Z})$, $\mathcal{M} = R^1 f_*(\mathcal{O}_E \to \omega^1_{E/\Delta})$, and the filtration $(\mathcal{M}^p)_p$ as in 0.2.15. Define a filtration F on $\mathcal{O}^{\log}_\Delta \otimes_{\mathbf{Z}} H_{\mathbf{Z}}$ by

$$F^p = \mathcal{O}^{\log}_\Delta \otimes_{\tau^{-1}(\mathcal{O}_\Delta)} \tau^{-1}(\mathcal{M}^p) \subset \mathcal{O}^{\log}_\Delta \otimes_{\tau^{-1}(\mathcal{O}_\Delta)} \tau^{-1}(\mathcal{M}) = \mathcal{O}^{\log}_\Delta \otimes_{\mathbf{Z}} H_{\mathbf{Z}}.$$

Take $y \in \Delta^{\log}$ lying over $0 \in \Delta$. We consider the specializations $F(s)$ for $s \in \mathrm{sp}(y)$. Take a branch of $e_2 \in H_{\mathbf{Z}}$ at y and take the corresponding branch of $\log(q)$ at y. Then F^1_y is a free $\mathcal{O}^{\log}_{\Delta,y}$-module of rank 1 generated by $(2\pi i)^{-1}\log(q)e_1 + e_2$

(0.2.15 (2)). Since $\mathcal{O}^{\log}_{\Delta,y}$ is a polynomial ring in one variable $\log(q)$ over $\mathcal{O}_{\Delta,0}$, an element $s \in \mathrm{sp}(y)$ is determined by $s(\log(q)) \in \mathbf{C}$. The filtration $F(s)$ is described as: $F^0(s) = H_{\mathbf{C},y}$, $F^2(s) = 0$, and $F^1(s)$ is the one-dimensional $\mathbf{C}$-subspace of $H_{\mathbf{C},y}$ generated by $s((2\pi i)^{-1}\log(q))e_1 + e_2$.

If the imaginary part of $s((2\pi i)^{-1}\log(q))$ is > 0 (that is, if $|\exp(s(\log(q)))| < 1$), then $(H_{\mathbf{Z},y}, F(s))$ is a Hodge structure of weight 1, and for the antisymmetric pairing $\langle\ ,\ \rangle : H_{\mathbf{Q}} \times H_{\mathbf{Q}} \to \mathbf{Q}$ defined as $\langle e_2, e_1\rangle = 1$, $(H_{\mathbf{Z},y}, \langle\ ,\ \rangle_y, F(s))$ becomes a polarized Hodge structure.

0.2.19

The observation in 0.2.18 leads us to the notion of "logarithmic variation of polarized Hodge structure."

Let X be a logarithmically smooth fs logarithmic analytic space. A *logarithmic variation of polarized Hodge structure* (LVPH) *on* X *of weight* w is a triple $(H_{\mathbf{Z}}, \langle\ ,\ \rangle, F)$ consisting of a locally constant sheaf $H_{\mathbf{Z}}$ of free $\mathbf{Z}$-modules of finite rank on $X^{\log}$, a bilinear form $\langle\ ,\ \rangle : H_{\mathbf{Q}} \times H_{\mathbf{Q}} \to \mathbf{Q}$, and a decreasing filtration F of $\mathcal{O}^{\log}_X \otimes H_{\mathbf{Z}}$ by $\mathcal{O}^{\log}_X$-submodules which satisfy the following three conditions (1)–(3).

(1) There exist a locally free $\mathcal{O}_X$-module $\mathcal{M}$ and a decreasing filtration $(\mathcal{M}^p)_{p\in\mathbf{Z}}$ by $\mathcal{O}_X$-submodules of $\mathcal{M}$ such that $\mathcal{O}^{\log}_X \otimes_{\mathbf{Z}} H_{\mathbf{Z}} = \mathcal{O}^{\log}_X \otimes_{\tau^{-1}(\mathcal{O}_X)} \tau^{-1}(\mathcal{M})$ and $F^p = \mathcal{O}^{\log}_X \otimes_{\tau^{-1}(\mathcal{O}_X)} \tau^{-1}(\mathcal{M}^p)$ for all p, and such that $\mathcal{M}^p = \mathcal{M}$ for $p \ll 0$, $\mathcal{M}^p = 0$ for $p \gg 0$, and $\mathcal{M}^p/\mathcal{M}^{p+1}$ are locally free for all p.

(2) Let $x \in X$, and let $(f_j)_{1\le j\le n}$ be elements of $M_{X,x}$ that are not contained in $\mathcal{O}^{\times}_{X,x}$ such that $(f_j \bmod \mathcal{O}^{\times}_{X,x})_{1\le j\le n}$ generates the monoid $(M_X/\mathcal{O}^{\times}_X)_x$. Let $y \in \tau^{-1}(x) \subset X^{\log}$. Then if $s \in \mathrm{sp}(y)$ and if $\exp(s(\log(f_j)))$ are sufficiently near to 0 for all j, $(H_{\mathbf{Z},y}, \langle\ ,\ \rangle_y, F(s))$ is a polarized Hodge structure of weight w.

(3) $(d \otimes_{\mathbf{Z}} 1_{H_{\mathbf{Z}}})(F^p) \subset \omega^{1,\log}_X \otimes_{\mathcal{O}^{\log}_X} F^{p-1}$ for all p.

0.2.20

Elliptic curve (continued). The pair $(H_{\mathbf{Z}}, F)$ in 0.2.18 arising from $E \to \Delta$ is an LVPH on Δ. (The Griffiths transversality (3) in 0.2.19 is satisfied automatically.)

At a point of Δ^*, the fiber of $(H_{\mathbf{Z}}, F)$ is a polarized Hodge structure. However, at $0 \in \Delta$, the fiber of $(H_{\mathbf{Z}}, F)$ should be understood as a family $(F(s))_{s\in\mathrm{sp}(y)}$ for some fixed $y \in \tau^{-1}(0)$ and for varying s. As is explained in Section 0.4, this family is the so-called nilpotent orbit. This is expressed in the schema (2) in Introduction.

0.2.21

LVPH arising from geometry. By the weakly semistable reduction theorem of Abramovich and Karu [AK], any projective fiber space is modified, by alteration and birational modification, to a projective, toroidal morphism $f : Y \to X$ without horizontal divisor, which is equivalent to a projective, vertical, logarithmically smooth morphism (2.1.11; see also 0.2.13) with Coker $\big((M^{\mathrm{gp}}_X/\mathcal{O}^{\times}_X)_{f(y)} \to (M^{\mathrm{gp}}_Y/\mathcal{O}^{\times}_Y)_y\big)$ being torsion-free at any $y \in Y$.

Then by Kato-Matsubara-Nakayama [KMN], for any $m \in \mathbf{Z}$, a variation of polarized logarithmic Hodge structure $(H_{\mathbf{Z}}, \langle\ ,\ \rangle, F)$ of weight m on X is obtained in the following way.

$$H_{\mathbf{Z}} = R^m (f^{\log})_* \mathbf{Z}/(\text{torsion}),$$
$$\langle\ ,\ \rangle : H_{\mathbf{Q}} \times H_{\mathbf{Q}} \to \mathbf{Q} \ \text{ induced from an ample line bundle,}$$
$$\mathcal{M} = R^m f_*(\omega^\bullet_{Y/X}),$$
$$\mathcal{M}^p = R^m f_*(\omega^{\geq p}_{Y/X}) \hookrightarrow \mathcal{M}$$
$$F^p = \mathcal{O}_X^{\log} \otimes_{\tau^{-1}(\mathcal{O}_X)} \tau^{-1}(\mathcal{M}^p) \hookrightarrow \mathcal{O}_X^{\log} \otimes_{\tau^{-1}(\mathcal{O}_X)} \tau^{-1}(\mathcal{M}) = \mathcal{O}_X^{\log} \otimes_{\mathbf{Z}} H_{\mathbf{Z}}.$$

There are many other contributors: [F], [Kf2], [Ma1], [Ma2], [U3], etc.

This is a generalization of work of Steenbrink [St].

0.2.22

The nilpotent orbit theorem of Schmid [Sc] is interpreted as follows (see Theorem 2.5.14).

Let X be a logarithmically smooth, fs logarithmic analytic space, and let $U = \{x \in X \mid M_{X,x} = \mathcal{O}^\times_{X,x}\}$ be the open set of X consisting of all points at which the logarithmic structure is trivial. Let H be a VPH on U with unipotent local monodromy along $X - U$. Then H extends to a LVPH on X.

0.2.23

The theory of logarithmic structure was started in p-adic Hodge theory to construct the logarithmic crystalline cohomology theory for varieties with semistable reduction ([I1], [HK], etc.).

Usually, the theory over p-adic fields begins by following its analogue over $\mathbf{C}$. But in the theory of logarithmic structure, applications appeared first in p-adic Hodge theory. We hope that the theory of logarithmic structure will also be useful in Hodge theory.

0.3 GRIFFITHS DOMAINS AND MODULI OF PH

In [G1], Griffiths defined and studied classifying spaces D of polarized Hodge structures. We review the definition of D. We regard D as moduli of polarized Hodge structures by discarding the Griffiths transversality from VPH (0.3.5–0.3.7).

Fix

$$(w, (h^{p,q})_{p,q \in \mathbf{Z}}, H_0, \langle\ ,\ \rangle_0)$$

where w is an integer, $(h^{p,q})_{p,q\in\mathbf{Z}}$ is a family of non-negative integers such that $h^{p,q} = 0$ unless $p+q = w$, $h^{p,q} \neq 0$ for only finitely many (p, q), and such that $h^{p,q} = h^{q,p}$ for all p, q, H_0 is a free $\mathbf{Z}$-module of rank $\sum_{p,q} h^{p,q}$, and $\langle\ ,\ \rangle_0$ is a nondegenerate bilinear form $H_{0,\mathbf{Q}} \times H_{0,\mathbf{Q}} \to \mathbf{Q}$, which is symmetric if w is even and antisymmetric if w is odd.

DEFINITION 0.3.1 *The classifying space D of polarized Hodge structures of type $\Phi_0 = \big(w, (h^{p,q})_{p,q}, H_0, \langle\ ,\ \rangle_0\big)$ is the set of all decreasing filtrations F on $H_{0,\mathbf{C}} = \mathbf{C} \otimes_{\mathbf{Z}} H_0$ such that the triple $(H_0, \langle\ ,\ \rangle_0, F)$ is a polarized Hodge structure of weight w and of Hodge type $(h^{p,q})_{p,q}$.*

The "compact dual" $\check{D}$ *of D is defined to be the set of all decreasing filtrations F on $H_{0,\mathbf{C}}$ such that* $\dim_{\mathbf{C}}(F^p/F^{p+1}) = h^{p,w-p}$ *for all p that satisfy the Riemann-Hodge first bilinear relation* 0.1.8 (1).

The spaces D and $\check{D}$ have natural structures of analytic manifolds, and D is an open submanifold of $\check{D}$.

The space D is called the *Griffiths domain* and also the *period domain*.

0.3.2

Examples. (i) *Upper half plane.* Consider the case $w = 1$, $h^{1,0} = h^{0,1} = 1$ and $h^{p,q} = 0$ for other (p, q). Let H_0 be a free $\mathbf{Z}$-module of rank 2 with basis e_1, e_2, and define an antisymmetric $\mathbf{Z}$-bilinear form $\langle\ ,\ \rangle_0 : H_0 \times H_0 \to \mathbf{Z}$ by $\langle e_2, e_1\rangle_0 = 1$. Then $D \simeq \mathfrak{h}$, the upper half plane, where we identify a point $\tau \in \mathfrak{h}$ with $F(\tau) \in D$ defined by

$$F^0(\tau) = H_{0,\mathbf{C}}, \quad F^1(\tau) = \mathbf{C}(\tau e_1 + e_2), \quad F^2(\tau) = \{0\}.$$

In this case, $\check{D}$ is identified with $\mathbf{P}^1(\mathbf{C})$.

(ii) *Upper half space* (a generalization of Example (i)). Let $g \geq 1$ and consider the case $w = 1$, $h^{1,0} = h^{0,1} = g$ and $h^{p,q} = 0$ for other (p, q). Let H_0 be a free $\mathbf{Z}$-module with basis $(e_j)_{1\leq j\leq 2g}$ and define a $\mathbf{Z}$-bilinear form $\langle\ ,\ \rangle_0 : H_0 \times H_0 \to \mathbf{Z}$ by

$$(\langle e_j, e_k\rangle_0)_{j,k} = \begin{pmatrix} 0 & -1_g \\ 1_g & 0 \end{pmatrix}.$$

Then $D \simeq \mathfrak{h}_g$, the Siegel upper half space of degree g. Recall that $\mathfrak{h}_g$ is the space of all symmetric matrices over $\mathbf{C}$ of degree g whose imaginary parts are positive definite. We identify a matrix $\tau \in \mathfrak{h}_g$ with $F(\tau) \in D$ as follows:

$$F^0(\tau) = H_{0,\mathbf{C}}, \quad F^1(\tau) = \begin{pmatrix} \text{subspace of } H_{0,\mathbf{C}} \text{ spanned} \\ \text{by the column vectors of } \begin{pmatrix} \tau \\ 1_g \end{pmatrix} \end{pmatrix}, \quad F^2(\tau) = \{0\}.$$

(The symmetry of τ corresponds to the Riemann-Hodge first bilinear relation for $F(\tau)$ and the positivity of $\mathrm{Im}(\tau)$ corresponds to the second bilinear relation (cf. 0.1.8).)

(iii) *Example with* $w = 2$, $h^{2,0} = h^{0,2} = 2$, $h^{1,1} = 1$ (a special case of the example investigated in Section 12.2 where the weight w is shifted to 0). Let H_0 be a free $\mathbf{Z}$-module of rank 5 with basis $(e_j)_{1\leq j\leq 5}$, and let $\langle\ ,\ \rangle_0 : H_{0,\mathbf{Q}} \times H_{0,\mathbf{Q}} \to \mathbf{Q}$ be the bilinear form defined by

$$\left\langle \sum_{1\leq j\leq 5} c_j e_j, \sum_{1\leq j\leq 5} c'_j e_j \right\rangle_0 = -c_1c'_1 - c_2c'_2 - c_3c'_3 + c_4c'_5 + c_5c'_4 \quad (c_j, c'_j \in \mathbf{Q}).$$

Let

$$Q := \{(z_1 : z_2 : z_3) \in \mathbf{P}^2(\mathbf{C}) \mid z_1^2 + z_2^2 + z_3^2 = 0\},$$

$$X := \left\{ (z, a) \;\middle|\; \begin{array}{l} z = (z_1 : z_2 : z_3) \in Q,\ a \in (\sum_{j=1}^{3} \mathbf{C}e_j)/\mathbf{C}(z_1e_1 + z_2e_2 + z_3e_3), \\ a \notin \mathrm{Image}(\mathbf{R}e_1 + \mathbf{R}e_2 + \mathbf{R}e_3) \end{array} \right\}.$$

For $z \in Q$, define a decreasing filtration $F(z)$ of $H_{0,\mathbf{C}}$ by $F^p(z) = H_{0,\mathbf{C}}$ for $p \leq 0$, $F^p(z) = 0$ for $p \geq 3$; $F^2(z)$ is the two-dimensional $\mathbf{C}$-subspace generated by $z_1e_1 + z_2e_2 + z_3e_3$ and e_5, and $F^1(z)$ is the annihilator of F^2 with respect to $\langle\ ,\ \rangle_0$. For $a \in \sum_{j=1}^{3} \mathbf{C}e_j$, let $N_a : H_{0,\mathbf{C}} \to H_{0,\mathbf{C}}$ be the nilpotent $\mathbf{C}$-linear map defined by

$$N_a(e_5) = a, \quad N_a(b) = -\langle a, b\rangle_0 e_4 \text{ for } b \in \sum_{j=1}^{3} \mathbf{C}e_j, \quad N_a(e_4) = 0.$$

Then $a \mapsto N_a$ is $\mathbf{C}$-linear, and $N_aN_b = N_bN_a$ for any a, b. For $a, b \in \sum_{j=1}^{3} \mathbf{C}e_j$ and $z \in Q$, $\exp(N_a)F(z) = \exp(N_b)F(z)$ if and only if $a \equiv b \bmod \mathbf{C}(z_1e_1 + z_2e_2 + z_3e_3)$. We have

$$X \xrightarrow{\sim} D, \quad (z, a) \mapsto \exp(N_a)F(z).$$

The complex dimension of D is 3.

In this example, D is not a symmetric Hermitian domain.

0.3.3

Let $G_{\mathbf{Z}} = \mathrm{Aut}(H_0, \langle\ ,\ \rangle_0)$, and for $R = \mathbf{Q}, \mathbf{R}, \mathbf{C}$ let $G_R = \mathrm{Aut}_R(H_{0,R}, \langle\ ,\ \rangle_0)$ and $\mathfrak{g}_R = \mathrm{Lie}\, G_R = \{A \in \mathrm{End}_R(H_{0,R}) \mid \langle A(x), y\rangle_0 + \langle x, A(y)\rangle_0 = 0\ (\forall x, y \in H_{0,R})\}$.

Then D is a homogeneous space under the natural action of $G_{\mathbf{R}}$. For $\mathbf{r} \in D$, let $K_{\mathbf{r}}$ be the maximal compact subgroup of $G_{\mathbf{R}}$ consisting of all elements that preserve the Hodge metric $(\ ,\ )_{\mathbf{r}}$ associated with $\mathbf{r}$ (0.1.8). The isotropy subgroup $K'_{\mathbf{r}}$ of $G_{\mathbf{R}}$ at $\mathbf{r} \in D$ is contained in $K_{\mathbf{r}}$, but they need not coincide for general D (cf. [G1], and also [Sc]). The following conditions (1) and (2) are equivalent for any $\mathbf{r} \in D$:

(1) D is a symmetric Hermitian domain.
(2) $\dim(K_{\mathbf{r}}) = \dim(K'_{\mathbf{r}})$.

0.3.4

Examples. Upper half space (continued). In this case, we have, for $R = \mathbf{Q}, \mathbf{R}, \mathbf{C}$,

$$G_R = \mathrm{Sp}(g, R) = \{h \in \mathrm{GL}(2g, R) \mid {}^t h J_g h = J_g\},$$

where $J_g = \left(\begin{smallmatrix} 0 & -1_g \\ 1_g & 0 \end{smallmatrix}\right)$. The matrix $\left(\begin{smallmatrix} A & B \\ C & D \end{smallmatrix}\right) \in \mathrm{Sp}(g, \mathbf{R})$ acts on D by

$$F(\tau) \mapsto F(\tau'), \quad \tau' = (A\tau + B)(C\tau + D)^{-1}.$$

Let $\mathbf{r} = F(i1_g) \in D$. Then $K_{\mathbf{r}} = K'_{\mathbf{r}}$, and this group is isomorphic to the unitary group $U(g)$ by $A + iB \mapsto \begin{pmatrix} A & B \\ -B & A \end{pmatrix} \in K_{\mathbf{r}} \subset \mathrm{Sp}(g, \mathbf{R})$.

Example with $h^{2,0} = h^{0,2} = 2$, $h^{1,1} = 1$ (continued). As in 0.3.2 (iii), let $Q = \{(z_1 : z_2 : z_3) \in \mathbf{P}^2(\mathbf{C}) \mid z_1^2 + z_2^2 + z_3^2 = 0\}$. We have a homeomorphism

$$\theta : Q \xrightarrow{\sim} \mathbf{S}^2 = \left\{ \sum_{j=1}^{3} x_j e_j \in H_{0,\mathbf{R}} \;\middle|\; x_j \in \mathbf{R}, x_1^2 + x_2^2 + x_3^2 = 1 \right\}$$

characterized as follows. Let $z = (a_1 + ib_1 : a_2 + ib_2 : a_3 + ib_3) \in Q$ $(a_j, b_j \in \mathbf{R})$. Then $(a_j)_j$ and $(b_j)_j$ are orthogonal in $\mathbf{R}^3$ and have the same length. The characterization of θ is that for $\theta(z) = \sum_{j=1}^{3} c_j e_j \in \mathbf{S}^2$, $(c_j)_j$ is orthogonal to $(a_j)_j$ and $(b_j)_j$, and $\det((a_j)_j, (b_j)_j, (c_j)_j) > 0$.

For $v \in \mathbf{S}^2$, write

$$\mathbf{r}(v) = \exp(iN_v)F(\theta^{-1}(v)) \in D.$$

If we take a basis $(f_j)_{1 \le j \le 5}$ of $H_{0,\mathbf{Q}}$ given by $f_j := e_j$ $(j = 1, 2, 3)$, $f_4 := e_5 - \frac{1}{2}e_4$, $f_5 := e_5 + \frac{1}{2}e_4$, then $(\langle f_j, f_k \rangle_0)_{j,k} = \begin{pmatrix} -1_4 & 0 \\ 0 & 1 \end{pmatrix}$, and hence $G_{\mathbf{R}} \simeq O(1, 4, \mathbf{R})$. Furthermore, for $v \in \mathbf{S}^2$, for the Hodge decomposition $H_{0,\mathbf{C}} = \bigoplus_{p,q} H^{p,q}_{\mathbf{r}(v)}$ corresponding to $\mathbf{r}(v)$, $(f_j)_{1 \le j \le 4}$ is a $\mathbf{C}$-basis of $H^{2,0}_{\mathbf{r}(v)} \oplus H^{0,2}_{\mathbf{r}(v)}$, f_5 is a $\mathbf{C}$-basis of $H^{1,1}_{\mathbf{r}(v)}$, and $((f_j, f_k)_{\mathbf{r}(v)})_{j,k} = 1_5$. Hence the maximal compact subgroup $K_{\mathbf{r}(v)}$ is $O(4, \mathbf{R}) \times O(1, \mathbf{R})$ for the basis $(f_j)_{1 \le j \le 5}$ and is independent of v. For this basis, the isotropy subgroup $K'_{\mathbf{r}(e_2)}$ is the image of

$$U(2) \times O(1, \mathbf{R}) \to O(4, \mathbf{R}) \times O(1, \mathbf{R}), \quad (A + iB) \times (\pm 1) \mapsto \begin{pmatrix} A & B & 0 \\ -B & A & 0 \\ 0 & 0 & \pm 1 \end{pmatrix}.$$

We have $\dim_{\mathbf{R}}(K_{\mathbf{r}(v)}) = 6 > \dim_{\mathbf{R}}(K'_{\mathbf{r}(v)}) = 4$. (This shows, following 0.3.3, that D is not a symmetric Hermitian domain.)

0.3.5

Now we consider the moduli of polarized Hodge structures. We consider what functors D and $\Gamma \backslash D$, for torsion-free subgroups Γ of $G_{\mathbf{Z}}$, represent as analytic spaces. That is, we ask what are $\mathrm{Mor}_{\mathcal{A}}(\ , D)$ and $\mathrm{Mor}_{\mathcal{A}}(\ , \Gamma \backslash D)$ for the category $\mathcal{A}$ of analytic spaces.

We consider $\mathrm{Mor}_{\mathcal{A}}(\ , D)$ first.

For an analytic space X, a morphism $X \to D$ is identified with a decreasing filtration $F = (F^p)_{p \in \mathbf{Z}}$ on the $\mathcal{O}_X$-module $\mathcal{O}_X \otimes_{\mathbf{Z}} H_0$ having the following properties (1) and (2).

(1) $F^p = \mathcal{O}_X \otimes_{\mathbf{Z}} H_0$ for $p \ll 0$, $F^p = 0$ for $p \gg 0$, and the $\mathcal{O}_X$-modules F^p / F^{p+1} are locally free for all $p \in \mathbf{Z}$.

(2) For any $x \in X$, the fiber $(H_0, \langle\ ,\ \rangle_0, F(x))$ is a PH of weight w and of Hodge type $(h^{p,q})_{p,q}$.

Here the object $(\mathcal{O}_X \otimes_{\mathbf{Z}} H_0, F)$ need not satisfy the Griffiths transversality (3) in Section 0.1.10. Corresponding to the identity morphism $D \to D$, there is a universal Hodge filtration F on $\mathcal{O}_D \otimes_{\mathbf{Z}} H_0$, but this F need not satisfy the Griffiths transverality. Hence, for an analytic space X, we will consider an object $(H_{\mathbf{Z}}, \langle\ ,\ \rangle, F)$ which satisfies all conditions of variation of polarized Hodge structure in 0.1.10 except the Griffiths transversality (3). The correct name of such an object might be "the analytic family of polarized Hodge structures parametrized by X without the assumption of Griffiths transversality," but in this book, for simplicity, we will call this object just a *polarized Hodge structure on* X, or, simply, a PH on X.

For an analytic space X, by a PH on X of type $\Phi_0 = \big(w, (h^{p,q})_{p,q}, H_0, \langle\ ,\ \rangle_0\big)$, we mean a PH on X of weight w and of Hodge type $(h^{p,q})_{p,q}$ endowed with an isomorphism $(H_{\mathbf{Z}}, \langle\ ,\ \rangle) \simeq (H_0, \langle\ ,\ \rangle_0)$ of local systems on X. Here $(H_0, \langle\ ,\ \rangle_0)$ is considered as a constant sheaf on X. Let $\underline{\mathrm{PH}}_{\Phi_0}(X)$ be the set of all isomorphism classes of PH on X of type Φ_0. The above interpretation of $\mathrm{Mor}_{\mathcal{A}}(\ , D)$ is rewritten as follows.

Lemma 0.3.6 *We have an isomorphism* $\underline{\mathrm{PH}}_{\Phi_0} \simeq \mathrm{Mor}_{\mathcal{A}}(\ , D)$ *of functors from* $\mathcal{A}$ *to* (Sets).

If $H = (H_{\mathbf{Z}}, \langle\ ,\ \rangle, F)$ is a PH on X of type Φ_0 and $\varphi : X \to D$ is the corresponding morphism, $\varphi(x) \in D$ for $x \in X$ is nothing but the fiber $F(x)$ of F at x regarded as a filtration of $H_{0,\mathbf{C}}$ via the endowed isomorphism $(H_{\mathbf{Z},y}, \langle\ ,\ \rangle) \simeq (H_0, \langle\ ,\ \rangle_0)$.

Let Γ be a torsion-free subgroup of $G_{\mathbf{Z}}$. Then $\Gamma\backslash D$ is an analytic manifold. Let X be an analytic space and let $H = (H_{\mathbf{Z}}, \langle\ ,\ \rangle, F)$ be a PH on X. By a Γ-*level structure* of H, we mean a global section of the sheaf

$$\Gamma\backslash \underline{\mathrm{Isom}}((H_{\mathbf{Z}}, \langle\ ,\ \rangle), (H_0, \langle,\rangle_0)),$$

on X, where $(H_0, \langle\ ,\ \rangle_0)$ is considered as a constant sheaf on X.

A level structure appears as follows. Let X be a connected analytic space and let $H = (H_{\mathbf{Z}}, \langle\ ,\ \rangle, F)$ be a PH on X of weight w and of Hodge type $(h^{p,q})$, let $x \in X$, and define $(H_0, \langle\ ,\ \rangle_0)$ to be the stalk $(H_{\mathbf{Z},x}, \langle\ ,\ \rangle_x)$. Assume Γ contains the image of $\pi_1(X, x) \to G_{\mathbf{Z}}$. Then H has a unique Γ-level structure μ such that the germ μ_x is the germ of the identity map of H_0 modulo Γ.

Let $\Phi_1 = \big(w, (h^{p,q})_{p,q}, H_0, \langle\ ,\ \rangle_0, \Gamma\big)$. For an analytic space X, by a PH on X of type Φ_1, we mean a PH on X of weight w and of Hodge type $(h^{p,q})_{p,q}$ endowed with a Γ-level structure. Let $\underline{\mathrm{PH}}_{\Phi_1}(X)$ be the set of all isomorphism classes of PH on X of type Φ_1.

Lemma 0.3.7 *We have an isomorphism* $\underline{\mathrm{PH}}_{\Phi_1} \simeq \mathrm{Mor}_{\mathcal{A}}(\ , \Gamma\backslash D)$ *of functors from* $\mathcal{A}$ *to* (Sets).

This is deduced from Lemma 0.3.6 as follows. The isomorphism $\underline{\mathrm{PH}}_{\Phi_0} \simeq \mathrm{Mor}_{\mathcal{A}}(\ , D)$ in 0.3.6 preserves the actions of Γ. The functor $\mathrm{Mor}_{\mathcal{A}}(\ , \Gamma\backslash D) : \mathcal{A} \to$ (Sets) is identified with the quotient $\Gamma\backslash \mathrm{Mor}_{\mathcal{A}}(\ , D)$ in the category of sheaf functors, that is, for each object X of $\mathcal{A}$, $\mathrm{Mor}_{\mathcal{A}}(X, \Gamma\backslash D)$ coincides with the set of global sections of the sheaf on X associated with the presheaf $U \mapsto \Gamma\backslash \mathrm{Mor}_{\mathcal{A}}(U, D)$ on X (here U is an open set of X). Similarly, $\underline{\mathrm{PH}}_{\Phi_1}$ is identified with the quotient $\Gamma\backslash \underline{\mathrm{PH}}_{\Phi_0}$ in the category of sheaf functors. Hence $\underline{\mathrm{PH}}_{\Phi_1} \simeq \mathrm{Mor}_{\mathcal{A}}(\ , \Gamma\backslash D)$. □

There is a torsion-free subgroup of $G_{\mathbf{Z}}$ of finite index. More strongly, there is a neat subgroup of $G_{\mathbf{Z}}$ of finite index (see 0.4.1 below).

0.3.8

We emphasize again here that the moduli conditions for $\underline{\mathrm{PH}}_{\Phi_0}$ and $\underline{\mathrm{PH}}_{\Phi_1}$ do not contain the Griffiths transversality on X (0.1.10 (3)). Griffiths transversality should be understood as an important property of the period map, which is satisfied by period maps coming from geometry. Let X be an analytic manifold, let H be a PH on X of type Φ_0, and let $\varphi : X \to D$ be the morphism corrsponding to H (the period map of H). Then the following two conditions (1) and (2) are equivalent.

(1) H satisfies the Griffths transversality.
(2) The image of the morphism of tangent bundles $d\varphi : T_X \to T_D$ induced by φ is contained in the horizontal tangent bundle $T_D^h \subset T_D$.

Here the horizontal tangent bundle is defined by

$$\begin{aligned} T_D^h &= F^{-1}(\mathcal{E}nd_{\langle\,,\,\rangle}(H_{\mathcal{O}}))/F^0(\mathcal{E}nd_{\langle\,,\,\rangle}(H_{\mathcal{O}})) \\ &\subset T_D = \mathcal{E}nd_{\langle\,,\,\rangle}(H_{\mathcal{O}})/F^0(\mathcal{E}nd_{\langle\,,\,\rangle}(H_{\mathcal{O}})) \end{aligned}$$

where $\mathcal{E}nd_{\langle\,,\,\rangle}(H_{\mathcal{O}}) = \{A \in \mathcal{E}nd_{\mathcal{O}}(H_{\mathcal{O}}) \mid \langle Ax, y\rangle + \langle x, Ay\rangle = 0 \ (\forall\, x, y \in H_{\mathcal{O}})\}$ and $F(\mathcal{E}nd_{\langle\,,\,\rangle}(H_{\mathcal{O}}))$ is the filtration on $\mathcal{E}nd_{\langle\,,\,\rangle}(H_{\mathcal{O}})$ induced by the universal Hodge filtration on $H_{\mathcal{O}} = \mathcal{O}_D \otimes_{\mathbf{Z}} H_{\mathbf{Z}}$ ([G2]; also [Sc]).

In particular, by applying this equivalence to the identity morphism of D, we see that the universal PH on D satisfies the Griffiths transversality if and only if the tangent bundle of D coincides with the horizontal tangent bundle. These equivalent conditions are satisfied by Example (ii) (and hence by Example (i)) but not by Example (iii) in 0.3.2. In Example (iii) in 0.3.2, T_D^h is a vector bundle of rank 2 whereas T_D is of rank 3.

0.3.9

Let $f : Y \to X$ be a projective, smooth morphism of analytic manifolds. Fix a polarization of Y over X. Let $(H_{\mathbf{Z}}, \langle\,,\,\rangle, F)$ be the associated VPH (0.1.11). Assume that X is connected, fix a base point $x \in X$, and let $(H_0, \langle\,,\,\rangle_0) := (H_{\mathbf{Z},x}, \langle\,,\,\rangle_x)$, and assume that $\Gamma := \mathrm{Image}(\pi_1(X, x) \to \mathrm{Aut}(H_0, \langle\,,\,\rangle_0)$ is torsion-free. Let $\varphi : X \to \Gamma\backslash D$ be the associated period map (0.3.7). For the differential $d\varphi$ of the period map φ, Griffiths obtained the following commutative diagram:

$$\begin{array}{ccc} T_X & \xrightarrow{\ d\varphi\ } & \varphi^* T^h_{\Gamma\backslash D} = \mathrm{gr}_F^{-1}\, \mathcal{E}nd_{\langle\,,\,\rangle}(H_{\mathcal{O}}) \\ {\scriptstyle \text{K-S}}\downarrow & & \cap\downarrow \\ R^1 f_* T_{Y/X} & \xrightarrow{\text{via coupling}} & \bigoplus_p \mathcal{H}om_{\mathcal{O}_X}(R^{m-p} f_* \Omega^p_{Y/X}, R^{m-p+1} f_* \Omega^{p-1}_{Y/X}) \end{array}$$

where $T^h_{\Gamma\backslash D}$ is the horizontal tangent bundle in the tangent bundle $T_{\Gamma\backslash D}$, K-S on the left vertical arrow means the Kodaira-Spencer map, and the right vertical arrow is the canonical map (for details, see [G2]). The bottom horizontal arrow is often more computable than the top horizontal arrow. This gives a geometric presentation of the differential of the period map.

0.4 TOROIDAL PARTIAL COMPACTIFICATIONS OF $\Gamma\backslash D$ AND MODULI OF PLH

We discuss how to add points at infinity to D to construct a kind of toroidal partial compactification $\Gamma\backslash D_\Sigma$ of $\Gamma\backslash D$. Since "nilpotent orbits" appear in degenerations of Hodge structures, it is natural to add "nilpotent orbits" as points at infinity. We do this in 0.4.1–0.4.12. We describe the space $\Gamma\backslash D_\Sigma$ in 0.4.13–0.4.19. We then explain in 0.4.20–0.4.34 that this enlarged classifying space is a moduli space of "polarized logarithmic Hodge structures."

We fix $(w, (h^{p,q})_{p,q\in\mathbf{Z}}, H_0, \langle\ ,\ \rangle_0)$ as in Section 0.3.

0.4.1

We say a subgroup Γ of $G_\mathbf{Z}$ is *neat* if, for each $\gamma\in\Gamma$, the subgroup of $\mathbf{C}^\times$ generated by all the eigenvalues of γ is torsion-free. It is known that there exists a neat subgroup of $G_\mathbf{Z}$ of finite index (cf. [B]). A neat subgroup of $G_\mathbf{Z}$ is, in particular, torsion-free.

Let Γ be a neat subgroup of $G_\mathbf{Z}$. We will construct some toroidal partial compactifications of $\Gamma\backslash D$ by adding "nilpotent orbits" as points at infinity.

Definition 0.4.2 *A subset σ of $\mathfrak{g}_\mathbf{R} = \mathrm{Lie}\, G_\mathbf{R}$ (0.3.3) is called a* nilpotent cone *if the following conditions (1)–(3) are satisfied.*

(1) $\sigma = (\mathbf{R}_{\ge 0})N_1 + \cdots + (\mathbf{R}_{\ge 0})N_n$ *for some* $n\ge 1$ *and for some* $N_1,\dots, N_n\in\sigma$.
(2) *Any element of σ is nilpotent as an endomorphism of $H_{0,\mathbf{R}}$.*
(3) *$NN' = N'N$ for any $N, N'\in\sigma$ as endomorphisms of $H_{0,\mathbf{R}}$.*

A nilpotent cone is said to be rational *if we can take $N_1,\dots,N_n\in\mathfrak{g}_\mathbf{Q}$ in (1) above.*

0.4.3

For a nilpotent cone σ, a *face* of σ is a nonempty subset τ of σ satisfying the following two conditions.

(1) If $x, y\in\tau$ and $a\in\mathbf{R}_{\ge 0}$, then $x+y, ax\in\tau$.
(2) If $x, y\in\sigma$ and $x+y\in\tau$, then $x, y\in\tau$.

One can show that a face of a nilpotent cone (resp. rational nilpotent cone) is a nilpotent cone (resp. rational nilpotent cone), and that a nilpotent cone has only finitely many faces.

For example, let $N_j\in\mathfrak{g}_\mathbf{R}$ $(1\le j\le n)$ be mutually commutative nilpotent elements that are linearly independent over $\mathbf{R}$. Then $\sigma := \sum_{j=1}^n (\mathbf{R}_{\ge 0})N_j$ is a nilpotent cone, and the faces of σ are $\sum_{j\in J}(\mathbf{R}_{\ge 0})N_j$ for finite subsets J of $\{1,\dots,n\}$.

DEFINITION 0.4.4 *A fan in* $\mathfrak{g}_{\mathbf{Q}}$ *is a nonempty set* Σ *of rational nilpotent cones in* $\mathfrak{g}_{\mathbf{R}}$ *satisfying the following three conditions*:

(1) *If* $\sigma \in \Sigma$, *any face of* σ *belongs to* Σ.
(2) *If* $\sigma, \sigma' \in \Sigma$, $\sigma \cap \sigma'$ *is a face of* σ *and of* σ'.
(3) *Any* $\sigma \in \Sigma$ *is sharp. That is,* $\sigma \cap (-\sigma) = \{0\}$.

0.4.5

Examples. (i) Let

$$\Xi := \{(\mathbf{R}_{\geq 0})N \mid N \text{ is a nilpotent element of } \mathfrak{g}_{\mathbf{Q}}\}.$$

Then Ξ is a fan in $\mathfrak{g}_{\mathbf{Q}}$.

(ii) Let $\sigma \in \mathfrak{g}_{\mathbf{R}}$ be a sharp rational nilpotent cone. Then the set of all faces of σ is a fan in $\mathfrak{g}_{\mathbf{Q}}$.

0.4.6

Nilpotent orbits. Let σ be a nilpotent cone in $\mathfrak{g}_{\mathbf{R}} = \operatorname{Lie} G_{\mathbf{R}}$. For $R = \mathbf{R}, \mathbf{C}$, we denote by σ_R the R-linear span of σ in $\mathfrak{g}_R$.

DEFINITION 0.4.7 *Let* $\sigma = \sum_{1 \leq j \leq r} (\mathbf{R}_{\geq 0}) N_j$ *be a nilpotent cone. A subset* Z *of* $\check{D}$ *is called a* σ-nilpotent orbit *if there is* $F \in \check{D}$ *which satisfies* $Z = \exp(\sigma_{\mathbf{C}})F$ *and satisfies the following two conditions.*

(1) $NF^p \subset F^{p-1}$ $(\forall p, \forall N \in \sigma)$.
(2) $\exp(\sum_{1 \leq j \leq r} z_j N_j)F \in D$ *if* $z_j \in \mathbf{C}$ *and* $\operatorname{Im}(z_j) \gg 0$.

In this case, the pair (σ, Z) *is called a* nilpotent orbit.

DEFINITION 0.4.8 *Let* Σ *be a fan in* $\mathfrak{g}_{\mathbf{Q}}$. *We define the* space D_Σ of nilpotent orbits in the directions in Σ *by*

$$D_\Sigma := \{(\sigma, Z) \mid \sigma \in \Sigma,\ Z \subset \check{D} \text{ is a } \sigma\text{-nilpotent orbit}\}.$$

Note that we have the inclusion map

$$D \hookrightarrow D_\Sigma, \quad F \mapsto (\{0\}, \{F\}).$$

For a sharp rational nilpotent cone σ in $\mathfrak{g}_{\mathbf{R}}$, we denote $D_{\{\text{face of } \sigma\}}$ by D_σ. Then, for a fan Σ in $\mathfrak{g}_{\mathbf{Q}}$, we have $D_\Sigma = \cup_{\sigma \in \Sigma} D_\sigma$.

0.4.9

Upper half plane (continued). Let Ξ be as in 0.4.5. Then $D_\Xi = D \cup \mathbf{P}^1(\mathbf{Q})$. This is explained as follows. For $a \in \mathbf{P}^1(\mathbf{Q})$, let V_a be the one-dimensional $\mathbf{R}$-vector subspace of $H_{0,\mathbf{R}}$ corresponding to a, that is, $V_a = \mathbf{R}(ae_1 + e_2)$ if $a \in \mathbf{Q}$, and $V_\infty = \mathbf{R}e_1$.

For $a \in \mathbf{P}^1(\mathbf{Q})$, define a sharp rational nilpotent cone $\sigma_a \in \Xi$ by

$$\sigma_a = \{N \in \mathfrak{g}_{\mathbf{R}} \mid N(H_{0,\mathbf{R}}) \subset V_a,\, N(V_a) = \{0\},\\ \langle x, N(x)\rangle_0 \geq 0 \text{ for any } x \in H_{0,\mathbf{R}}\}.$$

We identify $a \in \mathbf{P}^1(\mathbf{Q})$ with the nilpotent orbit $(\sigma_a, Z_a) \in D_\Xi$ where $Z_a = \{F \in \check{D} \mid F^1 \neq V_{a,\mathbf{C}}\}$. For example,

$$\sigma_\infty = \begin{pmatrix} 0 & \mathbf{R}_{\geq 0} \\ 0 & 0 \end{pmatrix}, \quad Z_\infty = \mathbf{C} \subset \mathbf{P}^1(\mathbf{C}) = \check{D}.$$

Definition 0.4.10 *Let Σ be a fan in $\mathfrak{g}_\mathbf{Q}$ and let Γ be a subgroup of $G_\mathbf{Z}$.*

(i) *We say Γ is* compatible with Σ *if the following condition* (1) *is satisfied.*

(1) *If $\gamma \in \Gamma$ and $\sigma \in \Sigma$, then $\mathrm{Ad}(\gamma)(\sigma) \in \Sigma$. Here, $\mathrm{Ad}(\gamma)(\sigma) = \gamma\sigma\gamma^{-1}$.*
Note that, if Γ is compatible with Σ, Γ acts on D_Σ by

$$\gamma : (\sigma, Z) \mapsto (\mathrm{Ad}(\gamma)(\sigma), \gamma Z) \quad (\gamma \in \Gamma).$$

(ii) *We say Γ is* strongly compatible with Σ *if it is compatible with Σ and the following condition* (2) *is also satisfied. For $\sigma \in \Sigma$, define*

$$\Gamma(\sigma) := \Gamma \cap \exp(\sigma).$$

(2) *The cone σ is generated by* $\log \Gamma(\sigma)$, *that is, any element of σ can be written as a sum of $a \log(\gamma)$ ($a \in \mathbf{R}_{\geq 0}$, $\gamma \in \Gamma(\sigma)$).*

Note that $\Gamma(\sigma)$ is a sharp fs monoid and $\Gamma(\sigma)^{\mathrm{gp}} = \Gamma \cap \exp(\sigma_\mathbf{R})$.

0.4.11

Example. If $\Sigma = \Xi$ in 0.4.5 and Γ is of finite index in $G_\mathbf{Z}$, then Γ is strongly compatible with Σ. If $\Sigma = \Xi$ and Γ is just a subgroup of $G_\mathbf{Z}$, it is compatible with Σ but is not necessarily strongly compatible with Σ.

0.4.12

Assume that Γ and Σ are strongly compatible.

In Chapter 3, we will define a topology of $\Gamma\backslash D_\Sigma$ for which $\Gamma\backslash D$ is a dense open subset of $\Gamma\backslash D_\Sigma$ and which has the following property. Let $(\sigma, Z) \in D_\Sigma$, $F \in D$, and $N_j \in \mathfrak{g}_\mathbf{R}$ ($1 \leq j \leq n$), and assume $\sigma = \sum_{j=1}^n (\mathbf{R}_{\geq 0})N_j$. Then

$$\left(\exp\left(\sum_{j=1}^n z_j N_j\right) F \bmod \Gamma\right)$$
$$\to ((\sigma, Z) \bmod \Gamma) \quad \text{if } z_j \in \mathbf{C} \text{ and } \mathrm{Im}(z_j) \to \infty \ (\forall j).$$

Furthermore, in Chapter 3, we will introduce on $\Gamma\backslash D_\Sigma$ a structure of a local ringed space over $\mathbf{C}$ and also a logarithmic structure. In 0.4.13 (resp. 0.4.18), we describe what this local ringed structure looks like in the cases of Examples (i) and (ii) (resp. Example (iii)) in 0.3.2. If Γ is neat, the logarithmic structure M of $\Gamma\backslash D_\Sigma$ is an fs logarithmic structure and has the form $M = \{f \in \mathcal{O} \mid f$ is invertible on $\Gamma\backslash D\} \overset{\alpha}{\hookrightarrow} \mathcal{O}$ for the structure sheaf of rings $\mathcal{O}$ of $\Gamma\backslash D_\Sigma$.

0.4.13

Examples. Upper half plane (continued). For

$$\Gamma = \begin{pmatrix} 1 & \mathbf{Z} \\ 0 & 1 \end{pmatrix} \subset \mathrm{SL}(2, \mathbf{Z}), \quad \sigma = \sigma_\infty = \begin{pmatrix} 0 & \mathbf{R}_{\geq 0} \\ 0 & 0 \end{pmatrix},$$

we have a commutative diagram of analytic spaces

$$\begin{array}{ccc} \Delta^* & \simeq & \Gamma\backslash D \\ \cap & & \cap \\ \Delta & \simeq & \Gamma\backslash D_\sigma. \end{array}$$

Here the upper isomorphism sends $e^{2\pi i\tau} \in \Delta^*$ ($\tau \in \mathfrak{h} = D$) to ($\tau \bmod \Gamma$); the lower isomorphism extends the upper isomorphism by sending $0 \in \Delta$ to the class of the nilpotent orbit ((σ_∞, $\mathbf{C}$) mod Γ).

Next we consider a subgroup Γ of $G_{\mathbf{Z}}$ of finite index. Let $\sigma = \sigma_\infty$. Let $n \geq 1$ and let $\Gamma = \Gamma(n)$ be the kernel of $\mathrm{SL}(2, \mathbf{Z}) \to \mathrm{SL}(2, \mathbf{Z}/n\mathbf{Z})$. Then

$$\Gamma(\sigma) = \begin{pmatrix} 1 & n\mathbf{N} \\ 0 & 1 \end{pmatrix}, \quad \Gamma(\sigma)^{\mathrm{gp}} = \Gamma \cap \begin{pmatrix} 1 & \mathbf{Z} \\ 0 & 1 \end{pmatrix} = \begin{pmatrix} 1 & n\mathbf{Z} \\ 0 & 1 \end{pmatrix}.$$

Γ is neat if and only if $n \geq 3$. For $n \geq 3$, as is well known in the theory of modular curves, we have the following local description of $\Gamma\backslash D_\Xi$ at the boundary point ($\infty \bmod \Gamma$) in $\Gamma\backslash D_\Xi = \Gamma\backslash(\mathfrak{h} \cup \mathbf{P}^1(\mathbf{Q}))$. We have

$$\Delta \overset{\sim}{\to} \Gamma(\sigma)^{\mathrm{gp}}\backslash D_\sigma \xrightarrow{\text{local isom}} \Gamma\backslash D_\Xi,$$

where the first arrow is an isomorphism of analytic spaces which sends $q \in \Delta^*$ to ($\tau \bmod \Gamma(\sigma)^{\mathrm{gp}}$) with $q = e^{2\pi i\tau/n}$, and sends $0 \in \Delta$ to ((σ_∞, $\mathbf{C}$) mod $\Gamma(\sigma)^{\mathrm{gp}}$), and the second arrow is the canonical projection and is locally an isomorphism of analytic spaces.

Upper half space (continued). We consider the case $g = 2$, i.e., $D = \mathfrak{h}_2$. Let U be the open set of $\mathbf{C}^3$ defined by

$$U = \{(q_1, q_2, a) \in \Delta^2 \times \mathbf{C} \mid \text{if } q_1 q_2 \neq 0, \\ \text{then } \log(|q_1|)\log(|q_2|) > (2\pi\, \mathrm{Im}(a))^2\}.$$

Let $N_1, N_2 \in \mathfrak{g}_{\mathbf{R}} = \mathfrak{sp}(2, \mathbf{R})$ be the nilpotent elements defined by

$$N_1(e_3) = e_1,\ N_1(e_j) = 0 \text{ for } j \neq 3, \quad N_2(e_4) = e_2,\ N_2(e_j) = 0 \text{ for } j \neq 4.$$

Then

$$\exp(z_1N_1 + z_2N_2)F\begin{pmatrix} a & b \\ b & c \end{pmatrix} = F\begin{pmatrix} a + z_1 & b \\ b & c + z_2 \end{pmatrix}$$

for any $z_1, z_2, a, b, c \in \mathbf{C}$. Let

$$\sigma = (\mathbf{R}_{\geq 0})N_1 + (\mathbf{R}_{\geq 0})N_2 \subset \mathfrak{g}_{\mathbf{R}}.$$

For $\Gamma = \exp(\mathbf{Z}N_1 + \mathbf{Z}N_2) = 1 + \mathbf{Z}N_1 + \mathbf{Z}N_2$, Γ is strongly compatible with the fan $\{\{0\}, (\mathbf{R}_{\geq 0})N_1, (\mathbf{R}_{\geq 0})N_2, \sigma\}$ of all faces of σ, and we have the following isomorphism of analytic spaces. For $q_j = e^{2\pi i \tau_j}$,

$$\begin{aligned}
&\Delta^2 \times \mathbf{C} \supset U \simeq \Gamma\backslash D_\sigma, \\
&(q_1, q_2, a) \mapsto F\begin{pmatrix} \tau_1 & a \\ a & \tau_2 \end{pmatrix} \bmod \Gamma \quad (q_1 q_2 \neq 0), \\
&(0, q_2, a) \mapsto \left((\mathbf{R}_{\geq 0})N_1,\ F\begin{pmatrix} \mathbf{C} & a \\ a & \tau_2 \end{pmatrix}\right) \bmod \Gamma \quad (q_2 \neq 0), \\
&(q_1, 0, a) \mapsto \left(\mathbf{R}_{\geq 0}N_2,\ F\begin{pmatrix} \tau_1 & a \\ a & \mathbf{C} \end{pmatrix}\right) \bmod \Gamma \quad (q_1 \neq 0), \\
&(0, 0, a) \mapsto \left(\sigma,\ F\begin{pmatrix} \mathbf{C} & a \\ a & \mathbf{C} \end{pmatrix}\right) \bmod \Gamma.
\end{aligned}$$

More generally, for any strongly compatible pair (Γ, Σ) such that $\sigma \in \Sigma$ and Γ is neat and such that $\Gamma(\sigma)^{\mathrm{gp}} = \exp(\mathbf{Z}N_1 + \mathbf{Z}N_2)$, the above isomorphism $U \simeq \Gamma(\sigma)^{\mathrm{gp}}\backslash D_\sigma$ induces a morphism of analytic spaces $U \to \Gamma\backslash D_\Sigma$ which is locally an isomorphism.

0.4.14

We say that we are in the classical situation if D is a symmetric Hermitian domain and the tangent bundle of D coincides with the horizontal tangent bundle. Example (ii) (and hence Example (i)) in 0.3.2 belongs to the classical situation, but Example (iii) does not.

In the classical situation, $\Gamma\backslash D_\Sigma$ is a toroidal partial compactification as constructed by Mumford et al. [AMRT]. Under the assumption $D \neq \emptyset$, the classical situation is listed as follows.

Case (1) $w = 2t+1$, $h^{t+1,t} = h^{t,t+1} \geq 0$, $h^{p,q} = 0$ for other (p, q).
Case (2) $w = 2t$, $h^{t+1,t-1} = h^{t-1,t+1} \leq 1$, $h^{t,t} \geq 0$, $h^{p,q} = 0$ for other (p, q).

For general D, the space $\Gamma\backslash D_\Sigma$ is not necessarily a complex analytic space because it may have "slits" caused by "Griffiths transversality at the boundary". But still it has a kind of complex structure, period maps can be extended to $\Gamma\backslash D_\Sigma$, and infinitesimal calculus can be performed nicely. This was first observed by the simplest example in [U2].

In the terminology of this book, $\Gamma\backslash D_\Sigma$ is a "logarithmic manifold," as explained in 0.4.15–0.4.17 below.

0.4.15

Strong topology. The underlying local ringed space over $\mathbf{C}$ of $\Gamma\backslash D_\Sigma$ is not necessarily an analytic space in general. Sometimes, it can be something like

(1) $S := \{(x, y) \in \mathbf{C}^2 \mid x \neq 0\} \cup \{(0, 0)\} = \{(x, y) \in \mathbf{C}^2 \mid \text{if } x = 0, \text{ then } y = 0\}$

endowed with a topology that is stronger than the topology as a subspace of $\mathbf{C}^2$, called the "strong topology."

Let Z be an analytic space and S be a subset of Z. A subset U of S is open in the *strong topology of S in Z* if and only if, for any analytic space Y and any morphism $\lambda : Y \to Z$ such that $\lambda(Y) \subset S$, $\lambda^{-1}(U)$ is open on Y.

If S is a locally closed analytic subspace of Z, the strong topology coincides with the topology as a subspace of Z. However the strong topology of the set S in (1) is stronger than the topology as a subspace of $\mathbf{C}^2$. For example,

(2) Let $f : \mathbf{R}_{>0} \to \mathbf{R}_{>0}$ be a map such that for each integer $n \geq 1$, there exists $\varepsilon_n > 0$ for which $f(s) \leq s^n$ if $0 < s < \varepsilon_n$. (An example of $f(s)$ is $e^{-1/s}$.) Then, if $s > 0$ and $s \to 0$, $(f(s), s)$ converges to $(0, 0)$ for the topology of S as a subspace of $\mathbf{C}^2$, but it does not converge for the strong topology (see 3.1.3). (Roughly speaking, $(f(s), s)$ runs too near to the "bad line" $\{0\} \times \mathbf{C}$.)

0.4.16

Categories $\mathcal{A}$, $\mathcal{A}$(log), $\mathcal{B}$, $\mathcal{B}$(log). We define the categories

$$\mathcal{A} \subset \mathcal{B}, \quad \mathcal{A}(\log) \subset \mathcal{B}(\log)$$

as follows (cf. 3.2.4).

We denote by

$$\mathcal{A}, \quad \mathcal{A}(\log),$$

the category of analytic spaces and the category of fs logarithmic analytic spaces, respectively.

Let $\mathcal{B}$ be the category of local ringed spaces X over $\mathbf{C}$ (over $\mathbf{C}$ means that $\mathcal{O}_X$ is a $\mathbf{C}$-algebra) having the following property: X has an open covering $(U_\lambda)_\lambda$ such that, for each λ, there exists an isomorphism $U_\lambda \simeq S_\lambda$ of local ringed spaces over $\mathbf{C}$ for some subset S_λ of an analytic space Z_λ, where S_λ is endowed with the strong topology in Z_λ and with the inverse image of $\mathcal{O}_{Z_\lambda}$.

Let $\mathcal{B}$(log) be the category of objects of $\mathcal{B}$ endowed with an fs logarithmic structure.

0.4.17

Logarithmic manifolds. Our space $\Gamma \backslash D_\Sigma$ is a very special object in $\mathcal{B}$(log), called a "logarithmic manifold" (cf. Section 3.5).

We first describe the idea of the logarithmic manifold by using the example $S \subset \mathbf{C}^2$ in 0.4.15 (1). Let $Z = \mathbf{C}^2$ with coordinate functions x, y, and endow Z with the logarithmic structure M_Z associated with the divisor "$x = 0$." Then the sheaf ω^1_Z of logarithmic differential forms on Z (= the sheaf of differential forms with logarithmic poles along $x = 0$) is a free $\mathcal{O}_Z$-module with basis $(d\log(x), dy)$. For each $z \in Z$, let ω^1_z be the module of logarithmic differential forms on the point z which is regarded as an fs logarithmic analytic space endowed with the ring $\mathbf{C}$ and with the inverse image of M_Z. Then, if z does not belong to the part $x = 0$ of Z, z is just the usual point Spec(**C**) with the trivial logarithmic structure $\mathbf{C}^\times$, and $\omega_z^1 = 0$. If z is in the part $x = 0$, z is a point Spec(**C**) with the induced logarithmic structure $M_z = \bigsqcup_{n\geq 0} \mathbf{C}^\times x^n \simeq \mathbf{C}^\times \times \mathbf{N}$, and hence ω_z^1 is a one-dimensional **C**-vector space generated by $d\log(x)$. Thus ω_z^1 is not equal to the fiber of ω_Z^1 at z which is a two-dimensional **C**-vector space with basis $(d\log(x), dy)$. Now the above set S has a presentation

$$S = \{z \in Z \mid \text{the image of } yd\log(x) \text{ in } \omega_z^1 \text{ is zero}\}. \tag{1}$$

Recall that zeros of a holomorphic function on Z form a closed analytic subset of Z. Here we discovered that S is the set of "zeros" of the differential form $yd\log(x)$ on Z, but the meaning of "zero" is not that the image of $yd\log(x)$ in the fiber of ω_Z^1 is zero (the latter "zeros" form the closed analytic subset $y = 0$ of Z). The set "zeros in the new sense" of a differential form with logarithmic poles is the idea of a "logarithmic manifold."

The precise definition is as follows (cf. 3.5.7). By a *logarithmic manifold*, we mean a local ringed space over **C** endowed with an fs logarithmic structure which has an open covering $(U_\lambda)_\lambda$ with the following property: For each λ, there exist a logarithmically smooth fs logarithmic analytic space Z_λ, a finite subset I_λ of $\Gamma(Z_\lambda, \omega_{Z_\lambda}^1)$, and an isomorphism of local ringed spaces over **C** with logarithmic structures between U_λ and an open subset of

$$S_\lambda = \{z \in Z_\lambda \mid \text{the image of } I_\lambda \text{ in } \omega_z^1 \text{ is zero}\}, \tag{2}$$

where S_λ is endowed with the strong topology in Z_λ and with the inverse images of $\mathcal{O}_{Z_\lambda}$ and M_{Z_λ}.

0.4.18

Example with $h^{2,0} = h^{0,2} = 2$, $h^{1,1} = 1$ (continued). Let Γ be a neat subgroup of $G_\mathbf{Z}$ of finite index. We give a local description of the space $\Gamma\backslash D_\Xi$ and observe that this space has a slit.

Fix $v \in \mathbf{S}^2 \cap (\sum_{j=1}^3 \mathbf{Q}e_j)$ and fix a nonzero element v' of $\sum_{j=1}^3 \mathbf{R}e_j$ which is linearly independent of v over **R**. Define $\ell \in \mathbf{Q}_{>0}$ by $\Gamma \cap \exp(\mathbf{Q}N_v) = \exp(\ell\mathbf{Z}N_v)$. Let

$$U = \{(q, a, z) \in \mathbf{C}^2 \times Q \mid (q, a, z) \text{ satisfies (1) and (2) below}\}.$$

(1) If $q \neq 0$ and $q = e^{2\pi i\tau/\ell}$, then $\langle \mathrm{Im}(\tau)v + \mathrm{Im}(a)v', \theta(z)\rangle_0 < 0$ (for θ, see 0.3.4).
(2) If $q = 0$, then $\theta(z) = v$.

Endow U with the strong topology in $Z := \mathbf{C}^2 \times Q$, with the sheaf of rings $\mathcal{O}_Z|_U$, and with the inverse image of the logarithmic structure of Z associated to the divisor $q = 0$. By the above condition (2), U has a slit and it is not an analytic space. But U is a logarithmic manifold.

Let $\sigma = (\mathbf{R}_{\geq 0})N_v$. We have morphisms of local ringed spaces over **C** with logarithmic structures

$$U \to \Gamma(\sigma)^{\mathrm{gp}}\backslash D_\sigma \to \Gamma\backslash D_\Xi$$

which are locally isomorphisms, and the left of which sends $(q, a, z) \in U$ to the class of $\exp(\tau N_v + aN_{v'})F(z) = \exp(\tau N_v)\exp(aN_{v'})F(z)$ if $q \neq 0$ and $q = e^{2\pi i\tau/\ell}$, and to the class of $(\sigma, \exp(\sigma_{\mathbf{C}})\exp(aN_{v'})F(z))$ if $q = 0$ and $\theta(z) = v$.

The above local description of $\Gamma\backslash D_\Xi$ is obtained from Proposition 12.2.5. Note that e_1, e_2 in Section 12.2 are e_4, e_5 here, respectively. The slit in $\Gamma\backslash D_\Xi$ corresponding to the slit in U appears by "small Griffiths transversality," as explained in 0.4.29 below.

Theorem 0.4.19 (cf. Theorem A in Section 4.1) *Let Σ be a fan in $\mathfrak{g}_{\mathbf{Q}}$ and let Γ be a neat subgroup of $G_{\mathbf{Z}}$ which is strongly compatible with Σ.*

(i) *Then $\Gamma\backslash D_\Sigma$ is a logarithmic manifold. It is a Hausdorff space.*
(ii) *For any $\sigma \in \Sigma$, the canonical projection $\Gamma(\sigma)^{\mathrm{gp}}\backslash D_\sigma \to \Gamma\backslash D_\Sigma$ is locally an isomorphism of logarithmic manifolds.*

The Hausdorff property of $\Gamma\backslash D_\Sigma$ is by virtue of the strong topology. Proposition 12.3.6 gives an example such that $\Gamma\backslash D_\Sigma$ is not Hausdorff if we use a naive topology that is weaker than the strong topology.

Now we consider a polarized logarithmic Hodge structure and its moduli.

0.4.20

For an object X of $\mathcal{B}(\log)$, a ringed space $(X^{\log}, \mathcal{O}_X^{\log})$ is defined just as in the case of fs logarithmic analytic spaces (see Section 2.2). It is described locally as follows. Assume that the logarithmic structure of X is induced from a homomorphism $\mathcal{S} \to \mathcal{O}_X$ with $\mathcal{S}$ an fs monoid. Let $Z = \mathrm{Spec}(\mathbf{C}[\mathcal{S}])_{\mathrm{an}}$. Then

$$X^{\log} = X \times_Z Z^{\log} = X \times_{\mathrm{Hom}(\mathcal{S}, \mathbf{C}^{\mathrm{mult}})} \mathrm{Hom}(\mathcal{S}, \mathbf{R}^{\mathrm{mult}}_{\geq 0} \times \mathbf{S}^1), \quad \mathcal{O}_X^{\log} = \mathcal{O}_X \otimes_{\mathcal{O}_Z} \mathcal{O}_Z^{\log}.$$

0.4.21

Let X be an object of $\mathcal{B}(\log)$. A *prepolarized logarithmic Hodge structure (pre-PLH) on X of weight w* is a triple $(H_{\mathbf{Z}}, \langle\ ,\ \rangle, F)$ consisting of a locally constant sheaf $H_{\mathbf{Z}}$ of free $\mathbf{Z}$-modules of finite rank on $X^{\log}$, a bilinear form $\langle\ ,\ \rangle : H_{\mathbf{Q}} \times H_{\mathbf{Q}} \to \mathbf{Q}$, and a decreasing filtration F on $\mathcal{O}_X^{\log} \otimes_{\mathbf{Z}} H_{\mathbf{Z}}$ by $\mathcal{O}_X^{\log}$-submodules which satisfy the following condition (1).

(1) There exist a locally free $\mathcal{O}_X$-module $\mathcal{M}$ and a decreasing filtration $(\mathcal{M}^p)_{p\in\mathbf{Z}}$ by $\mathcal{O}_X$-submodules of $\mathcal{M}$ such that $\mathcal{M}^p = \mathcal{M}$ for $p \ll 0$, $\mathcal{M}^p = 0$ for $p \gg 0$, and $\mathcal{M}^p/\mathcal{M}^{p+1}$ are locally free for all p, and such that $\mathcal{O}_X^{\log} \otimes_{\mathbf{Z}} H_{\mathbf{Z}} = \mathcal{O}_X^{\log} \otimes_{\tau^{-1}(\mathcal{O}_X)} \tau^{-1}(\mathcal{M})$ and $F^p = \mathcal{O}_X^{\log} \otimes_{\tau^{-1}(\mathcal{O}_X)} \tau^{-1}(\mathcal{M}^p)$ for all p. Furthermore, the annihilator of F^p with respect to $\langle\ ,\ \rangle$ coincides with F^{w+1-p} for any p.

We give two remarks concerning pre-PLH.

(i) If $H = (H_{\mathbf{Z}}, \langle\ ,\ \rangle, F)$ is a pre-PLH on X, the $\mathcal{O}_X$-modules $\mathcal{M}$ and $\mathcal{M}^p$ in (1) above are determined by H as

$$\mathcal{M} = \tau_*(\mathcal{O}_X^{\log} \otimes_{\mathbf{Z}} H_{\mathbf{Z}}), \quad \mathcal{M}^p = \tau_*(F^p).$$

This follows from Proposition 2.2.10.

(ii) (Proposition 2.3.3 (ii)) If $(H_{\mathbf{Z}}, \langle\ ,\ \rangle, F)$ is a pre-PLH on X, then for any $x \in X$ and for any $y \in X^{\log}$ lying over x, the action of $\pi_1(x^{\log})$ on $H_{\mathbf{Z},y}$ is unipotent.

0.4.22

Let X be an object of $\mathcal{B}(\log)$ and let $H = (H_{\mathbf{Z}}, \langle\ ,\ \rangle, F)$ be a pre-PLH on X. We call H a *polarized logarithmic Hodge structure* (*PLH*) *on* X if for each $x \in X$, it satisfies the following two conditions.

Positivity on x. Let y be any element of $X^{\log}$ lying over x. Take a finite family f_j $(1 \le j \le n)$ of elements of $M_{X,x}$ which do not belong to $\mathcal{O}_{X,x}^\times$ such that the monoid $(M_X/\mathcal{O}_X^\times)_x$ is generated by the images of f_j. Then, if $s \in \mathrm{sp}(y)$ and if $\exp(s(\log(f_j)))$ are sufficiently near to 0, $(H_{\mathbf{Z},y}, \langle\ ,\ \rangle_y, F(s))$ is a polarized Hodge structure.

Griffiths transversality on x. $(d \otimes 1_{H_{\mathbf{Z}}})(F^p|_{x^{\log}}) \subset \omega_x^{1,\log} \otimes_{\mathcal{O}_x^{\log}} (F^{p-1}|_{x^{\log}})$ for any p.

Here $\mathcal{O}_x^{\log}$ and $\omega_x^{1,\log}$ are those of the point $x = \mathrm{Spec}(\mathbf{C})$ endowed with the inverse image of M_X, and $F|_{x^{\log}}$ denotes the module-theoretic inverse image of F under the morphism of ringed spaces $(x^{\log}, \mathcal{O}_x^{\log}) \to (X^{\log}, \mathcal{O}_X^{\log})$.

In other words, PLH is a pre-PLH whose pullback to the fs logarithmic point x for any $x \in X$ satisfies the conditions in 0.2.19 for an LVPH (we take x as X in 0.2.19 here). Although x is not logarithmically smooth unless the logarithmic structue of x is trivial, the conditions in 0.2.19 make sense when we replace X there by x.

In the case that X is a logarithmically smooth, fs logarithmic analytic space, the validity of the above Griffiths transversality on x for all $x \in X$ (we call this the *small Griffiths transversality*) is much weaker than the Griffiths transversality (3) in 0.2.19 (we call this the *big Griffiths transversality*). In fact if the logarithmic structure of X is trivial (that is, $M_X = \mathcal{O}_X^\times$), $\omega_x^{1,\log} = 0$ for any point $x \in X$, and hence the small Griffiths transversality is an empty condition. Hence an LVPH on X is a PLH on X, but a PLH on X is not necessarily an LVPH on X.

0.4.23

In 0.4.23–0.4.25, we will see that the notion of a "nilpotent orbit" is nothing but a "PLH on an fs logarithmic point".

Let x be an fs logarithmic point. Then $x^{\log} \simeq \mathrm{Hom}(M_x^{\mathrm{gp}}/\mathcal{O}_x^\times, \mathbf{S}^1)$ and hence $\pi_1(x^{\log}) \simeq \mathrm{Hom}(M_x^{\mathrm{gp}}/\mathcal{O}_x^\times, \mathbf{Z})$. Let $\pi_1^+(x^{\log}) \subset \pi_1(x^{\log})$ be the part corresponding to the part $\mathrm{Hom}(M_x/\mathcal{O}_x^\times, \mathbf{N}) \subset \mathrm{Hom}(M_x^{\mathrm{gp}}/\mathcal{O}_x^\times, \mathbf{Z})$. Then $\pi_1^+(x^{\log})$ is an fs monoid.

Proposition 0.4.24 (cf. Propositions 2.5.1, and 2.5.5) *Let x be an fs logarithmic point and let $H = (H_{\mathbf{Z}}, \langle\ ,\ \rangle, F)$ be a pre-PLH on x. Let $y \in x^{\log}$.*

(i) *Let $h_j \in \mathrm{Hom}(M_x^{\mathrm{gp}}/\mathcal{O}_x^\times, \mathbf{Z})$ $(1 \le j \le n)$, let $\gamma_j \in \pi_1(x^{\log})$ be the element corresponding to h_j, and let $N_j : H_{\mathbf{Q},y} \to H_{\mathbf{Q},y}$ be the logarithm of the (unipotent) action of γ_j on $H_{\mathbf{Q},y}$. Let $z_j \in \mathbf{C}$ $(1 \le j \le n)$, $s_0 \in \mathrm{sp}(y)$, and let s be the element of $\mathrm{sp}(y)$ characterized by*

$$s((2\pi i)^{-1}\log(f)) - s_0((2\pi i)^{-1}\log(f)) = \sum_{j=1}^{n} z_j h_j(f) \quad \textit{for any} \quad f \in M_x^{\mathrm{gp}}$$

(see 0.2.17). Then

$$F(s) = \exp\left(\sum_{j=1}^{n} z_j N_j\right) F(s_0).$$

(ii) *Let $(\gamma_j)_{1\le j\le n}$ be a finite family of generators of the monoid $\pi_1^+(x^{\log})$ and let $N_j : H_{\mathbf{Q},y} \to H_{\mathbf{Q},y}$ be the logarithm of the action of γ_j on $H_{\mathbf{Q},y}$. Fix $s \in \mathrm{sp}(y)$. Then H satisfies the positivity on x in 0.4.22 if and only if the following condition is satisfied:*

$$\left(H_{\mathbf{Z},y}, \langle\ ,\ \rangle_y, \exp\left(\sum_{j=1}^{n} z_j N_j\right) F(s)\right) \textit{ is a PH if } \mathrm{Im}(z_j) \gg 0 \ (\forall j).$$

(iii) *Let N_j $(1 \le j \le n)$ be as in* (ii) *and let $s \in \mathrm{sp}(y)$. Then H satisfies the Griffiths transversality on x in 0.4.22 if and only if*

$$N_j F^p(s) \subset F^{p-1}(s) \textit{ for any } j \textit{ and } p.$$

0.4.25

With the notation in (ii) in 0.4.24, let $\sigma = \sum_{j=1}^{n}(\mathbf{R}_{\ge 0})N_j$. By (i) of 0.4.24, $\{F(s)\}_{s\in \mathrm{sp}(y)}$ is an $\exp(\sigma_{\mathbf{C}})$-orbit. By (ii) and (iii) of 0.4.24, this $\exp(\sigma_{\mathbf{C}})$-orbit is a σ-nilpotent orbit if and only if H is a PLH on x. In other words,

$$(\text{a PLH on an fs logarithmic point}) = (\text{a nilpotent orbit}).$$

Hence if H is a PLH on an object X of $\mathcal{B}(\log)$, for each $x \in X$, the pullback $H(x)$ of H to the fs logarithmic point x is regarded as a nilpotent orbit. This fact is presented in schema (2) in Introduction.

0.4.26

We generalize the functor $\underline{\mathrm{PH}}_{\Phi_1} : \mathcal{A} \to (\mathrm{Sets})$ in 0.3.7 to the logarithmic case.

Let Γ be a neat subgroup of $G_{\mathbf{Z}}$ and let Σ be a fan in $\mathfrak{g}_{\mathbf{Q}}$. Assume that Γ and Σ are strongly compatible (0.4.10). Let $\Phi = \big(w, (h^{p,q})_{p,q}, H_0, \langle\ ,\ \rangle_0, \Gamma, \Sigma\big)$.

For a PLH $H = (H_{\mathbf{Z}}, \langle\ ,\ \rangle, F)$ on X, by a Γ-level structure on H, we mean a global section of the sheaf $\Gamma\backslash\, \underline{\mathrm{Isom}}((H_{\mathbf{Z}}, \langle\ ,\ \rangle), (H_0, \langle\ ,\ \rangle_0))$ on $X^{\log}$.

We define a functor $\underline{\mathrm{PLH}}_{\Phi} : \mathcal{B}(\log) \to (\mathrm{Sets})$ as follows. For $X \in \mathcal{B}(\log)$, let $\underline{\mathrm{PLH}}_{\Phi}(X)$ be the set of all isomorphism classes of PLH on X of type $(w, (h^{p,q})_{p,q \in \mathbf{Z}})$ endowed with a Γ-level structure μ satisfying the following condition:

For any $x \in X$, any $y \in x^{\log}$ and a lifting $\tilde{\mu}_y : (H_{\mathbf{Z},y}, \langle\ ,\ \rangle_y) \xrightarrow{\sim} (H_0, \langle\ ,\ \rangle_0)$ of the germ of μ at y, we have the following (1) and (2).

(1) There exists $\sigma \in \Sigma$ such that

$$\mathrm{Image}\left(\pi_1^+(x^{\log}) \to \mathrm{Aut}(H_{\mathbf{Z},y}) \xrightarrow{\text{by } \tilde{\mu}_y} Aut(H_0)\right) \subset \exp(\sigma).$$

(2) For the smallest such $\sigma \in \Sigma$ and for $s \in \mathrm{sp}(y)$, $\exp(\sigma_{\mathbf{C}})\tilde{\mu}_y(F(s))$ is a σ-nilpotent orbit.

Note that the Griffiths transversality that is required for PLH is only the Griffiths transversality. But this definition fits well the moduli problem.

Now the precise form of Theorem for Subject I in Introduction is stated as follows.

Theorem 0.4.27 (cf. Theorem B in Section 4.2) *Let Γ be a neat subgroup of $G_{\mathbf{Z}}$ and let Σ be a fan in $\mathfrak{g}_{\mathbf{Q}}$. Assume that Γ and Σ are strongly compatible.*

(i) *The logarithmic manifold $\Gamma\backslash D_{\Sigma}$ represents the functor $\underline{\mathrm{PLH}}_{\Phi} : \mathcal{B}(\log) \to$ (Sets), that is, there exists an isomorphism $\varphi : \underline{\mathrm{PLH}}_{\Phi} \xrightarrow{\sim} \mathrm{Mor}(\ , \Gamma\backslash D_{\Sigma})$.*

(ii) *For any local ringed space Z over $\mathbf{C}$ with a logarithmic structure (which need not be fs) and for any morphism of functors $h : \underline{\mathrm{PLH}}_{\Phi}|_{\mathcal{A}(\log)} \to \mathrm{Mor}(\ , Z)|_{\mathcal{A}(\log)}$ (where $|_{\mathcal{A}(\log)}$ denotes the restrictions to $\mathcal{A}(\log)$), there exists a unique morphism $f : \Gamma\backslash D_{\Sigma} \to Z$ such that $h = (f \circ \varphi)|_{\mathcal{A}(\log)}$, where f is regarded as a morphism $\mathrm{Mor}(\ , \Gamma\backslash D_{\Sigma}) \to \mathrm{Mor}(\ , Z)$.*

0.4.28

For $X \in \mathcal{B}(\log)$ and for $H = (H_{\mathbf{Z}}, \langle\ ,\ \rangle, F, \mu) \in \underline{\mathrm{PLH}}_{\Phi}(X)$, the morphism $\varphi_H : X \to \Gamma\backslash D_{\Sigma}$ corresponding to H is called the associated *period map*, which is set-theoretically given by sending $x \in X$ to the Γ-equivalence class of the nilpotent orbit $(\sigma, \exp(\sigma_{\mathbf{C}})\tilde{\mu}_y(F(s)))$ at x (which is independent of the choices of y, $\tilde{\mu}_y$, and s) in 0.4.26 (2). Note that this map is an extension of the classical period map. If U is an open set on which the logarithmic structure of X is trivial (that is, $M_X|_U = \mathcal{O}_U^{\times}$), then the restriction $(H_{\mathbf{Z}}, \langle\ ,\ \rangle, F, \mu)|_U$ is a PH on U with a Γ-level structure, and the period map of H is an extension of the period map $U \to \Gamma\backslash D$.

Theorem 0.4.27 (ii) characterizes $\Gamma\backslash D_{\Sigma}$ as the universal object among the targets of period maps from objects of $\mathcal{A}(\log)$ into local ringed spaces over $\mathbf{C}$ with logarithmic structures. This indicates that the topology of $\Gamma\backslash D_{\Sigma}$, its ringed space structure, and the logarithmic structure, that we define in this book are in fact intrinsic structures (not artificial ones) determined by this universality.

0.4.29

The reason that the moduli space $\Gamma\backslash D_{\Sigma}$ of PLH is not necessarily an analytic space but a logarithmic manifold is as follows. We also explain why slits and the strong topology naturally appear.

Any PLH of type Φ on an fs logarithmic analytic space X (2.5.8) comes, locally on X, from a universal pre-PLH H_σ on some logarithmically smooth, fs logarithmic analytic space $\check{E}_\sigma$ (3.3.2) for some $\sigma \in \Sigma$ by pulling back via a morphism $X \to \check{E}_\sigma$ (cf. Sections 3.3 and 8.2). Let

$$\tilde{E}_\sigma = \{x \in \check{E}_\sigma \mid \text{the inverse image of } H_\sigma \text{ on } x \text{ satisfies Griffiths transversality}\},$$
$$\supset E_\sigma = \{x \in \check{E}_\sigma \mid \text{the inverse image of } H_\sigma \text{ on } x \text{ is a PLH}\}.$$

Locally on $\check{E}_\sigma$, $\tilde{E}_\sigma$ is the zeros in the new sense (0.4.17) of a finite family of differential forms on $\check{E}_\sigma$ (see Proposition 3.5.10). Hence $\tilde{E}_\sigma$ can have slits. (Note that the Griffiths transversality of the inverse image of H_σ on $x \in \tilde{E}_\sigma$ is the small Griffiths transversality (0.4.22).) Furthermore, as in Theorem A (i) stated in Section 4.1, we have

$$E_\sigma \text{ is open in } \tilde{E}_\sigma \text{ for the strong topology of } \tilde{E}_\sigma \text{ in } \check{E}_\sigma.$$

This openness is not true in general if we use the topology of $\tilde{E}_\sigma$ as a subset of $\check{E}_\sigma$ (12.3.10). Consequently, E_σ is a logarithmic manifold.

For a neat subgroup Γ of $G_{\mathbf{Z}}$ and a fan Σ in $\mathfrak{g}_{\mathbf{Q}}$ that are strongly compatible, the local shape of $\Gamma\backslash D_\Sigma$ is similar to that of E_σ. More precisely, $\Gamma\backslash D_\Sigma$ is covered by the images of morphisms $\Gamma(\sigma)^{\mathrm{gp}}\backslash D_\sigma \to \Gamma\backslash D_\Sigma$ $(\sigma \in \Sigma)$ which are locally isomorphisms, and E_σ is a $\sigma_{\mathbf{C}}$-torsor over $\Gamma(\sigma)^{\mathrm{gp}}\backslash D_\sigma$, where $\sigma_{\mathbf{C}}$ is the $\mathbf{C}$-vector space spanned by σ (Theorem A (iii) and (iv) in Section 4.1). Thus, slits, the strong topology, and logarithmic manifolds naturally appear in the moduli of PLH.

In the nonlogarithmic case where Σ consists of one element $\{0\}$ and $\Gamma = \{1\}$, $\check{E}_{\{0\}} = \check{D}$ with the universal H, and we have $\tilde{E}_{\{0\}} = \check{D}$ and $E_{\{0\}} = D$.

Example with $h^{2,0} = h^{0,2} = 2$, $h^{1,1} = 1$ (continued). In this example, the fact that the slit "if $q = 0$ then $v = \theta(z)$" appears from the small Griffiths transversality is explained as follows. Let $U \subset \mathbf{C}^2 \times Q$ be as in 0.4.18. The pullback on U of the universal PLH on $\Gamma\backslash D_\Xi$ extends to a pre-PLH $H = (H_{\mathbf{Z}}, \langle\ ,\ \rangle, F)$ on the fs logarithmic analytic space $\mathbf{C}^2 \times Q$ whose logarithmic structure is defined by the divisor $\{(q, a, z) \in \mathbf{C}^2 \times Q \mid q = 0\}$. For $x = (0, a, z) \in \mathbf{C}^2 \times Q$, the inverse image of H on x satisfies Griffiths transversality if and only if $v = \pm\theta(z)$.

One of the motivations of the dream of Griffiths to enlarge D was the hope of extending the period map of VPH to the boundary. Concerning this, we have the following result.

THEOREM 0.4.30 (Theorem 4.3.1) *Let X be a connected, logarithmically smooth, fs logarithmic analytic space and let $U = X_{\mathrm{triv}} = \{x \in X \mid M_{X,x} = \mathcal{O}_{X,x}^\times\}$ be the open subspace of X consisting of all points of X at which the logarithmic structure of X is trivial. Let H be a variation of polarized Hodge structure on U with unipotent local monodromy along $X - U$. Fix a base point $u \in U$ and let $(H_0, \langle\ ,\ \rangle_0) = (H_{\mathbf{Z},u}, \langle\ ,\ \rangle_u)$. Let Γ be a subgroup of $G_{\mathbf{Z}}$ which contains the global monodromy* *group* $\mathrm{Image}(\pi_1(U, u) \to G_{\mathbf{Z}})$ *and assume* Γ *is neat. Let* $\varphi : U \to \Gamma\backslash D$ *be the associated period map.*

(i) *Assume that* $X - U$ *is a smooth divisor. Then the period map* φ *extends to a morphism* $X \to \Gamma\backslash D_\Sigma$ *of logarithmic manifolds for some fan* Σ *in* $\mathfrak{g}_{\mathbf{Q}}$ *that is strongly compatible with* Γ.

(ii) *For any* $x \in X$, *there exist an open neighborhood* W *of* x, *a logarithmic modification* W' *of* W, *a commutative subgroup* Γ' *of* Γ, *and a fan* Σ *in* $\mathfrak{g}_{\mathbf{Q}}$ *that is strongly compatible with* Γ' *such that the period map* $\varphi|_{U\cap W}$ *lifts to a morphism* $U \cap W \to \Gamma'\backslash D$ *and extends to a morphism* $W' \to \Gamma'\backslash D_\Sigma$ *of logarithmic manifolds.*

Here in (ii) a logarithmic modification is a special kind of proper morphism $W' \to W$ which is an isomorphism over $U \cap W$ (3.6.12). This theorem can be deduced from the nilpotent orbit theorem of Schmid and some results for fans.

Note that (i) can be applied to $X = \Delta$.

Elliptic curves (continued). In theorem 0.4.30, consider the case where $X = \Delta$, $U = \Delta^*$, and H is the LVPH on Δ in 0.2.18. Fix a branch of e_2 in $H_{\mathbf{Z},u} = H_0 = \mathbf{Z}e_1 + \mathbf{Z}e_2$. The image Γ of $\pi_1(U, u) \to G_{\mathbf{Z}}$ is isomorphic to $\mathbf{Z}$ and is generated by the element γ such that $\gamma(e_1) = e_1$, $\gamma(e_2) = e_1 + e_2$. The classical period map $\Delta^* \to \Gamma\backslash D = \Gamma\backslash\mathfrak{h}$ extends to the period map $\Delta \to \Gamma\backslash D_\sigma$ where $\sigma = \sigma_\infty$ (0.4.13), which coincides with the isomorphism $\Delta \simeq \Gamma\backslash D_\sigma$ in 0.4.13.

0.4.31

Infinitesimal period maps. Let $f : Y \to X$ be a projective, logarithmically smooth (2.1.11), vertical morphism of logarithmically smooth fs logarithmic analytic spaces with connected X. Assume, for any $y \in Y$, that $\mathrm{Coker}\big((M_X^{\mathrm{gp}}/\mathcal{O}_X^\times)_{f(y)} \to (M_Y^{\mathrm{gp}}/\mathcal{O}_Y^\times)_y\big)$ is torsion-free. Let $(H_{\mathbf{Z}}, \langle\ ,\ \rangle, F)$ be the associated LVPH of weight m on X as in 0.2.21.

Let Γ and Σ be a strongly compatible pair (0.4.10). Assume that Γ is neat and contains $\mathrm{Image}(\pi_1(X^{\log}) \to G_{\mathbf{Z}})$, and assume that we have the associated period map $\varphi : X \to \Gamma\backslash D_\Sigma$ (0.4.28). Note that, by Theorem 0.4.30 (ii), these assumptions will be fulfilled locally on X, if we allow a logarithmic modification of it.

Then as a generalization of 0.3.9, for the differential $d\varphi$ of the period map φ, we have the following commutative diagram:

$$\begin{array}{ccc} \theta_X & \xrightarrow{\ d\varphi\ } & \varphi^*\theta^h_{\Gamma\backslash D_\Sigma} = \mathrm{gr}^{-1}\,\mathcal{E}nd_{\langle\,,\,\rangle}(\mathcal{M}) \\ {\scriptstyle \text{K-S}}\big\downarrow & & \cap\big\downarrow \\ R^1 f_*\theta_{Y/X} & \xrightarrow{\text{via coupling}} & \bigoplus_p \mathcal{H}om_{\mathcal{O}_X}(R^{m-p} f_*\omega^p_{Y/X}, R^{m-p+1} f_*\omega^{p-1}_{Y/X}) \end{array}$$

where $\theta_{Y/X} := \mathcal{H}om_{\mathcal{O}_Y}(\omega^1_{Y/X}, \mathcal{O}_Y)$, and $\theta^h_{\Gamma\backslash D_\Sigma}$ is the horizontal logarithmic tangent bundle of the logarithmic tangent bundle $\theta_{\Gamma\backslash D_\Sigma}$, K-S on the left vertical arrow means the logarithmic version of the Kodaira-Spencer map, and the right vertical arrow is the canonical map (Section 4.4).

0.4.32

Note that the classifying space D for polarized Hodge structures on H^2 of surfaces of general type with $p_g \geq 2$, or on H^3 of Calabi-Yau threefolds, is not classical in the sense of 0.4.14. By the construction in the present book, we can now talk about the extended period maps and their differentials associated with degenerations of all complex projective manifolds.

0.4.33

Moduli of PLH with coefficients. We can generalize the above theorems of the moduli of PLH to the moduli of PLH with coefficients (see Chapter 11). Let A be a finite-dimensional semisimple $\mathbf{Q}$-algebra endowed with a map $A \to A, a \mapsto a^\circ$, satisfying

$$(a+b)^\circ = a^\circ + b^\circ, \quad (ab)^\circ = b^\circ a^\circ \quad (a, b \in A).$$

By a *polarized logarithmic Hodge structure with coefficients in* A (*A-PLH*) we mean a PLH $(H_{\mathbf{Z}}, \langle\ ,\ \rangle, F)$ endowed with a ring homomorphism $A \to \mathrm{End}_{\mathbf{Q}}(H_{\mathbf{Q}})$ satisfying

$$\langle ax, y\rangle = \langle x, a^\circ y\rangle \quad (a \in A, x, y \in H_{\mathbf{Q}}).$$

The theorems 0.4.19 and 0.4.27 can be generalized to the moduli $\Gamma\backslash D_\Sigma^A$ of A-PLH (11.1.7, 11.3.1).

0.4.34

In the classical situation 0.4.14, in the work [AMRT], they constructed a fan Σ which is strongly compatible with $G_{\mathbf{Z}}$ such that $G_{\mathbf{Z}}\backslash D_\Sigma$ is compact. In our general situation, it can often happen that $\Gamma\backslash D_\Sigma$ is not locally compact for any Σ such that $D_\Sigma \neq D$. However, we can define the notion of a complete fan (a sufficiently big fan, roughly speaking) such that, in the classical situation, Σ is complete if and only if $G_{\mathbf{Z}}\backslash D_\Sigma$ is compact (see Section 12.6). If Σ is complete and is strongly compatible with Γ, the classical period map $U \to \Gamma\backslash D$ in 0.4.30 always extends globally to a morphism $X' \to \Gamma\backslash D_\Sigma$ of logarithmic manifolds for some logarithmic modification $X' \to X$ (Theorem 12.6.6).

One problem which we cannot solve in this book is that of finding a complete fan in general.*

In Example 0.3.2 (iii) (see also 0.3.4, 0.4.18, and 0.4.29), the fan Ξ in 0.4.5 is complete. But $\Gamma\backslash D_\Xi$ is not compact, not even locally compact, since it has slits.

0.5 FUNDAMENTAL DIAGRAM AND OTHER ENLARGEMENTS OF D

We fix $(w, (h^{p,q})_{p,q\in\mathbf{Z}}, H_0, \langle\ ,\ \rangle_0)$ as in Section 0.3. Let D be the classifying space of polarized Hodge structures, i.e., a Griffiths domain, as in 0.3.1.

---

*See the end of section 12.7.

To prove the main theorems 0.4.19 and 0.4.27, as already mentioned in the Introduction, we need to construct the fundamental diagram (3) in Introduction and to study all the spaces and their relations there. Roughly speaking, in this fundamental diagram, the construction of the four spaces in the right-hand side is based on arithmetic theory of algebraic groups, and that of the four spaces in the left-hand side is based on Hodge theory. They are joined by the central continuous map $D^{\sharp}_{\Sigma,\mathrm{val}} \to D_{\mathrm{SL}(2)}$, which is a geometric interpretation of the SL(2)-orbit theorem of Cattan-Kaplan-Schmid [CKS]. We give an overview of our results concerning these spaces.

The organization of Section 0.5 is as follows. In 0.5.1, we describe the rough ideas of all the enlargements of D in the fundamental diagram. In 0.5.2, in the case of Example (i) in 0.3.2 (the case of the upper half plane), we give the complete descriptions of all the enlargements of D, other than D_{Σ} which was already described in Section 0.4. After that we explain each of these enlargements (other than D_{Σ}) one by one in the general case; $D^{\sharp}_{\Sigma}$ in 0.5.3–0.5.6, D_{BS} in 0.5.7–0.5.10, $D_{\mathrm{SL}(2)}$ in 0.5.11–0.5.18, and the "valuative spaces" completing the fundamental diagram in 0.5.19–0.5.29. In particular, explicit descriptions of the central bridge $D^{\sharp}_{\Sigma,\mathrm{val}} \to D_{\mathrm{SL}(2)}$ in Examples (ii) and (iii) in 0.3.2 are given in 0.5.26 and 0.5.27, respectively. In 0.5.30, we overview $\flat$-spaces, related to the work of Cattani and Kaplan [CK1].

0.5.1

First we give some general observations.

Recall that D_{Σ} (Σ is a fan in $\mathfrak{g}_{\mathbf{Q}}$) is the set of nilpotent orbits (σ, Z), where $\sigma \in \Sigma$ and Z is an $\exp(\sigma_{\mathbf{C}})$-orbit in $\check{D}$ satisfying a certain condition (0.4.7).

(i) *Space* $D^{\sharp}_{\Sigma}$. The space $D^{\sharp}_{\Sigma}$ (Σ is a fan in $\mathfrak{g}_{\mathbf{Q}}$) is a set of pairs (σ, Z) where $\sigma \in \Sigma$ and Z is an $\exp(i\sigma_{\mathbf{R}})$-orbit in $\check{D}$ satisfying a certain condition (see 0.5.3 below). The space $D^{\sharp}_{\Sigma}$ has a natural topology, D is a dense open subset of $D^{\sharp}_{\Sigma}$, and, roughly speaking, the element $(\sigma, Z) \in D^{\sharp}_{\Sigma}$ is the limit point of elements of Z which "run in the direction of degeneration conducted by σ." The space $D^{\sharp}_{\Sigma}$ is covered by open subsets $D^{\sharp}_{\sigma} = \{(\sigma', Z) \in D^{\sharp}_{\Sigma} \mid \sigma' \subset \sigma\}$, where σ runs over elements of Σ.

(ii) *Space* D_{BS}. The space D_{BS} is a set of pairs (P, Z) where P is a $\mathbf{Q}$-parabolic subgroup of $G_{\mathbf{R}}$ and Z is a subset of D satisfying a certain condition (see 0.5.7 below). The set Z is a torus orbit in the following sense. For $(P, Z) \in D_{\mathrm{BS}}$, there is an associated homomorphism of algebraic groups $s : (\mathbf{R}^{\times})^n \to P$ over $\mathbf{R}$ such that Z is an $s(\mathbf{R}^n_{>0})$-orbit in D. The space D_{BS} has a natural topology, D is a dense open subset of D_{BS}, and, roughly speaking, the element $(P, Z) \in D_{\mathrm{BS}}$ is the limit point of elements of Z which "run in the direction of degeneration conducted by P." The space D_{BS} is covered by open subsets $D_{\mathrm{BS}}(P) = \{(P', Z) \in D_{\mathrm{BS}} \mid P' \supset P\}$, where P runs over all $\mathbf{Q}$-parabolic subgroups of G.

(iii) *Space* $D_{\mathrm{SL}(2)}$. The space $D_{\mathrm{SL}(2)}$ is a set of pairs (W, Z) where W is a compatible family $(W^{(j)})_{1\le j\le r}$ (i.e., distributive families in [K]; see 5.2.12) of rational weight filtrations $W^{(j)} = (W^{(j)}_k)_{k\in\mathbf{Z}}$ on $H_{0,\mathbf{R}}$ and Z is a subset of D satisfying a certain condition (see 0.5.11–0.5.13 below). The set Z is a torus orbit in the following sense. For $(W, Z) \in D_{\mathrm{SL}(2)}$, there is an associated homomorphism of algebraic groups over $\mathbf{R}$

$$s : (\mathbf{R}^{\times})^r \to G_{W,\mathbf{R}} = \{g \in G_{\mathbf{R}} \mid gW_k^{(j)} = W_k^{(j)} \text{ for all } j, k\}$$

such that Z is an $s(\mathbf{R}_{>0})^n$-orbit in D. The space $D_{\mathrm{SL}(2)}$ has a natural topology, D is a dense open subset of $D_{\mathrm{SL}(2)}$, and, roughly speaking, the element $(W, Z) \in D_{\mathrm{SL}(2)}$ is the limit point of elements of Z which "run in the direction of degeneration conducted by W." The space $D_{\mathrm{SL}(2)}$ is covered by open subsets $D_{\mathrm{SL}(2)}(W) = \{(W', Z) \in D_{\mathrm{SL}(2)} \mid W' \text{ is a "subfamily" of } W\}$, where W runs over all compatible families of rational weight filtrations of $H_{0,\mathbf{R}}$.

(iv) The *other four spaces*. The other four spaces are spaces of "valuative" orbits which are located over D_{Σ}, $D_{\Sigma}^{\sharp}$, $D_{\mathrm{SL}(2)}$, and D_{BS}, respectively. These upper spaces in the fundamental diagram (3) in Introduction are obtained from the lower spaces as the limits when the directions of degenerations are divided into narrower and narrower directions. We can say also that the vertical arrows in that diagaram are projective limits of kinds of blow-ups.

0.5.2

Upper half plane (continued). In the easiest case $D = \mathfrak{h}$, the sets $D_{\Xi}^{\sharp}$, $D_{\mathrm{SL}(2)}$, and D_{BS} are described as follows.

First we describe $D_{\Xi}^{\sharp}$. Recall that $\Xi = \{\{0\}, \sigma_a \ (a \in \mathbf{P}^1(\mathbf{Q}))\}$ (0.4.9). Recall that

$$\sigma_{\infty} = \begin{pmatrix} 0 & \mathbf{R}_{\geq 0} \\ 0 & 0 \end{pmatrix}.$$

The space $D_{\Xi}^{\sharp}$ is covered by open sets $D_{\sigma}^{\sharp}$ for $\sigma \in \Xi$. The space $D_{\sigma}^{\sharp}$ for $\sigma = \{0\}$ is identified with D ($F \in D$ is identified with the pair (σ, Z) with $\sigma = \{0\}$ and $Z = \{F\}$). The complement $D_{\sigma_{\infty}}^{\sharp} - D$ is the set of all pairs (σ_{∞}, Z) where Z is a subset of $\mathbf{C} \subset \check{D} = \mathbf{P}^1(\mathbf{C})$ of the form $x + i\mathbf{R}$ for some $x \in \mathbf{R}$. This is a set of all $\exp(i\sigma_{\infty,\mathbf{R}})$-orbits in $\mathbf{C}$. We have a homeomorphism

$$D_{\sigma_{\infty}}^{\sharp} \simeq \{x + iy \mid x \in \mathbf{R},\ 0 < y \leq \infty\}, \quad (\sigma_{\infty}, x + i\mathbf{R}) \mapsto x + i\infty,$$

which extends the identity map of D. Hence $(\sigma_{\infty}, x + i\mathbf{R})$ is the limit of $x + iy \in D$ $(y > 0)$ for $y \to \infty$. Let $a \in \mathbf{P}^1(\mathbf{Q})$ and let g be any element of $\mathrm{SL}(2, \mathbf{Q})$ such that $a = g \cdot \infty$. Then $D_{\sigma_a}^{\sharp} - D$ is the set of all pairs (σ_a, Z) where Z is a subset of $\check{D} = \mathbf{P}^1(\mathbf{C})$ such that $(\sigma_{\infty}, g^{-1}(Z)) \in D_{\sigma_{\infty}}^{\sharp}$. We have a homeomorphism $D_{\sigma_{\infty}}^{\sharp} \xrightarrow{\sim} D_{\sigma_a}^{\sharp}$, $(\sigma_{\infty}, Z) \mapsto (\sigma_a, g(Z))$.

We describe D_{BS}. A $\mathbf{Q}$-parabolic subgroup of $G_{\mathbf{R}} = \mathrm{SL}(2, \mathbf{R})$ is either $G_{\mathbf{R}}$ itself or P_a $(a \in \mathbf{P}^1(\mathbf{Q}))$ defined by

$$P_a = \{g \in \mathrm{SL}(2, \mathbf{R}) \mid ga = a\} = \{g \in \mathrm{SL}(2, \mathbf{R}) \mid gV_a = V_a\}$$

where V_a is as in 0.4.9. For example,

$$P_{\infty} = \left\{ \begin{pmatrix} a & b \\ 0 & d \end{pmatrix} \;\middle|\; a, b, d \in \mathbf{R},\ ad = 1 \right\}.$$

The space D_{BS} is covered by open sets $D_{\mathrm{BS}}(P)$ for $P = G_{\mathbf{R}}, P_a$ ($a \in \mathbf{P}^1(\mathbf{Q})$). The space $D_{\mathrm{BS}}(G_{\mathbf{R}})$ is identified with D ($F \in D$ is identified with the pair $(G_{\mathbf{R}}, Z)$ with $Z = \{F\}$). The complement $D_{\mathrm{BS}}(P_\infty) - D$ is the set of all pairs (P_∞, Z) where Z is a subset of $\mathfrak{h} = D$ of the form $x + i\mathbf{R}_{>0}$ for some $x \in \mathbf{R}$. We have a homeomorphism

$$D_{\mathrm{BS}}(P_\infty) \simeq \{x + iy \mid x \in \mathbf{R}, 0 < y \leq \infty\}, \quad (P_\infty, x + i\mathbf{R}_{>0}) \mapsto x + i\infty,$$

which extends the identity map of D. Hence $(P_\infty, x + i\mathbf{R}_{>0})$ is the limit of $x + iy \in D$ ($y > 0$) for $y \to \infty$. Let $a \in \mathbf{P}^1(\mathbf{Q})$ and let g be any element of $\mathrm{SL}(2, \mathbf{Q})$ such that $a = g \cdot \infty$. Then $D_{\mathrm{BS}}(P_a) - D$ is the set of all pairs (P_a, Z) where Z is a subset of D such that $(P_\infty, g^{-1}(Z)) \in D_{\mathrm{BS}}(P_\infty)$. We have a homeomorphism $D_{\mathrm{BS}}(P_\infty) \xrightarrow{\sim} D_{\mathrm{BS}}(P_a)$, $(P_\infty, Z) \mapsto (P_a, g(Z))$.

We describe $D_{\mathrm{SL}(2)}$. For $a \in \mathbf{P}^1(\mathbf{Q})$, let $W(a)$ be the increasing filtration on $H_{0,\mathbf{R}}$ defined by

$$W_1(a) = H_{0,\mathbf{R}}, \quad W_0(a) = W_{-1}(a) = V_a, \quad W_{-2}(a) = 0.$$

For example,

$$W_1(\infty) = H_{0,\mathbf{R}} \supset W_0(\infty) = W_{-1}(\infty) = \mathbf{R}e_1 \supset W_{-2}(\infty) = 0.$$

The space $D_{\mathrm{SL}(2)}$ is covered by the open subsets $D_{\mathrm{SL}(2)}(W(a))$ where $W(a)$ now denotes the family of weight filtrations consisting of the single member $W(a)$. The space $D_{\mathrm{SL}(2)}(\emptyset)$ for the empty family $\emptyset$ is identified with D ($F \in D$ is identified with the pair $(\emptyset, Z)$ with $Z = \{F\}$). The complement $D_{\mathrm{SL}(2)}(W(\infty)) - D$ is the set of all pairs $(W(\infty), Z)$ where Z is a subset of $\mathfrak{h} = D$ of the form $x + i\mathbf{R}_{>0}$ for some $x \in \mathbf{R}$. We have a homeomorphism

$$D_{\mathrm{SL}(2)}(W(\infty)) \simeq \{x + iy \mid x \in \mathbf{R}, 0 < y \leq \infty\}, \quad (W(\infty), x + i\mathbf{R}_{>0}) \mapsto x + i\infty,$$

which extends the identity map of D. Hence $(W(\infty), x + i\mathbf{R}_{>0})$ is the limit of $x + iy \in D$ ($y > 0$) for $y \to \infty$. Let $a \in \mathbf{P}^1(\mathbf{Q})$ and let g be any element of $\mathrm{SL}(2, \mathbf{Q})$ such that $a = g \cdot \infty$. Then $D_{\mathrm{SL}(2)}(W(a)) - D$ is the set of all pairs $(W(a), Z)$ where Z is a subset of D such that $(W(\infty), g^{-1}(Z)) \in D_{\mathrm{SL}(2)}(W(\infty))$. We have a homeomorphism $D_{\mathrm{SL}(2)}(W(\infty)) \xrightarrow{\sim} D_{\mathrm{SL}(2)}(W(a))$, $(W(\infty), Z) \mapsto (W(a), g(Z))$.

The valuative spaces in this case are naturally identified with the spaces under them respectively in the fundamental diagram. That is, the canonical maps $D^\sharp_{\Xi,\mathrm{val}} \to D^\sharp_\Xi$, $D_{\mathrm{BS,val}} \to D_{\mathrm{BS}}$, $D_{\mathrm{SL}(2),\mathrm{val}} \to D_{\mathrm{SL}(2)}$ are homeomorphisms and the canonical map $D_{\Xi,\mathrm{val}} \to D_\Xi$ is bijective.

The identity map of D extends to $G_{\mathbf{Q}}$-equivariant homeomorphisms $D^\sharp_\Xi \simeq D_{\mathrm{SL}(2)} \simeq D_{\mathrm{BS}}$, which induce homeomorphisms $D^\sharp_{\sigma_a} \simeq D_{\mathrm{SL}(2)}(W(a)) \simeq D_{\mathrm{BS}}(P_a)$ for each $a \in \mathbf{P}^1(\mathbf{Q})$ described as

$$(\sigma_a, Z') \leftrightarrow (W(a), Z) \leftrightarrow (P(a), Z), \quad Z' = \exp(i\sigma_{a,\mathbf{R}})Z, \quad Z = Z' \cap D.$$

Thus the fundamental diagram in this case becomes like (4) in Introduction.

0.5.3

Space $D^{\sharp}_{\Sigma}$. In 0.5.3–0.5.6, we consider the space $D^{\sharp}_{\Sigma}$ which is on the left-hand side (the Hodge side) of the fundamental diagram, next to the space D_{Σ} of nilpotent orbits considered in Section 0.4.

A *nilpotent i-orbit* is a pair (σ, Z) consisting of a nilpotent cone $\sigma = \sum_{1\le j\le r}(\mathbf{R}_{\ge 0})N_j$ and a subset $Z \subset \check{D}$ which satisfy, for some $F \in Z$,

$$\begin{cases} Z = \exp(i\sigma_{\mathbf{R}})F, \\ NF^p \subset F^{p-1} \quad (\forall p, \forall N \in \sigma), \\ \exp\left(\sum_{1\le j\le r} iy_jN_j\right) F \in D \quad (\forall\, y_j \gg 0). \end{cases}$$

Let Σ be a fan in $\mathfrak{g}_{\mathbf{Q}}$. As a set, we define

$$D^{\sharp}_{\Sigma} := \{(\sigma, Z) \text{ nilpotent } i\text{-orbit} \mid \sigma \in \Sigma,\ Z \subset \check{D}\}.$$

Note that we have the inclusion map $D \hookrightarrow D^{\sharp}_{\Sigma}$, $F \mapsto (\{0\}, \{F\})$. There is a canonical surjection $D^{\sharp}_{\Sigma} \to D_{\Sigma}$, $(\sigma, Z) \mapsto (\sigma, \exp(\sigma_{\mathbf{C}})Z)$. For a rational nilpotent cone σ in $\mathfrak{g}_{\mathbf{R}}$, we denote $D^{\sharp}_{\{\text{face of } \sigma\}}$ by $D^{\sharp}_{\sigma}$. Then, for a fan Σ in $\mathfrak{g}_{\mathbf{Q}}$, we have $D^{\sharp}_{\Sigma} = \bigcup_{\sigma\in\Sigma} D^{\sharp}_{\sigma}$.

0.5.4

In Chapter 3, we will define a topology of $D^{\sharp}_{\Sigma}$ that has the following property. Let $(\sigma, Z) \in D^{\sharp}_{\Sigma}$, let $N_j \in \mathfrak{g}_{\mathbf{Q}}$ $(1 \le j \le n)$, $F \in Z$, and assume $\sigma = \sum_{j=1}^{n}(\mathbf{R}_{\ge 0})N_j$. Then

$$\exp\left(\sum_{j=1}^{n} iy_jN_j\right) F \to (\sigma, Z) \quad \text{if } y_j \in \mathbf{R} \text{ and } y_j \to \infty.$$

Theorem 0.5.5 (Theorem A in Section 4.1)

(i) *The space* $D^{\sharp}_{\Sigma}$ *is Hausdorff.*

(ii) *Assume that* Γ *is strongly compatible with* Σ. *Then* $\Gamma\backslash D^{\sharp}_{\Sigma}$ *is Hausdorff.*

(iii) *Assume that* Γ *is strongly compatible with* Σ *and is neat. Then the canonical projection* $D^{\sharp}_{\Sigma} \to \Gamma\backslash D^{\sharp}_{\Sigma}$ *is a local homeomorphism.*

(iv) *Assume that* Γ *is strongly compatible with* Σ *and is neat. Then we have a canonical homeomorphism*

$$\Gamma\backslash D^{\sharp}_{\Sigma} \simeq (\Gamma\backslash D_{\Sigma})^{\log}$$

which is compatible with the projections to $\Gamma\backslash D_{\Sigma}$.

By (iv), for Γ as in (iv), the canonical map $\Gamma\backslash D^{\sharp}_{\Sigma} \to \Gamma\backslash D_{\Sigma}$ is proper, and the fibers are products of finite copies of $\mathbf{S}^1$.

0.5.6

Here we give local descriptions of $D^\sharp_\Sigma \to \Gamma\backslash D^\sharp_\Sigma \to \Gamma\backslash D_\Sigma$ for Example (i), Example (ii) ($g = 2$), and Example (iii) in 0.3.2 (for some choices of Σ and Γ).

Upper half plane (continued). Let $\Gamma = \left(\begin{smallmatrix} 1 & \mathbf{Z} \\ 0 & 1 \end{smallmatrix}\right)$, $\sigma = \sigma_\infty$. Then we have a commutative diagram of topological spaces:

$$\begin{array}{ccc} \{x + iy \mid x \in \mathbf{R},\ 0 < y \leq \infty\} & \simeq & D^\sharp_\sigma \\ \downarrow & & \downarrow \\ \Delta^{\log} & \simeq & \Gamma\backslash D^\sharp_\sigma \\ \downarrow & & \downarrow \\ \Delta & \simeq & \Gamma\backslash D_\sigma. \end{array}$$

Here the lower horizontal isomorphism is that in 0.4.13, the upper horizontal isomorphism is the one described in 0.5.2, and the upper left vertical arrow sends $\tau = x + iy$ $(0 < y < \infty)$ to $e^{2\pi i\tau} \in \Delta^* \subset \Delta^{\log}$, and $x + i\infty$ to $(0, e^{2\pi ix}) \in \Delta^{\log} = |\Delta| \times \mathbf{S}^1$.

Upper half space (continued). Let $g = 2$ and $D = \mathfrak{h}_2$. Let U be the open set of $\Delta^2 \times \mathbf{C}$ defined in 0.4.13, and let $\Gamma = \exp(\mathbf{Z}N_1 + \mathbf{Z}N_2) = 1 + \mathbf{Z}N_1 + \mathbf{Z}N_2$, $\sigma = (\mathbf{R}_{\geq 0})N_1 + (\mathbf{R}_{\geq 0})N_2$. We describe $D^\sharp_\sigma$ and $\Gamma\backslash D^\sharp_\sigma$. We have a commutative diagram of topological spaces:

$$\begin{array}{ccccc} (|\Delta| \times \mathbf{R})^2 \times \mathbf{C} & \supset & \tilde{U}^{\log} & \simeq & D^\sharp_\sigma \\ \downarrow & & \downarrow & & \downarrow \\ (|\Delta| \times \mathbf{S}^1)^2 \times \mathbf{C} & \supset & U^{\log} & \simeq & \Gamma\backslash D^\sharp_\sigma \\ \downarrow & & \downarrow & & \downarrow \\ \Delta^2 \times \mathbf{C} & \supset & U & \simeq & \Gamma\backslash D_\sigma. \end{array}$$

Here the upper left vertical arrow is induced by $\mathbf{R} \to \mathbf{S}^1$, $x \mapsto e^{2\pi ix}$, the lower left vertical arrow is induced by $|\Delta| \times \mathbf{S}^1 \to \Delta$, $(r, u) \mapsto ru$, and $\tilde{U}^{\log}$ is the inverse image of U in $(|\Delta| \times \mathbf{R})^2 \times \mathbf{C}$. The space $U^{\log}$ is identified with the inverse image of U in $(|\Delta| \times \mathbf{S}^1)^2 \times \mathbf{C}$. The inclusions $\supset$ in this diagram are open immersions. Let $r_j = e^{-2\pi y_j}$. The upper horizontal isomorphism of this diagram sends $((r_1, x_1), (r_2, x_2), a) \in \tilde{U}^{\log}$ to

$$\begin{array}{ll} F\begin{pmatrix} x_1 + iy_1 & a \\ a & x_2 + iy_2 \end{pmatrix} & \text{if } r_1 r_2 \neq 0, \\ \left((\mathbf{R}_{\geq 0})N_1,\ F\begin{pmatrix} x_1 + i\mathbf{R} & a \\ a & x_2 + iy_2 \end{pmatrix}\right) & \text{if } r_1 = 0 \text{ and } r_2 \neq 0, \\ \left((\mathbf{R}_{\geq 0})N_2,\ F\begin{pmatrix} x_1 + iy_1 & a \\ a & x_2 + i\mathbf{R} \end{pmatrix}\right) & \text{if } r_1 \neq 0 \text{ and } r_2 = 0, \\ \left(\sigma,\ F\begin{pmatrix} x_1 + i\mathbf{R} & a \\ a & x_2 + i\mathbf{R} \end{pmatrix}\right) & \text{if } r_1 = r_2 = 0. \end{array}$$

Example with $h^{2,0} = h^{0,2} = 2$, $h^{1,1} = 1$ (continued). Let the notation be as in 0.4.18. Let Γ be a neat subgroup of $G_\mathbf{Z}$ of finite index. We have a commutative diagram of topological spaces

$$\begin{array}{ccccc} ((\mathbf{R}_{\geq 0}) \times \mathbf{R}) \times \mathbf{C} \times Q & \supset & \tilde{U}^{\log} & \to & D_{\Xi}^{\sharp} \\ \downarrow & & \downarrow & & \downarrow \\ ((\mathbf{R}_{\geq 0}) \times \mathbf{S}^1) \times \mathbf{C} \times Q & \supset & U^{\log} & \to & \Gamma\backslash D_{\Xi}^{\sharp} \\ \downarrow & & \downarrow & & \downarrow \\ \mathbf{C} \times \mathbf{C} \times Q & \supset & U & \to & \Gamma\backslash D_{\Xi} \end{array}$$

in which the three horizontal arrows are local homeomorphisms. Here the upper left vertical arrow is induced by $\mathbf{R} \to \mathbf{S}^1$, $x \mapsto e^{2\pi i x}$, the lower left vertical arrow is induced from $(\mathbf{R}_{\geq 0}) \times \mathbf{S}^1 \to \mathbf{C}$, $(r, u) \mapsto ru$, and $\tilde{U}^{\log}$ is the inverse image of U in $((\mathbf{R}_{\geq 0}) \times \mathbf{R}) \times \mathbf{C} \times Q$. The space $U^{\log}$ is identified with the inverse image of U in $((\mathbf{R}_{\geq 0}) \times \mathbf{S}^1) \times \mathbf{C} \times Q$. Recall that U is endowed with the strong topology. The spaces $U^{\log}$ and $\tilde{U}^{\log}$ are endowed here with the topologies as fiber products by left squares. The upper horizontal arrow sends $(r, x, a, z) \in \tilde{U}^{\log}$ $(r \neq 0)$ to $\exp((x+iy)N_v + aN_{v'})F(z)$, where $r = e^{-2\pi y/\ell}$ with ℓ, v, and v' as in 0.4.18, and $(0, x, a, \theta^{-1}(v))$ to $((\mathbf{R}_{\geq 0})N_v, \exp(i\mathbf{R}N_v)\exp(xN_v + aN_{v'})F(\theta^{-1}(v)))$.

0.5.7

Space D_{BS}. In 0.5.7–0.5.10, we consider the space D_{BS} which is on the right-hand side (algebraic group side) of the fundamental diagram.

D_{BS} is a real manifold with corners.

The definition of D_{BS} will be reviewed in Section 5.1. Here we give an explicit presentation of the open set $D_{\mathrm{BS}}(P)$ of D_{BS}, for simplicity, under the assumptions that the **Q**-parabolic subgroup P of $G_{\mathbf{R}}$ is an **R**-minimal parabolic subgroup of $G_{\mathbf{R}}$ and that the largest **R**-split torus in the center of P/P_u (P_u is the unipotent radical of P) is **Q**-split. In this case, $D_{\mathrm{BS}}(P)$ is described by using the Iwasawa decomposition of $G_{\mathbf{R}}$.

Upper half plane (continued). We first observe the easiest case $D = \mathfrak{h}$. In this case, we have a homeomorphism

$$\begin{pmatrix} 1 & \mathbf{R} \\ 0 & 1 \end{pmatrix} \times (\mathbf{R}_{>0}) \times \mathrm{SO}(2, \mathbf{R}) \simeq G_{\mathbf{R}} = \mathrm{SL}(2, \mathbf{R}),$$

$$(g, t, k) \mapsto gs(t)k, \quad \text{where } s(t) = \begin{pmatrix} 1/t & 0 \\ 0 & t \end{pmatrix}.$$

This is an Iwasawa decomposition of $\mathrm{SL}(2, \mathbf{R})$. By $\mathrm{SL}(2, \mathbf{R})/\mathrm{SO}(2, \mathbf{R}) \xrightarrow{\sim} \mathfrak{h}$, $g \mapsto g \cdot i$, this Iwasawa decomposition induces a homeomorphism

$$\mathbf{R} \times \mathbf{R}_{>0} \xrightarrow{\sim} \mathfrak{h} = D, \quad (x, t) \mapsto \begin{pmatrix} 1 & x \\ 0 & 1 \end{pmatrix} s(t) \cdot i = x + t^{-2} i.$$

This homeomorphism extends to a homeomorphism

$$\mathbf{R} \times \mathbf{R}_{\geq 0} \simeq D_{\mathrm{BS}}(P_\infty), \quad (x, 0) \mapsto (P_\infty, x + i\mathbf{R}_{>0}).$$

In general, let K be a maximal compact subgroup of $G_{\mathbf{R}}$ and let P be an $\mathbf{R}$-minimal parabolic subgroup of $G_{\mathbf{R}}$. Denote by P_u the unipotent radical of P. Then we have a homeomorphism (Iwasawa decomposition)

$$P_u \times (\mathbf{R}^n_{>0}) \times K \simeq G_{\mathbf{R}}, \quad (g, t, k) \mapsto gs(t)k,$$

for a unique pair (n, s) where $n \geq 0$ is an integer and s is a homomorphism $(\mathbf{R}^\times)^n \to P$ of $\mathbf{R}$-algebraic groups satisfying the following conditions (1)–(3).

(1) The composition $(\mathbf{R}^\times)^n \xrightarrow{s} P \to P/P_u$ induces an isomorphism from $(\mathbf{R}^\times)^n$ onto the largest $\mathbf{R}$-split torus in the center of P/P_u.

(2) The Cartan involution $G_{\mathbf{R}} \xrightarrow{\sim} G_{\mathbf{R}}$ associated with K (see below) sends $s(t)$ to $s(t)^{-1}$.

(3) For $t = (t_j)_{1 \leq j \leq n} \in (\mathbf{R}^\times)^n$, $|t_j| < 1$ for any j if and only if all eigenvalues of $\mathrm{Ad}(s(t))$ on $\mathrm{Lie}(P_u)$ have absolute values > 1.

Here the Cartan involution associated with a maximal compact subgroup K of $G_{\mathbf{R}}$ is the unique homomorphism $\iota : G_{\mathbf{R}} \to G_{\mathbf{R}}$ of $\mathbf{R}$-algebraic groups such that $\iota^2 = \mathrm{id}$ and such that $K = \{g \in G_{\mathbf{R}} \mid \iota(g) = g\}$. For $\mathbf{r} \in D$, the Cartan involution of $G_{\mathbf{R}}$ associated with $K_{\mathbf{r}}$ coincides with the map $g \mapsto C_{\mathbf{r}} g C_{\mathbf{r}}^{-1}$ where $C_{\mathbf{r}}$ is the operator in 0.1.8 (2).

Now let P be a $\mathbf{Q}$-parabolic subgroup of $G_{\mathbf{R}}$ which is an $\mathbf{R}$-minimal parabolic subgroup of $G_{\mathbf{R}}$ such that the largest $\mathbf{R}$-split torus in the center of P/P_u is $\mathbf{Q}$-split. Let $\mathbf{r} \in D$, let $K_{\mathbf{r}}$ be the maximal compact subgroup of $G_{\mathbf{R}}$ corresponding to $\mathbf{r}$ (0.3.3), and consider the Iwasawa decomposition $P_u \times (\mathbf{R}^n_{>0}) \times K_{\mathbf{r}} \simeq G_{\mathbf{R}}$ with respect to $(P, K_{\mathbf{r}})$ satisfying (1)–(3) as above. It induces a homeomorphism

$$P_u \times (\mathbf{R}^n_{>0}) \times (K_{\mathbf{r}}/K'_{\mathbf{r}}) \simeq D, \quad (g, t, k) \mapsto gs(t)k \cdot \mathrm{r}.$$

This extends to a homeomorphism

$$P_u \times (\mathbf{R}^n_{\geq 0}) \times (K_{\mathbf{r}}/K'_{\mathbf{r}}) \simeq D_{\mathrm{BS}}(P).$$

The element of $D_{\mathrm{BS}}(P)$ corresponding to $(g, 0, k) \in P_u \times (\mathbf{R}^n_{\geq 0}) \times (K_{\mathbf{r}}/K'_{\mathbf{r}})$ ($g \in P_u$, $k \in K_{\mathbf{r}}$) coincides with the pair (P, Z) where $Z = \{gs(t)k \cdot \mathbf{r} \mid t \in \mathbf{R}^n_{>0}\}$. More generally, the element of $D_{\mathrm{BS}}(P)$ corresponding to (g, t, k) ($g \in P_u$, $t \in \mathbf{R}^n_{\geq 0}$, $k \in K_{\mathbf{R}}$) coincides with the element (Q, Z) of $D_{\mathrm{BS}}(P)$ where Q is the $\mathbf{Q}$-parabolic subgroup of $G_{\mathbf{R}}$ containing P which corresponds to the subset $J = \{j \mid 1 \leq j \leq n,\ t_j \neq 0\}$ of the set $\{1. \ldots, n\}$ (there is a bijection between the set of all subsets of $\{1, \ldots, n\}$ and the set of all $\mathbf{Q}$-parabolic subgroups of $G_{\mathbf{R}}$ containing P), and $Z = \{gs(t')k \cdot \mathbf{r} \mid t' \in \mathbf{R}^n_{>0},\ t'_j = t_j \text{ if } j \in J\} \subset D$.

Thus D_{BS} is understood by the theory of algebraic groups, rather than by Hodge theory.

THEOREM 0.5.8

(i) *D_{BS} is a real manifold with corners. For any $p \in D_{\mathrm{BS}}$, there are an open neighborhood U of p in D_{BS}, integers $m, n \geq 0$, and a homeomorphism*

$$U \simeq \mathbf{R}^m \times \mathbf{R}^n_{\geq 0}$$

which sends p to $(0, 0)$. The point p belongs to D if and only if $n = 0$.

(ii) *For any subgroup* Γ *of* $G_{\mathbf{Z}}$, $\Gamma\backslash D_{\mathrm{BS}}$ *is Hausdorff.*
(iii) *If* Γ *is of finite index in* $G_{\mathbf{Z}}$, $\Gamma\backslash D_{\mathrm{BS}}$ *is compact.*
(iv) *If* Γ *is a neat subgroup of* $G_{\mathbf{Z}}$, *the projection* $D_{\mathrm{BS}} \to \Gamma\backslash D_{\mathrm{BS}}$ *is a local homeomorphism.*

The definition of D_{BS} and the proof of this theorem were given in [KU2] and [BJ] independently. The definition of D_{BS} is a modification of the definition in [BS] of the original Borel-Serre space $\mathcal{X}_{\mathrm{BS}}$, which is an enlargement of the symmetric Hermitian space $\mathcal{X}$ of all maximal compact subgroups of $G_{\mathbf{R}}$. There is a canonical surjection $D_{\mathrm{BS}} \to \mathcal{X}_{\mathrm{BS}}$ which sends $F \in D$ to $K_F \in \mathcal{X}$. The proof of the above theorem is a reduction to the similar properties of the original Borel-Serre space proved in [BS].

In 0.5.9 and 0.5.10 below, we give local descriptions of D_{BS} for Example (ii) ($g = 2$) and Example (iii) in 0.3.2, respectively.

0.5.9

Upper half space (continued). Let $g = 2$ and $D = \mathfrak{h}_2$. Let P be the $\mathbf{Q}$-parabolic subgroup of $G_{\mathbf{R}}$ consisting of elements that preserve the $\mathbf{R}$-subspaces $\mathbf{R}e_1$, $\mathbf{R}e_1 + \mathbf{R}e_2$, and $\mathbf{R}e_1 + \mathbf{R}e_2 + \mathbf{R}e_4$ of $H_{0,\mathbf{R}}$. This is an $\mathbf{R}$-minimal parabolic subgroup of $G_{\mathbf{R}}$. There is a homeomorphism $\mathbf{R}^4 \simeq P_u$ (not an isomorphism of groups, for P_u is noncommutative). Let $s : (\mathbf{R}^\times)^2 \to P$ be the homomorphism of algebraic groups given by

$$s(t)e_1 = t_1^{-1}t_2^{-1}e_1, \quad s(t)e_2 = t_2^{-1}e_2, \quad s(t)e_4 = t_2e_4, \quad s(t)e_3 = t_1t_2e_3.$$

Put $\mathbf{r} = F\left(\begin{smallmatrix} i & 0 \\ 0 & i \end{smallmatrix}\right)$. Then we have $K_{\mathbf{r}} = K'_{\mathbf{r}} = \mathrm{Sp}(2, \mathbf{R}) \cap O(4, \mathbf{R}) \simeq U(2)$. We have a homeomorphism $P_u \times (\mathbf{R}^\times)^2 \simeq P$, $(g, t) \mapsto gs(t)$. We have a homeomorphism (Iwasawa decomposition)

$$P_u \times (\mathbf{R}^2_{>0}) \times K_{\mathbf{r}} \simeq G_{\mathbf{R}}, \quad (g, t, k) \mapsto gs(t)k,$$

which satisfies the conditions (1)–(3) in 0.5.7. This induces a homeomorphism

$$P_u \times \mathbf{R}^2_{>0} \simeq D, \quad (g, t) \mapsto gs(t) \cdot \mathbf{r},$$

which extends to a homeomorphism

$$P_u \times \mathbf{R}^2_{\geq 0} \simeq D_{\mathrm{BS}}(P).$$

0.5.10

Example with $h^{2,0} = h^{0,2} = 2$, $h^{1,1} = 1$ (continued). We use the notation in 0.3.4. Let $G^\circ_{\mathbf{R}}$ be the kernel of the determinant map $G_{\mathbf{R}} \to \{\pm 1\}$ (for the action on $H_{0,\mathbf{R}}$), and let P be the $\mathbf{Q}$-parabolic subgroup of $G_{\mathbf{R}}$ consisting of all elements of $G^\circ_{\mathbf{R}}$ which preserve the subspaces $\mathbf{R}e_4$ and $\sum_{j=1}^4 \mathbf{R}e_j$ of $H_{0,\mathbf{R}}$. This is an $\mathbf{R}$-minimal parabolic subgroup of $G_{\mathbf{R}}$. We have an isomorphism of $\mathbf{R}$-algebraic groups $\mathbf{R}^3 \xrightarrow{\sim} P_u$, $a \mapsto \exp(N_a)$. Let $s : \mathbf{R}^\times \to P$ be the homomorphism of $\mathbf{R}$-algebraic groups defined by

$$s(t)e_j = e_j \quad (1 \le j \le 3), \quad s(t)e_4 = t^{-1}e_4, \quad s(t)e_5 = te_5.$$

Then we have a homeomorphism $P_u \times \mathbf{R}^\times \xrightarrow{\sim} P$, $(g, t) \mapsto gs(t)$.

Let $v \in \mathbf{S}^2$. We have a homeomorphism (Iwasawa decomposition)

$$P_u \times (\mathbf{R}_{>0}) \times K_{\mathbf{r}(v)} \xrightarrow{\sim} G_{\mathbf{R}}, \quad (g, t, k) \mapsto gs(t)k$$

which satisfies the conditions (1)–(3) in 0.5.7. We have a homeomorphism

$$\{\pm 1\} \times \mathbf{S}^2 \xrightarrow{\sim} K_{\mathbf{r}(v)} \cdot \mathbf{r}(v), \quad (\pm 1, v') \mapsto s(\pm 1) \cdot \mathbf{r}(v').$$

Hence this Iwasawa decomposition induces a homeomorphism

$$\mathbf{R}^3 \times (\mathbf{R}_{>0}) \times \{\pm 1\} \times \mathbf{S}^2 \xrightarrow{\sim} D, \quad (a, t, \pm 1, v) \mapsto \exp(N_a)s(\pm t) \cdot \mathbf{r}(v),$$

which extends to a homeomorphism

$$\mathbf{R}^3 \times (\mathbf{R}_{\ge 0}) \times \{\pm 1\} \times \mathbf{S}^2 \xrightarrow{\sim} D_{\mathrm{BS}}(P).$$

In this example, all $\mathbf{Q}$-parabolic subgroups of $G_{\mathbf{R}}$ other than $G_{\mathbf{R}}^\circ$ are conjugate to P under $G_{\mathbf{Q}}$, and hence D_{BS} is covered by open sets $D_{\mathrm{BS}}(gPg^{-1})$ $(g \in G_{\mathbf{Q}})$ each of which is homeomorphic to $D_{\mathrm{BS}}(P)$ via the homeomorphism that extends $g^{-1} : D \to D$.

0.5.11

The space $D_{\mathrm{SL}(2)}$. In general, D_Σ and D_{BS} are still far from each other in nature. We find an intermediate existence $D_{\mathrm{SL}(2)}$ to connect them. We consider this space $D_{\mathrm{SL}(2)}$ in 0.5.11–0.5.18. Hodge theory and algebraic group theory are unified on this space. This unification is based on a fundamental property of the SL(2)-action on the upper half plane $\mathfrak{h}$:

$$\exp\left(iy\begin{pmatrix}0 & 1\\ 0 & 0\end{pmatrix}\right)(0) = \begin{pmatrix}\sqrt{y} & 0\\ 0 & 1/\sqrt{y}\end{pmatrix}(i).$$

When $y > 0$ varies, the left-hand side produces a nilpotent i-orbit, while the right-hand side produces a torus orbit in the Borel-Serre space.

We define $D_{\mathrm{SL}(2)}$ as follows.

A pair (ρ, φ), consisting of a homomorphism $\rho : \mathrm{SL}(2, \mathbf{C})^r \to G_{\mathbf{C}}$ of algebraic groups which is defined over $\mathbf{R}$ and a holomorphic map $\varphi : \mathbf{P}^1(\mathbf{C})^r \to \check{D}$, is called an SL(2)-*orbit of rank* r if it satisfies the following conditions (1)–(4) ([CKS, Chapter 4], [KU2, Chapter 3]):

(1) $\varphi(gz) = \rho(g)\varphi(z)$ for all $g \in \mathrm{SL}(2, \mathbf{C})^r$ and all $z \in \mathbf{P}^1(\mathbf{C})^r$.
(2) The Lie algebra homomorphism $\rho_* : \mathfrak{sl}(2, \mathbf{C})^{\oplus r} \to \mathfrak{g}_{\mathbf{C}}$ is injective.
(3) $\varphi(\mathfrak{h}^r) \subset D$.
(4) Let $z \in \mathfrak{h}^r$, let $F_z^\bullet(\mathfrak{sl}(2, \mathbf{C})^{\oplus r})$ be the Hodge filtration of $\mathfrak{sl}(2, \mathbf{C})^{\oplus r}$ induced by the Hodge filtration of $(\mathbf{C}^2)^{\oplus r}$ corresponding to z, and let $F_{\varphi(z)}^\bullet(\mathfrak{g}_{\mathbf{C}})$ be the Hodge filtration of $\mathfrak{g}_{\mathbf{C}}$ induced by the Hodge filtration $\varphi(z)$ of $H_{0,\mathbf{C}}$. Then $\rho_* : \mathfrak{sl}(2, \mathbf{C})^{\oplus r} \to \mathfrak{g}_{\mathbf{C}}$ sends $F_z^p(\mathfrak{sl}(2, \mathbf{C})^{\oplus r})$ into $F_{\varphi(z)}^p(\mathfrak{g}_{\mathbf{C}})$ for any p.

Here in (4), for $F \in D$, the Hodge filtration on $\mathfrak{g}_{\mathbf{C}}$ induced by F is defined as

$$F_F^p(\mathfrak{g}_{\mathbf{C}}) = \{h \in \mathfrak{g}_{\mathbf{C}} \mid h(F^s) \subset F^{s+p}\ (\forall s)\}.$$

The Hodge filtration of $(\mathbf{C}^2)^{\oplus r} = \bigoplus_{j=1}^r (\mathbf{C}e_{1j} \oplus \mathbf{C}e_{2j})$ corresponding to z is defined as $F^0(z) = (\mathbf{C}^2)^{\oplus r}$, $F^1(z) = \bigoplus_{j=1}^r \mathbf{C}(z_j e_{1j} + e_{2j})$, and $F^2(z) = 0$.

Let $\mathbf{i} = (i, \ldots, i) \in \mathfrak{h}^r$. Then, if the condition (1) is satisfied, (3) is satisfied if and only if $\varphi(\mathbf{i}) \in D$, and (4) is satisfied if and only if ρ_* sends $F_{\mathbf{i}}^p(\mathfrak{sl}(2, \mathbf{C})^{\oplus r})$ into $F_{\varphi(\mathbf{i})}^p(\mathfrak{g}_{\mathbf{C}})$ for any p.

0.5.12

For an SL(2)-orbit (ρ, φ) of rank r, let $N_j \in \mathfrak{g}_{\mathbf{R}}$ be the image under ρ_* of $\left(\begin{smallmatrix} 0 & 1 \\ 0 & 0 \end{smallmatrix}\right) \in \mathfrak{sl}(2, \mathbf{R})$ in the jth factor. Let

$$W^{(j)} = W(N_1 + \cdots + N_j) \quad (1 \le j \le r),$$

where $W(N)$ for a nilpotent linear operator N denotes the monodromy weight filtration associated with N (Deligne [D5]; see 5.2.4). The family $W = (W^{(j)})_{1 \le j \le r}$ is called the *family of weight filtrations associated with* (ρ, φ).

Definition 0.5.13 ([KU2, 3.6], 5.2.6) *Two* SL(2)*-orbits* (ρ_1, φ_1) *and* (ρ_2, φ_2) *of rank r are equivalent if there exists* $(t_1, \ldots, t_r) \in \mathbf{R}^r_{>0}$ *such that*

$$\rho_2 = \mathrm{Int}\left(\rho_1\left(\begin{pmatrix} t_1^{-1} & 0 \\ 0 & t_1 \end{pmatrix}, \ldots, \begin{pmatrix} t_r^{-1} & 0 \\ 0 & t_r \end{pmatrix}\right)\right) \circ \rho_1,$$

$$\varphi_2 = \rho_1\left(\begin{pmatrix} t_1^{-1} & 0 \\ 0 & t_1 \end{pmatrix}, \ldots, \begin{pmatrix} t_r^{-1} & 0 \\ 0 & t_r \end{pmatrix}\right) \cdot \varphi_1.$$

Here $\mathrm{Int}(g)$ *means the inner automorphism by g.*

Define $D_{\mathrm{SL}(2),r}$ *to be the set of all equivalence classes of* SL(2)*-orbits* (ρ, φ) *of rank r whose associated family of weight filtrations is defined over* **Q**.

Define $D_{\mathrm{SL}(2)} = \bigsqcup_{r \ge 0} D_{\mathrm{SL}(2),r}$ *where* $D_{\mathrm{SL}(2),0} = D$.

For an SL(2)-orbit (ρ, φ) of rank r, the family W of weight filtrations associated with (ρ, φ) and the set $Z = \{\varphi(iy_1, \ldots, iy_r) \mid y_j > 0\ (1 \le j \le r)\}$ are determined by the class $[\rho, \varphi]$ of (ρ, φ) in $D_{\mathrm{SL}(2)}$. Conversely, $[\rho, \varphi]$ is determined by the pair (W, Z) (see [KU2, 3.10]). We will denote $[\rho, \varphi] \in D_{\mathrm{SL}(2)}$ also by (W, Z).

0.5.14

If the condition (2) in 0.5.11 is omitted, a pair (ρ, φ) is called an SL(2)*-orbit in r variables.*

For an SL(2)-orbit (ρ, φ) in n variables, there exists a unique SL(2)-orbit (ρ', φ') such that for some $J \subset \{1, \ldots, n\}$, $(\rho, \varphi) = (\rho', \varphi') \circ \pi_J$, where $\pi_J : (\mathrm{SL}(2, \mathbf{C}) \times \mathbf{P}^1(\mathbf{C}))^n \to (\mathrm{SL}(2, \mathbf{C}) \times \mathbf{P}^1(\mathbf{C}))^r$ is the projection to the J-factor and such that (ρ', φ') is an SL(2)-orbit of rank $r := \sharp(J)$. We denote the point $[\rho', \varphi']$ of $D_{\mathrm{SL}(2)}$ also by $[\rho, \varphi]$.

The notion of the SL(2)-orbit generalizes the (H_1)-homomorphism in the context of equivariant holomorphic maps of symmetric domains (cf. [Sa2, II § 8]).

In the classical situation (0.4.14), the Satake-Baily-Borel compactification of $\Gamma\backslash D$ for a subgroup Γ of $G_{\mathbf{Z}}$ of finite index is a quotient of $\Gamma\backslash D_{\mathrm{SL}(2)}$, and is philosophically close to $\Gamma\backslash D_{\mathrm{SL}(2)}$.

0.5.15

In 5.2.13, we will review the definition of the topology of $D_{\mathrm{SL}(2)}$ given in [KU2]. In this topology, for $[\rho, \varphi] \in D_{\mathrm{SL}(2)}$, we have

$$\varphi(iy_1, \dots, iy_n) \to [\rho, \varphi] \quad \text{if } y_j \in \mathbf{R}_{>0} \text{ and } y_j/y_{j+1} \to \infty \ (y_{n+1} \text{ denotes } 1).$$

Theorem 0.5.16 (5.2.16, 5.2.15)

(i) *Let $p \in D_{\mathrm{SL}(2)}$ be an element of rank r. Then there are an open neighborhood U of p in $D_{\mathrm{SL}(2)}$, a finite dimensional vector space V over $\mathbf{R}$, $\mathbf{R}$-vector subspaces V_J of V given for each subset J of the set $\{1, \dots, r\}$, satisfying $V_J \supset V_{J'}$ if $J \subset J' \subset \{1, \dots, r\}$ and $V_\emptyset = V$, and a homeomorphism*

$$U \simeq \{(a, t) \in V \times \mathbf{R}^r_{\geq 0} \mid a \in V_J \text{ where } J = \{j \mid t_j = 0\}\}$$

which sends p to $(0, 0)$.

(ii) *For any subgroup Γ of $G_{\mathbf{Z}}$, $\Gamma\backslash D_{\mathrm{SL}(2)}$ is Hausdorff.*

(iii) *If Γ is a neat subgroup of $G_{\mathbf{Z}}$, the projection $D_{\mathrm{SL}(2)} \to \Gamma\backslash D_{\mathrm{SL}(2)}$ is a local homeomorphism.*

0.5.17

We consider the relation between D_{BS} and $D_{\mathrm{SL}(2)}$.

In D_{BS}, the direction of degeneration is determined by a parabolic subgroup of $G_{\mathbf{R}}$. On the other hand, in $D_{\mathrm{SL}(2)}$, it is determined by a family of weight filtrations.

Let $[\rho, \varphi] \in D_{\mathrm{SL}(2)}$ be an element of rank r and let $W = (W^{(j)})_{1\leq j\leq r}$ be the family of weight filtrations associated with (ρ, φ).

Let $G^\circ_{\mathbf{R}}$ be the kernel of the determinant map $G_{\mathbf{R}} \to \{\pm 1\}$, and let $G^\circ_{W,\mathbf{R}}$ be the subgroup of $G^\circ_{\mathbf{R}}$ consisting of all elements which preserve $W^{(j)}_k$ for any j, k. If $r = 1$, i.e., if W consists of one weight filtration, $G^\circ_{W,\mathbf{R}}$ is a $\mathbf{Q}$-parabolic subgroup of $G_{\mathbf{R}}$. Let $D_{\mathrm{SL}(2),\leq 1}$ be the part of $D_{\mathrm{SL}(2)}$ consisting of all elements of rank ≤ 1. Then $D_{\mathrm{SL}(2),\leq 1}$ is an open set of $D_{\mathrm{SL}(2)}$. The identity map of D is extended to a continuous map $D_{\mathrm{SL}(2),\leq 1} \to D_{\mathrm{BS}}$ which has the form $(W, Z) \mapsto (P, Z')$, where $P = G^\circ_{W,\mathbf{R}}$ and Z' is a certain subset of D containing Z (5.1.5).

Even for $r \geq 2$, in the case $h^{p,q} = 0$ for any $(p, q) \neq (1, 0), (0, 1)$ (the case $D = \mathfrak{h}_g$), the associated family $W = (W^{(j)})_{1\leq j\leq r}$ of weight filtrations is so simple that all filters in this family are linearly ordered,

$$0 = W^{(j)}_{-2} \subset W^{(1)}_{-1} \cdots \subset W^{(r)}_{-1} \subset W^{(r)}_0 \subset \cdots \subset W^{(1)}_0 \subset W^{(j)}_1 = H_{\mathbf{Q}},$$

for any j, i.e., they form a single long filtration. With one exception below, the same is true for other classical situations in 0.4.14. Hence W and P are related directly by $P = G^{\circ}_{W,\mathbf{R}}$, and we have $D_{\mathrm{SL}(2)}(W) = D_{\mathrm{BS}}(G^{\circ}_{W,\mathbf{R}})$, $D_{\mathrm{SL}(2)} = D_{\mathrm{BS}}$ (which also coincides with the classical Borel-Serre space $\mathcal{X}_{\mathrm{BS}}$) as in (5) in Introduction (cf. 12.1.2, [KU2, 6.7]).

Exceptional Case ([KU2, 6.7]). The weight w is even, rank $H_0 = 4$, and there exists a $\mathbf{Q}$-basis $(e_j)_{1\le j\le 4}$ of $H_{0,\mathbf{Q}}$ such that $\langle e_j, e_k\rangle_0 = 1$ if $j+k=5$, and $=0$ otherwise.

In general, we have the following criterion.

Criterion ([KU2, 6.3]). *The following are equivalent.*

(i) *The identity map of D extends to a continuous map $D_{\mathrm{SL}(2)} \to D_{\mathrm{BS}}$.*
(ii) *At any point of $D_{\mathrm{SL}(2)}$, the filters $W^{(j)}_k$ which appear in the associated family $W = (W^{(j)})_j$ of weight filtrations are linearly ordered by inclusion.*

For examples with no continuous extension $D_{\mathrm{SL}(2)} \to D_{\mathrm{BS}}$ of the identity map of D, see [KU2, 6.10] and 12.4.7.

0.5.18

We have the following criterion for the local compactness of $D_{\mathrm{SL}(2)}$.

Criterion (Theorem 10.1.6). *Let $p = [\rho, \varphi] \in D_{\mathrm{SL}(2)}$. The following* (i) *and* (ii) *are equivalent.*

(i) *There exists a compact neighborhood of p in $D_{\mathrm{SL}(2)}$.*
(ii) *The following conditions* (1) *and* (2) *hold.*

(1) *All filters $W^{(j)}_k$ appeared in the associated compatible family $W = (W^{(j)})_j$ of weight filtrations at p are linearly ordered by inclusion.*
(2) $\mathrm{Lie}(K_{\mathbf{r}}) \subset \mathrm{Lie}(G_{W,\mathbf{R}}) + \mathrm{Lie}(K'_{\mathbf{r}})$, *where* $\mathbf{r} = \varphi(\mathbf{i})$ *and* $G_{W,\mathbf{R}} = \{g \in G_{\mathbf{R}} \mid gW^{(j)}_k = W^{(j)}_k\ (\forall j, \forall k)\}$.

By this criterion, $D_{\mathrm{SL}(2)}$ for the example with $h^{2,0} = h^{0,2} = 2$, $h^{1,1} = 1$ in 0.3.2 (iii) is locally compact, and hence has no slit. We have $D_{\mathrm{SL}(2)} = D_{\mathrm{SL}(2),\le 1} \xrightarrow{\sim} D_{\mathrm{BS}}$ in this case. But even for the examples of similar kind in 12.2.10, $D_{\mathrm{SL}(2)}$ can have slits in general.

0.5.19

Valuative spaces. In the general case, the family W of weight filtrations associated with $p \in D_{\mathrm{SL}(2)}$ becomes more complicated and we do not have a direct relation between the W and the parabolic subgroups P (Criterion in 0.5.17). We have to introduce the valuative spaces $D_{\mathrm{SL}(2),\mathrm{val}}$ and $D_{\mathrm{BS},\mathrm{val}}$ to relate the spaces $D_{\mathrm{SL}(2)}$ and D_{BS}. These are the projective limits of certain kinds of blow-ups of the respective spaces.

To relate the spaces D_Σ and $D_{\mathrm{SL}(2)}$, we also have to introduce the valuative space $D^\sharp_{\Sigma,\mathrm{val}}$ of $D^\sharp_\Sigma$. This is the projective limit of a kind of blow-up of $D^\sharp_\Sigma$ corresponding to rational subdivisions of the fan Σ. We have a continuous map $D^\sharp_{\Sigma,\mathrm{val}} \to D_{\mathrm{SL}(2)}$ which is a geometric interpretation of the SL(2)-orbit theorem [CKS] as in 0.5.24 below.

In all cases, the valuative spaces are projective limits over the corresponding original spaces so as to divide the directions of degenerations into narrower and narrower.

Theorem 0.5.20 *Let X be one of $D^\sharp_{\Sigma,\mathrm{val}}$, $D_{\mathrm{SL}(2),\mathrm{val}}$, $D_{\mathrm{BS},\mathrm{val}}$. Then*

(i) *X is Hausdorff.*
(ii) *Let Γ be a subgroup of $G_{\mathbf{Z}}$. In the case X is $D^\sharp_{\Sigma,\mathrm{val}}$, assume Γ is strongly compatible with Σ. Then $\Gamma\backslash X$ is Hausdorff. If furthermore Γ is neat, the canonical projection $X \to \Gamma\backslash X$ is a local homeomorphism.*

The definitions of the four valuative spaces in the fundamental diagram are given in Chapter 5. Here we just give in 0.5.22–0.5.23 explicit local descriptions of them in the case of Example (ii) with $g = 2$ in 0.3.2. (The cases of Examples (i) and (iii) are not interesting concerning valuative spaces, for, in these cases, the valuative spaces are identified with the spaces under them in the fundamental diagram.)

0.5.21

Example $(\mathbf{C}^2)_{\mathrm{val}}$ *and* $(\mathbf{R}^2_{\geq 0})_{\mathrm{val}}$. In 0.5.22 and 0.5.23, we give explicit descriptions of some valuative spaces in the case $D = \mathfrak{h}_2$. For this, we introduce here the spaces $(\mathbf{C}^2)_{\mathrm{val}}$ and $(\mathbf{R}^2_{\geq 0})_{\mathrm{val}}$ obtained as projective limits of blow-ups from $\mathbf{C}^2$ and $\mathbf{R}^2_{\geq 0}$, respectively. In general, for any object X of $\mathcal{B}(\log)$, we will define in Section 3.6 (see 3.6.18 and 3.6.23) a space X_{val} obtained from X by taking blow-ups along the logarithmic structure. The space $(\mathbf{C}^2)_{\mathrm{val}}$ is X_{val} for $X = \mathbf{C}^2$ which is endowed with the logarithmic structure associated with the normal crossing divisor $\mathbf{C}^2 - (\mathbf{C}^\times)^2$.

Let $X = X_0 = \mathbf{C}^2$, and let X_1 be the blow-up of X at the origin $(0, 0)$. Then $(\mathbf{C}^\times)^2 \subset X_1$, and the complement $X_1 - (\mathbf{C}^\times)^2$ is the union of three irreducible divisors C_0, C_1, C_∞ where C_0 is the closure of $\{0\} \times \mathbf{C}^\times$, C_∞ is the closure of $\mathbf{C}^\times \times \{0\}$, and C_1 is the inverse image of $(0, 0)$. Next let X_2 be the blow-up of X_1 at two points, the intersection of C_0 and C_1 and the intersection of C_1 and C_∞. Then $(\mathbf{C}^\times)^2 \subset X_2$, and the complement $X_2 - (\mathbf{C}^\times)^2$ is the union of five irreducible divisors C_0, $C_{1/2}$, C_1, C_2, and C_∞. Here $C_{1/2}$ is the inverse image of the intersection of C_0 and C_1 in X_1, C_2 is the inverse image of the intersection of C_1 and C_∞, and we denote the proper transformations of C_0, C_1, and C_∞ in X_1 simply by C_0, C_1, and C_∞, respectively. In this way, we have a sequence of blow-ups

$$\cdots \to X_3 \to X_2 \to X_1 \to X_0 = X,$$

where X_{n+1} is obtained from X_n by blow-up the intersections of different irreducible components of $X_n - (\mathbf{C}^\times)^2$. We define

$$(\mathbf{C}^2)_{\mathrm{val}} = \varprojlim_n X_n.$$

This $(\mathbf{C}^2)_{\text{val}}$ is obtained also as the inverse limit of the toric varieties [KKMS, Od] corresponding to finite rational subdivisions of the cone $\mathbf{R}^2_{\geq 0}$ in $\mathbf{R}^2$. For example, the above X_2 is the toric variety corresponding to the finite subdivision of $\mathbf{R}^2_{\geq 0}$ consisting of the subcones $\{0\}$, σ_0, $\sigma_{0,1/2}$, $\sigma_{1/2}$, $\sigma_{1/2,1}$, σ_1, $\sigma_{1,2}$, σ_2, $\sigma_{2,\infty}$, and σ_∞ of $\mathbf{R}^2_{\geq 0}$. Here σ_s ($s = 0,\ 1/2,\ 1,\ 2$) is the half line $\{(x, sx) \mid x \in \mathbf{R}_{\geq 0}\}$ of slope s, σ_∞ is the half line $\{0\} \times \mathbf{R}_{\geq 0}$, and $\sigma_{s,t}$ is the cone generated by σ_s and σ_t. The open subvariety of X_2 corresponding to the cone σ_s is $X_2 - \cup_{s' \neq s} C_{s'}$, and the open subvariety of X_2 corresponding to the cone $\sigma_{s,t}$ is $X_2 - \cup_{s' \neq s,t} C_{s'}$. Let q_1 and q_2 be the coordinate functions of $\mathbf{C}^2$. Then for a finite rational subdivision of $\mathbf{R}^2_{\geq 0}$, the corresponding toric variety is the union of the open subvarieties $\operatorname{Spec}(\mathbf{C}[P(\sigma)])_{\text{an}}$ for cones σ in this subdivision, where

$$P(\sigma) = \{q_1^m q_2^n \mid (m, n) \in \mathbf{Z},\ am + bn \geq 0\ \ \forall (a, b) \in \sigma\}.$$

The projection $f : (\mathbf{C}^2)_{\text{val}} \to \mathbf{C}^2$ is proper and surjective, and f induces a homeomorphism $f^{-1}(\mathbf{C}^2 - \{(0,0)\}) \xrightarrow{\sim} \mathbf{C}^2 - \{(0,0)\}$. We regard $\mathbf{C}^2 - \{(0,0)\}$ as an open subspace of $(\mathbf{C}^2)_{\text{val}}$ via f^{-1}. The complement $f^{-1}((0,0))$ is described as

$$\begin{aligned} f^{-1}((0,0)) = \{(0,0)_s \mid s \in [0,\infty],\ s \notin \mathbf{Q}_{>0}\} \\ \cup \{(0,0)_{s,z} \mid s \in \mathbf{Q}_{>0},\ z \in \mathbf{P}^1(\mathbf{C})\}. \end{aligned}$$

Here if $s \in \mathbf{Q}_{>0}$ and $z \in \mathbf{C}^\times$ and if s is expressed as m/n with $m, n \in \mathbf{Z}$, $m > 0$, $n > 0$ and $\operatorname{GCD}(m, n) = 1$, then $(0,0)_{s,z}$ is the unique point of $f^{-1}((0,0))$ at which both $q_1^m q_2^{-n}$ and $q_1^{-m} q_2^n$ are holomorphic and the value of $q_1^m q_2^{-n}$ is z. For any finite rational subdivision of $\mathbf{R}^2_{\geq 0}$ containing the half line $\sigma = \{(x, sx) \mid x \in \mathbf{R}_{\geq 0}\}$ of slope s, the map from $(\mathbf{C}^2)_{\text{val}}$ to the toric variety corresponding to this subdivision induces a bijection from $\{(s, z) \mid z \in \mathbf{C}^\times\}$ onto the fiber over $(0,0) \in \mathbf{C}^2$ of the open subvariety corresponding to σ. For $s \in \mathbf{Q}_{>0}$, $(0,0)_{s,0}$ (resp. $(0,0)_{s,\infty}$) is the limit of $(0,0)_{s,z}$ ($z \in \mathbf{C}^\times$) for $z \to 0$ (resp. $z \to \infty$). The point $(0,0)_0$ (resp. $(0,0)_\infty$) is the limit of $(0, z) \in \mathbf{C}^2$ (resp. $(z, 0) \in \mathbf{C}^2$) ($z \in \mathbf{C}^\times$) for $z \to 0$. Finally, $(0,0)_s$ for $s \in \mathbf{R}_{>0} - \mathbf{Q}_{>0}$ is the unique point of $f^{-1}((0,0))$ at which $q_1^m q_2^{-n}$ ($m, n \in \mathbf{Z}$, $m > 0$, $n > 0$) is holomorphic if and only if $m/n > s$ and $q_1^{-m} q_2^n$ is holomorphic if and only if $s > m/n$. A point of $f^{-1}((0,0))$ has the form $(0,0)_s$ ($s \in [0,\infty] - \mathbf{Q}_{>0}$) if and only if for any $n \geq 0$, its image in X_n is the intersection of two different irreducible components of the divisor $X_n - (\mathbf{C}^\times)^2$.

These points of $f^{-1}((0,0))$ are characterized as the limits of points of $(\mathbf{C}^\times)^2 \subset (\mathbf{C}^2)_{\text{val}}$ as follows. If $s \in [0,\infty] - \mathbf{Q}_{>0}$, $(q_1, q_2) \in (\mathbf{C}^\times)^2$ converges to $(0,0)_s$ if and only if $(q_1, q_2) \to (0,0)$ and $\log(|q_2|)/\log(|q_1|) \to s$. If $s \in \mathbf{Q}_{>0}$ and $z \in \mathbf{P}^1(\mathbf{C})$, and if s is expressed as m/n with $m, n \in \mathbf{Z}$, $m > 0$, $n > 0$ and $\operatorname{GCD}(m, n) = 1$, $(q_1, q_2) \in (\mathbf{C}^\times)^2$ converges to $(0,0)_{s,z}$ if and only if $(q_1, q_2) \to (0,0)$, $\log(|q_2|)/\log(|q_1|) \to s$, and $q_1^m/q_2^n \to z$.

Let $(\mathbf{R}^2_{\geq 0})_{\text{val}} \subset (\mathbf{C}^2)_{\text{val}}$ be the closure of the subset $\mathbf{R}^2_{\geq 0} - \{(0,0)\}$ (regarded as a subset of $\mathbf{C}^2 - \{(0,0)\} \subset (\mathbf{C}^2)_{\text{val}}$). That is, $(\mathbf{R}^2_{\geq 0})_{\text{val}}$ is the union of $\mathbf{R}^2_{\geq 0} - \{(0,0)\}$ and the part of $f^{-1}((0,0))$ consisting of elements $(0,0)_s$ for $s \in [0,\infty]$ such that $s \notin \mathbf{Q}_{>0}$, and elements $(0,0)_{s,z}$ with $s \in \mathbf{Q}_{>0}$ and with $z \in \mathbf{R}_{\geq 0} \cup \{\infty\} \subset \mathbf{P}^1(\mathbf{C})$. The canonical projection $(\mathbf{R}^2_{\geq 0})_{\text{val}} \to \mathbf{R}^2_{\geq 0}$ is proper and surjective. The inverse image of $(0,0)$ in $(\mathbf{R}^2_{\geq 0})_{\text{val}}$ is regarded as a very long totally ordered set by the following rule: $(0,0)_s < (0,0)_{s',z} < (0,0)_{s',z'} < (0,0)_{s''}$ if $0 \le s < s' < s'' \le \infty$, $s \notin \mathbf{Q}_{>0}$, $s' \in \mathbf{Q}_{>0}$, $s'' \notin \mathbf{Q}_{>0}$, $0 \le z < z' \le \infty$. Closed intervals form a base of closed sets in this totally ordered set.

0.5.22

Upper half space (continued). Let $g = 2$ and $D = \mathfrak{h}_2$. We have the following commutative diagrams of topological spaces in which all inclusions are open immersions.

$$\begin{array}{ccccc} (\Delta^2)_{\mathrm{val}} \times \mathbf{C} & \supset & U_{\mathrm{val}} & \simeq & \Gamma\backslash D_{\sigma,\mathrm{val}} \\ \downarrow & & \downarrow & & \downarrow \\ \Delta^2 \times \mathbf{C} & \supset & U & \simeq & \Gamma\backslash D_\sigma . \end{array} \qquad (1)$$

$$\begin{array}{ccccc} (|\Delta|^2)_{\mathrm{val}} \times \mathbf{R}^2 \times \mathbf{C} & \supset & \tilde{U}^{\log}_{\mathrm{val}} & \simeq & D^\sharp_{\sigma,\mathrm{val}} \\ \downarrow & & \downarrow & & \downarrow \\ |\Delta|^2 \times \mathbf{R}^2 \times \mathbf{C} & \supset & \tilde{U}^{\log} & \simeq & D^\sharp_\sigma . \end{array} \qquad (2)$$

$$\begin{array}{ccc} P_u \times (\mathbf{R}^2_{\ge 0})_{\mathrm{val}} & \simeq & D_{\mathrm{BS,val}}(P) \\ \downarrow & & \downarrow \\ P_u \times \mathbf{R}^2_{\ge 0} & \simeq & D_{\mathrm{BS}}(P). \end{array} \qquad (3)$$

Here in (1), $(\Delta^2)_{\mathrm{val}} \subset (\mathbf{C}^2)_{\mathrm{val}}$ is the inverse image of Δ^2 under $(\mathbf{C}^2)_{\mathrm{val}} \to \mathbf{C}^2$, U and Γ are as in 0.4.13, and U_{val} denotes the inverse image of $U \subset \Delta^2 \times \mathbf{C}$ in $(\Delta^2)_{\mathrm{val}} \times \mathbf{C}$. In (2), $(|\Delta|^2)_{\mathrm{val}} \subset (\mathbf{R}^2_{\ge 0})_{\mathrm{val}}$ is the inverse image of $|\Delta|^2$ under $(\mathbf{R}_{\ge 0})^2_{\mathrm{val}} \to \mathbf{R}^2_{\ge 0}$, $\tilde{U}^{\log}$ is as in 0.5.6, and $\tilde{U}^{\log}_{\mathrm{val}}$ denotes the inverse image of $\tilde{U}^{\log}$ in $(|\Delta|^2)_{\mathrm{val}} \times \mathbf{R}^2 \times \mathbf{C}$. In (3), P is as in 0.5.9, and $D_{\mathrm{BS,val}}(P)$ denotes the inverse image of $D_{BS}(P)$ in $D_{\mathrm{BS,val}}$.

The lower rows of these diagrams are those obtained in 0.4.13, 0.5.6, and 0.5.9, respectively.

The first diagram is obtained as follows. We identify σ with the cone $\mathbf{R}^2_{\ge 0}$ via $\mathbf{R}^2_{\ge 0} \simeq \sigma$, $(a_1, a_2) \mapsto a_1 N_1 + a_2 N_2$. For a rational subdivision S of $\mathbf{R}^2_{\ge 0}$, we have the corresponding subdivision of σ. If $B(S)$ denotes the toric variety corresponding to S with a proper birational morphism $B(S) \to \mathbf{C}^2$, we have an isomorphism $U(S) \simeq \Gamma \setminus D_\sigma(S)$, where $U(S)$ is the inverse image of U in $B(S) \times \mathbf{C}$ and $\Gamma \setminus D_\sigma(S)$ is a blow-up of $\Gamma \setminus D_\sigma$ corresponding to this subdivision of σ. The upper row of the diagram (1) is obtained as the projective limit of $B(S) \times \mathbf{C} \supset U(S) \simeq \Gamma\backslash D_\sigma(S)$.

Roughly speaking, in the first diagram, we are dividing the direction of degeneration $\exp(z_1 N_1 + z_2 N_2)$ with $\mathrm{Im}(z_1), \mathrm{Im}(z_2) \to \infty$ into narrower and narrower directions. A narrow direction that appears here is the direction $\exp(z_1(N_1 + sN_2) + z_2(N_1 + s'N_2))$ with $\mathrm{Im}(z_1), \mathrm{Im}(z_2) \to \infty$ for some $s, s' \in \mathbf{Q}_{\ge 0}$ or the direction $\exp(z_1(N_1 + sN_2) + z_2 N_2)$ with $\mathrm{Im}(z_1), \mathrm{Im}(z_2) \to \infty$ for some $s \in \mathbf{Q}_{>0}$. When the directions become infinitely narrow, we obtain points at infinity in $\Gamma\backslash D_{\sigma,\mathrm{val}}$.

The second diagram is obtained in a similar manner. For the third diagram, see [KU2, 2.14]. Note that $D_{\mathrm{SL}(2),\mathrm{val}} = D_{\mathrm{BS,val}}$ in this case [KU2, 6.7].

0.5.23

Upper half space (continued). Let the notation be as in 0.5.22 and consider $\exp(iy_1N_1 + iy_2N_2)F(0) \in D$ for $y_1, y_2 \in \mathbf{R}_{>0}$. We observe how this point converges or diverges when y_1 and y_2 move in special ways. This point corresponds to $(e^{-2\pi y_1}, e^{-2\pi y_2}, 0, 0, 0) \in \tilde{U}^{\log}_{\mathrm{val}}$ in the diagram (2), and to $(1, (y_2/y_1)^{1/2}, (1/y_2)^{1/2}) \in P_u \times (\mathbf{R}_{>0})^2$ in the diagram (3). From this, we have

(i) When $t \to \infty$, $\exp(it(2+\sin(t))N_1 + itN_2)F(0)$ converges in $D^\sharp_\sigma$ to the image of $(0, 0, 0, 0, 0) \in \tilde{U}^{\log}$, but diverges in $D^\sharp_{\sigma,\mathrm{val}}$, D_{BS}, $D_{\mathrm{BS},\mathrm{val}}$.

We show here the divergence in D_{BS}. The corresponding point is $p(t) := (1, (2+\sin(t))^{-1/2}, t^{-1/2}) \in P_u \times \mathbf{R}^2_{\geq 0}$, and

$$p(t) = \begin{cases} (1, 2^{-1/2}, t^{-1/2}) \to (1, 2^{-1/2}, 0) \in P_u \times \mathbf{R}^2_{\geq 0} \\ \qquad \text{when } t = \pi n,\ n = 1, 2, 3, \ldots, \\ (1, 1, t^{-1/2}) \to (1, 1, 0) \in P_u \times \mathbf{R}^2_{\geq 0} \\ \qquad \text{when } t = 2\pi n - \pi/2,\ n = 1, 2, 3, \ldots. \end{cases}$$

Hence $p(t)$ diverges in D_{BS}.

This (i) shows that the image of $(0, 0, 0, 0, 0) \in \tilde{U}^{\log}$ in $D^\sharp_\sigma$ has no neighborhood V such that the inclusion map $V \cap D \to D$ extends to a continuous map $V \to D_{\mathrm{SL}(2)} = D_{\mathrm{BS}}$.

We also see at the end of 0.5.26, concerning a map in the converse direction, that the image of $(1, 0, 0) \in P_u \times \mathbf{R}^2_{\geq 0}$ in D_{BS} has no neighborhood V such that the inclusion map $V \cap D \to D$ extends to a continuous map $V \to D^\sharp_\sigma$, and even that it has no neighborhood V such that the canonical map $V \cap D \to \Gamma\backslash D$ extends to a continuous map $V \to \Gamma\backslash D_\sigma$.

Thus, to connect the world of D_σ and $D^\sharp_\sigma$ to the world of $D_{\mathrm{SL}(2)} = D_{\mathrm{BS}}$, we have to climb to the valuative space $D^\sharp_{\sigma,\mathrm{val}}$.

Similarly, by using the topological natures of $|\Delta|^2_{\mathrm{val}}$ and $(\mathbf{R}^2_{\geq 0})_{\mathrm{val}}$ explained in 0.5.21, we can show

(ii) When $t \to \infty$, $\exp(it^{c(2+\sin(t))}N_1 + itN_2)F(0)$, for a fixed $c > 1$ converges in $D^\sharp_{\sigma,\mathrm{val}}$ to the image of $((0, 0)_0, 0, 0, 0) \in \tilde{U}^{\log}_{\mathrm{val}}$, but diverges in $D_{\mathrm{BS},\mathrm{val}}$. Hence the image of $((0, 0)_0, 0, 0, 0)$ in $D^\sharp_{\sigma,\mathrm{val}}$ has no neighborhood V such that the inclusion map $V \cap D \to D$ extends to a continuous map $V \to D_{\mathrm{BS},\mathrm{val}}$. When $t \to \infty$, $\exp(i(t+\sin(t))N_1 + itN_2)F(0)$ converges in $D_{\mathrm{BS},\mathrm{val}}(P)$ to the image of $(1, (0, 0)_{1,1}) \in P_u \times (\mathbf{R}^2_{\geq 0})_{\mathrm{val}}$, but diverges in $D^\sharp_{\sigma,\mathrm{val}}$. Hence the image of $(1, (0, 0)_{1,1})$ in $D_{\mathrm{BS},\mathrm{val}}(P)$ has no neighborhood V such that the inclusion map $V \cap D \to D$ extends to a continuous map $V \to D^\sharp_{\sigma,\mathrm{val}}$.

Thus, the views of the infinity of various enlargements of D in the fundamental diagram are rather different from each other.

(iii) When $y_2 \to \infty$ and $y_1/y_2 \to \infty$, $\exp(iy_1N_1 + iy_2N_2)F(0)$ converges in $D^\sharp_{\sigma,\mathrm{val}}$ to the image of $((0, 0)_0, 0, 0, 0) \in \tilde{U}^{\log}_{\mathrm{val}}$, and also converges in $D_{\mathrm{BS}}(P)$

to the image of $(1, 0, 0) \in P_u \times \mathbf{R}^2_{\geq 0}$. As is explained in 0.5.26 below, these two limit points at infinity are related by the following continuous map $\psi : D^\sharp_{\sigma,\mathrm{val}} \to D_{\mathrm{SL}(2)}$.

0.5.24

SL(2)*-orbit Theorem and continuous map* $D^\sharp_{\Sigma,\mathrm{val}} \to D_{\mathrm{SL}(2)}$. The SL(2)-orbit theorem in several variables in [CKS] is interpreted as the relation between $D^\sharp_{\Sigma,\mathrm{val}}$ and $D_{\mathrm{SL}(2)}$.

Let $N_1, \dots, N_n \in \mathfrak{g}_{\mathbf{R}}$ be mutually commutative nilpotent elements, and $F \in D$. Assume that $(N_1, \dots, N_n, F)$ generates a nilpotent orbit, i.e., for $\sigma = \sum_{1\leq j\leq n}(\mathbf{R}_{\geq 0})N_j$, $\exp(\sigma_{\mathbf{C}})F$ is a σ-nilpotent orbit (0.4.7, 1.3.7). Cattani, Kaplan, and Schmid [CKS] defined an SL(2)-orbit (ρ, φ) in n variables associated to the family $(N_1, \dots, N_n, F)$ (cf. Section 6.1). Here the order of $N_1, \dots, N_n$ is important. They showed that two maps $\exp(\sum_{j=1}^n iy_jN_j)F$ and $\varphi(iy_1, \dots, iy_n)$ into D behave asymptotically when $y_j/y_{j+1} \to \infty$ (y_{n+1} means 1). Our geometric interpretation of the SL(2)-orbit Theorem is as follows.

There is a unique continuous map $\psi : D^\sharp_{\Sigma,\mathrm{val}} \to D_{\mathrm{SL}(2)}$ *which extends the identity map of* D (*we will prove this in* Chapter 6), $\exp(\sum_{j=1}^n iy_jN_j)F$ *converges in* $D^\sharp_{\Sigma,\mathrm{val}}$ *when* $y_j/y_{j+1} \to \infty$ $(1 \leq j \leq n)$ (*we saw this in* 0.5.23 (iii) *in a special case*), *and* ψ *sends this limit point to* $[\rho, \varphi]$.

This continuous map $\psi : D^\sharp_{\Sigma,\mathrm{val}} \to D_{\mathrm{SL}(2)}$ is the most important bridge in the fundamental diagram (3) in Introduction, which joins the four spaces D_Σ, $D_{\Sigma,\mathrm{val}}$, $D^\sharp_\Sigma$, and $D^\sharp_{\Sigma,\mathrm{val}}$ of orbits under nilpotent groups in the left-hand side and the four spaces $D_{\mathrm{SL}(2)}$, $D_{\mathrm{SL}(2),\mathrm{val}}$, D_{BS}, and $D_{\mathrm{BS},\mathrm{val}}$ of orbits under tori in the right-hand side.

0.5.25

Fundamental Diagram. We thus have the diagram that relates D_Σ and D_{BS} ((3) in Introduction; see also 5.0.1):

$$
\begin{array}{ccccccc}
 & & & & D_{\mathrm{SL}(2),\mathrm{val}} & \hookrightarrow & D_{\mathrm{BS},\mathrm{val}} \\
 & & & & \downarrow & & \downarrow \\
\Gamma\backslash D_{\Sigma,\mathrm{val}} & \leftarrow & D^\sharp_{\Sigma,\mathrm{val}} & \rightarrow & D_{\mathrm{SL}(2)} & & D_{\mathrm{BS}} \\
\downarrow & & \downarrow & & & & \\
\Gamma\backslash D_\Sigma & \leftarrow & D^\sharp_\Sigma & & & &
\end{array}
$$

where all maps are continuous, and all vertical maps are proper surjective.

The theorems on these spaces introduced in this Section 0.5 are proved by starting at D_{BS} and moving in this diagram from the right to the left. The spaces D_{BS}, $D_{\mathrm{BS},\mathrm{val}}$, $D_{\mathrm{SL}(2),\mathrm{val}}$, and $D_{\mathrm{SL}(2)}$ were already studied in [KU2]. By using the results on these spaces (which are reviewed in Chapter 5), in this book, we prove the results on the other spaces.

We now give explicit descriptions of $\psi : D^{\sharp}_{\Sigma,\mathrm{val}} \to D_{\mathrm{SL}(2)}$ in some examples. In the case of Example (i) (the case $D = \mathfrak{h}$) in 0.3.2, for any fan Σ in $\mathfrak{g}_{\mathbf{Q}}$, ψ is just the canonical map $D^{\sharp}_{\Sigma,\mathrm{val}} = D^{\sharp}_{\Sigma} \subset D^{\sharp}_{\Xi} \xrightarrow{\sim} D_{\mathrm{SL}(2)}$ in 0.5.2. In the following 0.5.26 and 0.5.27, we consider the cases of Example (ii) with $g = 2$ and Example (iii) in 0.3.2, respectively.

0.5.26

Upper half space (continued). Let $g = 2$ and $D = \mathfrak{h}_2$. In this case, $D_{\mathrm{SL}(2)} = D_{\mathrm{BS}}$. Let N_1, N_2, and σ be as before (0.4.13). The triple $(N_1, N_2, F(0))$ generates a σ-nilpotent orbit, and the associated SL(2)-orbit in two variables (ρ, φ) is given (see 6.1.4) by

$$\rho\left(\begin{pmatrix} a & b \\ c & d \end{pmatrix}, \begin{pmatrix} e & f \\ g & h \end{pmatrix}\right) = \begin{pmatrix} a & 0 & b & 0 \\ 0 & e & 0 & f \\ c & 0 & d & 0 \\ 0 & g & 0 & h \end{pmatrix}, \qquad \varphi(z, w) = F\begin{pmatrix} z & 0 \\ 0 & w \end{pmatrix}.$$

That is, the continuous map $\psi : D^{\sharp}_{\sigma,\mathrm{val}} \to D_{\mathrm{SL}(2)}$ sends the limit of $\exp(iy_1N_1 + iy_2N_2)F(0)$ for $y_1/y_2,\ y_2 \to \infty$ in $D^{\sharp}_{\sigma,\mathrm{val}}$ (see 0.5.23) to $[\rho, \varphi] \in D_{\mathrm{SL}(2)}$. Furthermore, (ρ, φ) is of rank 2, and "the N_1 and N_2 of ρ" defined in 0.5.12 coincide with N_1 and N_2 here, respectively. Let $W = (W^{(1)}, W^{(2)})$ be the family of weight filtrations associated with $[\rho, \varphi] \in D_{\mathrm{SL}(2)}$, that is, $W^{(1)} = W(N_1)$ and $W^{(2)} = W(N_1 + N_2)$. Then

$$0 = W^{(1)}_{-2} \subset W^{(1)}_{-1} = \mathbf{R}e_1 \subset W^{(1)}_{0} = \mathbf{R}e_1 + \mathbf{R}e_2 + \mathbf{R}e_4 \subset W^{(1)}_{1} = H_{0,\mathbf{R}},$$
$$0 = W^{(2)}_{-2} \subset W^{(2)}_{-1} = \mathbf{R}e_1 + \mathbf{R}e_2 = W^{(2)}_{0} \subset W^{(2)}_{1} = H_{0,\mathbf{R}}.$$

We have $G_{W,\mathbf{R}} = P$, $D_{\mathrm{SL}(2)}(W) = D_{\mathrm{BS}}(P)$. The element $[\rho, \varphi] \in D_{\mathrm{SL}(2)}$ is also written as $(W, Z) \in D_{\mathrm{SL}(2)}$ with $Z = F\left(\begin{smallmatrix} i\mathbf{R}_{>0} & 0 \\ 0 & i\mathbf{R}_{>0} \end{smallmatrix}\right)$, and is also written as $(P, Z) \in D_{\mathrm{BS}}$ with the same Z.

In the homeomorphism $\tilde{U}^{\log}_{\mathrm{val}} \simeq D^{\sharp}_{\sigma,\mathrm{val}}$ in 0.5.22, the above limit point of $D^{\sharp}_{\sigma,\mathrm{val}}$ is the image of $((0,0)_0, 0, 0, 0) \in \tilde{U}^{\log}_{\mathrm{val}} \subset (|\Delta|^2)_{\mathrm{val}} \times \mathbf{R}^2 \times \mathbf{C}$. In the homeomorphism $P_u \times \mathbf{R}^2_{\geq 0} \simeq D_{\mathrm{BS}}(P)$, the above $[\rho, \varphi] \in D_{\mathrm{SL}(2)} = D_{\mathrm{BS}}$ is the image of $(1, 0, 0)$.

We give an explicit description of $\psi : D^{\sharp}_{\sigma,\mathrm{val}} \to D_{\mathrm{SL}(2)}$ for some open neighborhood of the above limit point of $D^{\sharp}_{\sigma,\mathrm{val}}$. Let $(\tilde{U}^{\log}_{\mathrm{val}})'$ be the open set of $\tilde{U}^{\log}_{\mathrm{val}}$ consisting of all points (p, x_1, x_2, a) ($p \in (|\Delta|^2)_{\mathrm{val}}$, $x_1, x_2 \in \mathbf{R}$, $a \in \mathbf{C}$) such that $p \neq (0,0)_\infty$ and such that $p \neq (r, 0)$ for any $r \in |\Delta| - \{0\}$. Then $(\tilde{U}^{\log}_{\mathrm{val}})'$ contains the point $((0,0)_0, 0, 0, 0)$ of $\tilde{U}^{\log}_{\mathrm{val}}$. Define $N_3, N_4 \in \mathrm{Lie}(P_u)$ by

$$N_3 := e_{23} + e_{14} = \begin{pmatrix} 0 & 0 & 0 & 1 \\ 0 & 0 & 1 & 0 \\ 0 & 0 & 0 & 0 \\ 0 & 0 & 0 & 0 \end{pmatrix}, \qquad N_4 := e_{43} - e_{12} = \begin{pmatrix} 0 & -1 & 0 & 0 \\ 0 & 0 & 0 & 0 \\ 0 & 0 & 0 & 0 \\ 0 & 0 & 1 & 0 \end{pmatrix}.$$

We have a homeomorphism

$$\mathbf{R}^4 \xrightarrow{\sim} P_u, \quad (x_j)_{1\le j\le 4} \mapsto \exp\left(\sum\nolimits_{j=1}^{3} x_jN_j\right)\exp(x_4N_4).$$

The restriction of $\psi : D^\sharp_{\sigma,\mathrm{val}} \to D_{\mathrm{SL}(2)}(W) = D_{\mathrm{BS}}(P)$ to $(\tilde{U}^{\log}_{\mathrm{val}})'$ is explicitly described by the commutative diagram

$$\begin{array}{ccccc} (|\Delta|^2)_{\mathrm{val}} \times \mathbf{R}^2 \times \mathbf{C} \supset & (\tilde{U}^{\log}_{\mathrm{val}})' & \to & D^\sharp_{\sigma,\mathrm{val}} \\ & & & \downarrow \psi \\ & \downarrow & & D_{\mathrm{SL}(2)}(W) \\ & & & \| \\ & P_u \times \mathbf{R}^2_{\ge 0} & \xrightarrow{\sim} & D_{\mathrm{BS}}(P). \end{array}$$

Here the left vertical arrow sends $p = (r, x_1, x_2, x_3 + iy_3) \in (\tilde{U}^{\log}_{\mathrm{val}})'$ with $r \in (|\Delta|^2)_{\mathrm{val}}$ and $x_1, x_2, x_3, y_3 \in \mathbf{R}$ to the following element of $P_u \times \mathbf{R}^2_{\ge 0}$.

(1) When the image of r in $|\Delta|^2$ is not $(0, 0)$, if we write $r = (e^{-2\pi y_1}, e^{-2\pi y_2})$ with $0 < y_1 \le \infty$ and $0 < y_2 < \infty$, p is sent to

$$\left(\exp\left(\sum\nolimits_{j=1}^{3} x_jN_j\right)\exp\left(-\frac{y_3}{y_2}N_4\right), \frac{y_2}{y_1}\left(1 - \frac{y_3^2}{y_1y_2}\right)^{-1}, \frac{1}{y_2}\right) \in P_u \times \mathbf{R}^2_{\ge 0}.$$

(2) When the image of r in $|\Delta|^2$ is $(0, 0)$ and r has the form $(0, 0)_s$ or $(0, 0)_{s,z}$, p is sent to

$$\left(\exp\left(\sum_{j=1}^{3} x_jN_j\right), s, 0\right) \in P_u \times \mathbf{R}^2_{\ge 0}.$$

Note that, when $p \in (\tilde{U}^{\log}_{\mathrm{val}})'$ as in (1) converges to a point of $(\tilde{U}^{\log}_{\mathrm{val}})'$ as in (2) whose r has the form $(0, 0)_s$ or $(0, 0)_{s,z}$, then x_1, x_2, x_3, and y_3 converge, $y_1, y_2 \to \infty$, and $y_2/y_1 \to s$, and hence the terms y_3/y_2 and $1/y_2$ in (1) converge to 0 and the term $\frac{y_2}{y_1}\left(1 - \frac{y_3^2}{y_1y_2}\right)^{-1}$ in (1) converges to s.

Let $\Gamma = \exp(\mathbf{Z}N_1 + \mathbf{Z}N_2) = 1 + \mathbf{Z}N_1 + \mathbf{Z}N_2$. We show that, for any neighborhood V of $[\rho, \varphi] \in D_{\mathrm{SL}(2)}(W) = D_{\mathrm{BS}}(P)$, there is no continuous map $V \to \Gamma\backslash D_\sigma$ that extends the projection $V \cap D \to \Gamma\backslash D$. In fact, for any $c \in \mathbf{R}$, the image $p_c \in D^\sharp_{\sigma,\mathrm{val}}$ of $((0, 0)_0, 0, 0, ic) \in (\tilde{U}^{\log}_{\mathrm{val}})'$ is sent by ψ to $[\rho, \varphi] \in D_{\mathrm{SL}(2)}$ which is independent of c. On the other hand, the image p'_c of p_c in $\Gamma\backslash D_\sigma$ is the class of the nilpotent orbit

$$\left(\sigma,\ \exp(\sigma_{\mathbf{C}})F\begin{pmatrix} 0 & ic \\ ic & 0 \end{pmatrix}\right)$$

modulo Γ, and we have $p'_c \ne p'_d$ if $c, d \in \mathbf{R}$ and $c \ne d$. Hence there is no V with $V \to \Gamma\backslash D_\sigma$ as above.

0.5.27

Example with $h^{2,0} = h^{0,2} = 2$, $h^{1,1} = 1$ (continued). In this example, the fundamental diagram becomes (see Criteria in 0.5.17 and in 0.5.18, and Theorem 10.1.6).

$$\begin{array}{ccccccc}
 & & & & D_{\mathrm{SL}(2),\mathrm{val}} & = & D_{\mathrm{BS},\mathrm{val}} \\
 & & & & \| & & \| \\
D_{\Xi,\mathrm{val}} & \leftarrow & D^{\sharp}_{\Xi,\mathrm{val}} & \rightarrow & D_{\mathrm{SL}(2)} & = & D_{\mathrm{BS}}. \\
\| & & \| & & & & \\
D_{\Xi} & \leftarrow & D^{\sharp}_{\Xi}. & & & &
\end{array}$$

Let W be the $\mathbf{Q}$-rational increasing filtration of $H_{0,\mathbf{R}}$ defined by $W_k = 0$ $(k \le -1)$, $W_0 = W_1 = \mathbf{R}e_4$, $W_2 = W_3 = \sum_{j=1}^{4} \mathbf{R}e_j$, $W_k = H_{0,\mathbf{R}}$ $(k \ge 4)$. On the other hand, let the $\mathbf{Q}$-parabolic subgroup P of $G_{\mathbf{R}}$ be as in 0.5.10. Then, we have $D_{\mathrm{SL}(2)}(W) = D_{\mathrm{BS}}(P)$ (cf. 0.5.10), and the map $\psi : D^{\sharp}_{\Xi} = D^{\sharp}_{\Xi,\mathrm{val}} \to D_{\mathrm{SL}(2)}$ is injective.

Let $v \in \mathbf{S}^2 \cap (\sum_{j=1}^{3} \mathbf{Q}e_j)$ and let $\sigma = (\mathbf{R}_{\ge 0})N_v$. Then $\psi : D^{\sharp}_{\Xi} \to D_{\mathrm{SL}(2)}$ sends $D^{\sharp}_{\sigma}$ into $D_{\mathrm{SL}(2)}(W)$. We consider the map $\psi : D^{\sharp}_{\sigma} = D^{\sharp}_{\sigma,\mathrm{val}} \to D_{\mathrm{SL}(2)}(W) = D_{\mathrm{BS}}(P)$.

Consider the exterior product $\times$ in $\sum_{j=1}^{3} \mathbf{R}e_j$, i.e., the bilinear map $(\sum_{j=1}^{3} \mathbf{R}e_j) \times (\sum_{j=1}^{3} \mathbf{R}e_j) \to \sum_{j=1}^{3} \mathbf{R}e_j$ characterized by

$$e_1 \times e_2 = e_3, \quad e_2 \times e_3 = e_1, \quad e_3 \times e_1 = e_2, \quad e_j \times e_k = -e_k \times e_j.$$

For $z \in Q$ and $u \in \sum_{j=1}^{3} \mathbf{C}e_j$ such that u is not contained in $\mathbf{C}z + \sum_{j=1}^{3} \mathbf{R}e_j$, we have $\exp(N_u)F(z) = \exp(N_b)s(t) \cdot \mathbf{r}(\theta(z))$ with $b = \mathrm{Re}(u) - \mathrm{Im}(u) \times \theta(z)$, $t = (-\langle \mathrm{Im}(u), \theta(z)\rangle_0)^{-1/2}$ (cf. Section 12.2).

Fix $v' \in \mathbf{S}^2$ which is orthogonal to v. From 0.5.6, 0.5.10, and 0.5.18, we have a commutative diagram

$$\begin{array}{ccccc}
((\mathbf{R}_{\ge 0}) \times \mathbf{R}) \times \mathbf{C} \times Q \supset & \tilde{U}^{\log} & \rightarrow & D^{\sharp}_{\sigma} \\
 & & & \downarrow \psi \\
 & \downarrow & & D_{\mathrm{SL}(2)}(W) \\
 & & & \| \\
 & (\sum_{j=1}^{3} \mathbf{R}e_j) \times (\mathbf{R}_{\ge 0}) \times \{\pm 1\} \times \mathbf{S}^2 & \overset{\sim}{\rightarrow} & D_{\mathrm{BS}}(P).
\end{array}$$

The left vertical arrow sends $(r, x, a, z) \in \tilde{U}^{\log}$ with $r \ne 0$ to $(b, t, 1, \theta(z))$ where $b = xv + \mathrm{Re}(a)v' - (yv + \mathrm{Im}(a)v') \times \theta(z), t = (-\langle yv + \mathrm{Im}(a)v', \theta(z)\rangle_0)^{-1/2}$ with $y \in \mathbf{R}$ defined by $r = e^{-2\pi y/\ell}$ (ℓ is as in 0.4.18), and sends $(0, x, a, \theta^{-1}(v)) \in \tilde{U}^{\log}$ to $(xv + \mathrm{Re}(a)v' + \mathrm{Im}(a)v \times v', 0, 1, v)$.

We show the following two results.

(i) If we embed $D^{\sharp}_{\Xi}$ in $D_{\mathrm{SL}(2)}$ by the injection ψ, the topology of $D^{\sharp}_{\Xi}$ does not coincide with the topology as a subspace of $D_{\mathrm{SL}(2)}$.

(ii) Let us call the topology of $\tilde{U}^{\log}$ as a subspace of $((\mathbf{R}_{\geq 0}) \times \mathbf{R}) \times \mathbf{C} \times Q$ the naive topology. Then the composition $\tilde{U}^{\log} \to D^{\sharp}_{\sigma} \xrightarrow{\psi} D_{\mathrm{SL}(2)}(W) = D_{\mathrm{BS}}(P)$ is not continuous for the naive topology.

Let $p \in D^{\sharp}_{\sigma}$ be the image of $(0, 0, 0, \theta^{-1}(v)) \in \tilde{U}^{\log}$. Take $c > 0$ and consider the elements $f(s) = (\exp(-2\pi/(\ell s^c)), 0, 0, \theta^{-1}((1-s^2)^{1/2}v + sv')) \in \tilde{U}^{\log}$ $(s > 0, s \to 0)$. In $\tilde{U}^{\log}$, when $s \to 0$, $f(s)$ converges to $(0, 0, 0, \theta^{-1}(v))$ for the naive topology. However, in U with the strong topology, the image $(\exp(-2\pi/(\ell s^c)), 0, \theta^{-1}((1-s^2)^{1/2}v + sv'))$ of $f(s)$ does not converge to the image $(0, 0, \theta^{-1}(v))$ of $(0, 0, 0, \theta^{-1}(v))$ (0.4.15 (2)). Hence the image of $f(s)$ in $D^{\sharp}_{\sigma}$ does not converge to p (0.5.6).

The image of $f(s)$ in $(\sum_{j=1}^{3} \mathbf{R}e_j) \times (\mathbf{R}_{\geq 0}) \times \{\pm 1\} \times \mathbf{S}^2$ is

$$(s^{1-c}v' \times v,\ s^{c/2}(1-s^2)^{-1/4},\ 1,\ (1-s^2)^{1/2}v + sv').$$

This converges to $(0, 0, 1, v)$ if $c < 1$, but does not converge if $c > 1$ (by the existence of the term s^{1-c}).

For $c < 1$, this proves (i) because the image of $f(s)$ in $D^{\sharp}_{\sigma}$ does not converge to p but the image of $f(s)$ in $D_{\mathrm{SL}(2)}$ converges to $\psi(p)$.

For $c > 1$, this proves (ii) because $f(s)$ converges to $(0, 0, 0, \theta^{-1}(v))$ for the naive topology but the image of $f(s)$ in $D_{\mathrm{SL}(2)}$ does not converge to the image $\psi(p)$ of $(0, 0, 0, \theta^{-1}(v))$.

0.5.28

For an object X of $\mathcal{B}(\log)$, we will define a logarithmic local ringed space X_{val} over X in 3.6.18 and 3.6.23, by using the projective limit of blow-ups along the logarithmic structure. Though X_{val} need not belong to $\mathcal{B}(\log)$, we can define the topological space $(X_{\mathrm{val}})^{\log}$ (we denote it as $X^{\log}_{\mathrm{val}}$) in the same way as before. If $X = \mathbf{C}^2$ with the logarithmic structure associated with the normal crossing divisor $\mathbf{C}^2 - (\mathbf{C}^{\times})^2$, then

$$X_{\mathrm{val}} = (\mathbf{C})^2_{\mathrm{val}}, \quad X^{\log}_{\mathrm{val}} = (\mathbf{R}^2_{\geq 0})_{\mathrm{val}} \times (\mathbf{S}^1)^2.$$

For a fan Σ in $\mathfrak{g}_{\mathbf{Q}}$ and for a neat subgroup Γ of $G_{\mathbf{Z}}$ which is strongly compatible with Σ, we have

$$(\Gamma \backslash D_{\Sigma})_{\mathrm{val}} = \Gamma \backslash D_{\Sigma,\mathrm{val}}, \quad (\Gamma \backslash D_{\Sigma})^{\log}_{\mathrm{val}} = \Gamma \backslash D^{\sharp}_{\Sigma,\mathrm{val}}.$$

(See 8.4.3.)

In the classical situation 0.4.14, except for the unique exceptional case 0.5.17, the fundamental diagram and the fact that $D_{\mathrm{SL}(2)} = D_{\mathrm{BS}}$ in this situation show that there is a unique continuous map

$$(\Gamma \backslash D_{\Sigma})^{\log}_{\mathrm{val}} \to \Gamma \backslash D_{\mathrm{BS}}$$

which extends the identity map of D. Chikara Nakayama (unpublished) proved that in the classical situation 0.4.14, if we take Σ and Γ such that $\Gamma \backslash D_{\Sigma}$ is a toroidal compactification of $\Gamma \backslash D$, then $\Gamma \backslash D_{\Sigma,\mathrm{val}}$ coincides with the projective limit of all toroidal compactifications of $\Gamma \backslash D$.

In the general situation, we have

THEOREM 0.5.29 *Let X be a connected, logarithmically smooth, fs logarithmic analytic space, and let $U = X_{\mathrm{triv}} = \{x \in X \mid M_{X,x} = \mathcal{O}_{X,x}^{\times}\}$ be the open subspace of X consisting of all points of X at which the logarithmic structure of X is trivial. Let H be a variation of polarized Hodge structure on U with unipotent local monodromy along $X - U$. Fix a base point $u \in U$ and let $(H_0, \langle\ ,\ \rangle_0) = (H_{\mathbf{Z},u}, \langle\ ,\ \rangle_u)$. Let Γ be a subgroup of $G_{\mathbf{Z}}$ which contains the global monodromy group* Image$(\pi_1(U, u) \to G_{\mathbf{Z}})$ *and assume Γ is neat. Then the associated period map $\varphi : U \to \Gamma\backslash D$ extends to a continuous map*

$$X_{\mathrm{val}}^{\log} \to \Gamma\backslash D_{\mathrm{SL}(2)}.$$

Here X_{val} is the projective limit of certain blow-ups of X at the boundary (see Section 3.6).

As is explained in Section 8.4, this theorem 0.5.29 is obtained from the period maps in 0.4.30 (ii) and the map $\psi : D_{\Sigma,\mathrm{val}}^{\sharp} \to D_{\mathrm{SL}(2)}$. (The period map in 0.4.30 (ii) is obtained only locally on X, but the composition globalizes.)

0.5.30

$\flat$-*spaces.* Cattani and Kaplan [CK1] generalized Satake-Baily-Borel compactifications of $\Gamma\backslash D$ for a symmetric Hermitian domain D, to the case where D is a Griffiths domain of weight 2 under certain assumptions, and showed that period maps from a punctured disc Δ^* extend over the unit disc Δ. This was the first successful attempt to enlarge D beyond the classical situation 0.4.14. In Chapter 9 and Section 10.4, we consider the relationship between our theory and their theory and discuss related subjects.

We define the quotient topological spaces $D_{\mathrm{BS}}^{\flat} := D_{\mathrm{BS}}/\sim$, $D_{\mathrm{BS,val}}^{\flat} := D_{\mathrm{BS,val}}/\sim$ divided by the action of the unipotent radical of the parabolic subgroup of $G_{\mathbf{R}}$ associated to each point (9.1.1).

In the case of D being symmetric Hermitian domain, $D_{\mathrm{BS}}^{\flat}$ was studied by Zucker in [Z1, Z4], and is called the "reductive Borel-Serre space" by him.

Similarly we have the quotient space $D_{\mathrm{SL}(2),\le 1}^{\flat}$ of the part $D_{\mathrm{SL}(2),\le 1}$ of $D_{\mathrm{SL}(2)}$ of points of rank ≤ 1 by the unipotent part $G_{W,\mathbf{R},u}$ of $G_{W,\mathbf{R}}$ associated to each point of $D_{\mathrm{SL}(2),\le 1}$ (cf. 5.2.6). The space D^* of Cattani and Kaplan in [CK1] (defined for special D) is essentially this $D_{\mathrm{SL}(2),\le 1}^{\flat}$ (9.1.5).

Let X be an analytic manifold and U be the complement of a smooth divisor on X. Assume that we are given a variation of polarized Hodge structure H on U with unipotent local monodromy which has a Γ-level structure for a neat subgroup Γ of $G_{\mathbf{Z}}$ of finite index, then the associated period map $U \to \Gamma\backslash D$ extends to a continuous map $X \to \Gamma\backslash D_{\mathrm{SL}(2),\le 1}^{\flat}$ (see Section 9.4).

This is obtained as the composition of $X \to \Gamma\backslash D_{\Xi}$ (0.4.30 (i)) and the continuous map $\Gamma\backslash D_{\Xi} \to \Gamma\backslash D_{\mathrm{SL}(2),\le 1}^{\flat}$ which is induced from $\psi : D_{\Xi}^{\sharp} = D_{\Xi,\mathrm{val}}^{\sharp} \to D_{\mathrm{SL}(2),\le 1} \subset D_{\mathrm{SL}(2)}$.

0.6 PLAN OF THIS BOOK

The plan of this book is as follows.

Chapters 1–3 are preliminaries to state the main results of the present book, Theorems A and B in Chapter 4. In Chapter 1, we define the sets D_Σ and $D^\sharp_\Sigma$. In Chapter 2, we describe the theory of polarized logarithmic Hodge structures. In Chapter 3, we discuss the strong topology, logarithmic manifolds, the spaces E_σ, $\tilde{E}_\sigma$, $\check{E}_\sigma$, the categories $\mathcal{B}$, $\mathcal{B}(\log)$, and other enlargements of the category of analytic spaces. In Chapter 4, we state Theorems A and B without proofs. Theorems 0.4.19 and 0.5.5 are contained in Theorem A, and Theorem 0.4.27 is contained in Theorem B. Theorem 0.5.8 is contained in 5.1.10 and Theorem 5.1.14, Theorem 0.5.16 is contained in Theorem 5.2.15 and Proposition 5.2.16, and Theorem 0.5.20 is contained in Theorems 7.3,2, 7.4.2, 5.1.14, and 5.2.15. We also discuss, in Chapter 4, extensions of period maps over boundaries, and infinitesimal properties of extended period maps.

In Chapters 5–8, we prove Theorems A and B by moving from the right to the left in fundamental diagram (3) in Introduction (also in 0.5.25). In Chapter 5, we review the spaces $D_{\mathrm{SL}(2)}$, D_{BS}, $D_{\mathrm{SL}(2),\mathrm{val}}$, and $D_{\mathrm{BS},\mathrm{val}}$ defined in [KU2], and then we define $D_{\Sigma,\mathrm{val}}$ and $D^\sharp_{\Sigma,\mathrm{val}}$. By using the work [CKS] of Cattani, Kaplan, and Schmid on SL(2)-orbits in several variables, in Chapters 5 and 6 we connect the spaces $D^\sharp_{\Sigma,\mathrm{val}}$ and $D_{\mathrm{SL}(2)}$ as in Fundamental diagram (3) in Introduction (also in 0.5.25). In Chapter 7, we prove Theorem A, and in Chapter 8, Theorem B.

In Chapters 9–12, we give complements, examples, generalizations, and open problems. In Chapter 9, we consider the relationship of the present work with the enlargements of D studied by Cattani and Kaplan [CK1]. In Chapter 10, we describe local structures of $D_{\mathrm{SL}(2)}$. In Chapter 11, we consider the moduli of PLHs with coefficients. Although the case with coefficients is more general than the case without coefficients, we have chosen not to show the coefficients everywhere in this book (which would make the notation too complicated), but to describe the theory without coefficients except in Chapter 11, where we show that the results with coefficients can be simply deduced from those without. In Chapter 12, we give examples and discuss open problems.

Corrections to Previous Work

We indicate three mistakes in our previous work [KU1, KU2].

(i) In [KU1, (5.2)], there is a mistake in the definition of the notion of polarized logarithmic Hodge structures of type Φ. This mistake and its correction are explained in 2.5.16.

(ii) In [KU2, Lemma 4.7], the definition of $B(U, U', U'')$ is written as $\{g\tilde{\rho}(t)k \cdot \mathbf{r} \mid \cdots\}$, which is wrong. The correct definition is $\{\tilde{\rho}(t)gk \cdot \mathbf{r} \mid \cdots\}$. This point will be explained in 5.2.17.

(iii) In [KU2, Remarks 3.15, and 3.16], we indicated that we would consider a space $D^\flat_{\mathrm{SL}(2)}$ in this book. However, we actually consider only a part $D^\flat_{\mathrm{SL}(2),\leq 1}$ of $D^\flat_{\mathrm{SL}(2)}$ (Chapter 9). We realized that $D^\flat_{\mathrm{SL}(2)}$ is not necessarily Hausdorff and seems not to be a good object to consider, but that the part $D^{\flat}_{\mathrm{SL}(2),\le 1}$ is Haussdorff and is certainly a nice object.

The present work was announced in [KU1] under the title "Logarithmic Hodge Structures and Classifying Spaces" and in [KU2] under the title "Logarithmic Hodge Structures and Their Moduli," but we have changed the title.

0.7 NOTATION AND CONVENTION

Throughout this book, we use the following notation and terminology.

0.7.1

As usual, **N**, **Z**, **Q**, **R**, and **C** mean the set of natural numbers $\{0, 1, 2, 3, \dots\}$, the set of integers, the set of rational numbers, the set of real numbers, and the set of complex numbers, respectively.

0.7.2

In this book, a *ring* is assumed to have the unit element 1, a *subring* shares 1, and a ring homomorphism respects 1.

0.7.3

Let L be a $\mathbf{Z}$-module. For $R = \mathbf{Q}, \mathbf{R}, \mathbf{C}$, we denote $L_R := R \otimes_{\mathbf{Z}} L$.

We fix a 4-tuple

$$\Phi_0 = (w, (h^{p,q})_{p,q\in\mathbf{Z}}, H_0, \langle\ ,\ \rangle_0)$$

where w is an integer, $(h^{p,q})_{p,q\in\mathbf{Z}}$ is a set of non-negative integers satisfying

$$\begin{cases} h^{p,q} = 0 \text{ for almost all } p, q, \\ h^{p,q} = 0 \text{ if } p+q \neq w, \\ h^{p,q} = h^{q,p} \text{ for any } p, q, \end{cases}$$

H_0 is a free $\mathbf{Z}$-module of rank $\sum_{p,q} h^{p,q}$, and $\langle\ ,\ \rangle_0$ is a $\mathbf{Q}$-rational nondegenerate $\mathbf{C}$-bilinear form on $H_{0,\mathbf{C}}$ which is symmetric if w is even and antisymmetric if w is odd.

It is easily checked that the associated classifying space D of Griffiths (1.2.1) is nonempty if and only if either w is odd or $w = 2t$ and the signature (a, b) of $(H_{0,\mathbf{R}}, \langle\ ,\ \rangle_0)$ satisfies

$$a \text{ (resp. } b) = \sum_j h^{t+j,t-j},$$

where j ranges over all even (resp. odd) integers.

Let

$$G_{\mathbf{Z}} := \mathrm{Aut}(H_0, \langle\ ,\ \rangle_0),$$

and for $R = \mathbf{Q}, \mathbf{R}, \mathbf{C}$, let

$$\begin{aligned} G_R &:= \mathrm{Aut}(H_{0,R}, \langle\ ,\ \rangle_0), \\ \mathfrak{g}_R &:= \mathrm{Lie}\, G_R \\ &= \{N \in \mathrm{End}_R(H_{0,R}) \mid \langle Nx, y\rangle_0 + \langle x, Ny\rangle_0 = 0\ (\forall x, \forall y \in H_{0,R})\}. \end{aligned}$$

Following [BS], a *parabolic subgroup* of $G_{\mathbf{R}}$ means a parabolic subgroup of $(G^\circ)_{\mathbf{R}}$, where G° denotes the connected component of G in the Zariski topology containing the unity. (Note that $G^\circ = G$ if w is odd, and $G^\circ = \{g \in G \mid \det(g) = 1\}$ if w is even.)

0.7.4

We refer to a complex *analytic space* as an analytic space for brevity. We use the definition of analytic space (due to Grothendieck) in which the structure sheaf $\mathcal{O}_X$ of an analytic space X can have nonzero nilpotent sections. Precisely speaking, in this definition, an analytic space means a local ringed space over $\mathbf{C}$ which is locally isomorphic to the ringed space $(V, \mathcal{O}_U/(f_1, \dots, f_m))$, where U is an open set of $\mathbf{C}^n$ for some $n \geq 0$, $f_1, \dots, f_m$ are elements of $\Gamma(U, \mathcal{O}_U)$ for some $m \geq 0$, and $V = \{p \in U \mid f_1(p) = \dots = f_m(p) = 0\}$.

We denote by

$$\mathcal{A}, \quad \mathcal{A}(\log)$$

the category of analytic spaces and the category of fs logarithmic analytic spaces, i.e., analytic spaces endowed with an fs logarithmic structure, respectively.

0.7.5

Throughout this book, *compact* spaces and *locally compact* spaces are already Hausdorff as in Bourbaki [Bn].

Throughout this book, *proper* means "proper in the sense of Bourbaki [Bn] and separated." Here, for a continuous map $f : X \to Y$, f is proper in the sense of Bourbaki [Bn] if and only if for any topological space Z the map $X \times_Y Z \to Z$ induced by f is closed. (For example, if X and Y are locally compact, f is proper if and only if for any compact subset K of Y, the inverse image $f^{-1}(K)$ is compact.) On the other hand, f is *separated* if and only if the diagonal map $X \to X \times_Y X$ is closed. That is, f is separated if and only if, for any $a, b \in X$ such that $f(a) = f(b)$, there are open sets U and V of X such that $a \in U$, $b \in V$, and $U \cap V = \emptyset$.

In particular, in this book, a topological space X is compact if and only if the map from X to the one point set is proper.

0.7.6

For a continuous map $f : X \to Y$ and a sheaf $\mathcal{F}$ on Y, we denote the inverse image of $\mathcal{F}$ on X by $f^{-1}(\mathcal{F})$, not by $f^*(\mathcal{F})$. This is to avoid confusion with the module-theoretic inverse image. For f a morphism of ringed spaces $(X, \mathcal{O}_X) \to (Y, \mathcal{O}_Y)$

and a sheaf $\mathcal{F}$ of $\mathcal{O}_Y$-modules on Y, we denote by $f^*(\mathcal{F})$ the module-theoretic inverse image $\mathcal{O}_X \otimes_{f^{-1}(\mathcal{O}_Y)} f^{-1}(\mathcal{F})$ on X.

0.7.7

Concerning monoids and cones, we use the following concise terminology in this book, for simplicity.

We call a commutative monoid just a *monoid*. A monoid is assumed (as usual) to have the neutral element 1. A *submonoid* is assumed to share 1, and a homomorphism of monoids is assumed to respect 1.

Concerning cones, as explained in Section 1.3, a convex cone in the sense of [Od] is called just a *cone* in this book. A convex polyhedral cone in [Od] is called a *finitely generated cone*. A strongly convex cone in [Od] is called a *sharp cone*.

In [Od], for a finitely generated cone σ (in our sense), the topological interior of σ in the vector space $\sigma_{\mathbf{R}}$ is called the relative interior of σ. (Here $\sigma_{\mathbf{R}}$ denotes the $\mathbf{R}$-vector space generated by σ.) We call the relative interior of σ just the *interior* of σ. The interior of σ coincides with the complement of the union of all proper faces of σ. (Here a proper face of σ is a face of σ that is different from σ.)

For an fs monoid $\mathcal{S}$ (0.2.11 and 2.1.4), similarly, we call the complement of the union of all proper faces of $\mathcal{S}$ the *interior* of $\mathcal{S}$.

Chapter One

Spaces of Nilpotent Orbits and Spaces of Nilpotent i-Orbits

We recall polarized Hodge structures in Section 1.1, and the classifying space D of polarized Hodge structures in Section 1.2 (cf. [G1]). In Section 1.3, we introduce the space of nilpotent orbits D_Σ and the space of nilpotent i-orbits $D_\Sigma^\sharp$ in the directions in Σ as enlargements of D. The space D_Σ is the main object in this book.

1.1 HODGE STRUCTURES AND POLARIZED HODGE STRUCTURES

Let w and $(h^{p,q}) = (h^{p,q})_{p,q\in\mathbf{Z}}$ be as in Section 0.7.

1.1.1

A *Hodge structure* of weight w and of Hodge type $(h^{p,q})$ is a pair $(H_{\mathbf{Z}}, F)$ consisting of a free $\mathbf{Z}$-module $H_{\mathbf{Z}}$ of rank $\sum_{p,q} h^{p,q}$ and of a decreasing filtration F on $H_{\mathbf{C}} = \mathbf{C}\otimes H_{\mathbf{Z}}$, called a *Hodge filtration*, which satisfies the following two conditions.

$$\dim_{\mathbf{C}} F^p/F^{p+1} = h^{p,w-p} \quad \text{for all } p. \tag{1}$$

$$H_{\mathbf{C}} = \bigoplus_p F^p \cap \bar{F}^{w-p}. \tag{2}$$

1.1.2

A *polarized Hodge structure* of weight w and of Hodge type $(h^{p,q})$ is a triple $(H_{\mathbf{Z}}, \langle\ ,\ \rangle, F)$ consisting of a Hodge structure $(H_{\mathbf{Z}}, F)$ and of a nondegenerate $\mathbf{Q}$-bilinear form $\langle\ ,\ \rangle$ on $H_{\mathbf{Q}} = \mathbf{Q}\otimes H_{\mathbf{Z}}$, symmetric for even w and antisymmetric for odd w, which satisfies the following two conditions.

(1) $\langle F^p, F^q\rangle = 0$ for $p+q > w$.
(2) The Hermitian form

$$H_{\mathbf{C}} \times H_{\mathbf{C}} \to \mathbf{C}, \quad (x, y) \mapsto \langle C_F(x), \bar{y}\rangle,$$

is positive definite.

Here $\langle\ ,\ \rangle$ is regarded as the natural extension to the $\mathbf{C}$-bilinear form, the overbar indicates complex conjugation with respect to $H_{\mathbf{Z}}$, and C_F is the Weil operator, which is defined by $C_F(x) := i^{p-q}x$ for $x \in F^p \cap \bar{F}^q$ with $p+q = w$. The condition (1) (resp. (2)) is called the *Riemann-Hodge first* (resp. *second*) *bilinear relation.*

1.2 CLASSIFYING SPACES OF HODGE STRUCTURES

Let $\Phi_0 = \big(w, (h^{p,q}), H_0, \langle\ ,\ \rangle_0\big)$ be as in Section 0.7.

Definition 1.2.1 *The* classifying space D of polarized Hodge structures of type $\Phi_0 = \big(w, (h^{p,q}), H_0, \langle\ ,\ \rangle_0\big)$ *is the set of all decreasing filtrations F on $H_{0,\mathbf{C}} = \mathbf{C} \otimes H_0$ such that the triple $(H_0, \langle\ ,\ \rangle_0, F)$ is a polarized Hodge structure of weight w and of Hodge type $(h^{p,q})$.*

1.2.2

The compact dual $\check{D}$ of D *is the set of all decreasing filtrations F on $H_{0,\mathbf{C}}$ such that the triple $(H_0, \langle\ ,\ \rangle_0, F)$ satisfies the conditions* 1.1.1 (1) *and* 1.1.2 (1).

1.2.3

Example. Let $g \geq 1$ and consider the case $w = 1$, $h^{1,0} = h^{0,1} = g$ and $h^{p,q} = 0$ otherwise. Let H_0 be a free $\mathbf{Z}$-module with basis $(e_j)_{1\leq j\leq 2g}$ and define a $\mathbf{Z}$-bilinear form $\langle\ ,\ \rangle_0 : H_0 \times H_0 \to \mathbf{Z} \subset \mathbf{Q}$ by

$$(\langle e_j, e_k\rangle_0)_{j,k} = \begin{pmatrix} 0 & -1_g \\ 1_g & 0 \end{pmatrix}.$$

Then $D \simeq \mathfrak{h}_g$, the Siegel upper half space of degree g. Recall that $\mathfrak{h}_g$ is the space of all symmetric matrices over $\mathbf{C}$ of degree g whose imaginary parts are positive definite (corresponding to the Riemann-Hodge first and second bilinear relations, respectively). We identify a matrix $\tau \in \mathfrak{h}_g$ with $F(\tau) \in D$ as follows:

$$F(\tau)^0 := H_{0,\mathbf{C}},\ F(\tau)^1 := \left(\begin{matrix}\text{subspace of } H_{0,\mathbf{C}} \text{ spanned} \\ \text{by the column vectors of}\end{matrix} \begin{pmatrix} \tau \\ 1_g \end{pmatrix}\right),\ F(\tau)^2 := \{0\}.$$

Furthermore, if $g = 1$ then $D \simeq \mathfrak{h}_1 =: \mathfrak{h}$, the Poincaré upper half plane, where we identify a point $\tau \in \mathfrak{h}$ with $F(\tau) \in D$ under

$$F(\tau)^0 = H_{0,\mathbf{C}},\ F(\tau)^1 = \mathbf{C}(\tau e_1 + e_2),\ F(\tau)^2 = \{0\}.$$

In this case, we have $\check{D} = \mathbf{P}^1(\mathbf{C})$, the space of all one-dimensional $\mathbf{C}$-subspaces of $H_{0,\mathbf{C}}$.

1.2.4

As in Section 0.7, let $G_{\mathbf{Z}} = \mathrm{Aut}(H_0, \langle\ ,\ \rangle_0)$, and, for $R = \mathbf{Q}, \mathbf{R}, \mathbf{C}$, let $G_R = \mathrm{Aut}(H_{0,R}, \langle\ ,\ \rangle_0)$ and $\mathfrak{g}_R = \mathrm{Lie}\, G_R$. Then, $\check{D}$ (resp. D) is homogeneous under $G_{\mathbf{C}}$ (resp. $G_{\mathbf{R}}$).

1.2.5

Example. In the example 1.2.3, we have, for $R = \mathbf{Q}, \mathbf{R}, \mathbf{C}$,

$$G_R = \mathrm{Sp}(g, R) = \{M \in \mathrm{GL}(2g, R) \mid {}^tM J_g M = J_g\},$$

where $J_g = \left(\begin{smallmatrix} 0 & -1_g \\ 1_g & 0 \end{smallmatrix}\right)$. $\left(\begin{smallmatrix} A & B \\ C & D \end{smallmatrix}\right) \in \mathrm{Sp}(g, \mathbf{R})$ acts on D by

$$F(\tau) \mapsto F(\tau'), \quad \tau' = (A\tau + B)(C\tau + D)^{-1}.$$

1.3 EXTENDED CLASSIFYING SPACES

Definition 1.3.1 *Let A be a finite-dimensional $\mathbf{R}$-vector space.*

A subset σ of A is called a cone *of A if it is nonempty and the following condition* (1) *is satisfied:*

(1) *If $x, y \in \sigma$ and $a \in \mathbf{R}_{\geq 0}$, then $x + y, ax \in \sigma$.*

A cone σ in A is said to be sharp *if $\sigma \cap (-\sigma) = \{0\}$.*

For a cone σ in A, we denote the $\mathbf{R}$-linear span of σ in A by $\sigma_{\mathbf{R}}$, and the $\mathbf{C}$-linear span of σ in $A_{\mathbf{C}} = \mathbf{C} \otimes_{\mathbf{R}} A$ by $\sigma_{\mathbf{C}}$.

A cone σ in A is said to be finitely generated *if the following condition* (2) *is satisfied:*

(2) *There is a finite family $(N_j)_{1 \leq j \leq n}$ of elements of σ such that $\sigma = (\mathbf{R}_{\geq 0})N_1 + \cdots + (\mathbf{R}_{\geq 0})N_n$.*

Definition 1.3.2 *Let A be as in* 1.3.1, *and let σ be a cone in A. A* face *of σ is a subcone τ of σ in A satisfying the following condition* (1):

(1) *If $x, y \in \sigma$ and $x + y \in \tau$, then $x, y \in \tau$.*

The following is known: If σ is a finitely generated cone in A, then σ has only finitely many faces, and any face of σ is finitely generated (cf. [Od]).

Definition 1.3.3 *Let A be as in* 1.3.1. *A* fan *in A is a nonempty set Σ of finitely generated sharp cones in A satisfying the following conditions* (1) *and* (2):

(1) *If $\sigma \in \Sigma$, any face of σ belongs to Σ.*
(2) *If $\sigma, \tau \in \Sigma$, then $\sigma \cap \tau$ is a face of σ and of τ.*

Definition 1.3.4 *Now let A be a finite-dimensional vector space over $\mathbf{Q}$.*

A finitely generated cone σ in $A_{\mathbf{R}} = \mathbf{R} \otimes_{\mathbf{Q}} A$ is said to be rational *if we can take $N_1, \ldots, N_n \in A$ in* 1.3.1 (2).

It is known that any face of a rational finitely generated cone is also rational (cf. [Od]).

A fan in $A_{\mathbf{R}}$ is said to be *rational* if all members of Σ are rational.

Let $\Phi_0 = (w, (h^{p,q}), H_0, \langle\ ,\ \rangle_0)$ be as in Section 0.7.

Definition 1.3.5 *A* nilpotent cone *in $\mathfrak{g}_{\mathbf{R}}$ is a finitely generated cone σ in $\mathfrak{g}_{\mathbf{R}}$ satisfying the following conditions* (1) *and* (2):

(1) *Any element of σ is nilpotent as an endomorphism of $H_{0,\mathbf{R}}$.*
(2) *$NN' = N'N$ for any $N, N' \in \sigma$ as endomorphisms of $H_{0,\mathbf{R}}$.*

Definition 1.3.6 *Let $\check{D}_{\mathrm{orb}}$ (resp. $\check{D}^{\sharp}_{\mathrm{orb}}$) be the set of all pairs (σ, Z) where σ is a nilpotent cone in $\mathfrak{g}_{\mathbf{R}}$ and Z is an $\exp(\sigma_{\mathbf{C}})$-orbit (resp. $\exp(i\sigma_{\mathbf{R}})$-orbit) in $\check{D}$, that is, Z is a subset of $\check{D}$ having the form $\exp(\sigma_{\mathbf{C}})F$ (resp. $\exp(i\sigma_{\mathbf{R}})F$) for some $F \in \check{D}$.*

Definition 1.3.7 *By a* nilpotent orbit (*resp.* nilpotent i-orbit), *we mean an element (σ, Z) of $\check{D}_{\mathrm{orb}}$ (resp. $\check{D}^{\sharp}_{\mathrm{orb}}$) satisfying the following conditions* (1) *and* (2) *for some $F \in Z$:*

(1) $NF^p \subset F^{p-1}$ $(\forall\, p \in \mathbf{Z},\ \forall\, N \in \sigma)$.
(2) *Write $\sigma = (\mathbf{R}_{\geq 0})N_1 + \cdots + (\mathbf{R}_{\geq 0})N_n$. Then $\exp\left(\sum_{1\leq j\leq n} z_j N_j\right)F \in D$ if $z_j \in \mathbf{C}$ and $\mathrm{Im}(z_j) \gg 0$.*

(In fact, if (1) (resp. (2)) is satisfied for some $F \in Z$, then it is satisfied for any $F \in Z$.)

If (σ, Z) is a nilpotent orbit (resp. nilpotent i-orbit), we say Z is a *σ-nilpotent orbit* (resp. *σ-nilpotent i-orbit*).

Definition 1.3.8 *In this book, by a* fan in $\mathfrak{g}_{\mathbf{Q}}$, *we mean a rational fan in $\mathfrak{g}_{\mathbf{R}}$ consisting of nilpotent cones.*

For a fan Σ in $\mathfrak{g}_{\mathbf{Q}}$, we define the space $D_\Sigma \subset \check{D}_{\mathrm{orb}}$ (*resp.* $D^{\sharp}_\Sigma \subset \check{D}^{\sharp}_{\mathrm{orb}}$) of nilpotent orbits (*resp.* nilpotent i-orbits) in the directions in Σ *by*

$$D_\Sigma = \{(\sigma, Z) \in \check{D}_{\mathrm{orb}} \mid (\sigma, Z) \textit{ is a nilpotent orbit and } \sigma \in \Sigma\},$$
$$D^{\sharp}_\Sigma = \{(\sigma, Z) \in \check{D}^{\sharp}_{\mathrm{orb}} \mid (\sigma, Z) \textit{ is a nilpotent } i\textit{-orbit and } \sigma \in \Sigma\}.$$

1.3.9

We have the inclusion maps

$$D \hookrightarrow D_\Sigma, \quad D \hookrightarrow D^{\sharp}_\Sigma; \quad F \mapsto (\{0\}, F),$$

and the canonical surjection

$$D^{\sharp}_\Sigma \to D_\Sigma, \quad (\sigma, Z) \mapsto (\sigma, \exp(\sigma_{\mathbf{C}})Z),$$

which is compatible with the above inclusion maps.

For a rational sharp nilpotent cone σ in $\mathfrak{g}_{\mathbf{R}}$, we denote

$$D_\sigma := D_{\{\text{face of } \sigma\}}, \quad D^{\sharp}_\sigma := D^{\sharp}_{\{\text{face of } \sigma\}}.$$

Then, for a fan Σ in $\mathfrak{g}_{\mathbf{Q}}$, we have

$$D_\Sigma = \bigcup_{\sigma\in\Sigma} D_\sigma, \quad D^{\sharp}_\Sigma = \bigcup_{\sigma\in\Sigma} D^{\sharp}_\sigma.$$

Definition 1.3.10 *Let Σ be a fan in $\mathfrak{g}_{\mathbf{Q}}$ and let Γ be a subgroup of $G_{\mathbf{Z}}$.*

(i) *We say that Γ is* compatible with Σ *if the following condition* (1) *is satisfied.*

(1) *If $\gamma \in \Gamma$ and $\sigma \in \Sigma$ then $\mathrm{Ad}(\gamma)(\sigma) \in \Sigma$.*

Note that, if Γ is compatible with Σ, Γ acts on D_Σ and on $D^{\sharp}_\Sigma$ by

$$(\sigma, Z) \mapsto (\mathrm{Ad}(\gamma)(\sigma), \gamma Z).$$

(ii) *We say that* Γ *is* strongly compatible with Σ *if it is compatible with* Σ *and the following condition* (2) *is satisfied. For* $\sigma \in \Sigma$, *define a sharp fs monoid* $\Gamma(\sigma)$ *by*

$$\Gamma(\sigma) = \Gamma \cap \exp(\sigma).$$

(2) *If* $\sigma \in \Sigma$, *any element of* σ *can be written as a sum of a* $\log(\gamma)$ ($a \in \mathbf{R}_{\geq 0}$, $\gamma \in \Gamma(\sigma)$).

1.3.11

Example. Let

$$\Xi := \{(\mathbf{R}_{\geq 0})N \mid N \text{ is a nilpotent element of } \mathfrak{g}_{\mathbf{Q}}\}.$$

Then Ξ is a fan in $\mathfrak{g}_{\mathbf{Q}}$. If Γ is of finite index in $G_{\mathbf{Z}}$, Γ is strongly compatible with Σ.

Chapter Two

Logarithmic Hodge Structures

In Section 2.1, we recall basic facts about logarithmic structures (cf. [Kk1]). In Section 2.2, we recall the ringed spaces $(X^{\log}, \mathcal{O}_X^{\log})$ introduced in [KkNc], with some generalizations. We study properties of local systems on $X^{\log}$ in Section 2.3. In Section 2.4, we introduce the notion of the polarized logarithmic Hodge structure, which is defined on the ringed space $(X^{\log}, \mathcal{O}_X^{\log})$. In Section 2.5, we observe the relationship of polarized logarithmic Hodge structures and nilpotent orbits. In particular, we interpret the nilpotent orbit theorem of Schmid in the language of polarized logarithmic Hodge structures. Further, we introduce the notion of the polarized logarithmic Hodge structure of type Φ and study the associated period map. Finally, in Section 2.6, we introduce the notion of logarithmic mixed Hodge structures.

2.1 LOGARITHMIC STRUCTURES

We review the theory of logarithmic structures of Fontaine and Illusie briefly for our later use (cf. [Kk1,I2]).

2.1.1

Prelogarithmic structures and logarithmic structures. Let X be a ringed space with structure sheaf $\mathcal{O}_X$. A *prelogarithmic structure* on X is a sheaf of monoids M together with a homomorphism $\alpha : M \to \mathcal{O}_X$, where $\mathcal{O}_X$ is regarded as a sheaf of monoids by multiplication. (See our terminology 0.7.7. In particular, a monoid in this book means a commutative monoid.)

A *logarithmic structure* on X is a prelogarithmic structure (M, α) on X that satisfies

$$\alpha^{-1}(\mathcal{O}_X^\times) \xrightarrow{\sim} \mathcal{O}_X^\times \quad \text{via } \alpha.$$

If (M, α) is a logarithmic structure on X, we regard $\mathcal{O}_X^\times$ as a subsheaf of M via this isomorphism.

The simplest example of a logarithmic structure on X is $M = \mathcal{O}_X^\times$ with α the inclusion map. This logarithmic structure is called the *trivial logarithmic structure*.

Let (M, α) be a prelogarithmic structure on X. The *associated logarithmic structure* $(\tilde{M}, \tilde{\alpha})$ is defined as the push-out $\tilde{M}$ of

$$\begin{array}{ccc} \alpha^{-1}(\mathcal{O}_X^\times) & \xrightarrow{\subset} & M \\ {\scriptstyle\alpha}\downarrow & & \\ \mathcal{O}_X^\times & & \end{array}$$

in the category of sheaves of monoids on X, together with the homomorphism $\tilde{\alpha} : \tilde{M} \to \mathcal{O}_X$ induced by $\alpha : M \to \mathcal{O}_X$ and the inclusion $\mathcal{O}_X^\times \hookrightarrow \mathcal{O}_X$. More explicitly, $\tilde{M}$ is the sheafification of the presheaf $(M \times \mathcal{O}_X^\times)/\sim$, where $(m, f) \sim (m', f')$ if and only if there exist $g_1, g_2 \in \alpha^{-1}(\mathcal{O}_X^\times)$ such that $mg_1 = m'g_2$ and $f\alpha(g_2) = f'\alpha(g_1)$.

A ringed space endowed with a logarithmic structure is called a *logarithmic ringed space*.

2.1.2

Standard example. Let X be a complex manifold, let Y be a divisor on X with normal crossings, and let

$$M := \{f \in \mathcal{O}_X \mid f \text{ is invertible outside } Y\} \subset \mathcal{O}_X.$$

Then, M with the inclusion map $\alpha : M \hookrightarrow \mathcal{O}_X$ is a logarithmic structure, and is called the *logarithmic structure on X associated with Y*.

2.1.3

Inverse image. Let $f : (X, \mathcal{O}_X) \to (Y, \mathcal{O}_Y)$ be a morphism of ringed spaces and let (M, α) be a logarithmic structure on Y. Then the sheaf-theoretic inverse image $f^{-1}M$ together with the composite morphism $f^{-1}M \to f^{-1}\mathcal{O}_Y \to \mathcal{O}_X$ form a prelogarithmic structure on X. The *inverse image* $f^*(M, \alpha)$ of (M, α) is defined as the logarithmic structure on X associated with the above prelogarithmic structure.

We have

$$f^{-1}(M_Y/\mathcal{O}_Y^\times) \xrightarrow{\sim} f^*(M_Y)/\mathcal{O}_X^\times.$$

2.1.4

Fs monoids. Here we briefly recall some facts about monoids. For a monoid $\mathcal{S}$, we denote by $\mathcal{S}^\times$ the group of all invertible elements of $\mathcal{S}$. By an *fs monoid*, we mean a monoid (commutative always, see 0.7.7) having the following three properties:

(1) $\mathcal{S}$ is finitely generated.
(2) If $a, b, c \in \mathcal{S}$ and $ab = ac$, then $b = c$. (Hence $\mathcal{S}$ is embedded in the group $\mathcal{S}^{\mathrm{gp}} = \{\frac{a}{b} \mid a, b \in \mathcal{S}\}$.)
(3) If $a \in \mathcal{S}^{\mathrm{gp}}$ and $a^n \in \mathcal{S}$ for some integer $n \geq 1$, then $a \in \mathcal{S}$.

(The terminology "fs monoid" comes from "finitely generated and saturated monoid." If the reader feels this terminology strange, we propose the terminology "integral cone.")

2.1.5

Fs logarithmic structures and charts. A logarithmic structure (M, α) on a ringed space X is *fs* if there exist an open covering $(U_\lambda)_\lambda$ of X and a family of pairs $(\mathcal{S}_\lambda, \theta_\lambda)_\lambda$ consisting of an fs monoid $\mathcal{S}_\lambda$, regarded as a constant sheaf on U_λ, and of a homomorphism $\theta_\lambda : \mathcal{S}_\lambda \to M|_{U_\lambda}$ of sheaves of monoids which induces an isomorphism $\tilde{\mathcal{S}}_\lambda \xrightarrow{\sim} M|_{U_\lambda}$. Here $\tilde{\mathcal{S}}_\lambda$ denotes the logarithmic structure associated with the prelogarithmic structure $\mathcal{S}_\lambda \to M|_{U_\lambda} \to \mathcal{O}_{U_\lambda}$. In this case, $(\mathcal{S}_\lambda, \theta_\lambda)$ is called a *chart* of $M|_{U_\lambda}$.

We give basic facts about fs logarithmic structures.

(i) If M is an fs logarithmic structure, then, for any $x \in X$, $(M_X/\mathcal{O}_X^\times)_x$ is a sharp fs monoid and $(M_X^{\rm gp}/\mathcal{O}_X^\times)_x = ((M_X/\mathcal{O}_X^\times)_x)^{\rm gp}$ is a finitely generated free abelian group.

(ii) If M is an fs logarithmic structure and $\mathcal{S} \to M$ is a chart, then for each $x \in X$ the induced map $\mathcal{S} \to (M_X/\mathcal{O}_X^\times)_x$ is surjective and $\mathcal{S}(x)^{\rm gp}\mathcal{S}/\mathcal{S}(x)^{\rm gp} \xrightarrow{\sim} (M_X/\mathcal{O}_X^\times)_x$ where $\mathcal{S}(x)$ is the face of $\mathcal{S}$ defined as the inverse image of $\mathcal{O}_{X,x}^\times$ under $\mathcal{S} \to M_{X,x}$.

(iii) If M is an fs logarithmic structure and $x \in X$, then for some open neighborhood U of x there is a chart $\mathcal{S} \to M|_U$ such that $\mathcal{S} \xrightarrow{\sim} (M_X/\mathcal{O}_X^\times)_x$. In particular, for an fs logarithmic structure M, locally there is a chart $\mathcal{S} \to M$ with $\mathcal{S}$ a sharp fs monoid.

(iv) If $f : X \to Y$ is a morphism of ringed spaces and (M, α) is an fs logarithmic structure on Y, the inverse image $f^*(M, \alpha)$ is also an fs logarithmic structure. If $\mathcal{S} \to M$ is a chart, the induced homomorphism $\mathcal{S} \to f^*M$ is also a chart.

A ringed space endowed with an fs logarithmic structure is called an *fs logarithmic ringed space*. In particular, an analytic space endowed with an fs logarithmic structure is called an *fs logarithmic analytic space*.

2.1.6

Examples.

(i) The logarithmic structure M in Example 2.1.2 is an fs logarithmic structure. In fact, let $\prod_{1\le j\le r} z_j = 0$ be a local equation of Y in X such that each $\{z_j = 0\}$ is smooth. Then we have a chart

$$\mathcal{S} := \mathbf{N}^r \xrightarrow{\theta} M, \quad (a(j))_{1\le j\le r} \mapsto \prod_{1\le j\le r} z_j^{a(j)}.$$

(ii) Let $\mathcal{S}$ be an fs monoid and let $\mathbf{C}[\mathcal{S}]$ be the monoid ring (i.e., semigroup ring) of $\mathcal{S}$ over $\mathbf{C}$. Then the toric variety $X := \operatorname{Spec}(\mathbf{C}[\mathcal{S}])_{\rm an}$ has the canonical fs logarithmic structure associated with the prelogarithmic structure $\mathcal{S} \to \mathbf{C}[\mathcal{S}] \subset \mathcal{O}_X$.

2.1.7

Logarithmic differential forms. For an analytic space X, let $\Omega^1_X := \Delta^*(\mathcal{I}/\mathcal{I}^2)$ be the sheaf of Kähler differentials on X, where $\mathcal{I}$ is the sheaf of ideals of $\mathcal{O}_{X\times X}$ defining the image of the diagonal morphism $\Delta : X \to X \times X$. For $f \in \mathcal{O}_X$, the class of $\mathrm{pr}_1^*(f) - \mathrm{pr}_2^*(f)$ in Ω^1_X is denoted by df.

Let X be an fs logarithmic analytic space. The sheaf of logarithmic differential 1-forms on X is defined by

$$\omega^1_X := \left(\Omega^1_X \oplus \left(\mathcal{O}_X \otimes_{\mathbf{Z}} M^{\mathrm{gp}}_X\right)\right)/N,$$

where N is the $\mathcal{O}_X$-submodule generated by

$$\{(-d\alpha(f), \alpha(f) \otimes f) \mid f \in M_X\}.$$

For $f \in M^{\mathrm{gp}}_X$, the image of $(0, 1 \otimes f)$ in ω^1_X is denoted by $d\log(f)$.

For a morphism $f : X \to Y$ of fs logarithmic analytic spaces, define

$$\omega^1_{X/Y} := \mathrm{Coker}(f^*\omega^1_Y \to \omega^1_X).$$

Let

$$\omega^q_{X/Y} := \bigwedge^q_{\mathcal{O}_X} \omega^1_{X/Y}$$

for $q \in \mathbf{N}$.

These sheaves $\omega^q_{X/Y}$ are coherent as $\mathcal{O}_X$-modules.

We have the *logarithmic de Rham complex* $(\omega^\bullet_{X/Y}, d)$ *of* X *over* Y, where the differential d is characterized by the following properties. This d is compatible with the differential d of the usual de Rham complex $(\Omega^\bullet_{X/Y}, d)$, $d(d\log(f)) = 0$ for $f \in M^{\mathrm{gp}}_X$, and $d(f \wedge g) = df \wedge g + (-1)^q f \wedge dg$ for $f \in \omega^q_{X/Y}$ and $g \in \omega^r_{X/Y}$ $(q, r \in \mathbf{Z})$.

In the case $Y = \mathrm{Spec}(\mathbf{C})$ with the trivial logarithmic structure, $\omega^1_{X/Y} = \omega^1_X$, and $(\omega^\bullet_{X/Y}, d)$ is denoted as $(\omega^\bullet_X, d)$.

2.1.8

Examples.

(i) In the standard example 2.1.2, ω^1_X is nothing but the sheaf $\Omega^1_X(\log(Y))$ of differential forms with logarithmic poles along Y.

(ii) If X is the toric variety $\mathrm{Spec}(\mathbf{C}[\mathcal{S}])_{\mathrm{an}}$ ($\mathcal{S}$ is an fs monoid) with the canonical logarithmic structure in 2.1.6 (ii), we have an isomorphism

$$\mathcal{O}_X \otimes_{\mathbf{Z}} \mathcal{S}^{\mathrm{gp}} \xrightarrow{\sim} \omega^1_X, \quad f \otimes g \mapsto f\, d\log(g).$$

In particular, ω^1_X is free in this case (cf. [Od, 3.1]).

2.1.9

By an *fs logarithmic point*, we mean an fs logarithmic analytic space whose underlying ringed space over $\mathbf{C}$ is Spec($\mathbf{C}$).

If x is an fs logarithmic point, there is an fs monoid $\mathcal{S}$ such that $\mathcal{S}^\times = \{1\}$ and such that the logarithmic structure M_x of x is $\mathcal{S} \times \mathbf{C}^\times$ with $\alpha : M_x \to \mathbf{C}$ which sends (f, u) ($f \in \mathcal{S}, u \in \mathbf{C}^\times$) to 0 if $f \neq 1$ and to u if $f = 1$. We have

$$\mathbf{C} \otimes_{\mathbf{Z}} \mathcal{S}^{\mathrm{gp}} \simeq \mathbf{C} \otimes_{\mathbf{Z}} (M_x^{\mathrm{gp}}/\mathcal{O}_x^\times) \xrightarrow{\sim} \omega_x^1, \quad c \otimes f \mapsto c d\log(f).$$

2.1.10

Fiber products. The category $\mathcal{A}(\log)$ of fs logarithmic analytic spaces has fiber products. For fs logarithmic analytic spaces X, Y, Z and for morphisms $X \to Z$ and $Y \to Z$, the fiber product $X \times_Z Y$ in this category is obtained as follows. Locally on X, Y, Z there are fs monoids $\mathcal{S}_j$ ($j = 1, 2, 3$), charts $\mathcal{S}_1 \to M_X$, $\mathcal{S}_2 \to M_Y$, $\mathcal{S}_3 \to M_Z$, and homomorphisms $\mathcal{S}_3 \to \mathcal{S}_1$ and $\mathcal{S}_3 \to \mathcal{S}_2$ for which the following diagram is commutative.

$$\begin{array}{ccccc} X & \longrightarrow & Z & \longleftarrow & Y \\ \downarrow & & \downarrow & & \downarrow \\ \mathrm{Spec}(\mathbf{C}[\mathcal{S}_1])_{\mathrm{an}} & \longrightarrow & \mathrm{Spec}(\mathbf{C}[\mathcal{S}_3])_{\mathrm{an}} & \longleftarrow & \mathrm{Spec}(\mathbf{C}[\mathcal{S}_2])_{\mathrm{an}} \end{array}$$

In this local situation, let $\mathcal{S}_4$ be the push-out of $\mathcal{S}_1 \leftarrow \mathcal{S}_3 \to \mathcal{S}_2$ in the category of monoids, let $\mathcal{S}_5$ be the image of $\mathcal{S}_4$ in the group $\mathcal{S}_4^{\mathrm{gp}} = \{ab^{-1} \mid a, b \in \mathcal{S}_4\}$ associated with $\mathcal{S}_4$ (note that $\mathcal{S}_4 \to \mathcal{S}_4^{\mathrm{gp}}$ need not be injective), and let $\mathcal{S}_6 = \{a \in \mathcal{S}_4^{\mathrm{gp}} = \mathcal{S}_5^{\mathrm{gp}} \mid a^n \in \mathcal{S}_5 \text{ for some } n \geq 1\}$. (This $\mathcal{S}_6$ is the push-out of $\mathcal{S}_1 \leftarrow \mathcal{S}_3 \to \mathcal{S}_2$ in the category of fs monoids.) Let $(X \times_Z Y)'$ be the fiber product of the underlying analytic spaces of X, Y, Z taken in the category $\mathcal{A}$ of analytic spaces. Then we have the induced morphism of analytic spaces $(X \times_Z Y)' \to \mathrm{Spec}(\mathbf{C}[\mathcal{S}_4])_{\mathrm{an}}$. The fiber product $X \times_Z Y$ in the category $\mathcal{A}(\log)$ is the fiber product $(X \times_Z Y)' \times_{\mathrm{Spec}(\mathbf{C}[\mathcal{S}_4])_{\mathrm{an}}} \mathrm{Spec}(\mathbf{C}[\mathcal{S}_6])_{\mathrm{an}}$ in the category $\mathcal{A}$ endowed with the inverse image of the canonical logarithmic structure of $\mathrm{Spec}(\mathbf{C}[\mathcal{S}_6])_{\mathrm{an}}$.

In general, the fiber product $X \times_Z Y$ in $\mathcal{A}(\log)$ is obtained by gluing this local construction.

The underlying analytic space of the fiber product $X \times_Z Y$ in $\mathcal{A}(\log)$ need not coincide with the fiber product in the category $\mathcal{A}$ of analytic spaces. For example, if X and Y denote copies of $\mathrm{Spec}(\mathbf{C}[T_1, T_2])_{\mathrm{an}}$ which is endowed with the logarithmic structure associated with $\mathbf{N}^2 \to \mathbf{C}[T_1, T_2]$, $(m, n) \mapsto T_1^m T_2^n$, and if $X \to Y$ denotes the morphism induced by $\mathbf{C}[T_1, T_2] \to \mathbf{C}[T_1, T_2]$, $T_1 \mapsto T_1 T_2$, $T_2 \mapsto T_2$, then the fiber product $X \times_Y X$ in the category $\mathcal{A}(\log)$ coincides with X (the diagonal morphism $X \to X \times_Y X$ is an isomorphism) although the fiber product $(X \times_Y X)'$ of underlying spaces taken in $\mathcal{A}$ does not coincide with the diagonal. Here $X \times_Y X = X$ because the morphism of $\mathrm{Mor}(\ \ , X) \to \mathrm{Mor}(\ \ , Y)$ of functors from $\mathcal{A}(\log)$ to (Sets) is injective. In fact, for an fs logarithmic analytic space S, the map $\mathrm{Mor}(S, X) \to \mathrm{Mor}(S, Y)$ is identified with the injective map from the set $\{(a, b) \mid a, b \in \Gamma(S, M_S)\}$ to itself given by $(a, b) \mapsto (ab, b)$.

To avoid confusion, when we consider in this book a fiber product in $\mathcal{A}(\log)$ whose underlying analytic space differs from the fiber product in $\mathcal{A}$, we will always state explicitly that we take the fiber product in $\mathcal{A}(\log)$.

The underlying analytic space of the fiber product $X \times_Z Y$ in $\mathcal{A}(\log)$ coincides with the fiber product in the category $\mathcal{A}$ of analytic spaces if one of the following conditions (1) and (2) is satisfied.

(1) At least one of the morphisms $X \to Z$, $Y \to Z$ is strict.

Here we say a morphism $X \to Z$ is *strict* if the logarithmic structure of X is the inverse image of that of Z via this morphism. If $X \to Z$ is strict, the fiber product $X \times_Z Y$ in $\mathcal{A}(\log)$ coincides with the usual fiber product in $\mathcal{A}$ endowed with the inverse image of the logarithmic structure of Y.

(2) The logarithmic structure of Z is trivial.

In general, the canonical map from the fiber product $X \times_Z Y$ in $\mathcal{A}(\log)$ to the fiber product $(X \times_Z Y)'$ is proper as a map of topological spaces, and the fibers of this map are finite. This fact is reduced to the case $X = \mathrm{Spec}(\mathbf{C}[\mathcal{S}_1])_{\mathrm{an}}$, $Y = \mathrm{Spec}(\mathbf{C}[\mathcal{S}_2])_{\mathrm{an}}$, $Z = \mathrm{Spec}(\mathbf{C}[\mathcal{S}_3])_{\mathrm{an}}$, $X \times_Z Y = \mathrm{Spec}(\mathbf{C}[\mathcal{S}_6])_{\mathrm{an}}$, $(X \times_Z Y)' = \mathrm{Spec}(\mathbf{C}[\mathcal{S}_4])_{\mathrm{an}}$ in the above argument.

2.1.11

For a morphism $f : X \to Y$ of fs logarithmic analytic spaces, we say f is *logarithmically smooth* if locally on X and Y there are charts $P \to M_X$ and $Q \to M_Y$ with P, Q fs monoids, and a homomorphism $Q \to P$ of monoids satisfying the following conditions (1)–(3).

(1) The following diagram is commutative.

$$\begin{array}{ccc} Q & \longrightarrow & P \\ \downarrow & & \downarrow \\ f^{-1}(M_Y) & \longrightarrow & M_X. \end{array}$$

(2) The map $Q \to P$ is injective.

(3) The morphism $X \to Y \times_{\mathrm{Spec}(\mathbf{C}[Q])_{\mathrm{an}}} \mathrm{Spec}(\mathbf{C}[P])_{\mathrm{an}}$ of analytic spaces is smooth.

An fs logarithmic analytic space X is said to be *logarithmically smooth* if the evident morphism $X \to \mathrm{Spec}(\mathbf{C})$ is logarithmically smooth for the trivial logarithmic structure of $\mathrm{Spec}(\mathbf{C})$. An fs logarithmic analytic space X is logarithmically smooth if and only if there are an open covering $(U_\lambda)_\lambda$ of X and an fs monoid $\mathcal{S}_\lambda$ for each λ

such that each U_λ is isomorphic to an open subset of $Z_\lambda := \mathrm{Spec}(\mathbf{C}[\mathcal{S}_\lambda])_{\mathrm{an}}$ endowed with the restrictions of $\mathcal{O}_{Z_\lambda}$ and M_{Z_λ}.

2.2 RINGED SPACES $(X^{\log}, \mathcal{O}_X^{\log})$

In [KkNc], a ringed space $(X^{\log}, \mathcal{O}_X^{\log})$ is constructed associated with an fs logarithmic analytic space X. (This space $X^{\log}$ was also considered independently in [KyNy].) Here we generalize the definition to logarithmic local ringed spaces X over $\mathbf{C}$ which belong to the category $\overline{\mathcal{A}}_1(\log)$ introduced below. The construction of $(X^{\log}, \mathcal{O}_X^{\log})$ in this generalized situation is just similar to that in [KkNc].

The reason why we generalize the definition of $(X^{\log}, \mathcal{O}_X^{\log})$ is the following. As is described in Chapter 0, the moduli spaces of polarized logarithmic Hodge structures, which are the main subjects of this book, need not belong to the category $\mathcal{A}(\log)$ of fs logarithmic analytic spaces. They belong to $\mathcal{B}(\log)$ (3.2.4 below), which seems to us the best category to work with in the theory of moduli of polarized logarithmic Hodge structures. We have

$$\mathcal{A}(\log) \subset \mathcal{B}(\log) \subset \overline{\mathcal{A}}_1(\log) \qquad \text{(see Section 3.2).}$$

For the definition of $(X^{\log}, \mathcal{O}_X^{\log})$, $\overline{\mathcal{A}}_1(\log)$ seems to be the most natural category.

2.2.1

First, let $\overline{\mathcal{A}}_1$ be the full subcategory of the category of local ringed spaces over $\mathbf{C}$ consisting of objects X satisfying the following condition (A_1).

(A_1) *For any open set U of X and for any $n \geq 0$, the canonical map*

$$\mathrm{Mor}(U, \mathbf{C}^n) \to \mathcal{O}(U)^n \tag{1}$$

is bijective.

Here $\mathbf{C}^n$ is regarded as an analytic space (and hence a local ringed space over $\mathbf{C}$) in the standard way, and Mor means the set of morphisms in the category of local ringed spaces over $\mathbf{C}$. The map (1) sends a morphism $f : U \to \mathbf{C}^n$ to $(f^*(z_j))_{1 \leq j \leq n} \in \mathcal{O}(U)^n$ where the z_j are the coordinate functions of $\mathbf{C}^n$.

For example, analytic spaces belong to $\overline{\mathcal{A}}_1$. Also, any topological space endowed with the sheaf of complex valued continuous functions belongs to $\overline{\mathcal{A}}_1$.

Let $\overline{\mathcal{A}}_1(\log)$ be the full subcategory of the category of fs logarithmic local ringed spaces over $\mathbf{C}$ consisting of objects whose underlying local ringed spaces satisfy (A_1). That is, an object of $\overline{\mathcal{A}}_1(\log)$ is an object of $\overline{\mathcal{A}}_1$ endowed with an fs logarithmic structure.

The condition (A_1) is equivalent to the following condition:

(A'_1) *For any open set U of X and any scheme Z locally of finite type over* $\mathbf{C}$, *the canonical map*

$$\mathrm{Mor}(U, Z_{\mathrm{an}}) \to \mathrm{Mor}(U, Z) \tag{2}$$

is bijective.

Here Z_{an} denotes the analytic space associated with Z, and the map (2) is induced from the canonical morphism $Z_{\mathrm{an}} \to Z$.

The equivalence of (A_1) and (A'_1) is shown as follows. (A'_1) implies (A_1) because

$$\mathcal{O}(U)^n = \mathrm{Mor}(U, \mathrm{Spec}(\mathbf{C}[T_1, \ldots, T_n])),$$
$$\mathrm{Spec}(\mathbf{C}[T_1, \ldots, T_n])_{\mathrm{an}} = \mathbf{C}^n.$$

Assume (A_1) is satisfied. To prove that (A'_1) is satisfied, we may assume Z is affine, and hence $Z = \mathrm{Spec}(\mathbf{C}[T_1, \ldots, T_n]/(f_1, \ldots, f_m))$ for some $n, m \geq 0$ and for some $f_1, \ldots, f_m \in \mathbf{C}[T_1, \ldots, T_n]$. Then

$$\mathrm{Mor}(U, Z_{\mathrm{an}}) = \{\varphi \in \mathrm{Mor}(U, \mathbf{C}^n) \mid f_1(\varphi) = \cdots = f_m(\varphi) = 0\},$$
$$\mathrm{Mor}(U, Z) = \{\varphi \in \mathcal{O}(U)^n \mid f_1(\varphi) = \cdots = f_m(\varphi) = 0\}.$$

This shows that (A'_1) is satisfied.

2.2.2

We give remarks on objects X of $\overline{\mathcal{A}}_1$.

First, for an open set U of X, the addition (resp. multiplication) of the ring $\mathcal{O}(U)$ coincides with

$$\mathcal{O}(U) \times \mathcal{O}(U) = \mathrm{Mor}(U, \mathbf{C}^2) \to \mathrm{Mor}(U, \mathbf{C}) = \mathcal{O}(U),$$

where the arrow is induced from the addition (resp. multiplication) $\mathbf{C} \times \mathbf{C} \to \mathbf{C}$. From this we see that, for each $x \in U$, the map

$$\mathrm{Map}(U, \mathbf{C}) \to \mathbf{C}, \quad f \mapsto f(x),$$

induces a ring homomorphism

$$\mathcal{O}(U) = \mathrm{Mor}(U, \mathbf{C}) \to \mathrm{Map}(U, \mathbf{C}) \to \mathbf{C},$$

which we denote also by $f \mapsto f(x)$, and, in the limit, a ring homomorphism over $\mathbf{C}$

$$\mathcal{O}_{X,x} \to \mathbf{C},$$

which we denote again by $f \mapsto f(x)$. The kernel of this homomorphism is the unique maximal ideal of $\mathcal{O}_{X,x}$. Hence

(1) For any $x \in X$, the residue field of the local ring $\mathcal{O}_{X,x}$ is $\mathbf{C}$.

Next, $\exp : \mathbf{C} \to \mathbf{C}^\times$ induces

$$\exp : \mathcal{O}_X = \mathrm{Mor}(\quad, \mathbf{C}) \xrightarrow{\exp} \mathrm{Mor}(\quad, \mathbf{C}^\times) = \mathcal{O}_X^\times.$$

We have an exact sequence of sheaves

(2) $$0 \to \mathbf{Z} \xrightarrow{2\pi i} \mathcal{O}_X \xrightarrow{\exp} \mathcal{O}_X^\times \to 1.$$

Lastly, let $\mathcal{S}$ be an fs monoid. Then for an object X of $\overline{\mathcal{A}}_1$ we have a canonical bijection

$$\mathrm{Mor}_{\overline{\mathcal{A}}_1}(X, \mathrm{Spec}(\mathbf{C}[\mathcal{S}])_{\mathrm{an}}) \simeq \mathrm{Hom}(\mathcal{S}, \Gamma(X, \mathcal{O}_X)),$$

where $\Gamma(X, \mathcal{O}_X)$ is regarded as a monoid by multiplication. In fact, for any local ringed space X over $\mathbf{C}$, a morphism of local ringed spaces $X \to \mathrm{Spec}(\mathbf{C}[\mathcal{S}])$ (without $(\ )_{\mathrm{an}}$) over $\mathbf{C}$ corresponds bijectively to a homomorphism $\mathbf{C}[\mathcal{S}] \to \Gamma(X, \mathcal{O}_X)$ of rings over $\mathbf{C}$, and hence bijectively to a homomorphism $\mathcal{S} \to \Gamma(X, \mathcal{O}_X)$. If X belongs to $\overline{\mathcal{A}}_1$, we can add $(\ )_{\mathrm{an}}$ to $\mathrm{Spec}(\mathbf{C}[\mathcal{S}])$. From this, we can deduce easily that for an object X of $\overline{\mathcal{A}}_1(\log)$ we have a canonical bijection

$$\mathrm{Mor}_{\overline{\mathcal{A}}_1(\log)}(X, \mathrm{Spec}(\mathbf{C}[\mathcal{S}])_{\mathrm{an}}) \simeq \mathrm{Hom}(\mathcal{S}, \Gamma(X, M_X)).$$

Here $\mathrm{Spec}(\mathbf{C}[\mathcal{S}])$ is endowed with the canonical logarithmic structure (2.1.6 (ii)).

2.2.3

Let X be an object of $\overline{\mathcal{A}}_1(\log)$.

As a set, $X^{\log}$ is defined to be the set of all pairs (x, h) consisting of a point $x \in X$ and an *argument function* h which is a homomorphism $M^{\mathrm{gp}}_{X,x} \to \mathbf{S}^1$ whose restriction to $\mathcal{O}^{\times}_{X,x}$ is $u \mapsto u(x)/|u(x)|$. Here $\mathbf{S}^1 := \{z \in \mathbf{C} \mid |z| = 1\}$. Note that the value $u(x) \in \mathbf{C}^{\times}$ is defined as explained in (2.2.2).

We endow $X^{\log}$ with the weakest topology satisfying the following conditions (1) and (2).

(1) The canonical map

$$\tau : X^{\log} \to X, \quad (x, h) \mapsto x,$$

is continuous.

(2) For any open set U of X and for any $f \in \Gamma(U, M^{\mathrm{gp}}_X)$, the map $\tau^{-1}(U) \to \mathbf{S}^1$, $(x, h) \mapsto h(f)$ is continuous.

The canonical map $\tau : X^{\log} \to X$ is proper and surjective. The surjectivity is clear and the properness is seen as follows. Locally on X, take a chart $\mathcal{S} \to M_X$. Then we have an injective map

$$X^{\log} \hookrightarrow X \times \mathrm{Hom}(\mathcal{S}^{\mathrm{gp}}, \mathbf{S}^1), \quad (x, h) \mapsto (x, h_{\mathcal{S}}),$$

where $h_{\mathcal{S}}$ denotes the composite map $\mathcal{S}^{\mathrm{gp}} \to M^{\mathrm{gp}}_{X,x} \to \mathbf{S}^1$. The image of this injection is closed, and the topology of $X^{\log}$ coincides with the topology as a closed subset of $X \times \mathrm{Hom}(\mathcal{S}^{\mathrm{gp}}, \mathbf{S}^1)$. Since $\mathrm{Hom}(\mathcal{S}^{\mathrm{gp}}, \mathbf{S}^1)$ is compact, the map τ is proper.

For $x \in X$, the inverse image $\tau^{-1}(x)$ is homeomorphic to $(\mathbf{S}^1)^r$ where r is the rank of the abelian group $(M^{\mathrm{gp}}_X/\mathcal{O}^{\times}_X)_x$.

2.2.4

We define the sheaf of rings $\mathcal{O}^{\log}_X$ on $X^{\log}$ as follows. We will define first the *sheaf of logarithms* $\mathcal{L} = \mathcal{L}_X$ of M^{gp}_X on $X^{\log}$, and will then define $\mathcal{O}^{\log}_X$ as a sheaf of $\tau^{-1}(\mathcal{O}_X)$-algebras generated by $\mathcal{L}$.

Let $\mathcal{L}$ be the fiber product of

$$\begin{array}{ccc} & & \tau^{-1}(M_X^{\mathrm{gp}}) \\ & & \downarrow \\ \mathrm{Cont}(\ \ , i\mathbf{R}) & \xrightarrow{\exp} & \mathrm{Cont}(\ \ , \mathbf{S}^1), \end{array}$$

where Cont($\ , T$), for a topological space T, denotes the sheaf on $X^{\log}$ of continuous maps to T, and $\tau^{-1}(M_X^{\mathrm{gp}}) \to \mathrm{Cont}(\ \ , \mathbf{S}^1)$ comes from the definition of $X^{\log}$. We define

$$\mathcal{O}_X^{\log} := (\tau^{-1}(\mathcal{O}_X) \otimes_{\mathbf{Z}} \mathrm{Sym}_{\mathbf{Z}}(\mathcal{L}))/\mathfrak{a}, \tag{1}$$

where $\mathrm{Sym}_{\mathbf{Z}}(\mathcal{L})$ denotes the symmetric algebra of $\mathcal{L}$ over $\mathbf{Z}$, and $\mathfrak{a}$ is the ideal of $\tau^{-1}(\mathcal{O}_X) \otimes_{\mathbf{Z}} \mathrm{Sym}_{\mathbf{Z}}(\mathcal{L})$ generated by the image of

$$\tau^{-1}(\mathcal{O}_X) \to \tau^{-1}(\mathcal{O}_X) \otimes_{\mathbf{Z}} \mathrm{Sym}_{\mathbf{Z}}(\mathcal{L}), \quad f \mapsto f \otimes 1 - 1 \otimes \iota(f).$$

Here the map $\iota : \tau^{-1}(\mathcal{O}_X) \to \mathcal{L}$ is the one induced by

$$\tau^{-1}(\mathcal{O}_X) \to \mathrm{Cont}(\ \ , i\mathbf{R}), \quad f \mapsto \tfrac{1}{2}(f - \bar{f}),$$

and

$$\tau^{-1}(\mathcal{O}_X) \xrightarrow{\exp} \tau^{-1}(\mathcal{O}_X^\times) \subset \tau^{-1}(M_X^{\mathrm{gp}}).$$

In the above, the overbar indicates complex conjugation. Note that $\exp : \mathcal{O}_X \to \mathcal{O}_X^\times$ is defined as explained in 2.2.2.

We denote the projection $\mathcal{L} \to \tau^{-1}(M_X^{\mathrm{gp}})$ by exp, and the inverse $\tau^{-1}(M_X^{\mathrm{gp}}) \to \mathcal{L}/(2\pi i\mathbf{Z})$ by log. Then we have a commutative diagram with exact rows:

$$\begin{array}{ccccccccc} 0 & \longrightarrow & \mathbf{Z} & \xrightarrow{2\pi i} & \tau^{-1}(\mathcal{O}_X) & \xrightarrow{\exp} & \tau^{-1}(\mathcal{O}_X^\times) & \longrightarrow & 1 \\ & & \| & & \iota\downarrow\cap & & \downarrow\cap & & \\ 0 & \longrightarrow & \mathbf{Z} & \xrightarrow{2\pi i} & \mathcal{L} & \xrightarrow{\exp} & \tau^{-1}(M_X^{\mathrm{gp}}) & \longrightarrow & 1. \end{array} \tag{2}$$

We have the evident morphism of ringed spaces over $\mathbf{C}$

$$\tau : (X^{\log}, \mathcal{O}_X^{\log}) \to (X, \mathcal{O}_X).$$

A morphism $f : X \to Y$ in $\overline{\mathcal{A}}_1(\log)$ induces a morphism

$$f^{\log} : (X^{\log}, \mathcal{O}_X^{\log}) \to (Y^{\log}, \mathcal{O}_Y^{\log})$$

of ringed spaces over $\mathbf{C}$ in the obvious way.

2.2.5

For $y \in X^{\log}$, the stalk $\mathcal{O}_{X,y}^{\log}$ is described as follows. Let $x = \tau(y) \in X$ and $r = \mathrm{rank}_{\mathbf{Z}}(M_X^{\mathrm{gp}}/\mathcal{O}_X^\times)_x$. Let $(\ell_j)_{1 \le j \le r}$ be a family of elements of $\mathcal{L}_y$ whose images in $(M_X^{\mathrm{gp}}/\mathcal{O}_X^\times)_x$ under exp form a system of free generators. Then we have an isomorphism of $\mathcal{O}_{X,x}$-algebras

$$\mathcal{O}_{X,x}[T_1,\ldots,T_r] \xrightarrow{\sim} \mathcal{O}^{\log}_{X,y}, \quad T_j \mapsto \ell_j.$$

Note that $\mathcal{O}^{\log}_{X,y}$ is *not* a local ring if $r \geq 1$.

2.2.6

For a morphism $f : X \to Y$ of fs logarithmic analytic spaces, we denote

$$\omega^{q,\log}_{X/Y} := \tau^*(\omega^q_{X/Y}) = \mathcal{O}^{\log}_X \otimes_{\tau^{-1}(\mathcal{O}_X)} \tau^{-1}(\omega^q_{X/Y})$$

for $q \in \mathbf{Z}$ (cf. 2.1.7). We have the *logarithmic de Rham complex* $(\omega^{\bullet,\log}_{X/Y}, d)$ *of* $X^{\log}$ *over* $Y^{\log}$, where the differential d is characterized by the following properties (1)–(3).

(1) This d is compatible with the differential d of $(\omega^\bullet_{X/Y}, d)$.
(2) For $f \in M^{\mathrm{gp}}_X$, d sends $\log(f) \in \mathcal{L}/(2\pi i\mathbf{Z}) \subset \mathcal{O}^{\log}_X/(2\pi i\mathbf{Z})$ to $d\log(f) \in \omega^1_{X/Y} \subset \omega^{1,\log}_{X/Y}$ (2.1.7).
(3) $d(f \wedge g) = df \wedge g + (-1)^q f \wedge dg$ for $f \in \omega^{q,\log}_{X/Y}$ and $g \in \omega^{r,\log}_{X/Y}$ $(q, r \in \mathbf{Z})$.

More explicitly, for $f \in \mathcal{O}_X$, $\ell \in \mathcal{L}$, $g = \exp(\ell) \in M^{\mathrm{gp}}_X$, and $\eta \in \omega^q_{X/Y}$, we have

$$d(f\ell^n \otimes \eta) = \ell^n \otimes df \wedge \eta + nf\ell^{n-1} \otimes d\log(g) \wedge \eta + f\ell^n \otimes d\eta \quad \text{for } n \geq 1,$$
$$d(f \otimes \eta) = df \wedge \eta + f d\eta.$$

2.2.7

Examples.

(i) Let $X := \mathbf{C}$ and let z be the coordinate of X. Let M be the fs logarithmic structure on X associated with the divisor $\{0\}$ (see 2.1.2). We can take a chart $\mathbf{N} \to M_X = \bigcup_{n\geq 0} \mathcal{O}^\times_X z^n$, $n \mapsto z^n$. We have an isomorphism

$$X^{\log} \xrightarrow{\sim} (\mathbf{R}_{\geq 0}) \times \mathbf{S}^1, \quad (x, h) \mapsto (|z(x)|,\ h(z)).$$

We have $\mathcal{O}^{\log}_X = \tau^{-1}(\mathcal{O}_X)[\log(z)]$ and $\omega^{1,\log}_X = \tau^{-1}(\mathcal{O}_X)[\log(z)]d\log(z)$.

(ii) More generally, for a toric variety $X = \mathrm{Spec}(\mathbf{C}[\mathcal{S}])_{\mathrm{an}}$ with the canonical logarithmic structure in 2.1.6 (ii), we have

$X \simeq \mathrm{Hom}(\mathcal{S}, \mathbf{C}^{\mathrm{mult}})$,
$X^{\log} \simeq \mathrm{Hom}(\mathcal{S}, \mathbf{R}^{\mathrm{mult}}_{\geq 0}) \times \mathrm{Hom}(\mathcal{S}^{\mathrm{gp}}, \mathbf{S}^1)$,
$\tau : X^{\log} \to X$ is identified with the map induced by $(\mathbf{R}_{\geq 0}) \times \mathbf{S}^1 \to \mathbf{C}$, $(r, u) \mapsto ru$,
$\mathcal{O}^{\log}_X = \tau^{-1}(\mathcal{O}_X)[\log(f);\ f \in \mathcal{S}] \subset j^{\log}_* \mathcal{O}_U$ $(j^{\log} : U = \mathrm{Spec}(\mathbf{C}[\mathcal{S}^{\mathrm{gp}}])_{\mathrm{an}} \hookrightarrow X^{\log})$,
$\mathcal{O}^{\log}_X \otimes_{\mathbf{Z}} \mathcal{S}^{\mathrm{gp}} \xrightarrow{\sim} \omega^{1,\log}_X, \quad f \otimes g \mapsto f\,d\log(g)$.

(iii) We have local descriptions of $X^{\log}$ and $\mathcal{O}_X^{\log}$ for a general object X of $\overline{\mathcal{A}}_1(\log)$ by using the example in (ii). Locally on X, take a chart $\mathcal{S} \to M_X$ with $\mathcal{S}$ an fs monoid. Then as is explained in 2.2.2, we have the corresponding morphism $f : X \to Z := \mathrm{Spec}(\mathbf{C}[\mathcal{S}])_{\mathrm{an}}$ of objects of $\overline{\mathcal{A}}_1(\log)$. This induces a continuous map $X^{\log} \to Z^{\log}$, and we have a homeomorphism

$$X^{\log} \xrightarrow{\sim} X \times_Z Z^{\log}.$$

Furthermore, we have an isomorphism

$$\mathcal{O}_X \otimes_{\mathcal{O}_Z} \mathcal{O}_Z^{\log} \xrightarrow{\sim} \mathcal{O}_X^{\log},$$

where we denote the inverse images of $\mathcal{O}_X$, $\mathcal{O}_Z$, and $\mathcal{O}_Z^{\log}$ on $X^{\log}$ simply by $\mathcal{O}_X$, $\mathcal{O}_Z$, and $\mathcal{O}_Z^{\log}$. We have

$$\mathcal{O}_X \otimes_{\mathcal{O}_Z} \omega_Z^q \xrightarrow{\sim} \omega_X^q \quad \text{on } X,$$
$$\mathcal{O}_X^{\log} \otimes_{\mathcal{O}_Z} \omega_Z^q \xrightarrow{\sim} \mathcal{O}_X^{\log} \otimes_{\mathcal{O}_Z^{\log}} \omega_Z^{q,\log} \xrightarrow{\sim} \omega_X^{q,\log} \quad \text{on } X^{\log}.$$

2.2.8

Let x be an fs logarithmic point (2.1.9). Then $x^{\log} \simeq (\mathbf{S}^1)^r$ where $r = \mathrm{rank}(M_x^{\mathrm{gp}}/\mathcal{O}_x^\times)$. The sheaf $\mathcal{O}_x^{\log}$ is a locally constant sheaf on $x^{\log}$ which is locally isomorphic to $\mathbf{C}[T_1, \ldots, T_r]$.

For an object X of $\overline{\mathcal{A}}_1(\log)$ and for $x \in X$, we usually regard x as an fs logarithmic point whose logarithmic structure is the inverse image of the logarithmic structure of X. We identify $x^{\log} = \tau^{-1}(x)$.

2.2.9

Let x be an fs logarithmic point and let $r = \mathrm{rank}(M_x^{\mathrm{gp}}/\mathcal{O}_x^\times)$. Since $x^{\log} \simeq (\mathbf{S}^1)^r$, the fundamental group $\pi_1(x^{\log})$ of $x^{\log}$ is isomorphic to $\mathbf{Z}^r$. We define a canonical isomorphism

$$\pi_1(x^{\log}) \simeq \mathrm{Hom}(M_x^{\mathrm{gp}}/\mathcal{O}_x^\times, \mathbf{Z}) \tag{1}$$

as follows. Taking a base point $h_0 \in x^{\log}$, $h_0 : M_x \to \mathbf{S}^1$, we have a homeomorphism

$$x^{\log} \xrightarrow{\sim} \mathrm{Hom}(M_x^{\mathrm{gp}}/\mathcal{O}_x^\times, \mathbf{S}^1), \quad h \mapsto hh_0^{-1}.$$

The exact sequence

$$0 \to \mathbf{Z} \xrightarrow{2\pi i} i\mathbf{R} \xrightarrow{\exp} \mathbf{S}^1 \to 1$$

induces an exact sequence

$$0 \to \mathrm{Hom}(M_x^{\mathrm{gp}}/\mathcal{O}_x^\times, \mathbf{Z}) \xrightarrow{2\pi i} \mathrm{Hom}(M_x^{\mathrm{gp}}/\mathcal{O}_x^\times, i\mathbf{R}) \xrightarrow{\exp} \mathrm{Hom}(M_x^{\mathrm{gp}}/\mathcal{O}_x^\times, \mathbf{S}^1) \to 1.$$

This shows that the r-dimensional $\mathbf{R}$-vector space $\mathrm{Hom}(M_x^{\mathrm{gp}}/\mathcal{O}_x^\times, i\mathbf{R})$ is the universal covering of the real torus $x^{\log} \simeq \mathrm{Hom}(M_x^{\mathrm{gp}}/\mathcal{O}_x^\times, \mathbf{S}^1)$ whose fundamental group is $\pi_1(x^{\log}) \simeq \mathrm{Hom}(M_x^{\mathrm{gp}}/\mathcal{O}_x^\times, \mathbf{Z})$, which acts on the universal covering as translations by $(2\pi i)$-multiples of integral points. Note that the last isomorphism is independent of the choice of h_0.

We define

$$\pi_1^+(x^{\log}) := \mathrm{Hom}(M_x/\mathcal{O}_x^\times, \mathbf{N}) \subset \pi_1(x^{\log}). \tag{2}$$

Then, $\pi_1(x^{\log}) = \pi_1^+(x^{\log})^{\mathrm{gp}}$.

For $f \in M_x = \Gamma(x, M_x)$, since logarithms of f in the sheaf $\mathcal{L}_x$ on $x^{\log}$ form a locally constant sheaf on $x^{\log}$, the fundamental group $\pi_1(x^{\log})$ acts on local sections $\log(f)$ of $\mathcal{L}_x$. For $\gamma \in \pi_1(x^{\log})$, we have

$$\gamma(\log(f)) = \log(f) - 2\pi i[\gamma, f] \quad (\text{not } \log(f) + 2\pi i[\gamma, f], \text{ cf. Appendix A1}),$$

where $[\ ,\ ] : \pi_1(x^{\log}) \times M_x^{\mathrm{gp}} \to \mathbf{Z}$ is the pairing given by the above isomorphism $\pi_1(x^{\log}) \simeq \mathrm{Hom}(M_x^{\mathrm{gp}}/\mathcal{O}_x^\times, \mathbf{Z})$.

For example, if $x = 0 \in \mathbf{C}$ where $\mathbf{C}$ is endowed with the logarithmic structure associated to the divisor $\{0\} \subset \mathbf{C}$, then $M_x/\mathcal{O}_x^\times \simeq \mathbf{N}$ is generated by the image of the coordinate function q of $\mathbf{C}$, and the generator of $\pi_1^+(x^{\log}) = \mathrm{Hom}(M_x/\mathcal{O}_x^\times, \mathbf{N}) \simeq \mathbf{N}$, which sends q to 1, is the class of the counterclockwise directed circle $\gamma : [0, 1] \to x^{\log} = \{0\} \times \mathbf{S}^1$, $\gamma(t) = (0, \exp(2\pi i t))$. We have $\gamma(\log(q)) = \log(q) - 2\pi i$ (not $\log(q) + 2\pi i$).

Some reader may feel that the sign here is strange. See Appendix A1 for a discussion of the sign problem.

PROPOSITION 2.2.10 *For any object X of $\overline{\mathcal{A}}_1(\log)$, we have*

$$\tau_*(\mathcal{O}_X^{\log}) = \mathcal{O}_X, \quad R^m\tau_*(\mathcal{O}_X^{\log}) = 0 \quad \text{for } m \geq 1.$$

Proof. (Proofs for fs logarithmic analytic spaces X are given in [Ma1, 4.6] and [IKN, (3.7) (3)].) Let $x \in X$. Replacing X by an open neighborhood of x in X, take a family $(q_j)_{1\leq j\leq r}$ of elements of $\Gamma(X, M_X^{\mathrm{gp}})$ whose image in $(M_X^{\mathrm{gp}}/\mathcal{O}_X^\times)_x$ is a $\mathbf{Z}$-basis of $(M_X^{\mathrm{gp}}/\mathcal{O}_X^\times)_x$. Consider the sheaf of rings on $X^{\log}$

$$R = \mathbf{C}[\log(q_1), \ldots, \log(q_r)] \subset \mathcal{O}_X^{\log}.$$

Here $\log(q_j)$ are determined only modulo $2\pi i \cdot \mathbf{Z}$, but the subsheaf R of $\mathcal{O}_X^{\log}$ is independent of the local choices of the branches of $\log(q_j)$ and hence defined globally. Then, by 2.2.5, the germ of the homomorphism $\mathcal{O}_X \bigotimes_{\mathbf{C}} R \to \mathcal{O}_X^{\log}$ at any $y \in X^{\log}$ lying over x is bijective. Let $\iota : \tau^{-1}(x) \hookrightarrow X^{\log}$ be the inclusion map. Then, by the proper base change theorem (Appendix A2),

$$\begin{aligned}(R^m\tau_*(\mathcal{O}_X^{\log}))_x &\simeq \left(R^m\tau_*\left(\mathcal{O}_X \otimes_{\mathbf{C}} R\right)\right)_x \\ &\simeq \mathcal{O}_{X,x} \otimes_{\mathbf{C}} (R^m\tau_*R)_x \simeq \mathcal{O}_{X,x} \otimes_{\mathbf{C}} H^m(\tau^{-1}(x), \iota^{-1}R)\end{aligned}$$

It is sufficient to prove that $H^0(\tau^{-1}(x), \iota^{-1}R) = \mathbf{C}$ and $H^m(\tau^{-1}(x), \iota^{-1}R) = 0$ for $m \geq 1$. Since $\iota^{-1}R$ is a locally constant sheaf on $\tau^{-1}(x) = x^{\log}$,

$$H^m(\tau^{-1}(x), \iota^{-1}R) \simeq H^m(\pi_1(x^{\log}), R_y)$$

where y is a point of $X^{\log}$ lying over x, R_y is the stalk of R at y, and $H^m(\pi_1(x^{\log}), R_y)$ is the mth cohomology of the module R_y over the group $\pi_1(x^{\log})$. Let $(\gamma_j)_{1\leq j\leq r}$ be a $\mathbf{Z}$-basis of the abelian group $\pi_1(x^{\log}) \simeq \mathrm{Hom}((M_X^{\mathrm{gp}}/\mathcal{O}_X^\times)_x, \mathbf{Z})$ which is dual to the $\mathbf{Z}$-basis $(q_j \bmod \mathcal{O}_{X,x}^\times)_j$ of $(M_X^{\mathrm{gp}}/\mathcal{O}_X^\times)_x$. Then $\gamma_j(\log(q_k)) = \log(q_k) - 2\pi i \delta_{jk}$ where δ_{jk} is the Kronecker symbol (see 2.2.9 and Appendix A1). For $0 \le j \le r$, let Γ_j be the subgroup of $\pi_1(x^{\log})$ generated by γ_k $(j < k \le r)$ and let $R_j = \mathbf{C}[\log(q_1), \ldots, \log(q_j)] \subset R_y$. Hence $\Gamma_0 = \pi_1(x^{\log})$ and $R_0 = \mathbf{C}$. We prove $R\Gamma(\Gamma_j, R_y) = R_j$ by downward induction on j. The case $j = r$ is clear (note that $\Gamma_r = \{1\}$). We go from j $(1 \le j \le r)$ to $j - 1$. Assume $R\Gamma(\Gamma_j, R_y) = R_j$. Then we have

$$\begin{aligned} R\Gamma(\Gamma_{j-1}, R_y) &= R\Gamma(\Gamma_{j-1}/\Gamma_j,\ R\Gamma(\Gamma_j, R_y)) \\ &= (\text{the complex } [\gamma_j - 1 : R_j \to R_j]). \end{aligned}$$

As is easily seen, $\gamma_j - 1 : R_j \to R_j$ is surjective and the kernel is R_{j-1}. □

2.3 LOCAL SYSTEMS ON $X^{\log}$

In this section, we consider local systems on $X^{\log}$ for objects X of $\overline{\mathcal{A}}_1(\log)$ (2.2.1). We prove that, if L is a locally constant sheaf on $X^{\log}$ of free $\mathbf{Z}$-modules of finite rank with "unipotent local monodromy" (2.3.1 below), then locally on X, L is embedded in

$$\mathcal{O}_X^{\log} \otimes L_0$$

in a special way, where L_0 is a stalk of L regarded as a constant sheaf.

From the next section, we will consider "logarithmic Hodge structures" and "polarized logarithmic Hodge structures" by putting Hodge filtrations on local systems on $X^{\log}$.

2.3.1

Let L be a locally constant sheaf on $X^{\log}$. For $x \in X$ and $y \in X^{\log}$ lying over x, we call the action of $\pi_1(x^{\log}) = \pi_1(\tau^{-1}(x))$ on L_y the *local monodromy* of L at y.

Assume that L is a locally constant sheaf of abelian groups on $X^{\log}$. We say the local monodromy of L is *unipotent* if the local monodromy of L at y is unipotent for any $y \in X^{\log}$.

Proposition 2.3.2 *Let X be an object of $\overline{\mathcal{A}}_1(\log)$, let A be a subring of $\mathbf{C}$, and let L be a locally constant sheaf on $X^{\log}$ of free A-modules of finite rank. Let $x \in X$, let y be a point of $X^{\log}$ lying over x, and assume that the local monodromy of L at y is unipotent. Let $(q_j)_{1 \le j \le n}$ be a finite family of elements of $M_{X,x}^{\mathrm{gp}}$ whose image in $(M_X^{\mathrm{gp}}/\mathcal{O}_X^\times)_x$ is a $\mathbf{Z}$-basis, and let $(\gamma_j)_{1 \le j \le n}$ be the dual $\mathbf{Z}$-basis of $\pi_1(x^{\log})$ in the duality in 2.2.9. Then if we replace X by some open neighborhood of x, we have an isomorphism of $\mathcal{O}_X^{\log}$-modules*

$$\nu : \mathcal{O}_X^{\log} \otimes_A L \xrightarrow{\sim} \mathcal{O}_X^{\log} \otimes_A L_0, \quad L_0 = \textit{the stalk } L_y,$$

where L_0 is regarded as a constant sheaf, satisfying the following condition (1). *Let*

$$N_j = \log(\gamma_j) : L_{0,\mathbf{Q}} \to L_{0,\mathbf{Q}},$$

lift q_j *in* $\Gamma(X, M_X^{\mathrm{gp}})$ *(by replacing* X *by an open neighborhood of* x*), and let*

$$\xi = \exp\left(\sum_{j=1}^{n}(2\pi i)^{-1}\log(q_j)\otimes N_j\right) : \mathcal{O}_X^{\log}\otimes_A L_0 \xrightarrow{\sim} \mathcal{O}_X^{\log}\otimes_A L_0.$$

Note that the operator $\xi = \exp(\sum_{j=1}^{n}(2\pi i)^{-1}\log(q_j)\otimes N_j)$ *depends on the local choices of the branches of* $\log(q_j)$*, but that the subsheaf* $\xi^{-1}(1\otimes L_0)$ *of* $\mathcal{O}_X^{\log}\otimes_A L_0$*, which we consider in* (1) *below, is independent of the choices and hence is defined globally on* $X^{\log}$*.*

(1) *The restriction of* ν *to* $L = 1\otimes L$ *induces an isomorphism of locally constant sheaves*

$$\nu : L \xrightarrow{\sim} \xi^{-1}(1\otimes L_0).$$

If we fix branches of the germs $\log(q_j)_y$ *at* y $(1 \le j \le n)$*, we can take an isomorphism* ν *satisfying* (1) *as above which satisfies furthermore the following condition* (2)*.*

(2) *The branch of* ξ_y *defined by the fixed branches of* $\log(q_j)_y$ *satisfies*

$$\nu_y(1\otimes v) = \xi_y^{-1}(1\otimes v) \quad \textit{for any } v \in L_0.$$

Proof. Let L' be the locally constant subsheaf $\xi^{-1}(1\otimes L_0)$ of $\mathcal{O}_X^{\log}\otimes L_0$. Fix a branch of $\log(q_j)_y$ at y for $1\le j\le n$, and let $\nu : L_y \to (L')_y$ be the isomorphism of A-modules $v \mapsto \xi_y^{-1}(1\otimes v)$ where ξ_y is defined by the fixed branches of $\log(q_j)_y$. Then ν preserves the local monodromy actions of $\pi_1(x^{\log})$ on these stalks of the locally constant sheaves L and L'. In fact, for $v\in L_0$ and for $1\le k\le n$,

$$\begin{aligned}
\gamma_k(\xi_y^{-1}(1\otimes v) \text{ in } L'_y) &= \gamma_k(\xi_y)^{-1}\cdot(1\otimes v)\\
&= \exp\left(-\left(\sum_{j=1}^{n}((2\pi i)^{-1}\log(q_j)_y - \delta_{jk})\otimes N_j\right)\right)\cdot(1\otimes v)\\
&= \xi_y^{-1}\exp(1\otimes N_k)(1\otimes v) = \xi_y^{-1}(1\otimes\gamma_k(v \text{ in } L_y))
\end{aligned}$$

(δ_{jk} is the Kronecker symbol). Hence there is a unique isomorphism $\nu : L|_{x^{\log}} \to L'|_{x^{\log}}$ between the pullbacks of L and L' to $x^{\log}$ which induces the above isomorphism ν on the stalks at y.

By the proper base change theorem (Appendix A2) applied to the proper map $\tau : X^{\log}\to X$ and to the sheaf $\mathcal{F}$ of isomorphisms from L to L' on $X^{\log}$, the isomorphism ν extends to an isomorphism $\nu : L \xrightarrow{\sim} L'$ if we replace X by some open neighborhood of x in X. This isomorphism ν induces an isomorphism of $\mathcal{O}_X^{\log}$-modules

$$\nu : \mathcal{O}_X^{\log}\otimes_A L \xrightarrow{\sim} \mathcal{O}_X^{\log}\otimes_A L' = \mathcal{O}_X^{\log}\otimes_A L_0. \qquad \square$$

Example. Let $f : E\to\Delta$ be the degenerating family of elliptic curves in 0.2.10 and consider the locally constant sheaf $L = R^1 f_*^{\log}(\mathbf{Z})$ on $\Delta^{\log}$. In 2.3.2, take

$X = \Delta$, $x = 0 \in \Delta$, $A = \mathbf{Z}$, and take the coordinate function q of Δ as q_1 ($n = 1$ in this situation). Then the element γ_1, which we denote here by γ, is the positive generator of $\pi_1(\Delta^{\log})$ (represented by a circle in Δ^* in the counterclockwise direction; cf. Appendix A1). As is explained in 0.2, L has a $\mathbf{Z}$-basis (e_1, e_2) locally (e_1 is defined globally but e_2 is determined by a local choice of the branch of $\log(q)$). Fix a branch of $\log(q)$ at y and take the corresponding $e_{2,y}$. We have

$$\gamma(e_{1,y}) = e_{1,y}, \ \gamma(e_{2,y}) = e_{1,y} + e_{2,y},$$
$$N(e_{1,y}) = 0, \ N(e_{2,y}) = e_{1,y}, \quad \text{where } N = \log(\gamma).$$

The $\mathcal{O}_X^{\log}$-module $\mathcal{O}_X^{\log} \otimes_{\mathbf{Z}} L$ has a global base $(1 \otimes e_1, \ \omega)$ as in 0.2.15. We have an isomorphism of $\mathcal{O}_X^{\log}$-modules

$$\nu : \mathcal{O}_X^{\log} \otimes L \xrightarrow{\sim} \mathcal{O}_X^{\log} \otimes L_0, \quad 1 \otimes e_1 \to 1 \otimes e_{1,y}, \quad \omega \mapsto e_{2,y}.$$

This ν has the property stated in Proposition 2.3.2 globally on Δ. In fact, since $\omega = (2\pi i)^{-1} \log(q) e_1 + e_2$, ν sends e_1 to $1 \otimes e_{1,y} = \xi^{-1}(1 \otimes e_1)$ and e_2 to $-(2\pi i)^{-1} \log(q) e_{1,y} + e_{2,y} = \xi^{-1}(1 \otimes e_{2,y})$.

PROPOSITION 2.3.3 *Let X be an object of $\overline{\mathcal{A}}_1(\log)$ and let L be a locally constant sheaf of finite-dimensional $\mathbf{C}$-vector spaces on $X^{\log}$.*

(i) *If the local monodromy of L is unipotent, the $\mathcal{O}_X$-module*

$$\mathcal{M} := \tau_*\left(\mathcal{O}_X^{\log} \otimes_{\mathbf{C}} L\right)$$

is locally free of finite rank, and we have an isomorphism

$$\mathcal{O}_X^{\log} \otimes_{\tau^{-1}(\mathcal{O}_X)} \tau^{-1}(\mathcal{M}) \xrightarrow{\sim} \mathcal{O}_X^{\log} \otimes_{\mathbf{C}} L.$$

(ii) *Conversely, assume that there are a locally free $\mathcal{O}_X$-module $\mathcal{M}$ of finite rank on X and an isomorphism of $\mathcal{O}_X^{\log}$-modules $\mathcal{O}_X^{\log} \otimes_{\tau^{-1}(\mathcal{O}_X)} \tau^{-1}(\mathcal{M}) \simeq \mathcal{O}_X^{\log} \otimes_{\mathbf{C}} L$. Then the local monodromy of L is unipotent and $\mathcal{M} \xrightarrow{\sim} \tau_*(\mathcal{O}_X^{\log} \otimes_{\mathbf{C}} L)$.*

Proof. We prove (i). By Proposition 2.3.2, locally on X, we have an isomorphism of $\mathcal{O}_X^{\log}$-modules $\mathcal{O}_X^{\log} \otimes_{\mathbf{C}} L \simeq \mathcal{O}_X^{\log} \otimes_{\mathbf{C}} L_0$ for some finite-dimensional $\mathbf{C}$-vector space L_0 regarded as a constant sheaf on $X^{\log}$. Hence

$$\tau_*\left(\mathcal{O}_X^{\log} \otimes_{\mathbf{C}} L\right) \simeq \tau_*\left(\mathcal{O}_X^{\log} \otimes_{\mathbf{C}} L_0\right) = \mathcal{O}_X \otimes_{\mathbf{C}} L_0$$

by proposition 2.2.10.

We prove (ii). The fact that $\mathcal{M} \xrightarrow{\sim} \tau_*(\mathcal{O}_X^{\log} \otimes_{\mathbf{C}} L)$ is proved as

$$\tau_*\left(\mathcal{O}_X^{\log} \otimes_{\mathbf{C}} L\right) \simeq \tau_*\left(\mathcal{O}_X^{\log} \otimes_{\tau^{-1}(\mathcal{O}_X)} \tau^{-1}(\mathcal{M})\right) = \tau_*(\mathcal{O}_X^{\log}) \otimes_{\mathcal{O}_X} \mathcal{M} = \mathcal{M}$$

by proposition 2.2.10. It remains to prove that the local monodromy of L is unipotent. For this we may assume that X is an fs logarithmic point. In this case $\tau^{-1}(\mathcal{M})$ is a constant sheaf, L is embedded in $\mathcal{O}_X^{\log} \otimes_{\mathbf{C}} \tau^{-1}(\mathcal{M})$, and

$(\gamma-1)^n(\log(q_1)\cdots\log(q_m))=0$ if $n>m$ for any $\gamma\in\pi_1(x^{\log})$ and any $q_1,\ldots,q_m\in M^{\mathrm{gp}}_{X,x}$. Hence the action of $\pi_1(x^{\log})$ in L is unipotent. □

Proposition 2.3.4 *Let X be an object of $\overline{\mathcal{A}}_1(\log)$ (2.2.1), let A be a subring of $\mathbf{C}$ containing $\mathbf{Q}$, and let L be a locally constant sheaf on $X^{\log}$ of free A-modules of finite rank. Assume that the local monodromy of L is unipotent.*

(i) *There exists a unique A-homomorphism*

$$\mathcal{N}:L\to(M^{\mathrm{gp}}_X/\mathcal{O}^\times_X)\otimes L$$

satisfying the following condition (1). Here we denote the inverse image of $M^{\mathrm{gp}}_X/\mathcal{O}^\times_X$ on $X^{\log}$ simply by $M^{\mathrm{gp}}_X/\mathcal{O}^\times_X$.

(1) *For any $x\in X$, any $y\in X^{\log}$ lying over x, and for any $\gamma\in\pi_1(x^{\log})$, if $h_\gamma:(M^{\mathrm{gp}}_X/\mathcal{O}^\times_X)_x\to\mathbf{Z}$ denotes the homomorphism corresponding to γ (2.2.9), the composition $L_y\xrightarrow{\mathcal{N}}(M^{\mathrm{gp}}_X/\mathcal{O}^\times_X)_x\otimes L_y\xrightarrow{h_\gamma}L_y$ coincides with the logarithm of the action of γ on L_y.*

(ii) *Assume that X is an fs logarithmic point $\{x\}$. Let*

$$\mathcal{N}':L\to\omega^1_x\otimes_A L$$

be the composition of $\mathcal{N}$ and the $\mathbf{C}$-linear map $(M^{\mathrm{gp}}_x/\mathcal{O}^\times_x)\otimes L\to\omega^1_x\otimes L$, $f\otimes v\mapsto(2\pi i)^{-1}d\log(f)\otimes v$, and let $1\otimes\mathcal{N}':\mathcal{O}^{\log}_x\otimes_A L\to\omega^{1,\log}_x\otimes_A L$ be the $\mathcal{O}^{\log}_x$-linear homomorphism induced by $\mathcal{N}'$. Let $\mathcal{M}:=H^0(x^{\log},\mathcal{O}^{\log}_x\otimes_A L)=\tau_*(\mathcal{O}^{\log}_X\otimes_A L)$. Then the restriction $\mathcal{M}\to\omega^1_x\otimes_{\mathbf{C}}\mathcal{M}$ of $d\otimes 1_L:\mathcal{O}^{\log}_x\otimes_A L\to\omega^{1,\log}\otimes_A L$ coincides with the restriction of $1\otimes\mathcal{N}'$ to $\mathcal{M}$.

Proof.

(i) We may assume that $L=\xi^{-1}(1\otimes L_0)$ with the notation in 2.3.2. Define $\mathcal{N}:L\to(M^{\mathrm{gp}}_X/\mathcal{O}^\times_X)\otimes L$ by

$$\mathcal{N}(\xi^{-1}(1\otimes v))=\sum_{j=1}^{n}q_j\otimes\xi^{-1}(1\otimes N_j(v))$$

(then $\mathcal{N}$ is independent of the choice of the branch of ξ). We show that $\mathcal{N}$ has the property stated in (i). Let $x'\in X$, y' being a point of $X^{\log}$ lying over x', $\gamma\in\pi_1((x')^{\log})$, and let $h_\gamma:(M^{\mathrm{gp}}_X/\mathcal{O}^\times_X)_{x'}\to\mathbf{Z}$ be the homomorphism corresponding to γ. Then, for $v\in L_0$ and for any branch of $\xi^{-1}_{y'}$ at y', the local monodromy action of γ on $L_{y'}$ satisfies $\log(\gamma)(\xi^{-1}_{y'}(1\otimes v))=(\log(\gamma)(\xi^{-1}_{y'}))(1\otimes v)$ and $\log(\gamma)(\xi^{-1}_{y'})=\sum_{j=1}^n h_\gamma(q_j)\xi^{-1}_{y'}(1\otimes N_j)$.

(ii) Since $\mathcal{O}^{\log}_x$ is the union of $\mathbf{C}$-subsheaves that are locally constant sheaves of finite dimensional $\mathbf{C}$-vector spaces, the collection of $\mathcal{N}$ of these subsheaves gives $\mathcal{N}:\mathcal{O}^{\log}_x\to(M^{\mathrm{gp}}_x/\mathcal{O}^\times_x)\otimes\mathcal{O}^{\log}_x$ and $\mathcal{N}':\mathcal{O}^{\log}_x\to\omega^{1,\log}_x$. It is easy to see that the last homomorphism coincides with $-d$. Since the local monodromy acts trivially on the images of $\mathcal{M}$ in the stalks of $\mathcal{O}^{\log}_x\otimes_A L$, ($\mathcal{N}'$ of $\mathcal{O}^{\log}_x)\otimes 1+1\otimes(\mathcal{N}'$ of L)

on $\mathcal{O}_x^{\log} \otimes_A L$ induces the zero map on $\mathcal{M} \subset \mathcal{O}_x^{\log} \otimes_A L$. Hence $-d \otimes 1_L + 1 \otimes \mathcal{N}'$ is zero on $\mathcal{M}$. □

PROPOSITION 2.3.5 (Ogus [Og, Theorem 3.1.2]) *Let X be a logarithmically smooth, fs logarithmic analytic space, let $U := X_{\mathrm{triv}}$ be the open set of X consisting of all points at which the logarithmic structure is trivial (that is, $U = \{x \in X \mid M_{X,x} = \mathcal{O}_{X,x}^\times\}$), and let $j^{\log} : U \hookrightarrow X^{\log}$ be the inclusion map. Then the restriction to U gives an equivalence from the category of locally constant sheaves on $X^{\log}$ to the category of locally constant sheaves on U. The inverse functor is given by $(j^{\log})_*$.*

If X is connected, $\pi_1(U, u) \xrightarrow{\sim} \pi_1(X^{\log}, u)$ for any $u \in U$.

Proof. It is sufficient to prove that for any $y \in X^{\log}$ and any neighborhood V of y, there is a contractible open neighborhood $W \subset V$ of y such that $W \cap U$ is contractible. We may assume $X = \mathrm{Spec}(\mathbf{C}[\mathcal{S}])_{\mathrm{an}}$ for some fs monoid $\mathcal{S}$ such that $\mathcal{S}^\times = \{1\}$, endowed with the canonical logarithmic structure, and $y = (0, 1) \in X^{\log} = \mathrm{Hom}(\mathcal{S}, \mathbf{R}_{\geq 0}^{\mathrm{mult}}) \times \mathrm{Hom}(\mathcal{S}^{\mathrm{gp}}, \mathbf{S}^1)$ where 0 denotes the homomorphism which sends all elements of $\mathcal{S} - \{1\}$ to $0 \in \mathbf{R}_{\geq 0}$ and 1 denotes the homomorphism which sends all elements of $\mathcal{S}^{\mathrm{gp}}$ to 1. Take a finite family $(a_j)_{1 \leq j \leq n}$ of elements of $\mathcal{S} - \{1\}$ which generates $\mathcal{S}$, and a sufficiently small contractible open neighborhood $C \subset \mathrm{Hom}(\mathcal{S}^{\mathrm{gp}}, \mathbf{S}^1)$ of $1 \in \mathrm{Hom}(\mathcal{S}^{\mathrm{gp}}, \mathbf{S}^1)$. If we take $c > 0$ sufficiently small and define

$$W := \{(r, u) \in \mathrm{Hom}(\mathcal{S}, \mathbf{R}_{\geq 0}^{\mathrm{mult}}) \times C \mid r(a_j) < c \ \ (\forall j)\} \subset X^{\log},$$

then W is a contractible open neighborhood of y such that $W \subset V$, and $W \cap U = \{(r, u) \in W \mid r(a_j) \neq 0 \ \ (\forall j)\}$ is contractible. □

In [KjNc], it is shown that, under the assumption of Proposition 2.3.5, $X^{\log}$ is a topological manifold with the boundary $X^{\log} - U$.

2.3.6

The *canonical extension of Deligne* in [D1] can be understood from our logarithmic point of view, as follows. Let X be a logarithmically smooth fs logarithmic analytic space. Let L be a locally constant sheaf of finite-dimensional $\mathbf{C}$-vector spaces on $U = X_{\mathrm{triv}}$ with unipotent local monodromy along X-U. Let L' be the unique locally constant sheaf on $X^{\log}$ whose restriction on U is L (2.3.5). Then the local monodromy of L' is unipotent in the sense of 2.3.1. The canonical extension of Deligne, $\mathcal{M}$, of $\mathcal{O}_U \otimes_{\mathbf{C}} L$ to X is $\tau_*(\mathcal{O}_X^{\log} \otimes_{\mathbf{C}} L')$. The canonical extension $\mathcal{M}$ has a connection with logarithmic poles $\nabla : \mathcal{M} \to \omega_X^1 \otimes_{\mathcal{O}_X} \mathcal{M}$. This ∇ is induced by $d \otimes 1 : \mathcal{O}_X^{\log} \otimes_{\mathbf{C}} L' \to \omega_X^{1,\log} \otimes_{\mathbf{C}} L'$.

2.3.7

The isomorphism ν in 2.3.2 appears locally on X depending on the local choice of $(q_j)_j$. Here we see that, in the case $X = \mathrm{Spec}(\mathbf{C}[\mathcal{S}])_{\mathrm{an}}$, a canonical ν exists globally on X.

Let $\mathcal{S}$ be an fs monoid, $X = \mathrm{Spec}(\mathbf{C}[\mathcal{S}])_{\mathrm{an}}$ with the canonical logarithmic structure, and let $U = X_{\mathrm{triv}} = \mathrm{Spec}(\mathbf{C}[\mathcal{S}^{\mathrm{gp}}])_{\mathrm{an}}$. Then $U = \mathrm{Hom}(\mathcal{S}^{\mathrm{gp}}, \mathbf{C}^\times)$, and via the exact sequence

$$0 \to \mathrm{Hom}(\mathcal{S}^{\mathrm{gp}}, \mathbf{Z}) \to \mathrm{Hom}(\mathcal{S}^{\mathrm{gp}}, \mathbf{C}) \to \mathrm{Hom}(\mathcal{S}^{\mathrm{gp}}, \mathbf{C}^\times) \to 0$$

(the third arrow is induced from $\mathbf{C} \to \mathbf{C}^\times$, $z \mapsto \exp(2\pi i z)$), $\mathrm{Hom}(\mathcal{S}^{\mathrm{gp}}, \mathbf{C})$ is regarded as a universal covering of U and the fundamental group of U is identified with $\mathrm{Hom}(\mathcal{S}^{\mathrm{gp}}, \mathbf{Z})$.

Let A be a subring of $\mathbf{C}$, let L be a locally constant sheaf on $X^{\log}$ of free A-modules of finite rank with unipotent local monodromy, and let L_0 be the stalk of L at the unit point $1 \in U = \mathrm{Hom}(\mathcal{S}^{\mathrm{gp}}, \mathbf{C}^\times)$ regarded as a constant sheaf on $X^{\log}$. Then there is a unique isomorphism of $\mathcal{O}_X^{\log}$-modules

$$\nu : \mathcal{O}_X^{\log} \otimes_A L \xrightarrow{\sim} \mathcal{O}_X^{\log} \otimes_A L_0$$

satisfying the following conditions (1) and (2) for *any* finite family $(q_j)_{1 \le j \le n}$ of elements of $\mathcal{S}^{\mathrm{gp}}$ which is a $\mathbf{Z}$-basis of $\mathcal{S}^{\mathrm{gp}}/(\text{torsion})$. Let $(\gamma_j)_{1 \le j \le n}$ be the $\mathbf{Z}$-basis of $\pi_1(U, 1) = \mathrm{Hom}(\mathcal{S}^{\mathrm{gp}}, \mathbf{Z})$ which is dual to $(q_j)_{1 \le j \le n}$, and let $N_j : L_{0,\mathbf{Q}} \to L_{0,\mathbf{Q}}$ be the logarithm of γ_j.

(1) $\nu(1 \otimes L) = \xi^{-1}(1 \otimes L_0)$ with $\nu = \exp(\sum_{j=1}^n (2\pi i)^{-1} \log(q_j) \otimes N_j)$.

(2) Let $\log(q_j)_{1,0}$ be the branch of the germ of $\log(q_j)$ at $1 \in U$ which has the value 0 at 1, and let $\xi_{1,0} = \exp(\sum_{j=1}^n (2\pi i)^{-1} \log(q_j)_{1,0} \otimes N_j)$. Then the map $\xi_{1,0} \circ \nu_y : 1 \otimes L_y \to 1 \otimes L_0$ is the identity map.

The proof is similar to that of 2.3.2. First fix $(q_j)_{1 \le j \le n}$. For the locally constant subsheaf $L' = \xi^{-1}(1 \otimes L_0)$ of $\mathcal{O}_X^{\log} \otimes L_0$, the isomorphism $\xi_{0,1}^{-1} : L_1 \to L'_1$ of stalks preserves the actions of $\pi_1(X^{\log}, 1) \simeq \pi_1(U, 1)$, and it is extended uniquely to an isomorphism $\nu : L \xrightarrow{\sim} L'$ on $X^{\log}$. This induces an isomorphism of $\mathcal{O}_X^{\log}$-modules $\nu : \mathcal{O}^{\log} \otimes_A L \xrightarrow{\sim} \mathcal{O}_X^{\log} \otimes_A L' = \mathcal{O}_X^{\log} \otimes_A L_0$. It is easy to check that ν is independent of the choice of $(q_j)_{1 \le j \le n}$.

2.3.8

The following variant of 2.3.7 appears in 2.5.15. Let $X = \Delta^n$, $U = (\Delta^*)^n$, and endow X with the logarithmic structure associated with the divisor $X - U$. Let L be a locally constant sheaf on $X^{\log}$ of free A-modules of finite rank and assume that the local monodromy of L along $X - U$ is unipotent. Let $p \in U$ and fix a lifting $\tilde{p} \in \mathfrak{h}^n$ of p ($\mathfrak{h}$ is the upper half plane) for the surjection $\mathfrak{h}^n \to (\Delta^*)^n$, $(z_j)_{1 \le j \le n} \mapsto (\exp(2\pi i z_j))_{1 \le j \le n}$. Let L_0 be the stalk L_p regarded as a constant sheaf on $X^{\log}$. Then there is a unique isomorphism of $\mathcal{O}_X^{\log}$-modules

$$\nu : \mathcal{O}_X^{\log} \otimes_A L \xrightarrow{\sim} \mathcal{O}_X^{\log} \otimes L_0$$

satisfying the following conditions (1) and (2).

(1) $\nu(1 \otimes L) = \xi^{-1}(1 \otimes L_0)$ where $\xi = \exp(\sum_{j=1}^n (2\pi i)^{-1} \log(q_j) \otimes N_j)$ with $(q_j)_{1 \le j \le n}$ the standard coordinate of Δ^n, $N_j = \log(\gamma_j)$ with $(\gamma_j)_{1 \le j \le n}$ the standard basis of $\pi_1(X^{\log}) \simeq \pi_1((\Delta^*)^n)$ (in the counterclockwise direction).

(2) The map $\xi_{p,0} \circ \nu_p : 1 \otimes L_p \xrightarrow{\sim} 1 \otimes L_0$ is the identity map where $\xi_{p,0}$ is the branch of ξ_p defined by the branch of $((2\pi i)^{-1} \log(q_j)_p)_{1 \le j \le n}$ whose values at p is $\tilde{p}$.

The proof is essentially the same as that of 2.3.7.

2.4 POLARIZED LOGARITHMIC HODGE STRUCTURES

In this section, X denotes an object of $\overline{\mathcal{A}}_1(\log)$ (2.2.1). Here we define the notion "polarized logarithmic Hodge structure (PLH) on X."

We will define the notion of a "logarithmic Hodge structure on X" later in Section 2.6.

Let w and $(h^{p,q}) = (h^{p,q})_{p,q \in \mathbf{Z}}$ be as in Section 0.7.

Definition 2.4.1 *A* prelogarithmic Hodge structure (pre-LH) *on X of weight w and of Hodge type $(h^{p,q})$ is a pair $(H_{\mathbf{Z}}, F)$ consisting of a locally constant sheaf of free $\mathbf{Z}$-modules $H_{\mathbf{Z}}$ of rank $\sum_{p,q} h^{p,q}$ on $X^{\log}$ and of a decreasing filtration F of the $\mathcal{O}_X^{\log}$-module $\mathcal{O}_X^{\log} \otimes H_{\mathbf{Z}}$, which satisfy the following condition.*

(1) *There exist an $\mathcal{O}_X$-module $\mathcal{M}$ on X and a decreasing filtration $(\mathcal{M}^p)_{p \in \mathbf{Z}}$ on $\mathcal{M}$ by $\mathcal{O}_X$-submodules such that $\mathcal{M}^p/\mathcal{M}^{p+1}$ is locally free of rank $h^{p,w-p}$ for each p, $\tau^*\mathcal{M} = \mathcal{O}_X^{\log} \otimes_{\mathbf{Z}} H_{\mathbf{Z}}$, $\tau^*(\mathcal{M}^p) = F^p$.*

Here τ^* is the module-theoretic inverse image $\mathcal{O}_X^{\log} \otimes_{\tau^{-1}(\mathcal{O}_X)} \tau^{-1}(\ )$.

Proposition 2.4.2 *Let $(H_{\mathbf{Z}}, F)$ be a pre-LH on X. Then the local monodromy of $H_{\mathbf{Z}}$ is unipotent.*

Proof. This follows from 2.3.3 (ii). □

Definition 2.4.3 *A* prepolarized logarithmic Hodge structure (pre-PLH) *on X of weight w and of Hodge type $(h^{p,q})$ is a triple $(H_{\mathbf{Z}}, \langle\ ,\ \rangle, F)$ consisting of a pre-LH $(H_{\mathbf{Z}}, F)$ and of a nondegenerate $\mathbf{Q}$-bilinear form $\langle\ ,\ \rangle$ on $H_{\mathbf{Q}} = \mathbf{Q} \otimes H_{\mathbf{Z}}$, symmetric for even w and antisymmetric for odd w, which satisfy the following condition.*

(1) *$\langle F^p, F^q \rangle = 0$ if $p + q > w$.*

Here $\langle\ ,\ \rangle$ is regarded as the natural extension to $\mathcal{O}_X^{\log}$-bilinear form.

For a morphism $Y \to X$ in $\overline{\mathcal{A}}_1(\log)$ and for a pre-LH (resp. pre-PLH) on X, its inverse image on Y, which is a pre-LH (resp. pre-PLH) on Y, is defined evidently.

Proposition 2.4.4

(i) *Let $\mathcal{H}$ be the category of pre-LH on X of weight w and of Hodge type $(h^{p,q})$, and let $\mathcal{H}'$ be the category of triples $(H_{\mathbf{Z}}, \mathcal{M}, \iota)$ where $H_{\mathbf{Z}}$ is a locally constant sheaf of free $\mathbf{Z}$-modules $H_{\mathbf{Z}}$ of rank $\sum_{p,q} h^{p,q}$ on $X^{\log}$, $\mathcal{M}$ is an $\mathcal{O}_X$-module on X endowed with a decreasing filtration $(\mathcal{M}^p)_p$ by $\mathcal{O}_X$-submodules such that $\mathcal{M}^p/\mathcal{M}^{p+1}$ is* *locally free of rank $h^{p,w-p}$ for each p, and ι is an isomorphism of $\mathcal{O}_X^{\log}$-modules $\tau^*(\mathcal{M}) \xrightarrow{\sim} \mathcal{O}_X^{\log} \otimes_{\mathbf{Z}} H_{\mathbf{Z}}$. Then we have an equivalence of categories*

$$\mathcal{H}' \xrightarrow{\sim} \mathcal{H}, \quad (H_{\mathbf{Z}}, \mathcal{M}, \iota) \mapsto (H_{\mathbf{Z}}, F), \quad F^p = \iota(\tau^*(\mathcal{M}^p)).$$

The inverse functor is given as $(H_{\mathbf{Z}}, F) \mapsto (H_{\mathbf{Z}}, \mathcal{M}, \iota)$ where $\mathcal{M} = \tau_*(\mathcal{O}_X^{\log} \otimes_{\mathbf{Z}} H_{\mathbf{Z}})$, $\mathcal{M}^p = \tau_*(F^p)$, and ι is the canonical isomorphism $\mathcal{O}_X^{\log} \otimes_{\tau^{-1}(\mathcal{O}_X)} \tau^{-1}(\mathcal{M}) \xrightarrow{\sim} \mathcal{O}_X^{\log} \otimes_{\mathbf{Z}} H_{\mathbf{Z}}$.

(ii) *Let $\mathcal{P}$ be the category of pre-PLH on X of weight w and of Hodge type $(h^{p,q})$, and let $\mathcal{P}'$ be the category of 4-tuples $(H_{\mathbf{Z}}, \langle\ ,\ \rangle, \mathcal{M}, \iota)$ where $(H_{\mathbf{Z}}, \mathcal{M}, \iota)$ is an object of $\mathcal{H}'$ and $\langle\ ,\ \rangle$ is a nondegenerate $\mathbf{Q}$-bilinear form on $H_{\mathbf{Q}} = \mathbf{Q} \otimes H_{\mathbf{Z}}$, symmetric for even w and antisymmetric for odd w, which satisfy the condition $\langle \tau^*\mathcal{M}^p, \tau^*\mathcal{M}^q \rangle = 0$ if $p+q > w$. Then we have an equivalence of categories*

$$\mathcal{P}' \xrightarrow{\sim} \mathcal{P}, \quad (H_{\mathbf{Z}}, \langle\ ,\ \rangle, \mathcal{M}, \iota) \mapsto (H_{\mathbf{Z}}, \langle\ ,\ \rangle, F), \quad F^p = \iota(\tau^*(\mathcal{M}^p)).$$

The inverse functor is given in a similar manner as in (i).

Proof. This follows from 2.4.2 and 2.3.3 (i). □

2.4.5

Let X be an fs logarithmic analytic space. Let $H = (H_{\mathbf{Z}}, F)$ be a pre-LH on X. Let

$$d \otimes 1 = d \otimes 1_{H_{\mathbf{C}}} : \mathcal{O}_X^{\log} \otimes_{\mathbf{C}} H_{\mathbf{C}} \to \omega_X^{1,\log} \otimes_{\mathbf{C}} H_{\mathbf{C}}.$$

We say that H satisfies the *Griffiths transversality over* X if

$$(d \otimes 1)(F^p) \subset \omega_X^{1,\log} \otimes_{\mathcal{O}_X^{\log}} F^{p-1} \quad \text{for all } p.$$

2.4.6

Let $x \in X$ (X is an object of $\overline{\mathcal{A}}_1(\log)$) and let $M_x \to \mathcal{O}_x = \mathbf{C}$ be the fs logarithmic structure on x induced from that on X. Let $(x^{\log}, \mathcal{O}_x^{\log})$ be the associated ringed space. For $y \in x^{\log} = \tau^{-1}(x)$, we define

$$\operatorname{sp}(y) := \operatorname{Hom}_{\mathbf{C}\text{-alg}}(\mathcal{O}_{x,y}^{\log}, \mathbf{C}). \tag{1}$$

For $s \in \operatorname{sp}(y)$, we have a well-defined homomorphism

$$\tilde{s} : M_x^{\mathrm{gp}} \to \mathbf{C}^\times, \quad a \mapsto \exp(s(\log(a))), \tag{2}$$

where $\log(a)$ is defined in $\mathcal{L}_y/(2\pi i\mathbf{Z})$ and its image in $\mathcal{O}_{x,y}^{\log}/(2\pi i\mathbf{Z})$ is also denoted by $\log(a)$ (see 2.2.4).

For a pre-LH $(H_{\mathbf{Z}}, F)$ on X, for $x \in X$, for $y \in x^{\log}$ and for $s \in \operatorname{sp}(y)$, we have a decreasing filtration

$$F(s) := \left(\mathbf{C} \otimes_{\mathcal{O}_{x,y}^{\log}} (F^p|_{x^{\log}})_y\right)_{p\in\mathbf{Z}}, \quad \text{with } s : \mathcal{O}_{x,y}^{\log} \to \mathbf{C}, \tag{3}$$

on the $\mathbf{C}$-vector space $H_{\mathbf{C},y} = \mathbf{C} \otimes_{\mathbf{Z}} H_{\mathbf{Z},y}$, where $F^p|_{x^{\log}} = \mathcal{O}_x^{\log} \otimes_{k^{-1}(\mathcal{O}_X^{\log})} k^{-1}(F^p)$ with $k : x^{\log} \hookrightarrow X^{\log}$.

DEFINITION 2.4.7 *Let x be an fs logarithmic point* (2.1.9). *A pre-PLH $(H_{\mathbf{Z}}, \langle\ ,\ \rangle, F)$ on x of weight w is called a* polarized logarithmic Hodge structure (PLH) *on x of weight w, if it satisfies the following two conditions.*

(1) *$(H_{\mathbf{Z}}, F)$ satisfies the Griffiths transversality over x* (*cf.* 2.4.5).
(2) (Positivity) *Let $y \in x^{\log}$ and $s \in \mathrm{sp}(y)$. If $\tilde{s} : M_x \to \mathbf{C}$ (see* 2.4.6 (2)) *is sufficiently near to the structure morphism of the logarithmic structure $\alpha : M_x \to \mathbf{C}$ in the topology of simple convergence of $\mathbf{C}$-valued functions, then $(H_{\mathbf{Z},y}, \langle\ ,\ \rangle_y, F(s))$ is a polarized Hodge structure of weight w in the usual sense* 1.1.2.

Note that the validity of the condition 2.4.7 (2) is independent of the choice of y. In fact, let $y, y' \in x^{\log}$. Since $\mathcal{O}_x^{\log}$ is a locally constant sheaf, a path from y to y' in $x^{\log}$ induces an isomorphism $\mathcal{O}_{x,y'}^{\log} \xrightarrow{\sim} \mathcal{O}_{x,y}^{\log}$ by rewinding the path (cf. Appendix A1). This induces a bijection $\mathrm{sp}(y) \to \mathrm{sp}(y')$ which preserves the corresponding morphisms $M_x^{\mathrm{gp}} \to \mathbf{C}^\times$ (see 2.4.6 (2)). Furthermore, if $(H_{\mathbf{Z}}, F)$ is a pre-LH on x, the path also induces an isomorphism $H_{\mathbf{Z},y'} \xrightarrow{\sim} H_{\mathbf{Z},y}$. The induced isomorphism $H_{\mathbf{C},y'} \xrightarrow{\sim} H_{\mathbf{C},y}$ sends $F(s')$ to $F(s)$, if $\mathrm{sp}(y) \to \mathrm{sp}(y')$ sends s to s'.

DEFINITION 2.4.8 *A pre-PLH $(H_{\mathbf{Z}}, \langle\ ,\ \rangle, F)$ on X is a* polarized logarithmic Hodge structure (PLH) *on X if, for any $x \in X$, the inverse image of $(H_{\mathbf{Z}}, \langle\ ,\ \rangle, F)$ on x is a PLH. Here we regard x as an fs logarithmic point endowed with the inverse image of the logarithmic structure of X.*

2.4.9

Assume that X is an fs logarithmic analytic space. Let $(H_{\mathbf{Z}}, F)$ be a pre-LH on X. We have two types of Griffiths transversality for $(H_{\mathbf{Z}}, F)$. One is the Griffiths transversality over X as in 2.4.5, which we call the *big Griffiths transversality*. The other is the condition that the inverse image of $(H_{\mathbf{Z}}, F)$ on each $x \in X$ satisfies the Griffiths transversality over x, which we call the *small Griffiths transversality*.

If the logarithmic structure of X is trivial, then, for any $x \in X$, the sheaf ω_x^1 of logarithmic 1-forms on x is 0 and hence any pre-LH on x satisfies the small Griffiths transversality automatically. For a pre-LH $(H_{\mathbf{Z}}, F)$ on X, the big Griffiths transversality is much stronger than the small Griffiths transversality.

When X is a smooth analytic space with trivial logarithmic structure, a PLH on X satisfying the big Griffiths transversality is nothing but a variation of polarized Hodge structure. So, when X is a logarithmically smooth fs logarithmic analytic space (2.1.11), we call a PLH on X satisfying the big Griffiths transversality a *logarithmic variation of polarized Hodge structure* (*LVPH*).

2.4.10

We already explained in 0.2.21 how LVPH appears in geometry. The LVPH in 0.2.21 is a higher direct image of a constant LVPH. Higher direct images of more general LVPHs are discussed in [KMN].

2.5 NILPOTENT ORBITS AND PERIOD MAPS

Here we show that the notion "PLH on an fs logarithmic point" is essentially equivalent to the notion "nilpotent orbit." We also discuss period maps of PLH. At the end of this section, we give a PLH-theoretic interpretation of the nilpotent orbit theorem of Schmid [Sc].

In this section, we mainly work on an object of $\overline{\mathcal{A}}_1(\log)$ (2.2.1).

PROPOSITION 2.5.1 *Let x be an fs logarithmic point* (2.1.9) *and let $H = (H_{\mathbf{Z}}, \langle\ ,\ \rangle, F)$ be a pre-PLH on x. Let $y \in x^{\log}$, let $s, s' \in \mathrm{sp}(y)$, and write the homomorphism*

$$M_x^{\mathrm{gp}}/\mathcal{O}_x^\times = M_x^{\mathrm{gp}}/\mathbf{C}^\times \to \mathbf{C}, \quad f \mapsto (2\pi i)^{-1}(s'(\log(f)) - s(\log(f))),$$

in the form $\sum_{j=1}^n z_j h_{\gamma_j}$ with $z_j \in \mathbf{C}$ and $\gamma_j \in \pi_1(x^{\log})$, where $h_{\gamma_j} : M_x^{\mathrm{gp}}/\mathcal{O}_x^\times \to \mathbf{Z}$ is the homomorphism corresponding to γ_j (2.2.9). *Then*

$$F(s') = \exp\left(\sum_{j=1}^n z_j N_j\right) F(s),$$

where $N_j : H_{\mathbf{Q},y} \to H_{\mathbf{Q},y}$ is the logarithm of the action of γ_j.

Proof. Take a family $(q_j)_{1\le j\le n}$ of elements of M_x^{gp} whose image in $M_x^{\mathrm{gp}}/\mathcal{O}_x^\times$ is a $\mathbf{Z}$-basis, and let $(\gamma_j)_{1\le j\le n}$ be the dual basis of $\pi_1(x^{\log})$. Then the above homomorphism $M_x/\mathcal{O}_x^\times \to \mathbf{C}$ has the form $\sum_{j=1}^n (2\pi i)^{-1}(s'(\log(q_j)) - s(\log(q_j)))h_{\gamma_j}$. Hence it is sufficient to prove

$$F(s') = \xi_y(s')\xi_y(s)^{-1}F(s) \tag{1}$$

where ξ is as in 2.3.2. Let the isomorphism $\nu : \mathcal{O}_x^{\log} \otimes H_{\mathbf{Z}} \xrightarrow{\sim} \mathcal{O}_x^{\log} \otimes H_{\mathbf{Z},y}$ be as in 2.3.2 and let $\xi_{y,0}$ be the branch of ξ_y for which $1 \otimes v = \nu^{-1} \circ \xi_{y,0}^{-1}(1 \otimes v)$ for any $v \in H_{\mathbf{Z},y}$. By 2.4.4, the filtration F has the form $\nu^{-1}(\mathcal{O}_x^{\log} \otimes_{\mathbf{C}} F_0)$ for some filtration F_0 on the $\mathbf{C}$-vector space $H_{\mathbf{C},y}$. Let $\nu(s) : H_{\mathbf{C},y} \to H_{\mathbf{C},y}$ be the isomorphism obtained from ν by applying s. Since $\nu^{-1}(s) \circ \xi_{y,0}^{-1}(s) : H_{\mathbf{C},y} \to H_{\mathbf{C},y}$ is the identity map, we have

$$F(s) = \nu(s)^{-1}F_0 = \xi_{0,y}(s)F_0 \tag{2}$$

and (1) follows from this. □

2.5.2

Γ-level structure. Let $\Phi_0 = (w, (h^{p,q}), H_0, \langle\ ,\ \rangle_0)$ be as in Section 0.7 and let Γ be a subgroup of $G_{\mathbf{Z}}$. Let X be an object of $\overline{\mathcal{A}}_1(\log)$ (2.2.1), and let $H = (H_{\mathbf{Z}}, \langle\ ,\ \rangle, F)$ be a pre-PLH on X of weight w and of Hodge type $(h^{p,q})$. A *Γ-level structure* on H is a global section of the sheaf

$$\Gamma\backslash \underline{\mathrm{Isom}}((H_{\mathbf{Z}}, \langle\ ,\ \rangle), (H_0, \langle\ ,\ \rangle_0))$$

on $X^{\log}$. Here H_0 is regarded as a constant sheaf on $X^{\log}$, $\underline{\mathrm{Isom}}$ is the sheaf of isomorphisms, and $\gamma \in \Gamma$ acts on $\underline{\mathrm{Isom}}(\cdots)$ by $h \mapsto \gamma \circ h$.

Example 1. *Assume that w is odd and $\langle\ ,\ \rangle_0$ is a perfect pairing $H_0 \times H_0 \to \mathbf{Z}$ of $\mathbf{Z}$-modules (i.e., $H_0 \xrightarrow{\sim} \mathrm{Hom}_{\mathbf{Z}}(H_0, \mathbf{Z})$ by this pairing), let $n \geq 1$, and let*

$$\Gamma = \{\gamma \in G_{\mathbf{Z}} \mid \gamma \text{ acts on } H_0/nH_0 \text{ trivially}\}.$$

Then, if $(H_{\mathbf{Z}}, \langle\ ,\ \rangle)$ has a Γ-level structure, $\langle\ ,\ \rangle$ is a perfect pairing $H_{\mathbf{Z}} \times H_{\mathbf{Z}} \to \mathbf{Z}$ of local systems of $\mathbf{Z}$-modules. If $\langle\ ,\ \rangle$ is a perfect pairing $H_{\mathbf{Z}} \times H_{\mathbf{Z}} \to \mathbf{Z}$ of local systems of $\mathbf{Z}$-modules, a Γ-level structure on $(H_{\mathbf{Z}}, \langle\ ,\ \rangle)$ is equivalent to an isomorphism of local systems $H_{\mathbf{Z}}/nH_{\mathbf{Z}} \simeq H_0/nH_0$ (usually called an n-level structure) which sends the pairing $(H_{\mathbf{Z}}/nH_{\mathbf{Z}}) \times (H_{\mathbf{Z}}/nH_{\mathbf{Z}}) \to \mathbf{Z}/n\mathbf{Z}$ induced by $\langle\ ,\ \rangle$ to the pairing $(H_0/nH_0) \times (H_0/nH_0) \to \mathbf{Z}/n\mathbf{Z}$ induced by $\langle\ ,\ \rangle_0$. (This follows from the surjectivity of $\mathrm{Sp}(g, \mathbf{Z}) \to \mathrm{Sp}(g, \mathbf{Z}/n\mathbf{Z})$.)

Example 2. *Let X be an object of $\mathcal{A}_1(\log)$, let H be a pre-PLH on X, let $x \in X$, let y be a point of $X^{\log}$ lying over x, and let $(H_0, \langle\ ,\ \rangle_0) = (H_{\mathbf{Z},y}, \langle\ ,\ \rangle_y)$. Let Γ be the image of the local monodromy action $\mathrm{Image}(\pi_1(x^{\log}) \to G_{\mathbf{Z}})$. Then, by 2.3.2, if we replace X by an open neighborhood of x, we have the isomorphism $\xi \circ \nu$ from $(H_{\mathbf{Z}}, \langle\ ,\ \rangle)$ to $(H_0, \langle\ ,\ \rangle_0)$ determined modulo the choices of the branches of ξ. This $\xi \circ \nu$ is regarded as a Γ-level structure of H.*

2.5.3

We define the period map associated with a pre-PLH endowed with a Γ-level structure. (See 2.5.10 for another more sophisticated period map.)

Let the notation be as in 2.5.2, and let $H = (H_{\mathbf{Z}}, \langle\ ,\ \rangle, F)$ be a pre-PLH on X of weight w and of Hodge type $(h^{p,q})$ endowed with a Γ-level structure.

Let $\check{D}_{\mathrm{orb}}$ be as in 1.3.6.

We define the *period map*

$$\check{\varphi} : X \to \Gamma\backslash\check{D}_{\mathrm{orb}}$$

associated with H. Here the quotient $\Gamma\backslash\check{D}_{\mathrm{orb}}$ is defined with respect to the action of Γ on $\check{D}_{\mathrm{orb}}$,

$$\gamma : (\sigma, Z) \mapsto (\mathrm{Ad}(\gamma)\sigma, \gamma Z) \quad (\gamma \in \Gamma, (\sigma, Z) \in \check{D}_{\mathrm{orb}}).$$

Note that this is only a set-theoretic map.

Let $x \in X$, and take $y \in X^{\log}$ lying over x, and let $\tilde{\mu}_y : (H_{\mathbf{Z},y}, \langle\ ,\ \rangle_y) \xrightarrow{\sim} (H_0, \langle\ ,\ \rangle_0)$ be a lifting of the germ μ_y of the Γ-level structure μ at y. Then, via $\tilde{\mu}_y$, the action of $\pi_1(x^{\log})$ on $H_{\mathbf{Z},y}$ defines a unipotent action of $\pi_1(x^{\log})$ on H_0 preserving $\langle\ ,\ \rangle_0$. Let σ be the cone in $\mathfrak{g}_{\mathbf{R}}$ generated by the logarithms of the actions of elements of $\pi_1^+(x^{\log})$ on H_0. By Proposition 2.5.1, the set $\{\tilde{\mu}_y(F(s)) \mid s \in \mathrm{sp}(y)\} \subset \check{D}$ of filtrations on $H_{0,\mathbf{C}}$ is an $\exp(\sigma_{\mathbf{C}})$-orbit. We define

$$\check{\varphi}(x) := ((\sigma, Z) \bmod \Gamma) \in \Gamma\backslash\check{D}_{\mathrm{orb}}.$$

In fact, $\check{\varphi}(x)$ is independent of the choices of y and $\tilde{\mu}_y$: If we fix y and change $\tilde{\mu}_y$, the element (σ, Z) of $\check{D}_{\mathrm{orb}}$ is changed by the action of Γ, and hence $((\sigma, Z) \bmod \Gamma)$ does not change. If we replace y by another element y' of $\tau^{-1}(x)$, a path from y to

y' sends $\tilde{\mu}_y$ to a lifting $\tilde{\mu}_{y'}$ of μ at y' and sends s to an element s' of $\mathrm{sp}(y')$, and we have $\tilde{\mu}_y(F(s)) = \tilde{\mu}_{y'}(F(s'))$.

The proof of the following lemma is straightforwards and we omit it.

Lemma 2.5.4 *Let X and H be as in 2.5.3, let X' be an object of $\overline{\mathcal{A}}_1(\log)$ over X, and let H' be the inverse image of H on X'. Consider the period maps*

$$\check{\varphi} : X \to \Gamma\backslash\check{D}_{\mathrm{orb}}, \quad \check{\varphi}' : X' \to \Gamma\backslash\check{D}_{\mathrm{orb}}$$

associated with H and H', respectively. Let $x' \in X'$ and let $x \in X$ be the image of x'.

(i) *Let*

$$\check{\varphi}(x) = ((\sigma, Z) \bmod \Gamma), \quad \check{\varphi}'(x) = ((\sigma', Z') \bmod \Gamma)$$

where $(\sigma', Z') \in \check{D}_{\mathrm{orb}}$ is defined by using a point $y' \in (X')^{\log}$ lying over x' and using a lifting $\tilde{\mu}'_{y'}$ of the germ of the inverse image μ' of μ on $(X')^{\log}$ at y', and $(\sigma, Z) \in \check{D}_{\mathrm{orb}}$ is defined by using the image $y \in X^{\log}$ of y' and using the unique lifting $\tilde{\mu}_y$ of μ_y which induces $\tilde{\mu}'_{y'}$. Then

$$\sigma' \subset \sigma, \quad \textit{and} \quad Z' \subset Z = \exp(\sigma_{\mathbf{C}})Z'.$$

(ii) *If $X' \to X$ is strict (i.e., the logarithmic structure of X' coincides with the inverse image of that of X), then $\check{\varphi}(x) = \check{\varphi}'(x')$.*

Proposition 2.5.5 *Let X be an object of $\overline{\mathcal{A}}_1(\log)$, let Γ be a subgroup of $G_{\mathbf{Z}}$, let H be a pre-PLH on X of weight w and of Hodge type $(h^{p,q})$ endowed with a Γ-level structure, and let $\check{\varphi} : X \to \Gamma\backslash\check{D}_{\mathrm{orb}}$ be the associated period map. Let $x \in X$ and write $\check{\varphi}(x) = ((\sigma, Z) \bmod \Gamma)$ with $(\sigma, Z) \in \check{D}_{\mathrm{orb}}$. Take $F \in Z$.*

(i) *The inverse image of H on the fs logarithmic point x satisfies the Griffiths transversality 2.4.7 (1) if and only if*

$$NF^p \subset F^{p-1} \quad \textit{for all } N \in \sigma \textit{ and for all } p.$$

(ii) *Write $\sigma = \sum_{j=1}^{n}(\mathbf{R}_{\geq 0})N_j$. Then the inverse image of H on x satisfies the positivity condition 2.4.7 (2) if and only if*

$$\exp\left(\sum_{j=1}^{n} z_j N_j\right) F \in D \quad \textit{for } \operatorname{Im}(z_j) \gg 0. \tag{1}$$

(iii) *The inverse image of H on x is a PLH if and only if (σ, Z) is a nilpotent orbit.*

Proof. Note that (iii) follows from (i) and (ii).

By Lemma 2.5.4 (ii), we may assume that X is the fs logarithmic point x. Take $y \in x^{\log}$.

We prove (i). Let $\mathcal{M} = \Gamma(x^{\log}, \mathcal{O}_x^{\log} \otimes_{\mathbf{Z}} H_{\mathbf{Z}}) = \tau_*(\mathcal{O}_x^{\log} \otimes_{\mathbf{C}} \mathcal{M})$, and let $\mathcal{M}^p = \Gamma(x^{\log}, F^p) \subset \mathcal{M}$. Then $\mathcal{O}_x^{\log} \otimes_{\mathbf{Z}} H_{\mathbf{Z}} = \mathcal{O}_x^{\log} \otimes_{\mathbf{C}} \mathcal{M}$ and $F^p = \mathcal{O}_x^{\log} \otimes_{\mathbf{C}}$

$\mathcal{M}^p$ (2.4.4). Hence

$$(d\otimes 1)(F^p)\subset \omega_x^{1,\log}\otimes F^{p-1}\ (\forall\, p) \iff \nabla(\mathcal{M}^p)\subset \omega_x^1\otimes_{\mathbf{C}}\mathcal{M}^{p-1}\ (\forall\, p),$$

where $\nabla:\mathcal{M}\to\omega_x^1\otimes_{\mathbf{C}}\mathcal{M}$ is the $\mathbf{C}$-linear map induced by $d\otimes 1$. Take $s\in \mathrm{sp}(y)$. Then the composition $\mathcal{M}\to\mathcal{O}_x^{\log}\otimes_{\mathbf{Z}} H_{\mathbf{Z}}\xrightarrow{s} H_{\mathbf{C},y}$ is bijective and induces $\mathcal{M}^p\xrightarrow{\sim} F^p(s)$ for all p. Furthermore, by 2.3.4 (ii), via this composite map, $\nabla:\mathcal{M}\to\omega_x^1\otimes_{\mathbf{C}}\mathcal{M}$ corresponds to $\mathcal{N}':H_{\mathbf{C},y}\to\omega_x^1\otimes_{\mathbf{C}}H_{\mathbf{C},y}$. Hence we have

$$\begin{aligned}\nabla(\mathcal{M}^p)\subset\omega_x^1\otimes_{\mathbf{C}}\mathcal{M}^{p-1}\ (\forall p) &\iff \mathcal{N}'(F^p(s))\subset\omega_x^1\otimes_{\mathbf{C}}F^{p-1}(s)\ (\forall\, p)\\ &\iff \mathcal{N}(F^p(s))\subset(M_x^{\mathrm{gp}}/\mathcal{O}_x^\times)\otimes F^{p-1}(s)\ (\forall\, p)\\ &\iff N(F^p(s))\subset F^{p-1}(s)\ (\forall N\in\sigma,\forall\, p).\end{aligned}$$

Here the second $\iff$ follows from $\mathbf{C}\otimes_{\mathbf{Z}}(M_x^{\mathrm{gp}}/\mathcal{O}_x^\times)\xrightarrow{\sim}\omega_x^1$ (2.1.9).

We prove (ii). Fix a finite subset $\{q_1,\dots,q_m\}$ of $M_x-\mathcal{O}_x^\times$ whose image in $M_x/\mathcal{O}_x^\times$ generates $M_x/\mathcal{O}_x^\times$. Fix $s_0\in\mathrm{sp}(y)$. Then, for $s\in\mathrm{sp}(y)$ varying, we have

$$\begin{aligned}&(\tilde{s}:M_x\to\mathbf{C}\text{ converges to }\alpha\text{ for the topology of simple convergence})\\ &\iff(\tilde{s}(q_j)\tilde{s}_0(q_j)^{-1}\to 0\text{ for any }j)\\ &\iff(\mathrm{Im}((2\pi i)^{-1}(s(\log(q_j))-s_0(\log(q_j))))\text{ tends to }\infty\text{ for any }j).\end{aligned}$$

On the other hand, take a finite family $(h_j)_{1\le j\le n}$ of elements of $\mathrm{Hom}(M_x/\mathcal{O}_x^\times,\mathbf{N})$ which generates the monoid $\mathrm{Hom}(M_x/\mathcal{O}_x^\times,\mathbf{N})$. Then $\sigma=\sum_{j=1}^n(\mathbf{R}_{\ge 0})N_j$ where N_j is the logarithm of the local monodromy of the element of $\pi_1^+(x^{\log})$ corresponding to h_j. By 2.5.1, we have $F(s)=\exp(\sum_{j=1}^n z_jN_j)F(s_0)$ for the element s of $\mathrm{sp}(y)$ characterized by $(2\pi i)^{-1}(s(\log(q_j))-s_0(\log(q_j)))=\sum_{j=1}^n z_jh_j(q_j)$ for $1\le j\le n$. Hence the proof of (ii) is reduced to the following lemma which we apply by taking $M_x/\mathcal{O}_x^\times$ as $\mathcal{S}$ and by taking the homomorphism $M_x^{\mathrm{gp}}/\mathcal{O}_x^\times\to\mathbf{R}$, $f\mapsto\mathrm{Im}((2\pi i)^{-1}(s(\log(f))-s_0(\log(f))))$ as φ. □

Lemma 2.5.6 *Let $\mathcal{S}$ be an fs monoid such that $\mathcal{S}^\times=\{1\}$, let q_j $(1\le j\le m)$ be elements of $\mathcal{S}-\{1\}$ which generate $\mathcal{S}$, and let $h_1,\dots,h_n$ be elements of* $\mathrm{Hom}(\mathcal{S},\mathbf{R}_{\ge 0}^{\mathrm{add}})$ *which generate the cone* $\mathrm{Hom}(\mathcal{S},\mathbf{R}_{\ge 0}^{\mathrm{add}})$. *For $c>0$, define subsets A_c and B_c of* $\mathrm{Hom}(\mathcal{S},\mathbf{R}_{\ge 0}^{\mathrm{add}})$ *by*

$$A_c=\left\{\sum_{j=1}^n y_jh_j\mid y_j\ge c\ (\forall\, j)\right\},\quad B_c=\{\varphi\in\mathrm{Hom}(\mathcal{S}^{\mathrm{gp}},\mathbf{R}^{\mathrm{add}})\mid\varphi(q_j)\ge c\ (\forall\, j)\}.$$

Then we have the following.

(i) *Fix $c>0$. Then $A_d\subset B_c$ for some $d>0$.*
(ii) *Fix $c>0$. Then $B_d\subset A_c$ for some $d>0$.*

Proof. We prove (i). Since $h_1,\dots,h_n$ generate the cone $\mathrm{Hom}(\mathcal{S},\mathbf{R}_{\ge 0}^{\mathrm{add}})$ and since $q_j\ne 1$ for any j, $(h_1+\cdots+h_n)(q_j)>0$ for any j. Hence there exist $d>0$ such that $(dh_1+\cdots+dh_n)(q_j)\ge c$ for any j. If $y_j\ge d$ for any j $(1\le j\le n)$, $(y_1h_1+\cdots+y_nh_n)(q_j)\ge(dh_1+\cdots+dh_n)(q_j)\ge c$ for any j.

We prove (ii). Let $d = \max\{(ch_1 + \cdots + ch_n)(q_j) \mid 1 \leq j \leq m\}$. Let $\varphi \in B_d$. Since $(\varphi - \sum_{j=1}^{n} ch_j)(q_j) \geq 0$ for any j, we have $\varphi - \sum_{j=1}^{n} ch_j \in \mathrm{Hom}(\mathcal{S}, \mathbf{R}^{\mathrm{add}}_{\geq 0})$. Hence $\varphi - \sum_{j=1}^{n} ch_j = r_1 h_1 + \cdots + r_n h_n$ for some $r_j \geq 0$. We have $\varphi = \sum_{j=1}^{n} (c + r_j) h_j \in A_c$. □

PROPOSITION 2.5.7 *Let x be an fs logarithmic point. Fix $y \in x^{\log}$ and $s \in \mathrm{sp}(y)$. Let $(w, (h^{p,q})_{p,q}, H_0, \langle\ ,\ \rangle_0)$ be as in Section 0.7, fix a unipotent action ρ of $\pi_1(x^{\log})$ on H_0 preserving $\langle\ ,\ \rangle_0$, and let σ be the cone in $\mathfrak{g}_{\mathbf{R}}$ generated by $\log(\rho(\pi_1^+(x^{\log})))$. Let $(H_{\mathbf{Z}}, \langle\ ,\ \rangle)$ be the local system on $x^{\log}$ whose stalk at y is $(H_0, \langle\ ,\ \rangle_0)$ with monodromy ρ, and let $\mathfrak{F}^+$ (resp. $\mathfrak{F}$) be the set of all decreasing filtrations F on $\mathcal{O}_x^{\log} \otimes H_{\mathbf{Z}}$ such that $(H_{\mathbf{Z}}, \langle\ ,\ \rangle, F)$ is a PLH (resp. pre-PLH) of weight w and of Hodge type $(h^{p,q})$.*

(i) *The map $\mathfrak{F} \to \check{D},\ F \mapsto F(s)$ is bijective.*

(ii) *The map in* (i) *induces a bijection from $\mathfrak{F}^+$ to the set of all $F \in \check{D}$ such that $(\sigma, \exp(\sigma_{\mathbf{C}})F)$ is a nilpotent orbit.*

Proof. Take $(q_j)_{1 \leq j \leq n}$ as in the proof of 2.5.1, fix a branch $\xi_{y,0}$ of ξ_y, and consider the associated isomorphism $\nu : \mathcal{O}_x^{\log} \otimes H_{\mathbf{Z}} \simeq \mathcal{O}_x^{\log} \otimes H_0$. Then we have a bijection $\mathfrak{F} \simeq \check{D}$ in which $F_0 \in \check{D}$ corresponds to $F = \nu^{-1}(\mathcal{O}_x^{\log} \otimes_{\mathbf{C}} F_0) \in \mathfrak{F}$. We have $F(s) = \xi_{y,0}(s)F_0$. Since $\xi_{y,0}(s)$ is an automorphism of $\check{D}$, we see that $\mathfrak{F} \to \check{D},\ F \mapsto F(s)$ is a bijection. Let $\Gamma = \rho(\pi_1(x^{\log})) \subset G_{\mathbf{Z}}$. The period map associated with the pre-PLH $H = (H_{\mathbf{Z}}, \langle\ ,\ \rangle, F)$ with the evident Γ-level structure sends x to $(\sigma, \exp(\sigma_{\mathbf{C}})F(s))$. By 2.5.5, H is a PLH if and only if $(\sigma, \exp(\sigma_{\mathbf{C}})F(s))$ is a nilpotent orbit. □

DEFINITION 2.5.8 *Let Σ be a fan in $\mathfrak{g}_{\mathbf{Q}}$ and let Γ be a subgroup of $G_{\mathbf{Z}}$ which is strongly compatible with Σ* (1.3.10). *Denote by*

$$\Phi := \big(w, (h^{p,q})_{p,q \in \mathbf{Z}}, H_0, \langle\ ,\ \rangle_0, \Gamma, \Sigma\big)$$

the 6*-tuple consisting of the* 4*-tuple $\big(w, (h^{p,q})_{p,q \in \mathbf{Z}}, H_0, \langle\ ,\ \rangle_0\big)$ as in Section 0.7 and of the above Σ and Γ.*

Let X be an object of $\overline{\mathcal{A}}_1(\log)$.

A polarized logarithmic Hodge structure on X of type Φ (PLH on X of type Φ, *or* PLH on X endowed with a Γ-level structure whose local monodromies are in the directions in Σ) *is a pre-PLH $(H_{\mathbf{Z}}, \langle\ ,\ \rangle, F)$ on X of weight w and of Hodge type $(h^{p,q})$, which is endowed with a Γ-level structure μ, satisfying the following condition* (1)*. Let $\check{\varphi} : X \to \Gamma\backslash\check{D}_{\mathrm{orb}}$ be the period map associated to H.*

(1) *For each $x \in X$, if we denote $\check{\varphi}(x) = ((\sigma, Z) \bmod \Gamma)$ with $(\sigma, Z) \in \check{D}_{\mathrm{orb}}$, then there is $\tau \in \Sigma$ such that $\sigma \subset \tau$. Furthermore, if we take the smallest such τ, $(\tau, \exp(\tau_{\mathbf{C}})Z)$ is a nilpotent orbit.*

PROPOSITION 2.5.9 *Let X be an object of $\overline{\mathcal{A}}_1(\log)$. Let Φ be as in* 2.5.8 *and let $(H_{\mathbf{Z}}, \langle\ ,\ \rangle, F, \mu)$ be a PLH of type Φ on X. Then $(H_{\mathbf{Z}}, \langle\ ,\ \rangle, F)$ is a PLH of weight w and of Hodge type $(h^{p,q})$.*

Proof. Let $\check{\varphi} : X \to \Gamma\backslash\check{D}_{\mathrm{orb}}$ be the period map associated with H, let $x \in X$, and write $\check{\varphi}(x) = ((\sigma, Z) \bmod \Gamma)$. Take the smallest $\tau \in \Sigma$ such that $\sigma \subset \tau$. Then

$(\tau, \exp(\tau_{\mathbf{C}})Z)$ is a nilpotent orbit. As σ contains a point of the interior (0.7.7) of τ, this implies that (σ, Z) is a nilpotent orbit. □

2.5.10

Let $H = (H_{\mathbf{Z}}, \langle\ ,\ \rangle, F, \mu)$ be a PLH of type Φ on X (2.5.8). We define maps

$$\varphi : X \to \Gamma\backslash D_{\Sigma}, \tag{1}$$

$$\varphi^{\log} : X^{\log} \to \Gamma\backslash D_{\Sigma}^{\sharp} \tag{2}$$

associated with H as below.

Let $\check{\varphi} : X \to \Gamma\backslash \check{D}_{\mathrm{orb}}$ be the period map associated with H in 2.5.3. Let $x \in X$, let $\check{\varphi}(x) = ((\sigma, Z) \bmod \Gamma)$, and let τ be the smallest element of Σ such that $\sigma \subset \tau$. We define

$$\varphi(x) := ((\tau, \exp(\tau_{\mathbf{C}})Z) \bmod \Gamma) \in \Gamma\backslash D_{\Sigma}. \tag{3}$$

This map φ is also called the *period map* associated with H. If there is a possibility of confusion with the period map $\check{\varphi}$, we call φ the *period map of H with respect to* Σ.

Later, in Chapter 3, we will endow $\Gamma\backslash D_{\Sigma}$ with the structure of a logarithmic local ringed space over $\mathbf{C}$, and φ will become the underlying map of a morphism of logarithmic local ringed spaces over $\mathbf{C}$. On the other hand, we consider the map $\check{\varphi}$ always just as a set-theoretic map.

We define the map $\varphi^{\log}$. Let $y \in X^{\log}$ and let x be the image of y in X. Let s be an element of $\mathrm{sp}(y)$ satisfying the following condition: The composite map $\mathcal{L}_y \to \mathcal{O}_{X,y}^{\log} \xrightarrow{s} \mathbf{C} \to \mathbf{C}/\mathbf{R} \simeq i\mathbf{R}$ coincides with the composite map $\mathcal{L}_y \to \mathrm{Cont}(x, i\mathbf{R}) = i\mathbf{R}$ (2.2.4). We define

$$\varphi^{\log}(y) := ((\tau, Z) \bmod \Gamma) \in \Gamma\backslash D_{\Sigma}^{\sharp}, \tag{4}$$

where Z is the τ-nilpotent i-orbit $\exp(i\tau_{\mathbf{R}})\tilde{\mu}_y(F(s))$. Then $\varphi^{\log}(y)$ is independent of the choices of s and $\tilde{\mu}_y$.

If $\varphi^{\log}(y) = ((\tau, Z) \bmod \Gamma)$, then $\varphi(x) = ((\tau, \exp(\tau_{\mathbf{C}})Z) \bmod \Gamma)$.

2.5.11

Let X be an object of $\overline{\mathcal{A}}_1(\log)$ (2.2.1), let Γ be a subgroup of $G_{\mathbf{Z}}$, and let H be a pre-PLH on X of weight w and of Hodge type $(h^{p,q})$ endowed with a Γ-level structure. By the *set of local monodromy cones of H in* $\mathfrak{g}_{\mathbf{R}}$, we mean the set of all nilpotent cones σ in $\mathfrak{g}_{\mathbf{R}}$ such that $((\sigma, Z) \bmod \Gamma)$ belongs to the image of the period map $X \to \Gamma\backslash \check{D}_{\mathrm{orb}}$ (2.5.3) for some $\exp(\sigma_{\mathbf{C}})$-orbit Z in $\check{D}$.

Proposition 2.5.12 *Let X be an object of $\overline{\mathcal{A}}_1(\log)$ and assume* $\mathrm{rank}_{\mathbf{Z}}(M_X^{\mathrm{gp}}/\mathcal{O}_X^{\times})_x \le 1$ *for all $x \in X$. Let H be a PLH on X of weight w and of Hodge type $(h^{p,q})$ endowed with a Γ-level structure. Let Σ' be the set of local monodromy cones of H in $\mathfrak{g}_{\mathbf{R}}$* (2.5.11) *and let $\Sigma = \Sigma' \cup \{\{0\}\}$. Then $\Sigma \subset \Xi$* (1.3.11)*, Σ is a fan in $\mathfrak{g}_{\mathbf{Q}}$ and is strongly compatible with Γ, the period map $\check{\varphi} : X \to \Gamma\backslash \check{D}_{\mathrm{orb}}$ of H in* 2.5.3 *takes values in $\Gamma\backslash D_{\Sigma}$, H is a PLH of type $(w, (h^{p,q}), H_0, \langle\ ,\ \rangle_0, \Gamma, \Sigma)$, and the period map $\varphi : X \to \Gamma\backslash D_{\Sigma}$ in* 2.5.10 (1) *coincides with $\check{\varphi}$.*

Proof. Straightforward. □

2.5.13

Schmid [Sc] proved the nilpotent orbit theorem which, roughly speaking, says the following: If X is an analytic manifold and Y is a divisor on X with normal crossings, then a variation of polarized Hodge structure on $X - Y$ yields a nilpotent orbit at each point of Y. By using the relationship between a PLH and a nilpotent orbit given in proposition 2.5.5, we can interpret the nilpotent orbit theorem as

THEOREM 2.5.14 *Let X be a logarithmically smooth, fs logarithmic analytic space, and let $U = X_{\mathrm{triv}}$ be the open set of X consisting of all points at which the logarithmic structure is trivial (that is, $U = \{x \in X \mid M_{X,x} = \mathcal{O}_{X,x}^\times\}$). Let $(H_{\mathbf{Z}}, \langle\ ,\ \rangle, F)$ be a variation of polarized Hodge structure on U of weight w and of Hodge type $(h^{p,q})$, and assume that the local monodromy of $H_{\mathbf{Z}}$ along $X - U$ is unipotent. Then, $(H_{\mathbf{Z}}, \langle\ ,\ \rangle, F)$ extends uniquely to a PLH on X of weight w and of Hodge type $(h^{p,q})$. This PLH satisfies the big Griffiths transversality* (2.4.9), *i.e., this is an LVPH in* 2.4.9.

Here "the local monodromy of $H_{\mathbf{Z}}$ along $X - U$ is unipotent" means that the local monodromy of the unique extension of $H_{\mathbf{Z}}$ to $X^{\log}$ as a locally constant sheaf is unipotent.

The reason why the nilpotent orbit theorem of Schmid implies theorem 2.5.14 is shown in [KMN]. Here we explain it in the following special case for simplicity.

2.5.15

Recalling the nilpotent orbit theorem of Schmid [Sc, Chapter 4], we explain the above Theorem 2.5.14 in the case $X = \Delta^n$ and $U = (\Delta^*)^n$.

Assume that a variation of polarized Hodge structure $H = (H_{\mathbf{Z}}, \langle\ ,\ \rangle, F)$ on $(\Delta^*)^n$ is given. Fix $p \in (\Delta^*)^n$, let $H_0 = H_{\mathbf{Z},p}$, and let $\langle\ ,\ \rangle_0 : H_0 \times H_0 \to \mathbf{Q}$ be the pairing induced by $\langle\ ,\ \rangle$.

Let

$$\mathfrak{h}^n \to (\Delta^*)^n, \quad (z_j)_j \mapsto (\exp(2\pi i z_j))_j.$$

Denote the pullback of $H_{\mathbf{Z}}$ to $\mathfrak{h}^n$ via this map by the same letter $H_{\mathbf{Z}}$. Fix $u \in \mathfrak{h}^n$ whose image in $(\Delta^*)^n$ is p. We have a unique isomorphism $\beta : H_{\mathbf{Z}} \xrightarrow{\sim} H_0$ on $\mathfrak{h}^n$ (H_0 is regarded as a constant sheaf) whose germ at u is the identity map from $H_{\mathbf{Z},u} = H_{\mathbf{Z},p}$ to $H_0 = H_{\mathbf{Z},p}$. For $z \in \mathfrak{h}^n$ with image q in $(\Delta^*)^n$, let $F(q)$ be the filtration on $H_{\mathbf{C},q}$ defined by F and let $\tilde{\Phi}(z) = \beta(F(q)) \in D$. Then $\tilde{\Phi} : \mathfrak{h}^n \to D$ is the associated period map, which is holomorphic.

Now assume that the monodromy $\gamma_j : H_0 \to H_0$ is unipotent for every $1 \leq j \leq n$. (γ_j is the jth standard generator of $\pi_1((\Delta^*)^n)$). Let $N_j = \log(\gamma_j) : H_{0,\mathbf{Q}} \to H_{0,\mathbf{Q}}$. Then

$$\tilde{\Phi}(z + 1_j) = \gamma_j \tilde{\Phi}(z) = \exp(N_j)\tilde{\Phi}(z)$$

for any $z \in \mathfrak{h}^n$ and $1 \le j \le n$, where $z + 1_j$ is the element of $\mathfrak{h}^n$ defined by $(z+1_j)_k = z_k$ for any $1 \le k \le n$ such that $k \ne j$, and $(z+1_j)_j = z_j + 1$. This shows that, if we define

$$\tilde{\Psi} : \mathfrak{h}^n \to \check{D}, \quad z \mapsto \exp\left(-\sum_{j=1}^{n} z_j N_j\right) \tilde{\Phi}(z),$$

then $\tilde{\Psi}$ descends to $(\Delta^*)^n$, that is, there is a unique holomorphic map $\Psi : (\Delta^*)^n \to \check{D}$ for which the following diagram is commutative:

$$\begin{array}{ccc} \mathfrak{h}^n & \xrightarrow{\tilde{\Psi}} & \check{D} \\ \downarrow & & \| \\ (\Delta^*)^n & \xrightarrow{\Psi} & \check{D}. \end{array}$$

Then the nilpotent orbit theorem of Schmid asserts the following (i) and (ii).

(i) Ψ extends to a holomorphic map $\Delta^n \to \check{D}$.
(ii) For any $q = (q_j)_{1 \le j \le n} \in \Delta^n$,

$$(\sigma, \exp(\sigma_{\mathbf{C}})\Psi(q)) \quad \text{with } \sigma = \sum_{1 \le j \le n, q_j = 0} (\mathbf{R}_{\ge 0}) N_j$$

is a nilpotent orbit.

We explain why the nilpotent orbit theorem implies that a VPH on $(\Delta^*)^n$ extends to an LVPH.

Let $X = \Delta^n$ with the logarithmic structure associated with the normal crossing divisor $X - U$ where $U = (\Delta^*)^n$. Then $H_{\mathbf{Z}}$ extends to $X^{\log}$ as a locally constant sheaf by 2.3.5. The morphism $\Psi : X \to \check{D}$ defines a filtration F_Ψ on $\mathcal{O}_X \otimes H_0$. Let

$$\nu : \mathcal{O}_X^{\log} \otimes H_{\mathbf{Z}} \simeq \mathcal{O}_X^{\log} \otimes H_0$$

be the isomorphism of $\mathcal{O}_X^{\log}$-modules given in 2.3.8. Let $F := \nu^{-1}(\mathcal{O}_X^{\log} \otimes_{\mathcal{O}_X} F_\Psi)$ be the filtration on $\mathcal{O}_X^{\log} \otimes H_{\mathbf{Z}}$. Then, the restriction of F to $(\Delta^*)^n$ coincides with the original Hodge filtration on $\mathcal{O}_U \otimes H_{\mathbf{Z}}$. This shows that the VPH in question extends to a pre-PLH on $X = \Delta^n$. By (ii) of the theorem of Schmid and by 2.5.5, this pre-PLH is a PLH. It satisfies the big Griffiths transversality because its restriction to $(\Delta^*)^n$ satisfies the big Griffiths transversality. Hence this PLH is an LVPH.

2.5.16

Correction. In [KU1, 5.5], there is a mistake in the definition of a PLH of type Φ. The correct definition is the one given in this Section 2.5. The mistake in [KU1, 5.5] is corrected as follows. In the last line of condition (i) in [KU1, 5.5], *form a σ-nilpotent orbit* must be corrected to *are contained in a σ-nilpotent orbit,* or, equivalently, to *together with their translations by* $\exp(\sigma_{\mathbf{C}})$ *form a σ-nilpotent orbit.*

2.6 LOGARITHMIC MIXED HODGE STRUCTURES

We give here a definition of logarithmic mixed Hodge structure, and define a logarithmic Hodge structure of weight w ($w \in \mathbf{Z}$) as a special case of logarithmic mixed Hodge structure.

In [D5, 1.8.15], Deligne illustrated how to formulate "good degenerations" of mixed Hodge structures, and Steenbrink and Zucker gave the precise definition in [SZ], [Z3] following the philosophy of Deligne. Our definition of logarithmic mixed Hodge structure below also follows the philosophy of Deligne and is similar to the definition of good degeneration of a mixed Hodge structure in [SZ], [Z3].

This section is not used in the rest of this book.

In this section, X denotes an object of $\overline{\mathcal{A}}_1(\log)$ (2.2.1).

Definition 2.6.1 *A* prelogarithmic mixed Hodge structure (pre-LMH) *on X is a triple $(H_{\mathbf{Z}}, W, F)$, consisting of a locally constant sheaf of free $\mathbf{Z}$-modules $H_{\mathbf{Z}}$ of finite rank on $X^{\log}$, a $\mathbf{Q}$-rational increasing filtration W of the sheaf $H_{\mathbf{R}}$ of $\mathbf{R}$-modules, and a decreasing filtration F of the $\mathcal{O}_X^{\log}$-module $\mathcal{O}_X^{\log} \otimes H_{\mathbf{Z}}$, which satisfies the following two conditions.*

(1) *For each $k \in \mathbf{Z}$, W_k is locally constant.*
(2) *There exist an $\mathcal{O}_X$-module $\mathcal{M}$ on X and a decreasing filtration $(\mathcal{M}^p)_{p\in\mathbf{Z}}$ of $\mathcal{M}$ by $\mathcal{O}_X$-submodules such that $\mathcal{M}$, $\mathcal{M}^p$, $\mathcal{M}/\mathcal{M}^p$ are locally free of finite rank, $\mathcal{O}_X^{\log} \otimes_{\mathbf{Z}} H_{\mathbf{Z}} = \tau^*(\mathcal{M})$ and $F^p = \tau^*(\mathcal{M}^p)$.*

Definition 2.6.2 *Let x be an fs logarithmic point* (2.1.9). *A pre-LMH $(H_{\mathbf{Z}}, W, F)$ on x is called a* logarithmic mixed Hodge structure (LMH) *on x if it satisfies the following conditions* (1) *and* (2).

(1) *$(H_{\mathbf{Z}}, F)$ satisfies the Griffiths transversality* (2.4.5).
(2) *There exists a family $(W(\mathcal{S}))_{\mathcal{S}}$ of $\mathbf{Q}$-rational increasing filtration $W(\mathcal{S})$ of the sheaf $H_{\mathbf{R}}$ given for each face $\mathcal{S}$ of the fs monoid $M_x/\mathcal{O}_x^\times$, satisfying the following conditions* (2.1)–(2.3).
   (2.1) *If $\mathcal{S} = M_x/\mathcal{O}_x^\times$, then $W(\mathcal{S}) = W$.*
   (2.2) *Let h be a homomorphism $M_x/\mathcal{O}_x^\times \to \mathbf{R}_{\geq 0}^{\mathrm{add}}$, where $\mathbf{R}_{\geq 0}^{\mathrm{add}}$ means $\mathbf{R}_{\geq 0}$ regarded as a monoid with respect to the addition. Let $y \in x^{\log}$ and let N_h be the image of h under the composite map*

$$\mathrm{Hom}(M_x/\mathcal{O}_x^\times, \mathbf{R}_{\geq 0}^{\mathrm{add}}) \subset \mathrm{Hom}(M_x^{\mathrm{gp}}/\mathcal{O}_x^\times, \mathbf{R})$$
$$\simeq \mathbf{R} \otimes_{\mathbf{Z}} \pi_1(x^{\log}) \xrightarrow{\log} \mathrm{End}_{\mathbf{R}}(H_{\mathbf{R},y}),$$

where log *is the logarithm of the action of $\pi_1(x^{\log})$ on $H_{\mathbf{R},y}$. Then, if $\mathcal{S}$ is a face of $M_x/\mathcal{O}_x^\times$ and $\mathcal{S} \subset \mathrm{Ker}(h)$, we have*

$$N_h(W(\mathcal{S})_{k,y}) \subset W(\mathcal{S})_{k-2,y}$$

for all $k \in \mathbf{Z}$. If $\mathcal{S}'$ is a face of $M_x/\mathcal{O}_x^\times$ and $\mathcal{S} = \mathcal{S}' \cap \mathrm{Ker}(h)$, then, for any integer k and any integer $l \geq 0$, we have an isomorphism

$$N_h^l : \mathrm{gr}_{k+l}^{W(\mathcal{S})}(\mathrm{gr}_k^{W(\mathcal{S}')})_y \xrightarrow{\sim} \mathrm{gr}_{k-l}^{W(\mathcal{S})}(\mathrm{gr}_k^{W(\mathcal{S}')})_y.$$

(2.3) *Fix* $y \in x^{\log}$ *and let* $s \in \mathrm{sp}(y)$. *If* $\tilde{s} : M_x \to \mathbf{C}$ *(see* 2.4.6*) is sufficiently near to the structure morphism of the logarithmic structure* $\alpha : M_x \to \mathbf{C}$ *in the topology of simple convergence of* $\mathbf{C}$*-valued functions, then for any face* $\mathcal{S}$ *of* $M_x/\mathcal{O}_x^\times$, $(H_{\mathbf{Z},y}, W(\mathcal{S})_y, F(s))$ *is a mixed Hodge structure in the usual sense.*

2.6.3

Remark. The family of filtrations $(W(\sigma))_\sigma$ satisfying 2.6.2 (2.1) and 2.6.2 (2.2) is unique if it exists, according to Deligne [D5, 1.6.3].

DEFINITION 2.6.4 *A pre-LMH* $(H_{\mathbf{Z}}, W, F)$ *on* X *is a* logarithmic mixed Hodge structure (LMH) on X *if, for any* $x \in X$*, the inverse image of* $(H_{\mathbf{Z}}, W, F)$ *on* x *is an LMH. Here we regard* x *as an fs logarithmic point with the inverse image of the logarithmic structure of* X.

DEFINITION 2.6.5 *Let* $w \in \mathbf{Z}$*. A* logarithmic Hodge structure (LH) on X of weight w *is an LMH* $(H_{\mathbf{Z}}, W, F)$ *on* X *such that* $W_w = H_{\mathbf{R}}$ *and* $W_{w-1} = 0$.

2.6.6

Let $w \in \mathbf{Z}$. By Cattani and Kaplan [CK2], a PLH on X of weight w is an LH on X of weight w.

Chapter Three

Strong Topology and Logarithmic Manifolds

In this chapter, we define a structure of a logarithmic local ringed space over $\mathbf{C}$ on $\Gamma\backslash D_\Sigma$ (see Section 3.4), where Σ is a fan in $\mathfrak{g}_\mathbf{Q}$ and Γ a subgroup of $G_\mathbf{Z}$ that is strongly compatible with Σ. For this, in Section 3.3, we define for each $\sigma \in \Sigma$ a subset E_σ of an fs logarithmic analytic space $\check{E}_\sigma$ (a logarithmic version of the subset D of $\check{D}$) and a map $E_\sigma \to \Gamma\backslash D_\Sigma$ whose image covers $\Gamma\backslash D_\Sigma$ when σ ranges in Σ. We endow E_σ with the so-called "strong topology" introduced in Section 3.1, and with the inverse images of $\mathcal{O}$ and M of $\check{E}_\sigma$, and then import these structures to $\Gamma\backslash D_\Sigma$. The spaces E_σ and $\Gamma\backslash D_\Sigma$ are not necessarily analytic spaces. In Section 3.2, we introduce various categories of various kinds of (logarithmic) generalized analytic spaces and consider relations between them. The spaces E_σ and the spaces $\Gamma\backslash D_\Sigma$, with Γ neat belong to a category $\mathcal{B}^*(\log)$ that appears in Section 3.2, and belong in fact to a much smaller category, the category of "logarithmic manifolds" introduced in Section 3.5. There, as a preparation for the infinitesimal extended period maps, we extend the infinitesimal calculus on analytic spaces to the category $\mathcal{B}(\log)$ that appears in Section 3.2 (see also 0.4.16) and that contains $\mathcal{B}^*(\log)$. In Section 3.6, we consider "logarithmic modifications" of objects of $\mathcal{B}(\log)$ (a generalization of the modification in the toric geometry arising from subdivision of a fan) which we will need in Section 4.3 to extend period maps to the boundary.

3.1 STRONG TOPOLOGY

Definition 3.1.1 *Let X be an analytic space and S a subset of X.*

(i) *The* weak topology *of S in X is the topology as a subspace of X. We denote this topological space by $S_{weak/X}$.*

(ii) *The* strong topology *of S in X is the strongest topology on S for which the map $\lambda : Y \to S$ is continuous for any analytic space Y and for any analytic morphism $\lambda : Y \to X$ with $\lambda(Y) \subset S$. We denote this topological space by $S_{\mathrm{str}/X}$. By definition, a subset U of $S_{\mathrm{str}/X}$ is open if and only if $\lambda^{-1}(U)$ is open in Y for any (Y, λ) as above.*

We have clearly

$$\text{(weak topology)} \leq \text{(strong topology)}.$$

By the resolution of singularity of Hironaka [H], [AHV], for any Hausdorff analytic space Y that is countable at infinity there are a smooth analytic space Y' and a proper morphism $Y' \to Y$. Hence we may assume the analytic spaces Y to be smooth in the above definition of strong topology.

Note also that if S is a locally closed analytic subspace of X then the strong topology on S coincides with the weak topology on S.

We give two examples 3.1.2 and 3.1.3 concerning the strong topology.

3.1.2

Example. Let S be a subset of the complex line $X := \mathbf{C}$. Then the strong topology of S is the one described as follows. Let S' be the interior of S in the usual topology of X, that is, an element s of S belongs to S' if and only if S contains some neighborhood of s in X in the usual topology of X. Then $S_{\mathrm{str}/X}$ is the disjoint union of the subspace S' of X and the discrete set $S - S'$.

Proof. It is sufficient to prove that, for a smooth analytic space Y and an analytic morphism $\lambda : Y \to X$ with $\lambda(Y) \subset S$, if $y \in Y$ and $\lambda(y) \in S - S'$, then $\lambda(y') = \lambda(y)$ for any y' in some neighborhood of y. Take an open neighborhood U of y in Y such that, for any $y' \in U$, there exists a connected complex smooth curve C on Y joining y and y'. If the restriction of λ to C is not a constant function, $\lambda : C \to X = \mathbf{C}$ is an open map and hence $\lambda(C)$ must be an open set in X. Since $\lambda(y) \in \lambda(C) \subset S$, this contradicts the assumption $\lambda(y) \in S - S'$. Hence the restriction of λ to C is constant and hence $\lambda(y') = \lambda(y)$ for any $y' \in U$. □

3.1.3

Example. Let $X := \mathbf{C}^2$ and let S be the subset $S := (\mathbf{C}^2 - (\{0\} \times \mathbf{C})) \cup \{(0, 0)\} \subset X$.

(i) The strong topology of S in X is described as follows.

(1) The topology on $S - \{(0, 0)\}$ as a subspace of $S_{\mathrm{str}/X}$ coincides with the one as a subspace of X.

(2) Let $\varepsilon = (\varepsilon_n)_{n \geq 1}$ be a sequence in $\mathbf{R}_{>0}$. Define

$$U_n(\varepsilon_n) := \{(x, y) \in S \mid |x| < \varepsilon_n,\ |y| < \varepsilon_n,\ |y|^n < |x|\},$$

and

$$U(\varepsilon) := \left(\bigcup_{n \geq 1} U_n(\varepsilon_n)\right) \cup \{(0, 0)\}.$$

Then the $U(\varepsilon)$, where ε runs over all sequences in $\mathbf{R}_{>0}$, form a fundamental system of neighborhoods of $(0, 0)$ in $S_{\mathrm{str}/X}$.

(ii) If $k \in \mathbf{Z}_{\geq 1}$ and $z \in \mathbf{C}$ converges to 0, then (z^k, z) converges to $(0, 0)$ in $S_{\mathrm{str}/X}$. However, if $f : \mathbf{R}_{>0} \to \mathbf{R}_{>0}$ is a map such that, for each $n \geq 1$, there is $\varepsilon_n > 0$ for which $f(s) \leq s^n$ if $0 < s < \varepsilon_n$, then $(f(s), s)$ $(s \to 0)$ converges to $(0, 0)$ for the weak topology of S but diverges for the strong topology of S. Examples of f are $f(s) = e^{-1/s}$, $f(s) = s^{1/s}$, etc.

(iii) $S_{\mathrm{str}/X}$ is not locally compact.

Proof. (i) (1) follows easily from the definition. For (i) (2), we will prove a more general result, Proposition 3.1.5 (ii), and deduce (i) (2) from it just after the end of the proof of 3.1.5.

We prove (ii). The convergence of (z^k, z) follows easily from (i) (2). The divergence of $(f(s), s)$ is deduced from (i) (2) as follows. For each $n \geq 1$, take $\varepsilon_n > 0$ such that $f(s) \leq s^n$ if $0 < s < \varepsilon_n$. Then $(f(s), s)$ does not belong to $U(\varepsilon)$ for any $s \in \mathbf{R}_{>0}$. Otherwise, $f(s) < \varepsilon_n$, $s < \varepsilon_n$, $s^n < f(s)$ for some $n \geq 1$, but this is impossible.

We prove (iii). Assume it is locally compact and let K be a compact neighborhood of $(0, 0)$ in $S_{\mathrm{str}/X}$. Then we have a set $U(\varepsilon)$ in (i) (2) such that $U(\varepsilon) \subset K$. Replacing ε_n by $\min(\frac{1}{2}, \varepsilon_1, \ldots, \varepsilon_n)$, we may assume $1 > \varepsilon_1 \geq \varepsilon_2 \geq \cdots$. Let $x_n = \varepsilon_n^n$, $y_n = \varepsilon_n^2$, $p_n = (x_n, y_n) \in U_n(\varepsilon_n) \subset U(\varepsilon)$ $(n \geq 1)$. Then p_n does not converge to $(0, 0)$ in the strong topology of S. Moreover, any cofinal subsequence of $(p_n)_n$ does not converge to $(0, 0)$ in the strong topology of S. In fact, let $\varepsilon'_n = \varepsilon_{2n}^{2n}$ $(n \geq 1)$. We show that $p_n \notin U(\varepsilon')$ for any n. If $p_n \in U_m(\varepsilon'_m)$, then $\varepsilon_n^{2m} = y_n^m < x_n = \varepsilon_n^n$. Hence $2m > n$ and $\varepsilon'_m = \varepsilon_{2m}^{2m} < \varepsilon_n^n = x_n$. But this conclusion $\varepsilon'_m < x_n$ contradicts $p_n \in U_m(\varepsilon'_m)$. On the other hand, since K is compact, some cofinal subsequence of $(p_n)_n$ must converge to some point p of K. Since it converges to p also in the weak topology, we have $p = (0, 0)$, a contradiction. □

3.1.4

In Proposition 3.1.5 (ii) below, we generalize Example 3.1.3 (i) to the strong topology of an "analytically constructible" subset in a complex analytic space.

A subset S of a complex analytic space X is said to be *analytically constructible* if the following condition (1) holds locally on X.

(1) There exist a finite family $(A_j)_{j \in J}$ of closed analytic subspaces A_j of X and a closed analytic subspace B_j of A_j for each j such that

$$S = \{x \in X \mid x \in B_j \text{ for any } j \in J \text{ such that } x \in A_j\}.$$

To state Proposition 3.1.5, we fix notation. Let X be a complex analytic space endowed with a metric $d : X \times X \to \mathbf{R}_{\geq 0}$ which is compatible with the analytic structure of X. That is, there are an open covering $(U_\lambda)_{\lambda \in \Lambda}$ of X, analytic immersions $\iota_\lambda : U_\lambda \to \mathbf{C}^{k(\lambda)}$ for some $k(\lambda) \geq 0$, and constants $c_\lambda, c'_\lambda > 0$ such that

$$c_\lambda |\iota_\lambda(x) - \iota_\lambda(y)| \leq d(x, y) \leq c'_\lambda |\iota_\lambda(x) - \iota_\lambda(y)| \quad (\forall \lambda \in \Lambda,\ \forall x, \forall y \in U_\lambda).$$

Here $|\ \ |$ denotes the usual Euclidean metric on $\mathbf{C}^{k(\lambda)}$.

For $x \in X$ and for a subset E of X, let

$$d(x, E) = \inf\{d(x, y) \mid y \in E\}.$$

Let S be a subset of X and let $(A_j)_{j \in J}$ be a finite family of closed analytic subspaces of X. For $s \in S$, for a function $n : J \to \mathbf{Z}_{>0}$, and for $\delta \in \mathbf{R}_{>0}$, let $U_n(s, \delta)$ be the set of all $x \in S$ satisfying the following conditions (2) and (3).

(2) $d(x, s) < \delta$.

(3) If $j \in J$ and if $s \in A_j$ and $x \notin A_j$, then $d(x, S \cap A_j)^{n(j)} < d(x, A_j)$.

Let $\mathcal{E}$ be the set of all families $(\varepsilon_n)_n$ of real numbers $\varepsilon_n > 0$ given for each function $n : J \to \mathbf{Z}_{>0}$. For $s \in S$ and for $\varepsilon = (\varepsilon_n)_n \in \mathcal{E}$, define

$$U(s, \varepsilon) = \bigcup_n U_n(s, \varepsilon_n) \subset S.$$

Proposition 3.1.5 *Let X be a complex analytic space endowed with a metric d that is compatible with the analytic structure of X, let S be a subset of X, and let $(A_j)_{j \in J}$ be a finite family of closed analytic subspaces of X. Let $\mathcal{E}$ and $U(s, \varepsilon) \subset S$ $(s \in S, \varepsilon \in \mathcal{E})$ be as above.*

(i) *If $s \in S$ and if $\varepsilon \in \mathcal{E}$, $U(s, \varepsilon)$ is an open neighborhood of s in S in the strong topology.*

(ii) *Assume that there exists a closed analytic subspace B_j of A_j for each $j \in J$ such that*

$$S = \{s \in X \mid s \in B_j \text{ if } j \in J \text{ and } s \in A_j\}.$$

Then, if $s \in S$ and if ε ranges over all elements of $\mathcal{E}$, $U(s, \varepsilon)$ form a fundamental system of neighborhoods of s in S in the strong topology.

Proof. We prove (i). As is easily seen, if $s, s' \in S, \varepsilon \in \mathcal{E}$ and $s' \in U(s, \varepsilon)$, then there exists $\varepsilon' \in \mathcal{E}$ such that $U(s', \varepsilon') \subset U(s, \varepsilon)$. Hence, for the proof of (i), it is sufficient to prove that, if U is a subset of S satisfying the following condition (1), then U is open in the strong topology of S.

(1) For any $s \in U$, there exists $\varepsilon \in \mathcal{E}$ such that $U(s, \varepsilon) \subset U$.

Assume $U \subset S$ satisfies (1). Let Y be an analytic space, and $\lambda : Y \to X$ be a morphism satisfying $\lambda(Y) \subset S$. We have to show that $\lambda^{-1}(U)$ is an open set on Y.

For an analytic space Y' and for a surjective morphism $Y' \to Y$ which is open or closed, we can replace Y by Y', because the topology on Y is the quotient of the topology on Y'. By the Hironaka resolution of singularities [H], [AHV, Theorem 5.3.1] together with the above remark, we may assume the following conditions (2) and (3).

(2) Y is smooth.

(3) For each $j \in J$, either $\lambda^{-1}(A_j) = Y$, or $\lambda^{-1}(A_j)$ is a divisor on Y with normal crossings.

Let $y \in Y$ and let $s = \lambda(y)$. It is enough to show that, if $z \in Y$ is sufficiently near y, then $\lambda(z) \in U$. Let J' be the subset of J consisting of all j such that $s \in A_j$ and $\lambda^{-1}(A_j)$ is a divisor on Y with normal crossings. For each $j \in J'$, let f_j be a generator of the ideal of $\mathcal{O}_{Y,y}$ whose zero coincides with $\lambda^{-1}(A_j)$ on some neighborhood of y, let $f_j = f_{j,1}^{e(j,1)} \cdots f_{j,l(j)}^{e(j,l(j))}$ be a prime decomposition in $\mathcal{O}_{Y,y}$, and let $e(j) := \sum_{1 \le k \le l(j)} e(j,k)$. Since the metric d is compatible with the analytic structure, if the values of $n : J \to \mathbf{Z}_{>0}$ are sufficiently large, then for any $j \in J'$, $|f_j(z)|^{n(j)} d(\lambda(z), A_j)^{-e(j)}$ converges to 0 when $z \in Y - \lambda^{-1}(A_j)$ converges to y. Take such a function n and take $\delta > 0$ such that $U_n(s, \delta) \subset U$. It is enough to prove that, if $z \in Y$ is sufficiently near y, then $\lambda(z) \in U_n(s, \delta)$. Hence it is sufficient to prove that, if $j \in J'$ and if $z \in Y - \lambda^{-1}(A_j)$ is sufficiently near y, then $d(\lambda(z), S \cap A_j)^{n(j)} < d(\lambda(z), A_j)$.

Choose local coordinates on Y at y in the form $(f_{j,1}, \dots, f_{j,l(j)}, t_{j,1}, \dots, t_{j,m(j)})$, let $|f_{j,k}(z)|$ be the minimal one among $|f_{j,1}(z)|, \dots, |f_{j,l(j)}(z)|$, and let w be the point on Y with coordinates

$$(f_{j,1}(z), \dots, f_{j,k-1}(z), 0, f_{j,k+1}(z), \dots, f_{j,l(j)}(z), t_{j,1}(z), \dots, t_{j,m(j)}(z)).$$

Note $\lambda(w) \in S \cap A_j$. There exists a constant $c > 0$ which is independent of z and satisfies

$$d(\lambda(z), \lambda(w))^{n(j)} \le c|f_{j,k}(z)|^{n(j)} \le c|f_j(z)|^{n(j)/e(j)} < d(\lambda(z), A_j).$$

Hence $d(\lambda(z), S \cap A_j)^{n(j)} < d(\lambda(z), A_j)$.

We prove (ii) of Proposition 3.1.5. For the same reason given at the beginning of the proof of (i), it is sufficient to prove that, if U is an open set of S in the strong topology, then U satisfies the condition (1). Fix $s \in U$ and $n : J \to \mathbf{Z}_{>0}$. It is sufficient to prove $U_n(s, \delta) \subset U$ for some $\delta > 0$. Working locally on X, we may assume that, for each $j \in J$, there exist $g_{j,1}, \dots, g_{j,a(j)}, h_{j,1}, \dots, h_{j,b(j)} \in \mathcal{O}_X(X)$ such that

$$A_j = \{x \in X \mid g_{j,k}(x) = 0 \ \ (1 \le k \le a(j))\},$$
$$B_j = \{x \in X \mid h_{j,l}(x) = 0 \ \ (1 \le l \le b(j))\}.$$

Since the metric d is compatible with the analytic structure, we may further assume that there exist integers $e, e' \ge 1$ such that

$$\max\{|h_{j,l}(x)|^e \mid 1 \le l \le b(j)\} \le d(x, B_j),$$
$$d(x, A_j)^{e'} \le \max\{|g_{j,k}(x)| \mid 1 \le k \le a(j)\}$$

for any $j \in J$ and any $x \in X$. Define a closed analytic subspace Y of $X \times \prod_{j \in J} \mathbf{C}^{a(j)b(j)}$ by

$$Y := \left\{ (x, z) \;\middle|\; \begin{array}{l} x \in X, z = (z_{j,k,l})_{j \in J, 1 \le k \le a(j), 1 \le l \le b(j)} \in \prod_{j \in J} \mathbf{C}^{a(j)b(j)}, \\ h_{j,l}(x)^{e(e'n(j)+1)} = \sum_{1 \le k \le a(j)} z_{j,k,l}\, g_{j,k}(x) \ \ (\forall j, \forall l) \end{array} \right\}. \tag{4}$$

Let $\lambda : Y \to X$ be the morphism $(x, z) \mapsto x$. Then $\lambda(Y) \subset S$. As is easily seen, there exists $\alpha = (\alpha_j)_{j \in J} \in \prod_{j \in J} \mathbf{C}^{a(j)b(j)}$ ($\alpha_j \in \mathbf{C}^{a(j)b(j)}$) such that $(s, \alpha) \in Y$ and such that, if $j \in J$ and $s \in A_j$, then $\alpha_j = 0$. Since U is open in the strong topology of S, $\lambda^{-1}(U)$ is a neighborhood of (s, α). Hence there exists $\delta > 0$ such that, if $(x, z) \in Y$ and if $d(x, s) < \delta$ and $|z - \alpha| < \delta$, then $x \in U$. Thus it is sufficient to prove that, for a given $j \in J$ and $c > 0$, there exists δ such that $0 < \delta \le c$ with the following property: If $x \in U_n(s, \delta)$, then there exists $z_j \in \mathbf{C}^{a(j)b(j)}$ such that $|z_j - \alpha_j| < c$ and satisfies

$$h_{j,l}(x)^{e(e'n(j)+1)} = \sum_{1 \le k \le a(j)} z_{k,l}\, g_{j,k}(x) \quad (\forall l). \tag{5}$$

The existence of such a δ is seen easily if $s \notin A_j$. Assume $s \in A_j$. We show that we can take $\delta = c$. Note that $\alpha_j = 0$ in this case. If $x \in A_j \cap U_n(s, \delta)$, we can take $z_j = 0$. Asume $x \in U_n(s, \delta)$ but $x \notin A_j$. Then we have $d(x, S \cap A_j)^{n(j)} < d(x, A_j)$ and $d(x, S \cap A_j) \leq d(x, s) < \delta$, and hence

$$\begin{aligned}&\max\{|h_{j,l}(x)|^{e(e'n(j)+1)} \mid 1 \leq l \leq b(j)\}\\ &\quad \leq d(x, B_j)^{e'n(j)+1} \leq d(x, S \cap A_j)^{e'n(j)+1} < \delta d(x, S \cap A_j)^{e'n(j)} < \delta d(x, A_j)^{e'}\\ &\quad \leq \delta \max\{|g_{j,l}(x)| \mid 1 \leq l \leq a(j)\}.\end{aligned}$$

Thus there exists $z_j \in \mathbf{C}^{a(j)b(j)}$ satisfying (5) and $|z_j| < \delta$. □

Proof of 3.1.3 (i) (2). We deduce 3.1.3 (i) (2) from Proposition 3.1.5 (ii). In 3.1.5, let $X = \mathbf{C}^2$, $S = (\mathbf{C}^2 - (\{0\} \times \mathbf{C})) \cup \{(0, 0)\}$, J be a one-point set $\{j\}$, $A_j = \{0\} \times \mathbf{C}$, and $B_j = \{(0, 0)\}$, $s = (0, 0)$. Then if we define $U_n(s, \delta)$ $(\delta > 0)$ in 3.1.4 with respect to the usual metric d of $\mathbf{C}^2$ defined by $d((x_1, y_1), (x_2, y_2)) = (|x_1 - x_2|^2 + |y_1 - y_2|^2)^{1/2}$, then $U_n(s, \delta)$ is contained in the $U_n(\delta)$ of 3.1.3. If we define $U_n(s, \delta)$ in 3.1.4 with respect to the metric $d((x_1, y_1), (x_2, y_2)) = (|x_1 - x_2| + |y_1 - y_2|)/2$, then $U_n(\delta)$ of 3.1.3 is contained in $U_{n+1}(s, \delta)$ when $\delta < 1/2$. Hence 3.1.3 (i) (2) follows from 3.1.5 (ii). □

The following result will be used in Chapters 6 and 7.

Proposition 3.1.6 *Let X, d, S be as in 3.1.5, and let A be a closed analytic subspace of X. Let $s \in S \cap A$ and let $(s_\lambda)_\lambda$ be a directed family of elements of $S \cap (X - A)$ that converges to s in the strong topology of S. Let $y_\lambda := |\log(d(s_\lambda, A))|$. (Note $y_\lambda \to \infty$.) Then, for any $k \geq 0$, we have*

$$y_\lambda^k d(s_\lambda, S \cap A) \to 0.$$

Proof. Note that, for each $n \geq 1$, there exists $\varepsilon_n > 0$ such that $t < |\log(t)|^{-(k+1)n}$ if $0 < t < \varepsilon_n$. Let $\varepsilon = (\varepsilon_n)_{n \geq 1}$. By Proposition 3.1.5 (i), $U(s, \varepsilon)$ is a neighborhood of s in the strong topology of S. Hence, if λ is large, there exists $n \geq 1$ such that $d(s_\lambda, s) < \varepsilon_n$ and $d(s_\lambda, S \cap A)^n < d(s_\lambda, A)$. Since $d(s_\lambda, A) \leq d(s_\lambda, s) < \varepsilon_n$, we have $d(s_\lambda, A) < y_\lambda^{-(k+1)n}$. Hence $d(s_\lambda, S \cap A) < d(s_\lambda, A)^{1/n} < y_\lambda^{-(k+1)}$ and this shows that $y_\lambda^k d(s_\lambda, S \cap A) < y_\lambda^{-1} \to 0$. □

3.1.7

We give basic remarks on strong topologies.

(i) Let X be an analytic space, let Y be a locally closed analytic subspace of X, and let S be a subset of Y. Then we have

$$S_{\mathrm{str}/Y} = S_{\mathrm{str}/X}.$$

(ii) Let X be an analytic space, let S be a subset of X, and let T be a locally closed subset of X contained in S. Then $T_{\mathrm{str}/X}$ is a topological subspace of $S_{\mathrm{str}/X}$.

These assertions are proved easily.

Proposition 3.1.8 *Let X and Y be analytic spaces and let $S \subset X$, $T \subset Y$ be subsets. Then, $(S \times T)_{\mathrm{str}/X\times Y}$ coincides with $S_{\mathrm{str}/X} \times T_{\mathrm{str}/Y}$ if either one of the following conditions* (1) *and* (2) *is satisfied.*

(1) *S and T are analytically constructible subsets of X and of Y, respectively.*
(2) $T = Y$.

We do not know whether $(S \times T)_{\mathrm{str}/X\times Y} = S_{\mathrm{str}/X} \times T_{\mathrm{str}/Y}$ holds in general.

Proof of Proposition 3.1.8. If (1) is satisfied, noticing that $S \times T$ is a constructible subset of $X \times Y$, we can deduce the conclusion of Proposition 3.1.8 easily from Proposition 3.1.5 (ii).

Next assume that (2) is satisfied. By working locally on Y, by the resolution of singularities of Hironaka [H], [AHV], we may assume that Y is an open set of $\mathbf{C}^n$ for some $n \geq 0$. Then, using an open immersion $Y \subset \mathbf{P}^n_{\mathbf{C}}$, we are reduced to the case $Y = \mathbf{P}^n_{\mathbf{C}}$. Thus we may assume that Y is compact. Then, since $S_{\mathrm{str}/X} \times Y$ is Hausdorff, it is sufficient to prove that the composite map

$$(S \times Y)_{\mathrm{str}/X\times Y} \to S_{\mathrm{str}/X} \times Y \to S_{\mathrm{str}/X}$$

is proper. The fiber of $(S \times Y)_{\mathrm{str}/X\times Y}$ on each $s \in S$ is Y and hence is compact. Thus (by [Bn, Ch. I, §10, Theorem 1]), it is sufficient to prove that $(S \times Y)_{\mathrm{str}/X\times Y} \to S_{\mathrm{str}/X}$ is a closed map. Let C be a closed subset of $(S \times Y)_{\mathrm{str}/X\times Y}$, and let C' be the image of C in S. Let A be an analytic space and let $\lambda : A \to X$ be a morphism such that $\lambda(A) \subset S$. It is sufficient to prove that $\lambda^{-1}(C')$ is closed in A. Let $\lambda_Y : A \times Y \to X \times Y$ be the morphism induced by λ. Then $\lambda_Y(A \times Y) \subset S \times Y$. Hence $\lambda_Y^{-1}(C)$ is closed in $A \times Y$. Since $A \times Y \to A$ is proper, the image of $\lambda_Y^{-1}(C)$ in A, which coincides with $\lambda^{-1}(C')$, is closed in A. □

Proposition 3.1.9 *Let $f : X \to Y$ be a morphism of analytic spaces and let S be a subset of Y. Then the topological space $f^{-1}(S)_{\mathrm{str}/X}$ coincides with the fiber product $X \times_Y S_{\mathrm{str}/Y}$.*

Proof. Via the embedding

$$\Gamma_f : X \to X \times Y, \quad x \mapsto (x, f(x)), \tag{1}$$

consider X as a locally closed analytic subspace of $X \times Y$. By 3.1.7 (i), $f^{-1}(S)_{\mathrm{str}/X} \simeq \Gamma_f(f^{-1}(S))_{\mathrm{str}/X\times Y}$. Since $\Gamma_f(f^{-1}(S))$ is the intersection of $X \times S$ and $\Gamma_f(X)$ in $X \times Y$, $\Gamma_f(f^{-1}(S))$ is locally closed in $X \times S$ in the weak topology of $X \times S$ in $X \times Y$, and hence is locally closed in $X \times S$ in the strong topology of $X \times S$ in $X \times Y$. Hence, by 3.1.7 (ii), $\Gamma_f(f^{-1}(S))_{\mathrm{str}/X\times Y}$ is a topological subspace of $(X \times S)_{\mathrm{str}/X\times Y}$ and the last space is $X \times S_{\mathrm{str}/Y}$ by Proposition 3.1.8 (2). This proves Proposition 3.1.9. □

Proposition 3.1.10 *Let X be an analytic space, let $(A_j)_{j\in J}$ be a finite family of closed analytic subspaces of X, and, for each $j \in J$, let B_j be a closed analytic subspace of A_j. Let*

$$S = \{x \in X \mid x \in B_j \textit{ for any } j \in J \textit{ such that } x \in A_j\},$$

and let S' be an open set of $S_{\mathrm{str}/X}$. Then

(i) *Assume that $X - A_j$ is dense in X for any $j \in J$. Let $x \in S'$. Then the following conditions* (1)–(4) *are equivalent.*

(1) *x has a compact neighborhood in $(S')_{\mathrm{str}/X}$.*
(2) *There exists an open neighborhood U of x in X such that $U \cap A_j = U \cap B_j$ for any $j \in J$.*
(3) *There exists an open neighborhood U of x in X such that $U \cap S$ is a locally closed analytic subset of X.*
(4) *There exists an open neighborhood U of x in X such that $U \cap S'$ is a locally closed analytic subset of X.*

(ii) *$(S')_{\mathrm{str}/X}$ is locally compact if and only if S' is a locally closed analytic subset of X.*

Proof. We prove (i). The implications (2) $\Rightarrow$ (3) $\Rightarrow$ (4) $\Rightarrow$ (1) are clear. It is sufficient to prove that (1) $\Rightarrow$ (2).

We first show that we may assume X is smooth and $A_j, B_j, \bigcup_{j\in J} A_j$, and $\bigcup_{j\in J} B_j$ are divisors with normal crossings on X. In fact, by Hironaka's resolution of singularities [H], [AHV], we may assume that there exists a smooth analytic space Y with a proper surjective morphism $f\colon Y \to X$ such that $f^{-1}(A_k)$, $f^{-1}(B_k)$, $f^{-1}(\bigcup_{j\in J} A_j)$, and $f^{-1}(\bigcup_{j\in J} B_j)$ are divisors with normal crossings on Y. If $(X, (A_j)_j, (B_j)_j, S', x)$ satisfies (1), then by Proposition 3.1.9, $(Y, (f^{-1}(A_j))_j, (f^{-1}(B_j))_j, f^{-1}(S'), y)$ satisfies (1) for any $y \in f^{-1}(x)$. We show that, if $(Y, (f^{-1}(A_j))_j, (f^{-1}(B_j))_j, f^{-1}(S'), y)$ satisfies (2) for any $y \in f^{-1}(x)$, then $(X, (A_j)_j, (B_j)_j, S', x)$ satisfies (2). In fact, we have an open neighborhood V of $f^{-1}(x)$ in Y such that $V \cap f^{-1}(A_j) = V \cap f^{-1}(B_j)$ for any $j \in J$, and since $C := f(Y - V)$ is closed in X by the properness of f, $U := X - C$ is an open neighborhood of x in X and satisfies $U \cap A_j = U \cap B_j$ for any $j \in J$.

Now we assume that X is smooth and $A_j, B_j, \bigcup_{j\in J} A_j$, and $\bigcup_{j\in J} B_j$ are divisors with normal crossings on X. By replacing X by an open neighborhood of x in X, we may assume that $X = \mathbf{C}^n$, $x = \mathbf{0} \in \mathbf{C}^n$, and $\bigcup_{j\in J} A_j$ is the set of zeros of $\prod_{1\le j\le r} z_j$ for some r where the z_j are the standard coordinates of $\mathbf{C}^n$. Assume (2) is not satisfied, that is, for some $k \in J$, $U \cap A_k \ne U \cap B_k$ for any open neighborhood U of x in X. We may assume that there exist $1 \le r_1 < r_2 \le r$ such that A_k is the set of zeros of $\prod_{1\le j\le r_1} z_j$ and B_k is the set of zeros of $\prod_{1\le j\le r_2} z_j$. Consider

$$Y := \{z \in \mathbf{C}^n \mid z_1 = \cdots = z_{r_1},\ z_{r_1+1} = \cdots = z_{r_2},\ z_j = 0 \text{ for } j > r_2\},$$

and identify Y with $\mathbf{C}^2$ via (z_{r_1}, z_{r_2}). Then $T := S \cap Y$ is either one of the following two cases:

$$(\mathbf{C}^2 - (\{0\} \times \mathbf{C})) \cup \{(0, 0)\}, \tag{a}$$

$$(\mathbf{C}^2 - ((\{0\} \times \mathbf{C}) \cup (\mathbf{C} \times \{0\}))) \cup \{(0, 0)\}. \tag{b}$$

If (1) is satisfied, then there should be a compact neighborhood of $(0, 0)$ in T for the strong topology of T in $\mathbf{C}^2$. This is a contradiction in case (a) by 3.1.3 (iii). By blowing up the origin of $\mathbf{C}^2$, Case (b) is reduced to Case (a). Thus, we have proved (1) $\Rightarrow$ (2) and hence (i).

We prove (ii). Let Y be the normalization of the reduced part of X. Since $Y \to X$ is proper and surjective, we are reduced to the case where X is normal. We may assume that X is connected. In this case, if $X - A_j$ is not dense, then $A_j = X$. By this argument, we are reduced to the case (i) where $X - A_j$ are dense for all j. □

3.2 GENERALIZATIONS OF ANALYTIC SPACES

In this section, we consider generalizations of the notion of analytic spaces, and compare the categories of various kinds of these generalized analytic spaces.

As we mentioned at the beginning of Section 2.2, we think that the category $\mathcal{B}(\log)$ is the best one to work with in the theory of moduli of polarized logarithmic Hodge structures. For the roles of other categories, see 3.2.6.

3.2.1

As in Section 0.7, let $\mathcal{A}$ (resp. $\mathcal{A}(\log)$) be the category of analytic spaces (resp. fs logarithmic analytic spaces). We have considered bigger categories

$$\overline{\mathcal{A}}_1 \supset \mathcal{A}, \quad \overline{\mathcal{A}}_1(\log) \supset \mathcal{A}(\log) \qquad (2.2.1).$$

Expanding these, we will consider full subcategories

$$\begin{array}{ccccc} \overline{\mathcal{A}}_1 & \supset & \overline{\mathcal{A}}_2 & \supset & \overline{\mathcal{A}} \\ & & \cup & & \cup \\ & & \mathcal{B} & \supset & \mathcal{B}^* \end{array} \tag{1}$$

of the category of local ringed spaces over $\mathbf{C}$, all of which, except $\mathcal{B}^*$, contain $\mathcal{A}$, and consider full subcategories

$$\begin{array}{ccccccc} \overline{\mathcal{A}}_1(\log) & \supset & \overline{\mathcal{A}}_2(\log) & \supset & \overline{\mathcal{A}}(\log) & \supset & \overline{\mathcal{A}(\log)} \\ & & \cup & & \cup & & \cup \\ & & \mathcal{B}(\log) & \supset & \mathcal{B}^*(\log) & = & \mathcal{B}^*(\log) \end{array} \tag{2}$$

of the category of logarithmic local ringed spaces over $\mathbf{C}$, all of which, except $\mathcal{B}^*(\log)$, contain $\mathcal{A}(\log)$. If $\mathcal{A}_{\text{red}}$ (resp. $\mathcal{A}_{\text{red}}(\log)$) denotes the full subcategory of $\mathcal{A}$ (resp. $\mathcal{A}(\log)$) consisting of all objects whose structure sheaf $\mathcal{O}$ is reduced (i.e., has no nonzero nilpotent local sections), then $\mathcal{B}^*$ (resp. $\mathcal{B}^*(\log)$) contains $\mathcal{A}_{\text{red}}$ (resp. $\mathcal{A}_{\text{red}}(\log)$). Among these categories, the categories with the letters $\mathcal{B}$ are defined in some concrete way, but the categories with the letters $\mathcal{A}$, other than the categories $\mathcal{A}$ and $\mathcal{A}(\log)$, are defined in some abstract way.

We will see that the spaces E_σ and the important spaces $\Gamma\backslash D_\Sigma$ belong to $\mathcal{B}^*(\log)$ and that, if Γ is neat, they belong in fact to a much smaller category, the category of "logarithmic manifolds" that will be introduced in Section 3.5.

First we define $\overline{\mathcal{A}}$ and $\overline{\mathcal{A}(\log)}$. We begin with general settings.

Definition 3.2.2 *Let $\mathcal{C}$ be a category and let $\mathcal{D}$ be a full subcategory of $\mathcal{C}$. For an object S of $\mathcal{C}$, let $h^S_{\mathcal{D}}$ be the contravariant functor*

$$\mathcal{D} \to (\text{Sets}), \quad Y \mapsto h^S_{\mathcal{D}}(Y) := \text{Mor}(Y, S).$$

Define the categorical closure *$\overline{\mathcal{D}}$ of $\mathcal{D}$ in $\mathcal{C}$ to be the full subcategory of $\mathcal{C}$ consisting of all objects S for which*

$$\text{Mor}(S, Z) \to \text{Mor}(h^S_{\mathcal{D}}, h^Z_{\mathcal{D}})$$

is bijective for any object Z of $\mathcal{C}$.

Then we have $\mathcal{D} \subset \overline{\mathcal{D}} \subset \mathcal{C}$. By definition, an object S of $\overline{\mathcal{D}}$ is determined, up to canonical isomorphisms, by the functor $h^S_{\mathcal{D}}$ (i.e., by morphisms from objects of $\mathcal{D}$).

For example, if $\mathcal{C}$ is the category of abelian groups and $\mathcal{D}$ is the category of finite abelian groups, then $\overline{\mathcal{D}}$ coincides with the category of torsion abelian groups.

Definition 3.2.3

(i) *In the case*

$$\mathcal{C} := (\textit{category of local ringed spaces over } \mathbf{C}) \supset \mathcal{D} := \mathcal{A}$$

in 3.2.2, *we denote $\overline{\mathcal{D}}$ by $\overline{\mathcal{A}}$.*

(ii) *In the case*

$$\mathcal{C} := (\textit{category of fs logarithmic local ringed spaces over } \mathbf{C}) \supset \mathcal{D} := \mathcal{A}(\log),$$

we denote $\overline{\mathcal{D}}$ by $\overline{\mathcal{A}(\log)}$.

(iii) *In the case*

$$\mathcal{C} := (\textit{category of logarithmic local ringed spaces over } \mathbf{C}) \supset \mathcal{D} := \mathcal{A}(\log),$$

we denote $\overline{\mathcal{D}}$ by $\overline{\mathcal{A}(\log)}'$, and we denote by $\overline{\mathcal{A}(\log)}'_{\text{fs}}$ the full subcategory of $\overline{\mathcal{A}(\log)}'$ consisting of objects whose logarithmic structures are fs.

We will show that $\overline{\mathcal{A}(\log)}'_{\text{fs}} = \overline{\mathcal{A}(\log)}$ (see Theorem 3.2.5 below).

3.2.4

We define the other categories in the diagrams (1) and (2) in 3.2.1.

Recall that $\overline{\mathcal{A}}_1$ is the category of local ringed spaces over $\mathbf{C}$ satisfying the condition 2.2.1 (A_1)

Let $\overline{\mathcal{A}}_2$ be the category of local ringed spaces X over $\mathbf{C}$ satisfying the condition 2.2.1 (A_1) and also the following condition (A_2).

(A_2) *A subset U of X is open if and only if, for any analytic space A and any morphism $\lambda : A \to X$, $\lambda^{-1}(U)$ is open in A.*

Let $\mathcal{B}$ be the category of local ringed spaces X over $\mathbf{C}$ which have an open covering $(U_\lambda)_\lambda$ satisfying the following condition (B).

(B) *For each λ, there exist an analytic space Z_λ and a subset S_λ of Z_λ such that, as local ringed spaces over* $\mathbf{C}$, *U_λ is isomorphic to S_λ which is endowed with the strong topology in Z_λ and the inverse image of $\mathcal{O}_{Z_\lambda}$.*

In [KU1], we called an object of $\overline{\mathcal{A}}$ a "generalized analytic space." Since an object of the category $\mathcal{B}$ is also a kind of a "generalized analytic space" and is also important in this book, we now propose to call an object of $\overline{\mathcal{A}}$ a *categorical generalized analytic space* and an object of $\mathcal{B}$ a *geometrical generalized analytic space*.

Let $\mathcal{B}^*$ be the category of local ringed spaces X over $\mathbf{C}$ which have an open covering $(U_\lambda)_\lambda$ satisfying the following condition (B*).

(B*) *For each λ, there exist a reduced analytic space Z_λ, a closed analytic subspace A_λ of Z_λ whose complement $Z_\lambda - A_\lambda$ is dense in Z_λ, and an injection $U_\lambda \to Z_\lambda$ such that, when we regard U_λ as a subset of Z_λ, the topology of U_λ coincides with the strong topology of U_λ in Z_λ, U_λ is open in $U_\lambda \cup (Z_\lambda - A_\lambda)$ in the strong topology of $U_\lambda \cup (Z_\lambda - A_\lambda)$ in Z_λ, and $\mathcal{O}_{U_\lambda}$ is isomorphic over $\mathbf{C}$ to the inverse image of $\mathcal{O}_{Z_\lambda}$.*

For $\mathcal{C} = \overline{\mathcal{A}}_1, \overline{\mathcal{A}}_2, \overline{\mathcal{A}}, \mathcal{B}, \mathcal{B}^*$, let $\mathcal{C}(\log)$ be the category of objects of $\mathcal{C}$ endowed with an fs logarithmic structure.

THEOREM 3.2.5 *We have inclusions in the diagrams* (1) *and* (2) *in* 3.2.1. *Furthermore, we have*

$$\overline{\mathcal{A}(\log)}'_{\mathrm{fs}} = \overline{\mathcal{A}(\log)}.$$

In theorem 3.2.5, the inclusions

$$\overline{\mathcal{A}}_2 \subset \overline{\mathcal{A}}_1, \quad \mathcal{B}^* \subset \mathcal{B}$$

and their "log versions" are clear. We prove $\mathcal{B} \subset \overline{\mathcal{A}}_2$ (and hence $\mathcal{B}(\log) \subset \overline{\mathcal{A}}_2(\log)$) in 3.2.10 below. The remaining inclusions

$$\mathcal{B}^* \subset \overline{\mathcal{A}} \subset \overline{\mathcal{A}}_2, \quad \mathcal{B}^*(\log) \subset \overline{\mathcal{A}(\log)} \subset \overline{\mathcal{A}}(\log)$$

and the coincidence $\overline{\mathcal{A}(\log)}'_{\mathrm{fs}} = \overline{\mathcal{A}(\log)}$ will be proved later in Section 8.3.

3.2.6

Among the above many categories, the categories $\overline{\mathcal{A}}_1(\log)$, $\overline{\mathcal{A}}_2(\log)$, $\mathcal{B}(\log)$, $\mathcal{B}^*(\log)$, and $\overline{\mathcal{A}(\log)}$ play the following special roles in this book.

First, as in Chapter 2, $\overline{\mathcal{A}}_1(\log)$ is the category of spaces on which we can define a PLH.

Next, $\overline{\mathcal{A}}_2(\log)$ is the category for which we have the theorem on moduli of PLHs

$$\underline{\mathrm{PLH}}_{\Phi,\overline{\mathcal{A}}_2(\log)} \simeq \mathrm{Mor}(\quad, \Gamma\backslash D_\Sigma) \quad \text{(theorem B in Section 4.2)},$$

which is more general than Theorem 0.4.27 (i) in Chapter 0. However, the definition of the category $\overline{\mathcal{A}}_2(\log)$ is too abstract, and we do not see clearly how big it is.

The full subcategory $\mathcal{B}(\log)$ of $\overline{\mathcal{A}}_2(\log)$ can be understood better, and this is the reason we have chosen to state Theorem 0.4.27 (i) in this category. Also, as in Section 3.5 below, $\mathcal{B}(\log)$ has a good theory of infinitesimal calculus. It has fiber products (3.5.1). We mainly work with $\mathcal{B}(\log)$ in this book.

Finally, it will be proved that our space $\Gamma\backslash D_\Sigma$ belongs to $\mathcal{B}^*(\log)$ and hence, by Theorem 3.2.5, to $\overline{\mathcal{A}(\log)} = \overline{\mathcal{A}(\log)}'_{\mathrm{fs}} \subset \overline{\mathcal{A}(\log)}'$. The facts that $\Gamma\backslash D_\Sigma \in \overline{\mathcal{A}(\log)}'$

and the isomorphism $\underline{\mathrm{PLH}}_{\Phi,\mathcal{A}(\log)} \simeq \mathrm{Mor}(\ \ ,\Gamma\backslash D_\Sigma)|_{\mathcal{A}(\log)}$ will show that $\Gamma\backslash D_\Sigma$ has the universal property stated in Theorem 0.4.27 (ii) (see 4.2.2).

3.2.7

We describe the differences among the above categories.

(1) Let Z be an analytic space and let X be a subset of Z. Endow X with a topology $\mathcal{T}$ such that

$$\text{(weak topology)} \le \mathcal{T} \le \text{(strong topology)},$$

and define $\mathcal{O}_X$ to be the inverse image of $\mathcal{O}_Z$. Then X belongs to $\overline{\mathcal{A}}_1$. It belongs to $\overline{\mathcal{A}}_2$ if and only if the topology of X is the strong topology.

Since $\overline{\mathcal{A}} \subset \overline{\mathcal{A}}_2$ by Theorem 3.2.5, this shows that if X in this (1) is a categorically generalized analytic space, then X is a geometrically generalized analytic space (cf. 3.2.4), and this shows that the strong topology occurs naturally when we consider categorically generalized analytic spaces.

For example, if $Z = \mathbf{C}$ and X is the interval $[0, 1] \subset \mathbf{R}$ with the weak topology in Z, or if $Z = \mathbf{C}^2$ and $X = (\mathbf{C}^2 - (\{0\} \times \mathbf{C})) \cup \{(0, 0)\}$ with the weak topology in Z (not as in 3.1.3 where we considered the strong topology), then the topology of X is different from the strong topology (see 3.1.2 and 3.1.3), and hence X belongs to $\overline{\mathcal{A}}_1$ but not to $\overline{\mathcal{A}}_2$.

(2) Let X be a one-point set and endow X with the ring $\mathbf{C}\{\{T\}\}$ (resp. $\mathbf{C}[[T]]$) of convergent power series (resp. formal power series). Then X belongs to $\mathcal{B}$ (resp. $\overline{\mathcal{A}}$) but not to $\overline{\mathcal{A}}$ (resp. $\mathcal{B}$).

(3) The authors do not know any example that gives the difference between $\overline{\mathcal{A}}(\log)$ and $\overline{\mathcal{A}(\log)}$.

These statements in 3.2.7 will be proved in 3.2.11 below.
We give preliminary lemmas for the proof of $\mathcal{B} \subset \overline{\mathcal{A}}_2$.

Lemma 3.2.8 *Let Z be an analytic space and let X be a subset of Z. Let $\mathcal{T}$ be a topology on X satisfying*

$$\textit{(weak topology)} \le \mathcal{T} \le \textit{(strong topology)}.$$

Regard X as a local ringed space over $\mathbf{C}$ with the topology $\mathcal{T}$ and with the inverse image of $\mathcal{O}_Z$.

(i) *If Y is a local ringed space over $\mathbf{C}$ satisfying the condition (A_2) in 3.2.4, then the canonical map*

$$\mathrm{Mor}(Y, X) \to \{\lambda \in \mathrm{Mor}(Y, Z) \mid \lambda(Y) \subset X\}$$

is bijective.

(ii) *X satisfies the condition (A_2) in 3.2.4 if and only if $\mathcal{T}$ coincides with the strong topology.*

Proof. (i) follows easily from the definitions, and (ii) follows from (i). □

Lemma 3.2.9 *Let X be a local ringed space over* **C**. *Assume that, for any $x \in X$, the local ring $\mathcal{O}_{X,x}$ is Noetherian and its residue field is* **C**. *Then, for any open set U of X, the canonical morphism* $\mathrm{Mor}(U, \mathbf{C}^n) \to \mathcal{O}(U)^n$ *is injective. (Here $\mathbf{C}^n$ is regarded as an analytic space.)*

Proof. Let $\varphi, \psi \in \mathrm{Mor}(U, \mathbf{C}^n)$ and assume $z_j \circ \varphi = z_j \circ \psi$ $(1 \le j \le n)$, where the z_j are the coordinate functions on $\mathbf{C}^n$. We prove $\varphi = \psi$. We have $\varphi(x) = \psi(x)$ for any $x \in U$. Put $y = \varphi(x) = \psi(x)$ and $Y = \mathbf{C}^n$. It is sufficient to prove that the local ring homomorphisms $\varphi_x^*, \psi_x^* : \mathcal{O}_{Y,y} \to \mathcal{O}_{X,x}$ coincide for any $x \in U$.

Since $\mathcal{O}_{X,x}$ is Noetherian, $\mathcal{O}_{X,x}$ is embedded into its completion $\hat{\mathcal{O}}_{X,x}$. Write $y = (a_1, \ldots, a_n) \in \mathbf{C}^n$. Then φ_x^* and ψ_x^* induce local ring homomorphisms

$$\hat{\varphi}_x^*, \hat{\psi}_x^* : \hat{\mathcal{O}}_{Y,y} = \mathbf{C}[[T_1, \ldots, T_n]] \to \hat{\mathcal{O}}_{X,x}, \tag{1}$$

where T_j denotes the image of $z_j - a_j$. Since $\mathbf{C}[[T_1, \ldots, T_n]]$ is generated topologically by $T_1, \ldots, T_n$ as a ring over **C** in the $(T_1, \ldots, T_n)$-adic topology, the homomorphisms $\hat{\varphi}_x^*$ and $\hat{\psi}_x^*$ in (1) are determined by the images of T_j $(1 \le j \le n)$. Hence $\hat{\varphi}_x^*$ and $\hat{\psi}_x^*$ are determined by the images of $z_j \circ \varphi$ and $z_j \circ \psi$ in $\hat{\mathcal{O}}_{X,x}$ $(1 \le j \le n)$, respectively. Hence $\varphi_x^* = \psi_x^*$. □

3.2.10

Proof of $\mathcal{B} \subset \overline{\mathcal{A}}_2$. Let X be an object of $\mathcal{B}$. Let $\mathrm{Mor}(\ \ , \mathbf{C}^n)$ be the sheaf on X defined by $U \mapsto \mathrm{Mor}(U, \mathbf{C}^n)$ for open sets U of X. By Lemma 3.2.8 (ii), the condition (A_2) in 3.2.4 is satisfied. Hence, to prove that X belongs to $\overline{\mathcal{A}}_2$, it is sufficient to show that the condition (A_1) in 2.2.1 is satisfied, that is, it is sufficient to show that $\mathrm{Mor}(\ \ , \mathbf{C}^n)_x \xrightarrow{\sim} \mathcal{O}_{X,x}^n$ for any $x \in X$. Since the question is local, we may assume that X is a subset of an analytic space Z and X is endowed with the strong topology in Z and with the inverse image of $\mathcal{O}_Z$. Then $\mathcal{O}_{X,x} = \mathcal{O}_{Z,x}$, and hence $\mathcal{O}_{X,x}$ is Noetherian. By Lemma 3.2.9, $\mathrm{Mor}(\ \ , \mathbf{C}^n)_x \to \mathcal{O}_{X,x}^n$ is injective. It remains to prove the surjectivity. But an element f of $\mathcal{O}_{X,x}^n = \mathcal{O}_{Z,x}^n$ defines a morphism from an open neighborhood of x in Z to $\mathbf{C}^n$, and it induces a morphism from an open neighborhood of x in X to $\mathbf{C}^n$ which induces f. □

3.2.11

Proofs of the statements in 3.2.7. Let X be as in 3.2.7 (1). Then the fact that X belongs to $\overline{\mathcal{A}}_1$ is proved in the same way as the proof of $\mathcal{B} \subset \overline{\mathcal{A}}_2$ in 3.2.10. The statement concerning $\overline{\mathcal{A}}_2$ follows from Lemma 3.2.8 (ii).

Next, as in 3.2.7 (2), let X be a one-point set with the ring $\mathbf{C}\{\{T\}\}$ (resp. $\mathbf{C}[[T]]$). For $n \ge 1$, let $S_n = \mathrm{Spec}(\mathbf{C}[T]/(T^n))$. Let S be the one-point set endowed with the ring $\mathbf{C}[[T]]$. For an analytic space Y and a morphism $Y \to X$ of local ringed spaces, since $T \in \mathcal{O}(X)$ is contained in the maximal ideal of the local ring of X at the unique point of X, the image of T in $\mathcal{O}_{Y,y}$ for any $y \in Y$ is contained in the maximal ideal of $\mathcal{O}_{Y,y}$. This shows that the image of T in $\mathcal{O}_Y$ is locally nilpotent. Hence the functor $h_{\mathcal{A}}^X$ is the sheaf on the category $\mathcal{A}$ associated to the presheaf $\varinjlim_n h_{\mathcal{A}}^{S_n}$. Hence, for any local ringed space Z over $\mathbf{C}$, we have

$$\begin{aligned}\mathrm{Mor}\,(h^X_{\mathcal{A}}, h^Z_{\mathcal{A}}) &= \mathrm{Mor}\,(\varinjlim_n h^{S_n}_{\mathcal{A}}, h^Z_{\mathcal{A}}) = \varprojlim_n \mathrm{Mor}\,(h^{S_n}_{\mathcal{A}}, h^Z_{\mathcal{A}}) \\ &= \varprojlim_n \mathrm{Mor}(S_n, Z) = \mathrm{Mor}(\varinjlim_n S_n, Z) = \mathrm{Mor}(S, Z).\end{aligned}$$

This shows that X does not belong (resp. belongs) to $\overline{\mathcal{A}}$. On the other hand, clearly X belongs (resp. does not belong) to $\mathcal{B}$. □

3.3 SETS E_σ AND $E^\sharp_\sigma$

In this section, let Σ be a fan in $\mathfrak{g}_{\mathbf{Q}}$ and let Γ be a subgroup of $G_{\mathbf{Z}}$ that is strongly compatible with Σ (cf. 1.3.10 (ii)). In the next section, we will endow $\Gamma\backslash D_\Sigma$ with the structure of a logarithmic local ringed space over $\mathbf{C}$. For this, we use a surjective map

$$\bigsqcup_{\sigma\in\Sigma} E_\sigma \to \Gamma\backslash D_\Sigma$$

where E_σ is a subset of a certain fs logarithmic analytic space $\check{E}_\sigma$, and we transport the structure on $\check{E}_\sigma$ of a logarithmic local ringed space to E_σ and then to $\Gamma\backslash D_\Sigma$ via this surjection. In this section, we define $\check{E}_\sigma$, E_σ, and the above surjection.

3.3.1

For an fs monoid $\mathcal{S}$, let

$$\mathcal{S}^\vee := \mathrm{Hom}(\mathcal{S}, \mathbf{N}),$$

where $\mathbf{N} := \mathbf{Z}_{\geq 0}$ is regarded as an additive monoid. Then $\mathcal{S}^\vee$ is an fs monoid. We have

$$\mathcal{S}/\mathcal{S}^\times \xrightarrow{\sim} (\mathcal{S}^\vee)^\vee.$$

For an fs monoid $\mathcal{S}$, a *face* of $\mathcal{S}$ is a submonoid $\mathcal{S}'$ of $\mathcal{S}$ having the following property: If $a, b \in \mathcal{S}$ and $ab \in \mathcal{S}'$, then $a, b \in \mathcal{S}'$. There is a bijection

$$\delta : \{\text{face of } \mathcal{S}\} \xrightarrow{\sim} \{\text{face of } \mathcal{S}^\vee\}, \quad \mathcal{S}' \mapsto \{h \in \mathcal{S}^\vee \mid h(\mathcal{S}') = \{0\}\}. \tag{1}$$

Let Σ and Γ be as at the beginning of this section. For $\sigma \in \Sigma$, $\Gamma(\sigma) = \Gamma \cap \exp(\sigma)$ (1.3.10 (2)) is an fs monoid with $\Gamma(\sigma)^\times = \{1\}$ and there is a bijection

$$\{\text{face of } \sigma\} \xrightarrow{\sim} \{\text{face of } \Gamma(\sigma)\}, \quad \tau \mapsto \Gamma(\tau). \tag{2}$$

3.3.2

We define the spaces toric_σ, torus_σ, and $\check{E}_\sigma$.

In the following, $\mathbf{C}^{\text{mult}}$ (resp. $\mathbf{R}^{\text{mult}}_{\geq 0}$) denotes the set $\mathbf{C}$ (resp. $\mathbf{R}_{\geq 0}$) regarded as a monoid with respect to the multiplication.

For $\sigma \in \Sigma$, denote

$$\begin{aligned}\text{toric}_\sigma &= \operatorname{Spec}(\mathbf{C}[\Gamma(\sigma)^\vee])_{\text{an}} = \operatorname{Hom}(\Gamma(\sigma)^\vee, \mathbf{C}^{\text{mult}}),\\ \text{torus}_\sigma &= \operatorname{Spec}(\mathbf{C}[\Gamma(\sigma)^{\vee\,\text{gp}}])_{\text{an}} = \operatorname{Hom}(\Gamma(\sigma)^{\vee\,\text{gp}}, \mathbf{C}^\times) = \mathbf{C}^\times \otimes \Gamma(\sigma)^{\text{gp}} \subset \text{toric}_\sigma,\end{aligned}$$

We denote

$$\check{E}_\sigma = \text{toric}_\sigma \times \check{D}.$$

(toric$_\sigma$, torus$_\sigma$, and $\check{E}_\sigma$ depend on Γ, not only on σ, and they should be written as toric$_{\Gamma(\sigma)}$, torus$_{\Gamma(\sigma)}$, and $\check{E}_{\Gamma(\sigma)}$, respectively. But we abbreviate them as above since Γ is usually clear.)

We endow toric$_\sigma$ with the canonical logarithmic structure in 2.1.6 (ii).

We endow $\check{E}_\sigma$ with the inverse image of the canonical logarithmic structure of the toric variety toric$_\sigma$. Since $\check{D}$ is smooth, $\check{E}_\sigma$ is a logarithmically smooth fs logarithmic analytic space.

We have

$$\pi_1(\text{toric}_\sigma^{\log}) \simeq \pi_1(\text{torus}_\sigma) \simeq \Gamma(\sigma)^{\text{gp}}.$$

We often identify these three groups.

Let $q \in \text{toric}_\sigma$. Then we have a face $\sigma(q)$ of σ defined as follows. We have a canonical injection $\pi_1(q^{\log}) \to \pi_1(\text{toric}_\sigma^{\log}) = \Gamma(\sigma)^{\text{gp}}$, and, if we regard $\pi_1(q^{\log})$ as a subgroup of $\Gamma(\sigma)^{\text{gp}}$ via this injection, we have

$$\pi_1^+(q^{\log}) = \pi_1(q^{\log}) \cap \Gamma(\sigma)$$

and $\pi_1^+(q^{\log})$ is a face of $\Gamma(\sigma)$. Let $\sigma(q)$ be the face of σ corresponding to the face $\pi_1^+(q^{\log})$ of $\Gamma(\sigma)$ via the bijection (2) in 3.3.1. The face of $\Gamma(\sigma)^\vee$ corresponding to $\pi_1^+(q^{\log})$ via the bijection (1) in 3.3.1 is $\{f \in \Gamma(\sigma)^\vee \mid f(q) \neq 0\}$. Here we regard $f \in \Gamma(\sigma)^\vee$ as a holomorphic function on $\operatorname{Spec}(\mathbf{C}[\Gamma(\sigma)^\vee])_{\text{an}}$ and so the value $f(q) \in \mathbf{C}$ of f at q is defined.

3.3.3

We define a canonical pre-PLH $H_\sigma = (H_{\sigma,\mathbf{Z}}, \langle\ ,\ \rangle_\sigma, F_\sigma)$ on $\check{E}_\sigma$ as follows.

Let $(H_{\sigma,\mathbf{Z}}, \langle\ ,\ \rangle_\sigma)$ be the local system on toric$_\sigma^{\log}$ whose stalk at the unit point $1 \in$ torus$_\sigma \subset$ toric$_\sigma^{\log}$ is $(H_0, \langle\ ,\ \rangle_0)$ with the action of $\pi_1(\text{toric}_\sigma^{\log})$ given by the canonical injection $\pi_1(\text{toric}_\sigma^{\log}) = \Gamma(\sigma)^{\text{gp}} \hookrightarrow G_\mathbf{Z}$. By 2.3.7, we have a canonical isomorphism of $\mathcal{O}^{\log}_{\text{toric}_\sigma}$-modules

$$\nu : \mathcal{O}^{\log}_{\text{toric}_\sigma} \otimes_\mathbf{Z} H_{\sigma,\mathbf{Z}} \simeq \mathcal{O}^{\log}_{\text{toric}_\sigma} \otimes_\mathbf{Z} H_0,$$

via which we identify these two $\mathcal{O}^{\log}_{\text{toric}_\sigma}$-modules. We also denote by $(H_{\sigma,\mathbf{Z}}, \langle\ ,\ \rangle_\sigma)$ the local system on $\check{E}_\sigma^{\log}$ which is the inverse image of $(H_{\sigma,\mathbf{Z}}, \langle\ ,\ \rangle_\sigma)$ via the projection

$\check{E}_\sigma^{\log} \to \text{toric}_\sigma^{\log}$. We have

$$\mathcal{O}_{\check{E}_\sigma}^{\log} \otimes_{\mathbf{Z}} H_{\sigma,\mathbf{Z}} = \mathcal{O}_{\check{E}_\sigma}^{\log} \otimes_{\mathbf{Z}} H_0$$

via the above identification. On the other hand, $\mathcal{O}_{\check{D}} \otimes_{\mathbf{Z}} H_0$ has the universal Hodge filtration. Via the projection $\check{E}_\sigma \to \check{D}$, it defines a filtration on $\mathcal{O}_{\check{E}_\sigma} \otimes_{\mathbf{Z}} H_0$ and hence a filtration F_σ on $\mathcal{O}_{\check{E}_\sigma}^{\log} \otimes_{\mathbf{Z}} H_0 = \mathcal{O}_{\check{E}_\sigma}^{\log} \otimes_{\mathbf{Z}} H_{\sigma,\mathbf{Z}}$ with which $H_\sigma := (H_{\sigma,\mathbf{Z}}, \langle\ ,\ \rangle_\sigma, F_\sigma)$ becomes a pre-PLH on $\check{E}_\sigma$.

This pre-PLH has the unique $\Gamma(\sigma)^{\text{gp}}$-level structure μ_σ whose germ at $1 \in \text{torus}_\sigma$ is the class of the identity map of H_0. With the identification $1 \otimes H_{\sigma,\mathbf{Z}} = \xi^{-1}(1 \otimes H_0)$ where ξ is as in 2.3.7 (we take $\Gamma(\sigma)^\vee$ as $\mathcal{S}$ in 2.3.7), this canonical $\Gamma(\sigma)^{\text{gp}}$-level structure is expressed as

$$\xi(1 \otimes v) = 1 \otimes \mu_\sigma(v) \quad \text{for } v \in H_{\sigma,\mathbf{Z}}. \tag{1}$$

3.3.4

We define subsets E_σ and $\tilde{E}_\sigma$ of $\check{E}_\sigma$. Let H_σ be the canonical pre-PLH on $\check{E}_\sigma$ defined above. We define E_σ to be the set of all points x of $\check{E}_\sigma$ such that the inverse image of H_σ to the fs logarithmic point x is a PLH. We define $\tilde{E}_\sigma$ to be the set of all points x of $\check{E}$ such that the inverse image of H_σ to x satisfies Griffiths transversality (the small Griffiths transversality). We have

$$E_\sigma \subset \tilde{E}_\sigma \subset \check{E}_\sigma.$$

3.3.5

We have a canonical surjective homomorphism

$$\begin{aligned} \mathbf{e} : \sigma_{\mathbf{C}} &\to \text{torus}_\sigma = \mathbf{C}^\times \otimes \Gamma(\sigma)^{\text{gp}}, \\ z\log(\gamma) &\mapsto \exp(2\pi i z) \otimes \gamma \quad (z \in \mathbf{C}, \gamma \in \Gamma(\sigma)^{\text{gp}}), \end{aligned} \tag{1}$$

whose kernel coincides with $\log(\Gamma(\sigma)^{\text{gp}})$.

Let $q \in \text{toric}_\sigma$, and let $\mathcal{S} = \Gamma(\sigma)^\vee$, $\mathcal{S}' = \{f \in \mathcal{S} \mid f(q) \neq 0\}$ where $f \in \mathcal{S}$ is regarded as a holomorphic function on toric_σ. By passing to the quotient, $\mathbf{e}$ induces a commutative diagram with surjective vertical arrows

$$\begin{array}{ccc} \sigma_{\mathbf{C}}/\log(\Gamma(\sigma)^{\text{gp}}) & \xrightarrow[\sim]{\mathbf{e}} & \text{torus}_\sigma = \text{Hom}(\mathcal{S}^{\text{gp}}, \mathbf{C}^\times) \\ \downarrow & & \downarrow \\ \sigma_{\mathbf{C}}/(\sigma(q)_{\mathbf{C}} + \log(\Gamma(\sigma)^{\text{gp}})) & \xrightarrow{\sim} & \text{Hom}((\mathcal{S}')^{\text{gp}}, \mathbf{C}^\times). \end{array} \tag{2}$$

We define the class of q in $\sigma_{\mathbf{C}}/(\sigma(q)_{\mathbf{C}} + \log(\Gamma(\sigma)^{\text{gp}}))$ to be the element that corresponds to the element $f \mapsto f(q)$ of $\text{Hom}((\mathcal{S}')^{\text{gp}}, \mathbf{C}^\times)$ via the lower isomorphism of this diagram (see 3.3.2 for $\sigma(q)$).

PROPOSITION 3.3.6 *The period map*

$$\check{\varphi} : \check{E}_\sigma \to \Gamma(\sigma)^{\mathrm{gp}} \backslash \check{D}_{\mathrm{orb}}$$

(2.5.3) *of the canonical pre-PLH H_σ on $\check{E}_\sigma$ with the canonical $\Gamma(\sigma)^{\mathrm{gp}}$-level structure μ_σ* (3.3.3) *is given by*

$$\check{\varphi}(q, F) = ((\sigma(q), \exp(\sigma(q)_{\mathbf{C}}) \exp(z) F) \bmod \Gamma(\sigma)^{\mathrm{gp}})$$

($q \in \mathrm{toric}_\sigma$, $F \in \check{D}$) *where z is any fixed element of $\sigma_{\mathbf{C}}$ whose image in $\sigma_{\mathbf{C}}/(\sigma(q)_{\mathbf{C}} + \log(\Gamma(\sigma)^{\mathrm{gp}}))$ is the class of q in the sense of* 3.3.5.

Proof. Take $y \in \check{E}_\sigma^{\log}$ lying over (q, F). Since the $\Gamma(\sigma)^{\mathrm{gp}}$-level structure μ_σ is given by $\mu_\sigma(v) = \xi(1 \otimes v)$ (3.3.3 (2)), we have $\tilde{\mu}_{\sigma,y}(F_\sigma(s)) = \xi(s)F$. Let $\mathcal{S} = \Gamma(\sigma)^\vee$, and let $\mathcal{S}' = \{f \in \mathcal{S} \mid f(q) \neq 0\}$ as in 3.3.5. For $s \in \mathrm{sp}(y)$, let $\tilde{s} : \mathcal{S}^{\mathrm{gp}} \to \mathbf{C}^\times$ be the homomorphism $f \mapsto \exp(s(\log(f)))$. When s ranges over $\mathrm{sp}(y)$, $\tilde{s}$ ranges over all homomorphisms $\mathcal{S}^{\mathrm{gp}} \to \mathbf{C}^\times$ whose restriction to $(\mathcal{S}')^{\mathrm{gp}}$ coincides with $f \mapsto f(q)$. On the other hand, when we regard $\tilde{s}$ as an element of torus_σ, we have $\tilde{s} = \mathbf{e}(\sum_{j=1}^n (2\pi i)^{-1} s(\log(q_j)) N_j)$ where $(q_j)_{1 \le j \le n}$ and $(N_j)_{1 \le j \le n}$ are as in 2.3.7 (we take $\Gamma(\sigma)^\vee$ as $\mathcal{S}$ in 2.3.7). Hence, as s ranges over $\mathrm{sp}(y)$, $(\sum_{j=1}^n (2\pi i)^{-1} s(\log(q_j)) N_j \bmod \log(\Gamma(\sigma)^{\mathrm{gp}}))$ ranges over all elements of $\sigma_{\mathbf{C}}/\log(\Gamma(\sigma)^{\mathrm{gp}})$ whose images in $\sigma_{\mathbf{C}}/(\sigma(q)_{\mathbf{C}} + \log(\Gamma(\sigma)^{\mathrm{gp}}))$ coincide with the class of q. This proves the result. □

COROLLARY 3.3.7 *Let $(q, F) \in \check{E}_\sigma$ and let z be an element of $\sigma_{\mathbf{C}}$ whose image in $\sigma_{\mathbf{C}}/(\sigma(q)_{\mathbf{C}} + \log(\Gamma(\sigma)^{\mathrm{gp}}))$ coincides with the class of q.*

(i) *$(q, F) \in E_\sigma$ if and only if $\exp(\sigma(q)_{\mathbf{C}}) \exp(z) F$ is a $\sigma(q)$-nilpotent orbit.*
(ii) *$(q, F) \in \tilde{E}_\sigma$ if and only if $N(F^p) \subset F^{p-1}$ for all $N \in \sigma(q)$ and for all p.*

Proof. This follows from 3.3.6 and 2.5.5. □

3.3.8

The inverse image of H_σ on E_σ is a PLH. By 3.3.6, it is a PLH of type $(w, (h^{p,q}), H_0, \langle\ ,\ \rangle_0, \Gamma(\sigma)^{\mathrm{gp}}, \{\text{face of } \sigma\})$. The period map $\check{\varphi} : E_\sigma \to \Gamma(\sigma)^{\mathrm{gp}} \backslash \check{D}_{\mathrm{orb}}$ takes values in $\Gamma(\sigma)^{\mathrm{gp}} \backslash D_\sigma$ and coincides with the period map $\varphi : E_\sigma \to \Gamma(\sigma)^{\mathrm{gp}} \backslash D_\sigma$ in 2.5.10.

3.3.9

For $\sigma \in \Sigma$, denote

$$|\mathrm{toric}|_\sigma = \mathrm{Hom}(\Gamma(\sigma)^\vee, \mathbf{R}_{\ge 0}^{\mathrm{mult}}) \subset \mathrm{toric}_\sigma,$$
$$|\mathrm{torus}|_\sigma = \mathrm{Hom}(\Gamma(\sigma)^{\vee\,\mathrm{gp}}, \mathbf{R}_{>0}) = \mathbf{R}_{>0} \otimes \Gamma(\sigma)^{\mathrm{gp}} \subset \mathrm{torus}_\sigma .$$

Define

$$\check{E}_\sigma^\sharp = |\mathrm{toric}|_\sigma \times \check{D} \subset \check{E}_\sigma, \quad E_\sigma^\sharp = E_\sigma \cap \check{E}_\sigma^\sharp.$$

3.3.10

We define a kind of a period map $\varphi^\sharp : E_\sigma^\sharp \to D_\sigma^\sharp$ which makes the following diagram commutative.

$$\begin{array}{ccc} E_\sigma^\sharp & \xrightarrow{\varphi^\sharp} & D_\sigma^\sharp \\ \downarrow & & \downarrow \\ E_\sigma & \xrightarrow{\varphi} & \Gamma(\sigma)^{\mathrm{gp}}\backslash D_\sigma. \end{array}$$

The injection

$$|\mathrm{toric}|_\sigma = \mathrm{Hom}(\Gamma(\sigma)^\vee, \mathbf{R}_{\geq 0}^{\mathrm{mult}})$$
$$\hookrightarrow \mathrm{toric}_\sigma^{\log} = \mathrm{Hom}(\Gamma(\sigma)^\vee, \mathbf{R}_{\geq 0}^{\mathrm{mult}}) \times \mathrm{Hom}(\Gamma(\sigma)^{\vee\,\mathrm{gp}}, \mathbf{S}^1)$$

sending $h \in \mathrm{Hom}(\Gamma(\sigma)^\vee, \mathbf{R}_{\geq 0}^{\mathrm{mult}})$ to $(h, 1)$, where 1 denotes the trivial homomorphism $\Gamma(\sigma)^{\vee\,\mathrm{gp}} \to \mathbf{S}^1$, induces an injection

$$\check{E}_\sigma^\sharp = |\mathrm{toric}|_\sigma \times \check{D} \hookrightarrow \mathrm{toric}_\sigma^{\log} \times \check{D} = \check{E}_\sigma^{\log}.$$

The restriction of $\xi = \exp(\sum_{j=1}^n (2\pi i)^{-1} \log(q_j) \otimes N_j)$ ($(q_j)_{1\leq j\leq n}$ and $(N_j)_{1\leq j\leq n}$ are as in 2.3.7 where we take $\Gamma(\sigma)^\vee$ as S of 2.3.7) to $|\mathrm{toric}|_\sigma$ has one global branch defined by taking the branch of $\log(q_j)$ whose restriction to $|\mathrm{torus}|_\sigma$ has only real values for each j. This branch of ξ on $|\mathrm{toric}|_\sigma$ is independent of the choice of the basis $(q_j)_{1\leq j\leq n}$ of $\Gamma(\sigma)^{\vee\,\mathrm{gp}}$. By using this ξ, define an isomorphism $H_{\sigma,\mathbf{Z}} \xrightarrow{\sim} H_0$ on $|\mathrm{toric}|_\sigma$ by $v \mapsto \xi(1 \otimes v) \in 1 \otimes H_0 = H_0$ $(v \in H_{\sigma,\mathbf{Z}})$.

Define $\varphi^\sharp : E_\sigma^\sharp \to D_\sigma^\sharp$ following the definition of $\varphi^{\log} : \check{E}_\sigma^{\log} \to \check{D}_\sigma^\sharp$ as in 2.5.10 but without modulo $\Gamma(\sigma)^{\mathrm{gp}}$ by using this isomorphism $H_{\sigma,\mathbf{Z}} \xrightarrow{\sim} H_0$ on $|\mathrm{toric}|_\sigma$ as $\tilde{\mu}_y$ in 2.5.10.

An explicit description of this map is

$$\varphi^\sharp(q, F) = (\sigma(q), \exp(i\sigma(q)_{\mathbf{R}}) \exp(iy)F),$$

where y is any element of $\sigma_{\mathbf{R}}$ such that the image of iy in $\sigma_{\mathbf{C}}/(\sigma(q)_{\mathbf{C}} + \sigma_{\mathbf{R}})$ coincides with the class of q (3.3.5).

3.3.11

Example. Let $\sigma \in \Xi$ with $\sigma \neq \{0\}$. Then, $\Gamma(\sigma) \simeq \mathbf{N}$, $\mathrm{toric}_\sigma = \mathbf{C}$, $|\mathrm{toric}|_\sigma = \mathbf{R}_{\geq 0}$, $\check{E}_\sigma = \mathbf{C} \times \check{D}$, $\check{E}_\sigma^\sharp = (\mathbf{R}_{\geq 0}) \times \check{D}$. Let γ be the generator of $\Gamma(\sigma)$ and let $N := \log(\gamma)$. Then we have

$$E_\sigma = \left\{ (q, F) \in \mathbf{C} \times \check{D} \;\middle|\; \begin{array}{l} \exp((2\pi i)^{-1} \log(q) N)F \in D \text{ if } q \neq 0, \text{ and} \\ \exp(\mathbf{C}N)F \text{ is a } \sigma\text{-nilpotent orbit if } q = 0 \end{array} \right\},$$

$$E_\sigma^\sharp = \left\{ (q, F) \in (\mathbf{R}_{\geq 0}) \times \check{D} \;\middle|\; \begin{array}{l} \exp((2\pi i)^{-1} \log(q) N)F \in D \text{ if } q \neq 0, \text{ and} \\ \exp(i\mathbf{R}N)F \text{ is a } \sigma\text{-nilpotent } i\text{-orbit if } q = 0 \end{array} \right\}.$$

Here the cases $q \neq 0$, $q = 0$ correspond to $\sigma(q) = \{0\}$, $\sigma(q) = \sigma$, respectively. The period map $\varphi : E_\sigma \to \Gamma(\sigma)^{\mathrm{gp}}\backslash D_\sigma$ (resp. $\varphi^\sharp : E_\sigma^\sharp \to D_\sigma^\sharp$) is

$$(q, F) \mapsto \exp((2\pi i)^{-1}\log(q)N)F \quad \text{if } q \neq 0,$$
$$(q, F) \mapsto (\sigma, \exp(\mathbf{C}N)F) \quad (\text{resp. } (\sigma, \exp(i\mathbf{R}N)F)) \quad \text{if } q = 0.$$

In the case of Example (i) in 0.3.2 where $D = \mathfrak{h}$ and $\check{D} = \mathbf{P}^1(\mathbf{C})$, for

$$\Gamma = \begin{pmatrix} 1 & \mathbf{Z} \\ 0 & 1 \end{pmatrix} \subset \mathrm{SL}(2, \mathbf{Z}), \quad \sigma = \begin{pmatrix} 0 & \mathbf{R}_{\geq 0} \\ 0 & 0 \end{pmatrix},$$

we have

$$E_\sigma = \{(q, z) \in \mathbf{C} \times \mathbf{C} \mid |qe^{2\pi i z}| < 1\}$$

and the map $\check{\varphi} : E_\sigma \to \Gamma(\sigma)^{\mathrm{gp}}\backslash D_\sigma = \Gamma\backslash D_\sigma$ is identified with $E_\sigma \to \Delta$, $(q, z) \mapsto qe^{2\pi i z}$ if we identify $\Gamma\backslash D_\sigma$ with Δ via the bijection that sends $q \in \Delta^* = \Delta - \{0\}$ to $(2\pi i)^{-1}\log(q) \in \mathbf{Z}\backslash\mathfrak{h} = \Gamma\backslash D$ and sends $0 \in \Delta$ to $((\sigma, \mathbf{C}) \bmod \Gamma) \in \Gamma\backslash D_\sigma$.

3.4 SPACES E_σ, $\Gamma\backslash D_\Sigma$, $E_\sigma^\sharp$, AND $D_\Sigma^\sharp$

Let Σ be a fan in $\mathfrak{g}_{\mathbf{Q}}$ and let Γ be a subgroup of $G_{\mathbf{Z}}$ that is strongly compatible with Σ (cf. 1.3.10 (ii)). Let $\sigma \in \Sigma$. We define structures of logarithmic local ringed spaces over $\mathbf{C}$ on E_σ, $\tilde{E}_\sigma$, and on $\Gamma\backslash D_\Sigma$, and introduce suitable topologies on $E_\sigma^\sharp$ and on $D_\Sigma^\sharp$.

3.4.1

First we consider E_σ and $\tilde{E}_\sigma$. Recall (3.3.2) that we endow $\check{E}_\sigma = \mathrm{toric}_\sigma \times \check{D}$ with the inverse image of the canonical fs logarithmic structure on toric_σ. We endow the subsets E_σ and $\tilde{E}_\sigma$ of $\check{E}_\sigma$, with the following structures of logarithmic local ringed spaces over $\mathbf{C}$. The topology is the strong topology in $\check{E}_\sigma$. The sheaf $\mathcal{O}$ of rings and the logarithmic structure M are the inverse images of $\mathcal{O}$ and M of $\check{E}_\sigma$, respectively.

Then the logarithmic structure of E_σ and that of $\tilde{E}_\sigma$ coincide with the inverse images of the canonical logarithmic structure of toric_σ. Note that the induced homomorphisms $\alpha : M_{E_\sigma} \to \mathcal{O}_{E_\sigma}$ and $\alpha : M_{\tilde{E}_\sigma} \to \mathcal{O}_{\tilde{E}_\sigma}$ are injective.

3.4.2

Next we consider $\Gamma\backslash D_\Sigma$. First, we endow $\Gamma(\sigma)^{\mathrm{gp}}\backslash D_\sigma$ with the quotient topology via the period map $E_\sigma \to \Gamma(\sigma)^{\mathrm{gp}}\backslash D_\sigma$ in 3.3.8. We then endow $\Gamma\backslash D_\Sigma$ with the strongest topology for which the maps $\Gamma(\sigma)^{\mathrm{gp}}\backslash D_\sigma \to \Gamma\backslash D_\Sigma$ are continuous for all $\sigma \in \Sigma$. We endow $\Gamma\backslash D_\Sigma$ with the following sheaf of rings $\mathcal{O}_{\Gamma\backslash D_\Sigma}$ over $\mathbf{C}$ and the following logarithmic structure $M_{\Gamma\backslash D_\Sigma}$. Let $\pi_\sigma : E_\sigma \to \Gamma(\sigma)^{\mathrm{gp}}\backslash D_\sigma \to \Gamma\backslash D_\Sigma$ be the composite map. For any open set U of $\Gamma\backslash D_\Sigma$ and for any $\sigma \in \Sigma$, let $U_\sigma := \pi_\sigma^{-1}(U)$ and define

$$\begin{aligned}&\mathcal{O}_{\Gamma\backslash D_\Sigma}(U) \text{ (resp. } M_{\Gamma\backslash D_\Sigma}(U))\\ &\quad := \{\text{map } f : U \to \mathbf{C} \mid f \circ \pi_\sigma \in \mathcal{O}_{E_\sigma}(U_\sigma) \text{ (resp. } \in M_{E_\sigma}(U_\sigma)) \ (\forall \sigma \in \Sigma)\}.\end{aligned}$$

Here, in the definition of $M_{\Gamma\backslash D_\Sigma}$, we identify M_{E_σ} with its image in $\mathcal{O}_{E_\sigma}$ via the injection $\alpha : M_{E_\sigma} \to \mathcal{O}_{E_\sigma}$.

3.4.3

We introduce the topology of $E_\sigma^\sharp$ as a subspace of E_σ (cf. 3.3.9). We introduce the quotient topology of $D_\sigma^\sharp$ via the surjection $E_\sigma^\sharp \to D_\sigma^\sharp$ in 3.3.10. We introduce on $D_\Sigma^\sharp$ the strongest topology for which the inclusion maps $D_\sigma^\sharp \to D_\Sigma^\sharp$ $(\sigma \in \Sigma)$ are continuous. Note that the surjection $D_\Sigma^\sharp \to \Gamma\backslash D_\Sigma$ (1.3.9) becomes continuous.

Proposition 3.4.4

(i) *Let $(\sigma, Z) \in D_\Sigma$, let $F \in Z$, and write $\sigma = \sum_{1\le j\le n} \mathbf{R}_{\ge 0} N_j$. Then*

$$((\sigma, Z) \bmod \Gamma) = \lim_{\substack{\mathrm{Im}(z_j)\to\infty\\ 1\le j\le n}} \left(\exp\left(\sum_{1\le j\le n} z_j N_j\right) F \bmod \Gamma\right) \quad \textit{in } \Gamma\backslash D_\Sigma.$$

(ii) *Let $(\sigma, Z) \in D_\Sigma^\sharp$, let $F \in Z$, and let N_j be as above. Then*

$$(\sigma, Z) = \lim_{\substack{y_j\to\infty\\ 1\le j\le n}} \exp\left(\sum_{1\le j\le n} iy_j N_j\right) F \quad \textit{in } D_\Sigma^\sharp.$$

Proof. Replacing F by $\exp\left(\sum_{1\le j\le n} ia_j N_j\right)F$ for some $a_j \gg 0$, we may assume that

$$\exp\left(\sum_{1\le j\le n} iy_j N_j\right) F \in D$$

for any $y_j \ge 0$ $(1 \le j \le n)$. Let $\mathbf{0} \in \mathrm{toric}_\sigma = \mathrm{Hom}(\Gamma(\sigma)^\vee, \mathbf{C}^{\mathrm{mult}})$ be the element that sends all nontrivial elements of $\Gamma(\sigma)^\vee$ to $0 \in \mathbf{C}$. Then, by the morphism of analytic spaces

$$\mathrm{toric}_\sigma \to \check{E}_\sigma, \quad q \mapsto (q, F),$$

a sufficiently small open neighborhood U of $\mathbf{0}$ in toric_σ is sent into E_σ, and hence gives a continuous map $U \to E_\sigma$ in the strong topology of E_σ in $\check{E}_\sigma$. When $\mathrm{Im}(z_j) \to \infty$ $(1 \le j \le n)$,

$$\mathbf{e}\left(\sum_{1\le j\le n} z_j N_j\right) \to \mathbf{0} \quad \text{in } \mathrm{toric}_\sigma, \tag{1}$$

where **e** is as in 3.3.5, and hence

$$\left(\mathbf{e}\left(\sum_{1\le j\le n} z_jN_j\right), F\right) \to (\mathbf{0}, F) \quad \text{in } E_\sigma \tag{2}$$

in the strong toplogy of E_σ. By taking the image of (2) under the continuous map $E_\sigma \to \Gamma\backslash D_\Sigma$, we have

$$\left(\exp\left(\sum_{1\le j\le n} z_jN_j\right) F \bmod \Gamma\right) \to ((\sigma, Z) \bmod \Gamma) \quad \text{in } \Gamma\backslash D_\Sigma,$$

proving (i).

Next, when $y_j \to \infty$ $(1 \le j \le n)$, we have, by (2),

$$\left(\mathbf{e}\left(\sum_{1\le j\le n} iy_jN_j\right), F\right) \to (\mathbf{0}, F) \quad \text{in } E_\sigma^\sharp. \tag{3}$$

By applying the continuous map $E_\sigma^\sharp \to D_\Sigma^\sharp$ to (3), we obtain

$$\exp\left(\sum_{1\le j\le n} iy_jN_j\right) F \to (\sigma, Z) \quad \text{in } D_\Sigma^\sharp. \qquad \square$$

3.4.5

Example. In the case of Example (i) in 0.3.2, by 3.4.2, the bijection between $\Gamma(\sigma)^{\mathrm{gp}}\backslash D_\sigma$ and Δ given in 3.3.11 is in fact an isomorphism $\Gamma(\sigma)^{\mathrm{gp}}\backslash D_\sigma \simeq \Delta$ of logarithmic local ringed spaces over **C** (cf. 0.4.13).

3.5 INFINITESIMAL CALCULUS AND LOGARITHMIC MANIFOLDS

We show that the usual infinitesimal calculus on analytic spaces is naturally extended to that on objects of the category $\mathcal{B}(\log)$ in 3.2.4. We consider "logarithmic manifolds" as a special kind of "logarithmically smooth" objects of $\mathcal{B}(\log)$.

Proposition 3.5.1

(i) *The category* $\mathcal{B}(\log)$ *has fiber products.*

(ii) *Let* $X \times_Z Y$ *be a fiber product in* $\mathcal{B}(\log)$ *and let* $X \times_Z^{cl} Y$ *be the fiber product of* $X \to Z \leftarrow Y$ *in the category of topological spaces. (Here* cl *means classical.) Consider the canonical continuous map* $f : X \times_Z Y \to X \times_Z^{cl} Y$. *This map* f *is bijective if either one of the following conditions* (1) *and* (2) *is satisfied.*

(1) *Either* $X \to Z$ *or* $Y \to Z$ *is strict (that is, either the logarithmic structure of* X *or that of* Y *coincides with the inverse image of that of* Z*).*
(2) *The logarithmic structure of* Z *is trivial.*

The map f is a homeomorphism if either one of the above conditions (1) *and* (2) *is satisfied and furthermore either one of the following conditions* (3) *and* (4) *is satisfied.*

(3) *Either one of X and Y is an fs logarithmic analytic space.*
(4) *Locally, X and Y are analytically constructible subsets of fs logarithmic analytic spaces X' and Y', respectively, X (resp. Y) is endowed with the strong topology in X' (resp. Y'), and $\mathcal{O}$ and M of X (resp. Y) are the pullbacks of those of X' (resp. Y').*

Proof. We prove (i). Let $p_1 : X \to Z$, $p_2 : Y \to Z$ be morphisms in $\mathcal{B}(\log)$. Working locally, we may assume that there are objects X', Y', and Z' of $\mathcal{A}(\log)$ (0.7.4) such that X (resp. Y, resp. Z) is a subset of X' (resp. Y', resp. Z') endowed with the strong topology and with the inverse images of the structure sheaf $\mathcal{O}$ and the logarithmic structure M of X' (resp. Y', resp. Z'), and morphisms $p_1' : X' \to Z'$ and $p_2' : Y' \to Z'$ in $\mathcal{A}(\log)$ such that the following diagram is commutative:

$$\begin{array}{ccccc} X & \xrightarrow{p_1} & Z & \xleftarrow{p_2} & Y \\ \downarrow & & \downarrow & & \downarrow \\ X' & \xrightarrow{p_1'} & Z' & \xleftarrow{p_2'} & Y'. \end{array}$$

Let $X' \times_{Z'} Y'$ be the fiber product in $\mathcal{A}(\log)$ (2.1.10), and let $S \subset X' \times_{Z'} Y'$ be the inverse image of $X \times_Z^{\mathrm{cl}} Y$ under $X' \times_{Z'} Y' \to X' \times_{Z'}^{\mathrm{cl}} Y'$ where $X' \times_{Z'}^{\mathrm{cl}} Y'$ denotes the fiber product of $X' \to Z' \leftarrow Y'$ in the category of analytic spaces (which is the fiber product also as a topological space). Endow S with the strong topology in $X' \times_{Z'} Y'$ and with the inverse images of $\mathcal{O}$ and M of $X' \times_{Z'} Y'$. We prove that S is the fiber product $X \times_Z Y$ in $\mathcal{B}(\log)$. For an object T of $\mathcal{B}(\log)$, by 3.2.8 and by $\mathcal{B}(\log) \subset \overline{\mathcal{A}}_2(\log)$ (3.2.10), we have the identifications

$$\begin{aligned} &\{\lambda : T \to S, \text{ a morphism}\} \\ &= \{\lambda : T \to X' \times_{Z'} Y', \text{ a morphism such that } \lambda(T) \subset S\} \\ &= \left\{ (\lambda_1, \lambda_2) \middle| \begin{array}{l} \lambda_1 : T \to X', \lambda_2 : T \to Y' \text{ morphisms}, \\ \lambda_1(T) \subset X, \lambda_2(T) \subset Y, p_1' \circ \lambda_1 = p_2' \circ \lambda_2 \end{array} \right\}, \\ &= \{(\lambda_1, \lambda_2) \mid \lambda_1 : T \to X, \lambda_2 : T \to Y \text{ morphisms}, p_1 \circ \lambda_1 = p_2 \circ \lambda_2\}. \end{aligned}$$

We prove (ii). Working locally, let X', Y', etc. be as in the proof of (i). Let $X \times Y$ be the product in $\mathcal{B}(\log)$ and let $X \times^{\mathrm{cl}} Y$ be the product in the category of topological spaces. The topology of $X \times Y$ is the strong topology in $X' \times Y'$. The canonical bijection $X \times Y \to X \times^{\mathrm{cl}} Y$ is continuous, and, by 3.1.8, it is a homeomorphism if one of the conditions (3) and (4) is satisfied. By Proposition 3.1.9 applied to the morphism $X' \times_{Z'} Y' \to X' \times Y'$ and to the subset $X \times Y$ of $X' \times Y'$, $X \times_Z Y$ coincides as a topological space with the fiber product of $X' \times_{Z'} Y' \to X' \times Y' \leftarrow X \times Y$ in the category of topological spaces. Let P be the fiber product of $X' \times_{Z'}^{\mathrm{cl}} Y' \to X' \times Y' \leftarrow X \times Y$ in the category of topological spaces. Then the canonical map $P \to X \times_Z^{\text{cl}} Y$ is bijective and continuous. If one of the conditions (1) and (2) is satisfied, locally, (p'_1, p'_2) also satisfies one of the conditions (1) and (2). Hence $X' \times_{Z'} Y' \xrightarrow{\sim} X' \times_{Z'}^{\text{cl}} Y'$ by 2.1.10, and this shows $X \times_Z Y \xrightarrow{\sim} P$ as topological spaces. Hence $X \times_Z Y \to X \times_Z^{\text{cl}} Y$ is bijective. If one of the conditions (3) and (4) is satisfied, $P \to X \times_Z^{\text{cl}} Y$ is a homeomorphism since $X \times Y \to X \times^{\text{cl}} Y$ is a homeomorphism. These results prove (ii). □

In this Section 3.5, we use direct products $X \times Y$ ($= X \times_{\text{Spec}(\mathbf{C})} Y$ with the trivial logarithmic structure on $\text{Spec}(\mathbf{C})$) in $\mathcal{B}(\log)$. In the next Section 3.6, we will use more general fiber products in $\mathcal{B}(\log)$.

3.5.2

Let X be an object of $\mathcal{B}(\log)$. We define the *sheaf* ω_X^1 *of logarithmic differential 1-forms* on X, imitating its definition in 2.1.7 for an fs logarithmic analytic space, by using the product $X \times X$ in 3.5.1.

Define $\omega_X^q := \bigwedge_{\mathcal{O}_X}^q \omega_X^1$ $(q \geq 0)$, $\omega_X^{q,\log} := \mathcal{O}_X^{\log} \otimes_{\tau^{-1}(\mathcal{O}_X)} \tau^{-1}(\omega_X^q)$, $d : \mathcal{O}_X \to \omega_X^1$, $d\log : M_X^{\text{gp}} \to \omega_X^1$, $d : \mathcal{O}_X^{\log} \to \omega_X^{1,\log}$ and the *logarithmic de Rham complexes* $\omega_X^\bullet$ and $\omega_X^{\bullet,\log}$, just as in 2.1.7.

Lemma 3.5.3 *Let Z be an object of $\mathcal{A}(\log)$, let X be a subset of Z, and endow X with the strong topology and with the inverse images of $\mathcal{O}_Z$ and M_Z. Then*

$$\omega_Z^q|_X \xrightarrow{\sim} \omega_X^q, \quad \omega_Z^{q,\log}|_{X^{\log}} \xrightarrow{\sim} \omega_X^{q,\log}, \quad (q \geq 0).$$

Proof. These are proved easily. □

Definition 3.5.4

(i) *Recall* (2.1.11) *that an object of* $\mathcal{A}(\log)$ *is* logarithmically smooth *if it is locally an open set of a toric variety endowed with the canonical logarithmic structure* (2.1.6 (ii)).

(ii) *An object of* $\mathcal{B}(\log)$ *is said to be* logarithmically smooth *if it is locally isomorphic to a subset of a logarithmically smooth object Z of* $\mathcal{A}(\log)$, *endowed with the strong topology in Z and with the inverse images of $\mathcal{O}_Z$ and M_Z.*

(iii) *Let X be a logarithmically smooth object of* $\mathcal{B}(\log)$. *We define the sheaf θ_X on X, called the* sheaf of logarithmic vector fields *on X, as the $\mathcal{O}_X$-dual of ω_X^1. (Note that ω_X^1 is locally free of finite rank by* 3.5.3.)

Proposition 3.5.5 *Let X and Y be objects of* $\mathcal{B}(\log)$. *Let $\mathcal{I}$ be an ideal of $\mathcal{O}_Y$ which is locally of finite type and which satisfies $\mathcal{I}^2 = 0$, and let Y_0 be the object of* $\mathcal{B}(\log)$ *whose underlying topological space is that of Y, and which is endowed with $\mathcal{O}_{Y_0} = \mathcal{O}_Y/\mathcal{I}$ and with the inverse image of M_Y. Let $f : Y_0 \to X$ be a morphism, and let P be the set of all morphisms $Y \to X$ that extend f.*

(i) *Assume that P is not empty. Then P is a principal homogeneous space under the group $Q := \text{Hom}_{\mathcal{O}_{Y_0}}(f^*(\omega_X^1), \mathcal{I})$.*

(ii) *If X is logarithmically smooth, then P is not empty locally on Y.*

Proof. We prove (i). We define an action of Q on the set P as follows. Let $\delta \in Q$. Then, for $g \in P$, $g' = \delta g \in P$ is the following extension $Y \to X$ of f. The homomorphism $g'^{*} : f^{-1}(\mathcal{O}_X) \to \mathcal{O}_Y$ is given by $a \mapsto g^{*}(a) + \delta(da)$, and the homomorphism $g'^{*} : f^{-1}(M_X) \to M_Y$ is given by $a \mapsto g^{*}(a)(1 + \delta(d\log(a)))$. We prove that, for any $g, g' \in P$, there exists a unique $\delta \in Q$ that sends g to g'. Let Ω^1_X be $\omega^1_{X'}$ where X' is the same as X as a local ringed space over $\mathbf{C}$ but the logarithmic structure of X' is trivial. Since the restriction of $(g', g) : Y \to X \times X$ to Y_0 coincides with the composite of $f : Y_0 \to X$ and the diagonal morphism $X \to X \times X$, and since $\mathcal{I}^2 = 0$, there is a unique homomorphism $u : f^{*}\Omega^1_X \to \mathcal{I}$ such that $g'^{*}(b) = g^{*}(b) + u(db)$ for any $b \in \mathcal{O}_X$. Furthermore, for $a \in f^{-1}(M_X^{\mathrm{gp}})$, $g^{*}(a) \in M_Y^{\mathrm{gp}}$ and $g'^{*}(a) \in M_Y^{\mathrm{gp}}$ have the same image in $M_{Y_0}^{\mathrm{gp}}$, and hence there exists a unique homomorphism $v : f^{-1}(M_X^{\mathrm{gp}}) \to \mathcal{I}$ such that $g'^{*}(a) = g^{*}(a)(1 + v(a))$ for $a \in f^{-1}(M_X^{\mathrm{gp}})$. For $a \in f^{-1}(M_X^{\mathrm{gp}})$, $\alpha(g'^{*}(a)) = \alpha(g^{*}(a)(1 + v(a)))$ shows that $u(d\alpha(a)) = \alpha(a)v(a)$. Since ω^1_X is the quotient of $\Omega^1_X \oplus (\mathcal{O}_X \otimes_{\mathbf{Z}} M_X^{\mathrm{gp}})$ by the $\mathcal{O}_X$-submodule generated by $\{(-d\alpha(a), \alpha(a) \otimes a) \mid a \in M_X\}$, we have a unique homomorphism $\delta : f^{*}(\omega^1_X) \to \mathcal{I}$ which induces u on $f^{*}\Omega^1_X$ and which satisfies $v(a) = \delta(d\log(a))$ for $a \in f^{-1}(M_X^{\mathrm{gp}})$.

We prove (ii). Because

$$\mathrm{Mor}(Y, \mathrm{Spec}(\mathbf{C}[\mathcal{S}])_{\mathrm{an}}) \simeq \mathrm{Hom}(\mathcal{S}, M_Y)$$

for any object Y of $\overline{\mathcal{A}}_1(\log)$ and any fs monoid $\mathcal{S}$, (ii) is reduced to the fact that, for an fs monoid $\mathcal{S}$, a homomorphism $\mathcal{S} \to M_{Y_0}$ lifts locally to a homomorphism $\mathcal{S} \to M_Y$. Since $M_Y^{\mathrm{gp}} \to M_{Y_0}^{\mathrm{gp}}$ is surjective and the kernel ($\simeq \mathcal{I}$) is divisible as an abelian group, $\mathcal{S}^{\mathrm{gp}} \to M_{Y_0}^{\mathrm{gp}}$ lifts to $\mathcal{S}^{\mathrm{gp}} \to M_Y^{\mathrm{gp}}$ locally. Any lifting $\mathcal{S}^{\mathrm{gp}} \to M_Y^{\mathrm{gp}}$ of $\mathcal{S} \to M_{Y_0}$ gives a homomorphism $\mathcal{S} \to M_Y$. □

3.5.6

For a logarithmically smooth object X of $\mathcal{B}(\log)$, we define the *logarithmic tangent bundle* T_X of X as the vector bundle associated with the sheaf θ_X. Here, for an object X of $\mathcal{B}(\log)$ and for a locally free $\mathcal{O}_X$-module $\mathcal{F}$ of finite rank on X, the vector bundle $V(\mathcal{F})$ associated with $\mathcal{F}$ is the object of $\mathcal{B}(\log)$ that represents the functor

$$Y \mapsto \{(f, a) \mid f : Y \to X,\ a \in \Gamma(Y, f^{*}\mathcal{F})\}$$

from $\mathcal{B}(\log)$ to the category of sets. The existence of $V(\mathcal{F})$ is shown as follows. We may work locally on X and hence we may assume $\mathcal{F} = \mathcal{O}_X^{\oplus r}$ for some $r \geq 0$. Then since $\mathcal{B}(\log) \subset \overline{\mathcal{A}}_1(\log)$ (3.2.10), the product $X \times \mathbf{C}^r$, as in 3.5.1, has the property of $V(\mathcal{F})$.

By 3.5.5, for a logarithmically smooth object X of $\mathcal{B}(\log)$, T_X represents the functor

$$Y \mapsto \mathrm{Mor}(Y[T]/(T^2), X) \qquad (1)$$

from $\mathcal{B}(\log)$ to the category of sets, where $Y[T]/(T^2)$ denotes the object of $\mathcal{B}(\log)$ whose underlying topological space is the same as Y, whose structure sheaf is

$\mathcal{O}_Y[T]/(T^2)$ (T is an indeterminate), and whose logarithmic structure is the inverse image of that of Y.

If X is a complex manifold with the trivial logarithmic structure, T_X coincides with the usual tangent bundle of X.

For a morphism $f : X \to X'$ of logarithmically smooth objects of $\mathcal{B}(\log)$, the canonical homomorphism $f^*(\omega^1_{X'}) \to \omega^1_X$ induces a morphism $T_X \to T_{X'}$, which we denote by df. This morphism df is also obtained from the above functorial interpretation (1) of the logarithmic tangent bundle.

DEFINITION 3.5.7 *By a* logarithmic manifold, *we mean a logarithmic local ringed space over* $\mathbf{C}$ *which has an open covering* $(U_\lambda)_\lambda$ *with the following property: For each* λ, *there exist a logarithmically smooth fs logarithmic analytic space* Z_λ, *a finite subset* I_λ *of* $\Gamma(Z_\lambda, \omega^1_{Z_\lambda})$, *and an isomorphism of logarithmic local ringed spaces over* $\mathbf{C}$ *between* U_λ *and an open set of*

$$S_\lambda = \{z \in Z_\lambda \mid \textit{the image of } I_\lambda \textit{ in } \omega^1_z \textit{ is zero}\}, \qquad (1)$$

where S_λ *is endowed with the strong topology in* Z_λ *and with the inverse images of* $\mathcal{O}_{Z_\lambda}$ *and* M_{Z_λ}.

We will see (Theorem A in Section 4.1 below) that E_σ and $\Gamma\backslash D_\Sigma$, for Γ neat and strongly compatible with Σ, are logarithmic manifolds.

The following 3.5.8 is seen easily.

PROPOSITION 3.5.8

(i) *A logarithmic manifold is a logarithmically smooth object of* $\mathcal{B}(\log)$.
(ii) *A logarithmic manifold is an object of the category* $\mathcal{B}^*(\log)$ *in* 3.2.4.

PROPOSITION 3.5.9 *Let* Z *be a logarithmically smooth fs logarithmic analytic space, let* I *be a finite subset of* $\Gamma(Z, \omega^1_Z)$, *and let* S *be the set of all* $z \in Z$ *such that the image of* I *in the sheaf* ω^1_z *of logarithmic differential* 1*-forms on the fs logarithmic point* z *is zero. Then* S *is an analytically constructible subset* (3.1.4) *of* Z.

Note that, by 3.5.9 and 3.1.5, the topology of a logarithmic manifold is well understood.

Proof of Proposition 3.5.9. We may assume that Z is an open set of the toric variety $\operatorname{Spec}(\mathbf{C}[\mathcal{S}])_{\mathrm{an}}$ for an fs monoid $\mathcal{S}$ and that the logarithmic structure of Z is induced from the canonical logarithmic structure of $\operatorname{Spec}(\mathbf{C}[\mathcal{S}])_{\mathrm{an}}$. We have $\mathcal{O}_Z \otimes_{\mathbf{Z}} \mathcal{S}^{\mathrm{gp}} \xrightarrow{\sim} \omega^1_Z$, $f \otimes g \mapsto f d\log(g)$ (2.1.8 (ii)). For each face $\mathcal{S}'$ of $\mathcal{S}$, let $A(\mathcal{S}')$ be the closed analytic subspace of Z defined by the ideal of $\mathcal{O}_Z$ generated by the image of the complement $\mathcal{S} - \mathcal{S}'$ of $\mathcal{S}'$. Let $J_{\mathcal{S}'}$ be the coherent ideal of $\mathcal{O}_{A(\mathcal{S}')}$ generated by the images of I under

$$\omega^1_Z \simeq \mathcal{O}_Z \otimes_{\mathbf{Z}} \mathcal{S}^{\mathrm{gp}} \to \mathcal{O}_{A(\mathcal{S}')} \otimes_{\mathbf{Z}} (\mathcal{S}^{\mathrm{gp}}/\mathcal{S}'^{\mathrm{gp}}) \xrightarrow{h} \mathcal{O}_{A(\mathcal{S}')},$$

where h ranges over all homomorphisms $h : \mathcal{S}^{\mathrm{gp}}/\mathcal{S}'^{\mathrm{gp}} \to \mathbf{Z}$. Let $B(\mathcal{S}')$ be the closed analytic subspace of $A(\mathcal{S}')$ defined by $J_{\mathcal{S}'}$. It is sufficient to prove that

$$S = \{z \in Z \mid z \in B(\mathcal{S}') \text{ for any face } \mathcal{S}' \text{ of } \mathcal{S} \text{ such that } z \in A(\mathcal{S}')\}. \qquad (1)$$

For $z \in Z$, let $\mathcal{S}(z) := \{f \in \mathcal{S} \mid f(z) \neq 0\}$ be the face of $\mathcal{S}$ corresponding to z, then, as is easily seen, we have

(1) For a face $\mathcal{S}'$ of $\mathcal{S}$, $z \in A(\mathcal{S}')$ if and only if $\mathcal{S}(z) \subset \mathcal{S}'$.
(2) $\mathbf{C} \otimes_{\mathbf{Z}} (\mathcal{S}^{\mathrm{gp}}/\mathcal{S}(z)^{\mathrm{gp}}) \xrightarrow{\sim} \omega^1_z$, $1 \otimes g \mapsto d\log(g)$.

These results (2) and (3) yield (1). □

We now prove that the space $\tilde{E}_\sigma \subset \check{E}_\sigma$ (3.3.4, 3.4.1) is a logarithmic manifold. Note that $\check{E}_\sigma$ is a logarithmically smooth fs logarithmic analytic space.

Proposition 3.5.10 *Let the notation be as in Section 3.3, and denote $\check{E}_\sigma$ (resp. $\tilde{E}_\sigma$) by Z (resp. S). Then, there exist an open covering $(U_\lambda)_\lambda$ of Z and a finite subset I_λ of $\Gamma\big(U_\lambda, \omega^1_{U_\lambda}\big)$ for each λ such that*

$$S \cap U_\lambda = \{z \in U_\lambda \mid \textit{the image of } I_\lambda \textit{ in } \omega^1_z \textit{ is zero}\}.$$

In particular, $\tilde{E}_\sigma$ is a logarithmic manifold.

Proof. Let $Y = \mathrm{toric}_\sigma$. Since $Y = \mathrm{Spec}(\mathbf{C}[\Gamma(\sigma)^\vee])_{\mathrm{an}}$, ω^1_Y is identified with $\mathcal{O}_Y \otimes_{\mathbf{Z}} (\Gamma(\sigma)^\vee)^{\mathrm{gp}}$ and, via $\log : \Gamma(\sigma)^{\mathrm{gp}} \to \sigma_{\mathbf{C}}$, it is identified with the $\mathcal{O}_Y$-dual of $\mathcal{O}_Y \otimes_{\mathbf{C}} \sigma_{\mathbf{C}}$. Hence, for each $q \in Y$, the fiber $\omega^1_Y(q)$ of ω^1_Y at q is identified with the dual $\mathbf{C}$-vector space of $\sigma_{\mathbf{C}}$. Furthermore, this identification induces an identification of the logarithmic differential module ω^1_q of the fs logarithmic point q, which is a quotient of $\omega^1_Y(q)$, with the dual $\mathbf{C}$-vector space of $\sigma(q)_{\mathbf{C}}$ (for $\sigma(q)$, see 3.3.2).

On the other hand, let

$$\mathcal{G} := \mathcal{O}_{\check{D}} \otimes_{\mathbf{C}} \sigma_{\mathbf{C}} \supset \mathcal{G}^{-1} := \{X \in \mathcal{G} \mid X F^p_{\mathrm{univ}} \subset F^{p-1}_{\mathrm{univ}}\ (\forall p \in \mathbf{Z})\},$$

where F_{univ} is the universal filtration of $\mathcal{O}_{\check{D}} \otimes_{\mathbf{Z}} H_0$ on $\check{D}$. Then, $\mathcal{G}$, $\mathcal{G}^{-1}$, and $\mathcal{G}/\mathcal{G}^{-1}$ are locally free $\mathcal{O}_{\check{D}}$-modules of finite rank. For $F \in \check{D}$, the fiber $\mathcal{G}^{-1}(F)$ of $\mathcal{G}^{-1}$ at F coincides with $\{X \in \sigma_{\mathbf{C}} \mid XF^p \subset F^{p-1}\ (\forall p \in \mathbf{Z})\}$. By 3.3.7 (ii), for $z = (q, F) \in \check{E}_\sigma$, $z \in S$ if and only if the map $\sigma(q)_{\mathbf{C}} \to \sigma_{\mathbf{C}}/\mathcal{G}^{-1}(F) = \mathcal{G}(F)/\mathcal{G}^{-1}(F)$ is the zero map.

Let $p_1 : Z \to Y$ and $p_2 : Z \to \check{D}$ be the projections, and define a locally free $\mathcal{O}_Z$-module $\mathcal{F}$ of finite rank by

$$\mathcal{F} := p_2^*(\mathcal{G}/\mathcal{G}^{-1}).$$

Then, the canonical projection $\mathcal{O}_Z \otimes_{\mathbf{C}} \sigma_{\mathbf{C}} \to \mathcal{F}$ defines a canonical global section of $p_1^*(\omega^1_Y) \otimes_{\mathcal{O}_Z} \mathcal{F}$. Hence we obtain a global section of $\omega^1_Z \otimes_{\mathcal{O}_Z} \mathcal{F}$, which we denote by η. For $z = (q, F) \in Z$, since $\omega^1_q \simeq \omega^1_z$, the canonical map $\sigma(q)_{\mathbf{C}} \to \sigma_{\mathbf{C}}/\mathcal{G}^{-1}(F)$ is identified with the image of η in $\omega^1_z \otimes_{\mathbf{C}} \mathcal{F}(z)$. Hence $z \in S$ if and only if the image of η in $\omega^1_z \otimes_{\mathbf{C}} \mathcal{F}(z)$ is zero.

If we have a basis $(e_j)_{j\in J}$ of $\mathcal{F}$ on an open set U of Z and if we write $\eta|_U = \sum_{j\in J}\eta_j\otimes e_j$ with η_j a section of $\omega_Z^1|_U$, then

$$S\cap U = \{z\in U \mid \text{the images of } \eta_j \text{ in } \omega_z^1 \text{ are zero for all } j\in J\}.$$

This proves 3.5.10. □

3.6 LOGARITHMIC MODIFICATIONS

The subject of this section is as follows.

Let $\mathcal{S}$ be an fs monoid, let X be the toric variety $\mathrm{Spec}(\mathbf{C}[\mathcal{S}])_{\mathrm{an}}$, and let U be the dense open set $\mathrm{Spec}(\mathbf{C}[\mathcal{S}^{\mathrm{gp}}])_{\mathrm{an}}$ of X. Then in the toric geometry [KKMS], for a finite rational subdivision Σ of the cone $\mathrm{Hom}(\mathcal{S}, \mathbf{R}^{\mathrm{add}}_{\geq 0})$, we have a modification $X(\Sigma)$ of X associated with Σ, with a proper morphism $f : X(\Sigma)\to X$ which induces an isomorphism $f^{-1}(U)\xrightarrow{\sim} U$. If I is an ideal of $\mathcal{O}_X$ generated by some elements of $\mathcal{S}$, the normalization of the blow-up of X with respect to I is isomorphic to $X(\Sigma)$ for some Σ.

In this section, we consider such a modification $X(\Sigma)$ ("logarithmic modification" 3.6.12) of an object X of $\mathcal{B}(\log)$. (See [Kk2, 9] for a similar theory for schemes.) This is important when we try to extend period maps to the boundary in Section 4.3. We also consider a space X_{val} defined by using projective limits of logarithmic modifications (3.6.18, 3.6.23). This space X_{val} will play important roles in later sections from Chapter 5, when we consider the "valuative spaces" in the fundamental diagram in Introduction.

PROPOSITION 3.6.1 *Let X be an object of $\mathcal{B}(\log)$ and J be a subset of $\Gamma(X, M_X^{\mathrm{gp}}/\mathcal{O}_X^\times)$ such that the submonoid of $\Gamma(X, M_X^{\mathrm{gp}}/\mathcal{O}_X^\times)$ generated by J has finite generators. (For example, any finite subset J of $\Gamma(X, M_X^{\mathrm{gp}}/\mathcal{O}_X^\times)$ satisfies this condition.) Then*

(i) *There is an object $X[J]$ of $\mathcal{B}(\log)$ over X having the following property. For any object Y of $\mathcal{B}(\log)$ over X, the set $\mathrm{Mor}_X(Y, X[J])$ of morphisms $Y\to X[J]$ over X is either a one-point set or an empty set, and it is nonempty if and only if the pullbacks of all elements of J on Y in $\Gamma(Y, M_Y^{\mathrm{gp}}/\mathcal{O}_Y^\times)$ belong to $\Gamma(Y, M_Y/\mathcal{O}_Y^\times)$.*

(ii) *If $a^{-1}\in\Gamma(X, M_X/\mathcal{O}_X^\times)$ for all $a\in J$, $X[J]$ is an open subobject of X in $\mathcal{B}(\log)$.*

Here we say an object U of $\mathcal{B}(\log)$ is an *open subobject* of an object X of $\mathcal{B}(\log)$ if U is an open set of X and $\mathcal{O}_U = \mathcal{O}_X|_U$ and $M_U = M_X|_U$.

The characterizing property of $X[J]$ in proposition 3.6.1 (i) shows that, if $\langle J\rangle$ denotes the submonoid of $\Gamma(X, M_X^{\mathrm{gp}}/\mathcal{O}_X^\times)$ generated by J, then $X[J] = X[\langle J\rangle]$.

3.6.2

Before we prove proposition 3.6.1, we consider a special case of it. Let $\mathcal{S}$ be an fs monoid, let $X = \mathrm{Spec}(\mathbf{C}[\mathcal{S}])_{\mathrm{an}}$ with the canonical logarithmic structure, and let $\tilde{J}$ be a subset of $\mathcal{S}^{\mathrm{gp}}$ that generates a finitely generated submonoid of $\mathcal{S}^{\mathrm{gp}}$. Let J be the image of $\tilde{J}$ in $\Gamma(X, M_X^{\mathrm{gp}}/\mathcal{O}_X^\times)$. Then $X[J] = \mathrm{Spec}(\mathbf{C}[\mathcal{S}'])_{\mathrm{an}}$, where $\mathcal{S}'$ is the smallest fs submonoid of $\mathcal{S}^{\mathrm{gp}}$ containing $\mathcal{S}$ and $\tilde{J}$, endowed with the canonical logarithmic structure. If $\mathcal{S}''$ denotes the submonoid of $\mathcal{S}^{\mathrm{gp}}$ generated by $\mathcal{S}$ and $\tilde{J}$, then $\mathcal{S}' = \{a \in \mathcal{S}^{\mathrm{gp}} \mid a^n \in \mathcal{S}'' \text{ for some } n \geq 1\}$, and $\mathrm{Spec}(\mathbf{C}[\mathcal{S}'])_{\mathrm{an}}$ coincides with the normalization of $\mathrm{Spec}(\mathbf{C}[\mathcal{S}''])_{\mathrm{an}}$.

3.6.3

Example. If $X = \mathrm{Spec}(\mathbf{C}[T_1, T_2])_{\mathrm{an}}$ with the logarithmic structure associated with $\mathbf{N}^2 \to \mathcal{O}_X$, $(m, n) \mapsto T_1^m T_2^n$, and if J is the one-point set $\{T_1/T_2 \bmod \mathcal{O}_X^\times\}$, then $X[J] = \mathrm{Spec}(\mathbf{C}[T_1/T_2, T_2])_{\mathrm{an}}$ with the logarithmic structure associated with $\mathbf{N}^2 \to \mathcal{O}_{X[J]}$, $(m, n) \mapsto (T_1/T_2)^m T_2^n$.

3.6.4

Proof of Proposition 3.6.1. We may assume that J is a finite set.

(i) It is sufficient to prove the existence of $X[J]$ locally on X. Locally on X, there are an fs monoid $\mathcal{S}$, a homomorphism $h : \mathcal{S} \to M_X$, and a finite subset $\tilde{J}$ of $\mathcal{S}^{\mathrm{gp}}$ such that $J = h(\tilde{J}) \bmod \mathcal{O}_X^\times$. Since $\mathcal{B}(\log) \subset \overline{\mathcal{A}}_1(\log)$ (3.2.10), we have a morphism $X \to \mathrm{Spec}(\mathbf{C}[\mathcal{S}])_{\mathrm{an}}$ corresponding to h (2.2.2). Let $\mathcal{S}'$ be the smallest fs submonoid of $\mathcal{S}^{\mathrm{gp}}$ containing $\mathcal{S}$ and $\tilde{J}$ and let X' be the fiber product $X \times_{\mathrm{Spec}(\mathbf{C}[\mathcal{S}])_{\mathrm{an}}} \mathrm{Spec}(\mathbf{C}[\mathcal{S}'])_{\mathrm{an}}$ in the category $\mathcal{B}(\log)$ (3.5.1). Then this X' has the property of $X[J]$ stated in Proposition 3.6.1.

(ii) Under the assumption of (ii), in the proof of (i), we can take $\tilde{J}$ such that $(\tilde{J})^{-1} \subset \mathcal{S}$. In this situation, $\mathrm{Spec}(\mathbf{C}[\mathcal{S}'])_{\mathrm{an}}$ is an open subobject of $\mathrm{Spec}(\mathbf{C}[\mathcal{S}])_{\mathrm{an}}$. Hence the fiber product X' becomes also an open subobject of X. □

3.6.5

Let X, J, and $X[J]$ be as in 3.6.1. Then the map of sheaves $\mathrm{Mor}(\ , X[J]) \to \mathrm{Mor}(\ , X)$ on the category $\mathcal{B}(\log)$ is injective by the characterizing property of $X[J]$ in Proposition 3.6.1 (i). However the map of the spaces $X[J] \to X$ is not necessarily injective. For example, in the example in 3.6.3, the map $X[J] \to X$ is described as $\mathbf{C}^2 \to \mathbf{C}^2$, $(x, y) \mapsto (xy, y)$ and hence not injective.

Proposition 3.6.6 *Let X be an object of $\mathcal{B}(\log)$ and I be a nonempty finite subset of $\Gamma(X, M_X^{\mathrm{gp}}/\mathcal{O}_X^\times)$. Then*

(i) *There is an object $B_I(X)$ of $\mathcal{B}(\log)$ over X (which we call the* logarithmic blow-up *of X with respect to I) having the following property. For any object Y of $\mathcal{B}(\log)$ over X, the set $\mathrm{Mor}_X(Y, B_I(X))$ is either a one-point set or an empty set, and it is nonempty if and only if, locally on Y, there is an element a of I such that the pullbacks of $a^{-1}b$ in $M_Y^{\mathrm{gp}}/\mathcal{O}_Y^\times$ for all $b \in I$ belong to $M_Y/\mathcal{O}_Y^\times$.*

(ii) *There is an object $B_I^*(X)$ of $\mathcal{B}(\log)$ over X having the following property. For any object Y of $\mathcal{B}(\log)$ over X, the set $\mathrm{Mor}_X(Y, B_I^*(X))$ is either a one-point set or an empty set, and it is nonempty if and only if for any $a, b \in I$, locally on Y, at least one of the pullbacks of $a^{-1}b$ and ab^{-1} in $M_Y^{\mathrm{gp}}/\mathcal{O}_Y^\times$ belong to $M_Y/\mathcal{O}_Y^\times$.*

(iii) *There is a morphism* $B_I^*(X) \to B_I(X)$ *over* X. *Furthermore,* $B_I^*(X) = B_{I'}(X)$ *for some nonempty finite subset* I' *of* $\Gamma(X, M_X^{gp}/\mathcal{O}_X^\times)$.

(iv) *Let* I' *be a nonempty subset of* $\Gamma(X, M_X^{gp}/\mathcal{O}_X^\times)$. *Then*

$$\mathrm{Mor}(\ , B_{I'}^*(X)) \subset \mathrm{Mor}(\ , B_I^*(X)) \quad in \quad \mathrm{Mor}(\ , X) \quad if\ I' \supset I,$$
$$\mathrm{Mor}(\ , B_I(X)) \cap \mathrm{Mor}(\ , B_{I'}(X)) = \mathrm{Mor}(\ , B_{II'}(X)) \quad in \quad \mathrm{Mor}(\ , X)$$

where $II' = \{aa' \mid a \in I, a' \in I'\}$.

(v) *Assume that* $X = \mathrm{Spec}(\mathbf{C}[\mathcal{S}])_{\mathrm{an}}$ *for an fs monoid* $\mathcal{S}$ *and that* $I \subset \Gamma(X, M_X^{gp}/\mathcal{O}_X^\times)$ *is the image of a nonempty finite subset* $\tilde{I}$ *of* $\mathcal{S}$. *Then, as an analytic space,* $B_I(X)$ *coincides with the normalization of the blow-up of* X *along the closed analytic subspace of* X *defined by the ideal of* $\mathcal{O}_X$ *generated by* $\tilde{I}$.

Proof. We prove (i). For $a \in I$, let $a^{-1}I = \{a^{-1}b \mid b \in I\}$ and $X[a^{-1}I]$ be as in 3.6.1. By 3.6.5, $\mathrm{Mor}(\ , X[a^{-1}I])$ is a subsheaf of the sheaf $\mathrm{Mor}(\ , X)$ on the category $\mathcal{B}(\log)$. Define the sheaf $\bigcup_{a\in I} \mathrm{Mor}(\ , X[a^{-1}I])$ as the union of the sheaves $\mathrm{Mor}(\ , X[a^{-1}I])$ in $\mathrm{Mor}(\ , X)$. For $a, b \in I$, $\mathrm{Mor}(\ , X[a^{-1}I]) \cap \mathrm{Mor}(\ , X[b^{-1}I])$ is represented by the open subobject $X[a^{-1}I][(a^{-1}b)^{-1}]$ of $X[a^{-1}I]$ in $\mathcal{B}(\log)$ (Proposition 3.6.1 (ii)), which is identified with the open subobject $X[b^{-1}I][(b^{-1}a)^{-1}]$ of $X[b^{-1}I]$ in $\mathcal{B}(\log)$. Hence the sheaf $\bigcup_{a\in I} \mathrm{Mor}(\ , X[a^{-1}I])$ is represented by an object $B_I(X) := \cup_{a\in I} X[a^{-1}I]$, the union of all $X[a^{-1}I]$, which we glue as open subobjects.

The assertion in (iv) concerning the product II' follows from the characterizing property of $B_I(X)$ in (i) easily. By this, if I' is the product of all sets $\{a, b\}$ with $a, b \in I$ in this sense, $B_{I'}(X)$ has the property of $B_I^*(X)$ stated in (ii). Hence we have (ii).

The assertions in (iii) and the remaining assertion in (iv) follow from the characterizing properties of $B_I(X)$ and $B_I^*(X)$ given in (i) and (ii).

The assertion (v) follows from 3.6.2. (We take the normalization, since $\mathrm{Spec}(\mathbf{C}[\mathcal{S}'])_{\mathrm{an}}$ is the normalization of $\mathrm{Spec}(\mathbf{C}[\mathcal{S}''])_{\mathrm{an}}$ in 3.6.2.) □

3.6.7

Let X be an object of $\mathcal{B}(\log)$, let $N_{\mathbf{Q}}$ be a finite-dimensional $\mathbf{Q}$-vector space, and assume that we are given an element

$$s \in \Gamma(X, M_X^{gp}/\mathcal{O}_X^\times) \otimes N_{\mathbf{Q}}.$$

We will consider an object $X(\Sigma)$ of $\mathcal{B}(\log)$ over X associated with a rational fan Σ in $N_{\mathbf{R}} = \mathbf{R} \otimes_{\mathbf{Q}} N_{\mathbf{Q}}$.

For an object Y of $\mathcal{B}(\log)$ over X and for a point $y \in Y$, we have a homomorphism

$$s_y : \pi_1(y^{\log}) \to N_{\mathbf{Q}} \subset N_{\mathbf{R}}$$

which is the germ in $(M_Y^{gp}/\mathcal{O}_Y^\times)_y \otimes N_{\mathbf{Q}} = \mathrm{Hom}(\pi_1(y^{\log}), N_{\mathbf{Q}})$ of the pullback of s in $\Gamma(Y, M_Y^{gp}/\mathcal{O}_Y^\times) \otimes N_{\mathbf{Q}}$.

Lemma 3.6.8 *Let* σ *be a finitely generated rational cone in* $N_{\mathbf{R}}$. *Then there is an object* $X(\sigma)$ *of* $\mathcal{B}(\log)$ *over* X *having the following property. For any object* Y *of*

$\mathcal{B}$(log) over X, the set $\mathrm{Mor}_X(Y, X(\sigma))$ *is either a one-point set or an empty set, and it is nonempty if and only if*

$$s_y(\pi_1^+(y^{\log})) \subset \sigma$$

for any $y \in Y$.

Proof. Take a finitely generated $\mathbf{Z}$-submodule $N_{\mathbf{Z}}$ of $N_{\mathbf{Q}}$ such that $N_{\mathbf{Q}} = \mathbf{Q} \otimes N_{\mathbf{Z}}$ and such that $s \in \Gamma(X, M_X^{\mathrm{gp}}/\mathcal{O}_X^\times) \otimes N_{\mathbf{Z}}$. Let

$$J(\sigma) = \{h(s) \mid h \in \mathrm{Hom}_{\mathbf{Z}}(N_{\mathbf{Z}}, \mathbf{Z}), h(\sigma) \subset \mathbf{R}_{\geq 0}\} \subset \Gamma(X, M_X^{\mathrm{gp}}/\mathcal{O}_X^\times),$$

where $h(s)$ denotes the image of s under $1 \otimes h : \Gamma(X, M_X^{\mathrm{gp}}/\mathcal{O}_X^\times) \otimes N_{\mathbf{Z}} \to \Gamma(X, M_X^{\mathrm{gp}}/\mathcal{O}_X^\times)$. Since $J(\sigma)$ is a finitely generated monoid, by 3.6.1 (i), we have $X[J(\sigma)]$. This $X[J(\sigma)]$ has the property of $X(\sigma)$ stated in Lemma 3.6.8. □

Lemma 3.6.9 *Let σ and τ be finitely generated rational cones in $N_{\mathbf{R}}$.*

(i) *If $\tau \subset \sigma$, then* $\mathrm{Mor}(\ , X(\tau)) \subset \mathrm{Mor}(\ , X(\sigma))$ *in* $\mathrm{Mor}(\ , X)$ *as sheaves on* $\mathcal{B}$(log).

(ii) $\mathrm{Mor}(\ , X(\sigma)) \cap \mathrm{Mor}(\ , X(\tau)) = \mathrm{Mor}(\ , X(\sigma \cap \tau))$ *in* $\mathrm{Mor}(\ , X)$.

(iii) *If τ is a face of σ, then $X(\tau)$ is an open subobject* (3.6.1) *of $X(\sigma)$.*

Proof. The assertions (i) and (ii) follow from 3.6.8.

We prove (iii). Take $N_{\mathbf{Z}}$ as in the proof of 3.6.8. Then $J(\tau)$ is generated by $J(\sigma)$ and the $h(s)^{-1}$ for the elements h of $\mathrm{Hom}_{\mathbf{Z}}(N_{\mathbf{Z}}, \mathbf{Z})$ satisfying $h(\sigma) \subset \mathbf{R}_{\geq 0}$ and $h(\tau) = 0$. Hence $X(\tau)$ is an open subobject of $X(\sigma)$ by 3.6.1 (ii). □

Proposition 3.6.10 *Let X be an object of $\mathcal{B}$(log), let $(N_{\mathbf{Q}}, s)$ be as in* 3.7.7, *and let Σ be a rational fan in $N_{\mathbf{R}}$. Then there is an object $X(\Sigma)$ of $\mathcal{B}$(log) over X having the following property. For any object Y of $\mathcal{B}$(log) over X, the set* $\mathrm{Mor}_X(Y, X(\Sigma))$ *is either a one-point set or an empty set, and it is nonempty if and only if, for each $y \in Y$, there is an element σ of Σ such that $s_y(\pi_1^+(y^{\log})) \subset \sigma$.*

Each $X(\sigma)$ ($\sigma \in \Sigma$) is an open subobject of $X(\Sigma)$, and $X(\Sigma) = \bigcup_{\sigma \in \Sigma} X(\sigma)$. We have $X(\sigma) = X(\{$face of $\sigma\})$.

Proof. This follows from 3.6.8 and 3.6.9. □

3.6.11

We describe the relation of $X(\Sigma)$ to toric varieties associated with fans ([KKMS]).

(i) Assume first that we are given a chart $h : \mathcal{S} \to M_X$ with $\mathcal{S}$ a sharp fs monoid, let $N_{\mathbf{Q}} = \mathrm{Hom}(\mathcal{S}^{\mathrm{gp}}, \mathbf{Q})$, and let $s \in \Gamma(X, M_X^{\mathrm{gp}}/\mathcal{O}_X^\times) \otimes N_{\mathbf{Q}}$ be the element corresponding to h. Assume that we are given a rational fan Σ in $N_{\mathbf{R}} = \mathrm{Hom}(\mathcal{S}, \mathbf{R}^{\mathrm{add}})$ such that $\sigma \subset \mathrm{Hom}(\mathcal{S}, \mathbf{R}^{\mathrm{add}}_{\geq 0})$ for any $\sigma \in \Sigma$. Let $V(\Sigma)$ be the toric variety constructed in [KKMS]. It is covered by open subspaces $V(\sigma) = \mathrm{Spec}(\mathbf{C}[\mathcal{S}(\sigma)])_{\mathrm{an}}$ where $\mathcal{S}(\sigma)$ is a submonoid of $\mathcal{S}^{\mathrm{gp}}$ consisting of all $f \in \mathcal{S}^{\mathrm{gp}}$ that are sent to $\mathbf{R}_{\geq 0}$ by all elements of σ. We endow $V(\Sigma)$ with a unique logarithmic structure whose restriction to each $V(\sigma)$ is the canonical logarithmic structure 2.1.6 (ii). The inclusions

$\mathcal{S} \subset \mathcal{S}(\sigma)$ induce a canonical morphism $V(\Sigma) \to \mathrm{Spec}(\mathbf{C}[\mathcal{S}])_{\mathrm{an}}$. We have

$$X(\Sigma) = X \times_{\mathrm{Spec}(\mathbf{C}[\mathcal{S}])_{\mathrm{an}}} V(\Sigma), \quad X(\sigma) = X \times_{\mathrm{Spec}(\mathbf{C}[\mathcal{S}])_{\mathrm{an}}} V(\sigma) \quad (\sigma \in \Sigma)$$

as objects of $\mathcal{B}(\log)$. These fiber products are taken here in $\mathcal{B}(\log)$ (3.5.1 (i)) but as topological spaces, they are the fiber products in the category of topological spaces by 3.5.1 (ii) (condition (1) there is satisfied because $X \to \mathrm{Spec}(\mathbf{C}[\mathcal{S}])_{\mathrm{an}}$ is strict, and condition (3) there is satisfied because $V(\Sigma)$ and $V(\sigma)$ are analytic spaces).

If Σ here is a finite subdivision of $\mathrm{Hom}(\mathcal{S}, \mathbf{R}^{\mathrm{add}}_{\geq 0})$ (this means that Σ is finite and $\bigcup_{\sigma \in \Sigma} \sigma = \mathrm{Hom}(\mathcal{S}, \mathbf{R}^{\mathrm{add}}_{\geq 0})$), the map $V(\Sigma) \to \mathrm{Spec}(\mathbf{C}[\mathcal{S}])_{\mathrm{an}}$ is proper surjective with connected fibers, and hence $X(\Sigma) \to X$ is proper surjective with connected fibers.

(ii) Next we consider the general situation. Let $(N_{\mathbf{Q}}, s)$ be as in 3.6.7 and let Σ be a rational fan in $N_{\mathbf{R}}$. Locally on X, there is a chart $h : \mathcal{S} \to M_X$ with $\mathcal{S}$ a sharp fs monoid (2.1.5 (iii)). Since $\mathcal{S}^{\mathrm{gp}} \to M_X^{\mathrm{gp}}/\mathcal{O}_X^\times$ is a surjective homomorphism of sheaves, locally on X, there is a homomorphism $a : \mathrm{Hom}(N_{\mathbf{Q}}, \mathbf{Q}) \to \mathbf{Q} \otimes \mathcal{S}^{\mathrm{gp}}$ such that the composition $h \circ a : \mathrm{Hom}(N_{\mathbf{Q}}, \mathbf{Q}) \to \mathbf{Q} \otimes (M_X^{\mathrm{gp}}/\mathcal{O}_X^\times)$ is the map induced by s. Let $b : \mathrm{Hom}(\mathcal{S}^{\mathrm{gp}}, \mathbf{R}) \to N_{\mathbf{R}}$ be the $\mathbf{R}$-homomorphism induced by a. Let Σ' be the rational fan in $\mathrm{Hom}(\mathcal{S}^{\mathrm{gp}}, \mathbf{R})$ consisting of $\mathrm{Hom}(\mathcal{S}, \mathbf{R}^{\mathrm{add}}_{\geq 0}) \cap b^{-1}(\sigma)$ $(\sigma \in \Sigma)$ and their faces. Then

$$X(\Sigma) = X(\Sigma') = X \times_{\mathrm{Spec}(\mathbf{C}[\mathcal{S}])_{\mathrm{an}}} V(\Sigma').$$

Here $X(\Sigma')$ is defined with respect to h.

From this, we see that the map $X(\Sigma) \to X$ is separated (0.7.5) in general. In fact, to show this we may work locally, and we are reduced to the well-known fact that $V(\Sigma')$ is Hausdorff.

Definition 3.6.12 *Let X be an object of $\mathcal{B}(\log)$ and let Y be an object of $\mathcal{B}(\log)$ over X. We say Y is a* logarithmic modification *of X if, locally on X, there are a chart $h : \mathcal{S} \to M_X$ with $\mathcal{S}$ a sharp fs monoid and a finite rational subdivision Σ of $\mathrm{Hom}(\mathcal{S}, \mathbf{R}^{\mathrm{add}}_{\geq 0})$ such that $Y \simeq X(\Sigma)$ over X. Here $X(\Sigma)$ is defined with respect to the chart h as in* 3.6.11 (i).

By 3.6.11, a logarithmic modification is proper and surjective with connected fibers as a map of topological spaces.

Lemma 3.6.13 *A logarithmic blow-up is a logarithmic modification.*

Proof. Let I be a nonempty finite subset of $\Gamma(X, M_X^{\mathrm{gp}}/\mathcal{O}_X^\times)$. Locally on X, we have a chart $\mathcal{S} \to M_X$ with $\mathcal{S}$ a sharp fs monoid and a finite subset $\tilde{I}$ of $\mathcal{S}^{\mathrm{gp}}$ whose image in $\Gamma(X, M_X^{\mathrm{gp}}/\mathcal{O}_X^\times)$ coincides with I. Let $N_{\mathbf{Q}} = \mathrm{Hom}(\mathcal{S}^{\mathrm{gp}}, \mathbf{Q})$ and let $s \in \Gamma(X, M_X^{\mathrm{gp}}/\mathcal{O}_X^\times) \otimes N_{\mathbf{Q}}$ be the induced element. For $a \in \tilde{I}$, let σ_a be the cone $\{h \in \mathrm{Hom}(\mathcal{S}, \mathbf{R}^{\mathrm{add}}_{\geq 0}) \mid h(b) \geq h(a)\ (\forall\, b \in \tilde{I})\}$. Then for $a, b \in \tilde{I}$, $\sigma_a \cap \sigma_b$ is a face of σ_a and also a face of σ_b. Hence we have a finite subdivision Σ of the cone $\mathrm{Hom}(\mathcal{S}, \mathbf{R}^{\mathrm{add}}_{\geq 0})$ consisting of σ_a $(a \in \tilde{I})$ and their faces. We have $B_I(X) = \bigcup_{a \in I} X(\sigma_a) = X(\Sigma)$. □

3.6.14

In 3.6.15 below, we give set-theoretic descriptions of $X(\Sigma)$ and $X(\Sigma)^{\log}$, and also those for logarithmic blow-ups.

For an object X of $\mathcal{B}(\log)$, we define sets $Q(X)$ and $Q'(X)$ as follows.

Let $Q_1(X)$ be the set of all pairs (x, P) where $x \in X$ and P is a submonoid of $(M_X^{gp}/\mathcal{O}_X^\times)_x$ such that $(M_X/\mathcal{O}_X^\times)_x \subset P$ and $P^\times \cap (M_X/\mathcal{O}_X^\times)_x = \{1\}$. Let $Q(X)$ be the set of all triples (x, P, h) where $(x, P) \in Q_1(X)$ and h is a homomorphism $(\tilde{P})^\times \to \mathbf{C}^\times$, where $\tilde{P}$ denotes the inverse image of P in $M_{X,x}^{gp}$ and $(\tilde{P})^\times$ denotes the group of all invertible elements of $\tilde{P}$, such that $h(f) = f(x)$ for $f \in \mathcal{O}_{X,x}^\times$.

Let $Q_1'(X)$ be the set of all pairs (x, σ) where $x \in X$ and σ is a submonoid of $\pi_1^+(x^{\log})$ which contains some point in the interior (0.7.7) of $\pi_1^+(x^{\log})$. We have a map

$$Q_1'(X) \to Q_1(X), \quad (x, \sigma) \mapsto (x, P(\sigma)),$$

where $P(\sigma)$ denotes the submonoid of $(M_X^{gp}/\mathcal{O}_X^\times)_x$ consisting of all elements f such that the homomorphism $\pi_1(x^{\log}) \to \mathbf{Z}$ corresponding to f (2.2.9) sends σ into $\mathbf{N}$. Let $Q'(X)$ be the set of all triples (x, σ, h) such that $(x, \sigma) \in Q_1'(X)$ and $(x, P(\sigma), h) \in Q(X)$.

We have an evident map

$$Q'(X) \to Q(X), \quad (x, \sigma, h) \mapsto (x, P(\sigma), h).$$

For an object Y of $\mathcal{B}(\log)$ over X, we have maps

$$q_Y' : Y \to Q'(X), \quad q_Y : Y \to Q(X)$$

defined as follows. The map q_Y is induced from q_Y' via $Q'(X) \to Q(X)$, and $q_Y'(y) = (x, \sigma, h)$, where x is the image of y in X, σ is the image of $\pi_1^+(y^{\log})$ in $\pi_1(x^{\log})$ (then $P := P(\sigma)$ coincides with the inverse image of $(M_Y/\mathcal{O}_Y^\times)_y$ under the map $(M_X^{gp}/\mathcal{O}_X^\times)_x \to (M_Y^{gp}/\mathcal{O}_Y^\times)_y$), and h is the composition $(\tilde{P})^\times \to \mathcal{O}_{Y,y}^\times \to \mathbf{C}^\times$ where the last arrow is $f \mapsto f(y)$. Forgetting h, we have maps

$$q_{Y,1}' : Y \to Q_1'(X), \quad q_{Y,1} : Y \to Q_1(X).$$

LEMMA 3.6.15 *Let X be an object of $\mathcal{B}(\log)$ and let Y be an object of $\mathcal{B}(\log)$ over X. Assume that we have one of the following four cases* (1)–(4).

(1) $Y = X[J]$ *for some subset J of $\Gamma(X, M_X^{gp}/\mathcal{O}_X^\times)$ which generates a finitely generated monoid* (3.6.1 (i)).

(2) $Y = B_I(X)$ *for some nonempty finite subset I of* $\Gamma(X, M_X^{gp}/\mathcal{O}_X^\times)$ (3.6.6 (i)).

(3) $Y = B_I^*(X)$ *for some nonempty finite subset I of* $\Gamma(X, M_X^{gp}/\mathcal{O}_X^\times)$ (3.6.6 (ii)).

(4) $Y = X(\Sigma)$ *for $(N_{\mathbf{Q}}, s)$ as in* 3.7.6 *and for a rational fan Σ in $N_{\mathbf{R}}$* (3.6.10).

Then

(i) *The maps $q_Y' : Y \to Q'(X)$ and $q_Y : Y \to Q(X)$* (3.6.14) *are injective.*

(ii) *For any $y \in Y$ with image x in X, the map $(M_X^{gp}/\mathcal{O}_X^\times)_x \to (M_Y^{gp}/\mathcal{O}_Y^\times)_y$ is surjective, and $q_Y'(y) = (x, \sigma, h) \in Q'(X)$ is recovered from $q_Y(y) = (x, P, h) \in Q(X)$ as $\sigma = \{\gamma \in \pi_1(x^{\log}) \mid h_\gamma(P) \subset \mathbf{N}\}$ where h_γ denotes the homomorphism $(M_X^{gp}/\mathcal{O}_X^\times)_x \to \mathbf{Z}$ corresponding to γ.*

In the following, by (i), *we regard* Y *as a subset of* $Q'(X)$ *via the injection* q'_Y *and also as a subset of* $Q(X)$ *via the injection* q_Y.

(iii) *In the case* (4), *as a subset of* $Q'(X)$, *we have*

$$Y = \{(x, \sigma, h) \in Q'(X) \mid \sigma \text{ is a face of } \pi_1^+(x^{\log}) \cap s_x^{-1}(\tau) \text{ for some } \tau \in \Sigma\}.$$

(iv) *In the case* (1), *as a subset of* $Q(X)$, *we have*

$$Y = \{(x, P, h) \in Q(X) \mid P = \langle (M_X/\mathcal{O}_X^\times)_x, J_x, (J'_x)^{-1}\rangle^{\text{sat}}\}$$

for some finite subset J' *of* J. *Here* $\langle (M_X/\mathcal{O}_X^\times)_x, J_x, (J'_x)^{-1}\rangle$ *denotes the submonoid of* $(M_X^{\text{gp}}/\mathcal{O}_X^\times)_x$ *generated by* $(M_X/\mathcal{O}_X^\times)_x$, *the stalk* J_x *of* J, *and the inverses of elements of the stalk* J'_x, *and* $\langle - \rangle^{\text{sat}}$ *means the saturation of* $\langle - \rangle$ *in* $(M_X^{\text{gp}}/\mathcal{O}_X^\times)_x$, *i.e.*,

$$\langle - \rangle^{\text{sat}} = \{a \in (M_X^{\text{gp}}/\mathcal{O}_X^\times)_x \mid a^n \in \langle - \rangle \text{ for some } n \geq 1\}.$$

In the case (2), *as a subset of* $Q(X)$, *we have*

$$Y = \left\{(x, P, h) \in Q(X) \;\middle|\; \begin{array}{l} P = \langle (M_X/\mathcal{O}_X^\times)_x, \{a^{-1}b \mid a \in I', b \in I\}\rangle^{\text{sat}} \\ \text{for some subset } I' \text{ of } I \end{array}\right\}.$$

In the case (3), *as a subset of* $Q(X)$, *we have*

$$Y = \left\{(x, P, h) \in Q(X) \;\middle|\; \begin{array}{l} P = \langle (M_X/\mathcal{O}_X^\times)_x, \{a^{-1}b \mid a, b \in I, f(a) \leq f(b)\}\rangle^{\text{sat}} \\ \text{for some map } f : I \to \mathbf{N} \end{array}\right\}.$$

(v) *The map* $Y^{\log} \to Y \times_X X^{\log}$ *is injective, and the image consists of all elements* $((x, P, h), (x, h'))$ $((x, P, h) \in Y \subset Q(X),\ (x, h') \in X^{\log})$ *such that* $h'(a) = h(a)/|h(a)|$ *for any* $a \in (\tilde{P})^\times$. *The topology of* $Y^{\log}$ *coincides with the topology as a subset of* $Y \times_X X^{\log}$.

(vi) *Let* Z *be an object of* $\mathcal{B}(\log)$ *over* X. *Then there is a morphism* $Z \to Y$ *over* X *if and only if, for any* $z \in Z$ *with* $q'_Z(z) = (x, \tau, f)$ (*resp.* $q_Z(z) = (x, Q, f)$), *there is* $(x, \sigma) \in q'_{Y,1}(Y)$ (*resp.* $(x, P) \in q_{Y,1}(Y)$) *such that* $\tau \subset \sigma$ (*resp.* $P \subset Q$). *If there is a morphism* $Z \to Y$ *over* X, *and* z, x, τ, f, σ, P *are as above, the image of* z *in* $Y \subset Q'(X)$ *and in* $Y \subset Q(X)$ *are described as* (x, σ', h) *and* (x, P', h), *respectively, where* σ' *is the smallest face of* σ *such that* $\tau \subset \sigma$, $P' = \langle P, (P \cap Q^\times)^{-1}\rangle$, *and* h *is the restriction of* f *to* $((P')^\sim)^\times$.

Proof. The assertions about the case (1) follow from the explicit construction of $X[J]$ in 3.6.4.

The assertions in (ii) for the cases (2), (3), and (4) are reduced to the case (1).

By (ii), the injectivity of $Y \to Q(X)$ is reduced to the injectivity of $Y \to Q'(X)$.

We prove the injectivity of $Y \to Q'(X)$ for the case (4). If $y_1, y_2 \in Y$ have the same image (x, σ, h) in $Q'(X)$, $s_{y_j}(\pi_1^+(y_j^{\log})) = s_x(\sigma)$ for $j = 1, 2$, and hence there is $\tau \in \Sigma$ such that $s_{y_j}(\pi_1^+(y_j^{\log})) \subset \tau$ for $j = 1, 2$. We have $y_1, y_2 \in X(\tau)$ and hence, by 3.6.8, we are reduced to the case (1).

The injectivity of $Y \to Q'(X)$ for the cases (2) and (3) follows from that for the case (4), since locally on X the cases (2) and (3) are special cases of (4) by 3.6.6 (iii) and 3.6.13.

The other assertions in 3.6.15 are reduced to the case (1). □

Definition 3.6.16 *For an abelian group L and for a submonoid V of L, we say V is* valuative *if $V \cup V^{-1} = L$.*

3.6.17

Example. A valuative submonoid of $\mathbf{Z}^2$ is one of the following types (i)–(iii). Let $\langle\ ,\ \rangle : \mathbf{Z}^2 \times \mathbf{R}^2 \to \mathbf{R}$ be the standard pairing $((x_1, x_2), (y_1, y_2)) \mapsto x_1 y_1 + x_2 y_2$.

(i) $\mathbf{Z}^2$.

(ii) V_ℓ for a half line $\ell = (\mathbf{R}_{\geq 0})v$ $(v \in \mathbf{R}^2 - \{(0,0)\})$, defined by

$$V_\ell = \{x \in \mathbf{Z}^2 \mid \langle x, \ell\rangle \subset \mathbf{R}_{\geq 0}\}.$$

(iii) $V_{\ell,\ell'}$ for a rational half line $\ell = (\mathbf{R}_{\geq 0})v$ $(v \in \mathbf{Q}^2 - \{(0,0)\})$ and for a half line $\ell' = (\mathbf{R}_{\geq 0})v'$ in the one-dimensional quotient vector space $\mathbf{R}^2/\mathbf{R}\ell$ $(v' \in \mathbf{R}^2/\mathbf{R}\ell - \{0\})$, defined by

$$V_{\ell,\ell'} = \{x \in \mathbf{Z}^2 \mid \langle x, \ell\rangle \subset \mathbf{R}_{\geq 0}.\ \text{If } \langle x, \ell\rangle = 0,\ \text{then } \langle x, \ell'\rangle \geq 0\}.$$

Definition 3.6.18 *Let X be an object of $\mathcal{B}(\log)$.*

As a set, X_{val} is the set of all elements (x, V, h) of $Q(X)$ (3.6.14) such that V is valuative in $(M_X^{\mathrm{gp}}/\mathcal{O}_X^\times)_x$. That is, X_{val} is the set of all triples (x, V, h) where $x \in X$, V is a valuative submonoid of $(M_X^{\mathrm{gp}}/\mathcal{O}_X^\times)_x$ containing $(M_X/\mathcal{O}_X^\times)_x$ such that $V^\times \cap (M_X/\mathcal{O}_X^\times)_x = \{1\}$, and h is a homomorphism $(\tilde{V})^\times \to \mathbf{C}^\times$, where $\tilde{V}$ denotes the inverse image of V in $M_{X,x}^{\mathrm{gp}}$ and $(\tilde{V})^\times$ denotes the group of all invertible elements of $\tilde{V}$, such that $h(f) = f(x)$ for all $f \in \mathcal{O}_{X,x}^\times$.

In 3.6.23 below, we will define a structure on X_{val} of a logarithmic local ringed space over X.

3.6.19

For a finite subset I of $\Gamma(X, M_X^{\mathrm{gp}}/\mathcal{O}_X^\times)$, we define a canonical map $X_{\mathrm{val}} \to B_I^*(X)$ as $(x, V, h) \mapsto (x, P, h')$ (3.6.6 (ii)), where

$$P = \langle (M_X/\mathcal{O}_X^\times)_x, V \cap \{a_x^{-1} b_x \mid a, b \in I\}\rangle^{\mathrm{sat}},$$

and h' is the restriction of h to $(\tilde{P})^\times$ (3.6.15 (iv) for the case (3)).

If I' is another nonempty finite subset of $\Gamma(X, M_X^{\mathrm{gp}}/\mathcal{O}_X^\times)$ such that $I' \supset I$, the map $X_{\mathrm{val}} \to B_I^*(X)$ coincides with the composition $X_{\mathrm{val}} \to B_{I'}^*(X) \to B_I^*(X)$. This follows from 3.6.15 (vi) for the case (3).

Proposition 3.6.20 *Assume that $\Gamma(X, M_X^{\mathrm{gp}}/\mathcal{O}_X^\times) \to (M_X^{\mathrm{gp}}/\mathcal{O}_X^\times)_x$ is surjective for any $x \in X$. Then the maps $X_{\mathrm{val}} \to B_I^*(X)$ in 3.6.19 induce a bijection*

$$X_{\mathrm{val}} \xrightarrow{\sim} \varprojlim_I B_I^*(X)$$

where I ranges over all non-empty finite subset of $\Gamma(X, M_X^{\mathrm{gp}}/\mathcal{O}_X^\times)$.

Proof. This follows from 3.6.15 (iv) and (vi), for the case (3). □

LEMMA 3.6.21 *Let* $(N_{\mathbf{Q}}, s)$ *be as in* 3.6.7. *Assume we are given a chart* $\mathcal{S} \to M_X$.

(i) *Let* Σ *be a finite rational fan in* $N_{\mathbf{R}}$. *Then, locally on* X, *there is a nonempty finite subset* $\tilde{I}$ *of* $\mathcal{S}^{\mathrm{gp}}$ *such that the fiber product* $X(\Sigma) \times_X B_I(X)$ *in the category* $\mathcal{B}(\log)$ *is an open subobject* (3.6.1)*of* $B_I(X)$. *Here* I *denotes the image of* $\tilde{I}$ *in* $\Gamma(X, M_X^{\mathrm{gp}}/\mathcal{O}_X^\times)$.

(ii) *Let* Y *be a logarithmic modification of* X. *Then, locally on* X, *there are a nonempty finite subset* $\tilde{I}$ *of* $\mathcal{S}^{\mathrm{gp}}$ *and a morphism* $B_I(X) \to Y$ *over* X. *Here* I *denotes the image of* $\tilde{I}$ *in* $\Gamma(X, M_X^{\mathrm{gp}}/\mathcal{O}_X^\times)$.

Proof.

(i) It is sufficient to treat the case $\Sigma = \{\text{face of } \sigma\}$ for some finitely generated sharp rational cone σ in $N_{\mathbf{R}}$. Take $N_{\mathbf{Z}} \subset N_{\mathbf{Q}}$ as in 3.6.7, and let I be a finite subset of $\Gamma(X, M_X^{\mathrm{gp}}/\mathcal{O}_X^\times)$ which generates $J(\sigma)$ (3.6.8) such that $1 \in I$. Locally on X, we have a finite subset $\tilde{I}$ of $\mathcal{S}^{\mathrm{gp}}$ whose image in $\Gamma(X, M_X^{\mathrm{gp}}/\mathcal{O}_X^\times)$ coincides with I. We have $X(\sigma) = X[I] = X[1^{-1}I]$ and this is an open subobject of $B_I(X)$.

(ii) By (i), locally on X, we have $\tilde{I}$ such that the fiber product $Y \times_X B_I(X)$ in $\mathcal{B}(\log)$ is an open subobject of $B_I(X)$. It is sufficient to prove that $Y \times_X B_I(X) = B_I(X)$. Since $Y \times_X B_I(X) \to B_I(X)$ is proper (3.6.12), for each $x \in X$, the fiber F_1 in $Y \times_X B_I(X)$ over x is a nonempty compact open subset of the fiber F_2 in $B_I(X)$ over x. Since F_2 is a Hausdorff space, F_1 is an open and closed subset of F_2. Since F_2 is connected by 3.6.12 and 3.6.13, we have $F_1 = F_2$. □

3.6.22

In general, for a directed ordered set Λ and for a projective system $(X_\lambda)_{\lambda \in \Lambda}$ of logarithmic local ringed spaces, the projective limit $\varprojlim_\lambda X_\lambda$ in the category of local ringed spaces exist. As a topological space, this projective limit P is the projective limit of the topological spaces X_λ. The structural sheaf $\mathcal{O}_P$ and the logarithmic structure M_P of P are given by

$$\mathcal{O}_P = \varinjlim_\lambda p_\lambda^{-1}(\mathcal{O}_{X_\lambda}), \quad M_P = \varinjlim_\lambda p_\lambda^{-1}(M_{X_\lambda})$$

where p_λ is projection $P \to X_\lambda$.

3.6.23

Let X be an object of $\mathcal{B}(\log)$. We endow X_{val} with the structure of a logarithmic local ringed space over $\mathbf{C}$ as follows.

First assume that there is a chart $\mathcal{S} \to M_X$. Then, by 3.6.6 (iii), 3.6.13, and 3.6.21 (ii), we have isomorphisms of logarithmic local ringed spaces

$$\varprojlim (\text{logarithmic modifications of } X) \simeq \varprojlim_I B_I^*(X) \simeq \varprojlim_{\tilde{I}} B_{\tilde{I}}^*(X), \tag{1}$$

where the transition morphisms in the first projective system are all morphisms over X, I in the second ranges over all nonempty finite subsets of $\Gamma(X, M_X^{\mathrm{gp}}/\mathcal{O}_X^\times)$, $\tilde{I}$ in the third ranges over all nonempty finite subsets of $\mathcal{S}^{\mathrm{gp}}$, and I in the third projective system denotes the image of $\tilde{I}$ in $\Gamma(X, M_X^{\mathrm{gp}}/\mathcal{O}_X^\times)$. By the existence of the chart, the assumption of Proposition 3.6.20 is satisfied. Hence, by Proposition 3.6.20, the underlying set of the projective limit in (1) above is identified with X_{val}. We endow X_{val} with the structure of a logarithmic local ringed space over X as the projective limit in (1).

In the case where X has a chart, by the presentation of X as the third projective limit in (1), we see that the logarithmic local ringed space U_{val} over X for an open set U of X is identified with the open subobject of X_{val} whose underlying set is the inverse image of U in X_{val}. Hence for a general object X of $\mathcal{B}(\log)$, globally on X_{val}, we have a structure of a logarithmic local ringed space over X which induces the logarithmic local ringed structure of U_{val} for any open subset U of X which has a chart.

Lemma 3.6.24 *The map $X_{\mathrm{val}} \to X$ is proper and surjective, and the fibers are connected.*

Proof. Since the problem is local on X, we may assume that X has a chart. In this case, since X_{val} is the projective limit of logarithmic modifications of X, 3.6.24 follows from 3.6.12. □

3.6.25

Example. Let $X = \mathbf{C}^2$ with the logarithmic structure given by the divisor $\mathbf{C}^2 - (\mathbf{C}^\times)^2$ with normal crossings. Then X_{val} coincides with the space $(\mathbf{C}^2)_{\mathrm{val}}$ in 0.5.21. Let $f : (\mathbf{C}^2)_{\mathrm{val}} \to \mathbf{C}^2$ be the projection. Let $x = (0, 0) \in \mathbf{C}^2$ and identify $(M_X/\mathcal{O}_X^\times)_x$ with $\mathbf{N}^2$ in the natural way. Then the point $(0, 0)_s$ of $f^{-1}(x)$ ($s \in \mathbf{R}_{>0} - \mathbf{Q}_{>0}$) in 0.5.21 is the point (x, V, h) of X_{val} (3.6.18) where $V = V_\ell$ with $\ell = (\mathbf{R}_{\geq 0})(1, s)$ in the notation of 3.6.17 and $h : (\tilde{V})^\times \to \mathbf{C}^\times$ is the evident one (note that $(\tilde{V})^\times = \mathcal{O}_{X,x}^\times$ in this case). The point $(0, 0)_0$ (resp. $(0, 0)_\infty$, resp. $(0, 0)_{s,0}$ with $s \in \mathbf{Q}_{>0}$, resp. $(0, 0)_{s,\infty}$ with $s \in \mathbf{Q}_{>0}$) of $f^{-1}(x)$ in 0.5.21 is the point (x, V, h) where $V = V_{\ell,\ell'}$ with $\ell = (\mathbf{R}_{\geq 0})(1, 0)$ (resp. $(\mathbf{R}_{\geq 0})(0, 1)$, resp. $(\mathbf{R}_{\geq 0})(1, s)$, resp. $(\mathbf{R}_{\geq 0})(1, s)$) and ℓ' is the image of $(\mathbf{R}_{\geq 0})(0, 1)$ (resp. $(\mathbf{R}_{\geq 0})(1, 0)$, resp. $(\mathbf{R}_{\geq 0})(1, 0)$, resp. $(\mathbf{R}_{\geq 0})(0, 1)$), and $h : (\tilde{V})^\times \to \mathbf{C}^\times$ is the evident one (note that $(\tilde{V})^\times = \mathcal{O}_{X,x}^\times$ in this case). Finally let $s \in \mathbf{Q}_{>0}$ and write $s = m/n$ with $m, n \in \mathbf{Z}, m, n > 0$, and $\mathrm{GCD}(m, n) = 1$. Then the point $(0, 0)_{s,z}$ with $s \in \mathbf{Q}_{>0}$ and $z \in \mathbf{C}^\times$ of $f^{-1}(x)$ in 0.5.21 is the point (x, V, h) of X_{val}, where $V = V_\ell$ with $\ell = (\mathbf{R}_{\geq 0})(1, s)$, $\tilde{V}$ in this case is generated by $\mathcal{O}_{X,x}^\times$ and q_1^m/q_2^n with (q_1, q_2) the standard coordinate of $\mathbf{C}^2$, and $h : (\tilde{V})^\times \to \mathbf{C}^\times$ is the homomorphism which sends $f \in \mathcal{O}_{X,x}^\times$ to $f(x)$ and q_1^m/q_2^n to z.

3.6.26

Let $X_{\mathrm{val}}^{\log}$ be the set of all pairs (x, h) where $x \in X_{\mathrm{val}}$ and h is a homomorphism $M_{X,x}^{\mathrm{gp}} \to \mathbf{S}^1$ such that $h(u) = u(x)/|u(x)|$ for any $u \in \mathcal{O}_{X_{\mathrm{val}},x}^\times$. Then we have a bijection

$$X_{\rm val}^{\log} \xrightarrow{\sim} \{((x, V, h), (x, h')) \in X_{\rm val} \times_X X^{\log} \mid \\ h'(f) = h(f)/|h(f)| \text{ for any } f \in (\tilde{V})^\times\}.$$

We endow $X_{\rm val}^{\log}$ with the topology as a subspace of $X_{\rm val} \times_X X^{\log}$. Since this is a closed subspace of $X_{\rm val} \times_X X^{\log}$, $X_{\rm val}^{\log}$ is proper over $X_{\rm val}$.

In the case X has a chart, we have a homeomorphism

$$X_{\rm val}^{\log} \xrightarrow{\sim} \varprojlim_Y Y^{\log}$$

where Y ranges over all logarithmic modifications of X. The bijectivity of this map follows from 3.6.15 (v), and the fact that it is a homeomorphism follows from the properness of $X_{\rm val}^{\log}$ and of $\varprojlim_Y Y^{\log}$ over X.

Lemma 3.6.27 *Let X be an object of $\mathcal{B}(\log)$, let $(N_{\mathbf{Q}}, s)$ be as in* 3.6.7, *and let Σ be a rational fan in $N_{\mathbf{R}}$.*

(i) *Via the morphism $X(\Sigma)_{\rm val} \to X_{\rm val}$ induced by $X(\Sigma) \to X$, $X(\Sigma)_{\rm val}$ is an open subobject* (3.6.1) *of $X_{\rm val}$.*

(ii) *As a subset of $X_{\rm val}$, $X(\Sigma)_{\rm val}$ is the set of all $(x, V, h) \in X_{\rm val}$ such that $P \subset V$ for some $(x, P, h) \in X(\Sigma) \subset Q(X)$.*

Proof.

(i) This is reduced to the case where Σ is finite and X has a chart, and then to 3.6.21 (i).

(ii) Write $Y = X(\Sigma)$. The inclusion map $Y_{\rm val} \to X_{\rm val}$ sends $(y, V_1, h_1) \in Y_{\rm val}$ (3.6.18) with $y = (x, P, h_2) \in Y \subset Q(X)$ (3.6.14) (P is the inverse image of $(M_Y/\mathcal{O}_Y^\times)_y$ under $(M_X^{\rm gp}/\mathcal{O}_X^\times)_x \to (M_Y^{\rm gp}/\mathcal{O}_Y^\times)_y$ and $h_2 : (\tilde{P})^\times \to \mathbf{C}^\times$ is induced from $h_1 : (\tilde{V}_1)^\times \to \mathbf{C}^\times$) to (x, V, h), where V is the inverse image of V_1 under $(M_X^{\rm gp}/\mathcal{O}_X^\times)_x \to (M_X^{\rm gp}/\mathcal{O}_X^\times)_x/P^\times \xrightarrow{\sim} (M_Y^{\rm gp}/\mathcal{O}_Y^\times)_y$, and $h : (\tilde{V})^\times \to \mathbf{C}^\times$ is the homomorphism induced by h_1. Hence, if $(x, V, h) \in X_{\rm val}$ is the image of $(y, V_1, h_1) \in Y_{\rm val}$ with $y = (x, P, h_2) \in Y \subset Q(X)$, then $P \subset V$ as desired. Conversely, let $(x, V, h) \in X_{\rm val}$ and assume that there is $(x, P, h_2) \in Y$ such that $P \subset V$. Let P' be the submonoid of $(M_X^{\rm gp}/\mathcal{O}_X^\times)_x$ generated by P and the inverses of elements of P contained in $V^\times$. Let $y := (x, P', h_2') \in Y \subset Q(X)$, where $h_2' : (\tilde{P}')^\times \to \mathbf{C}^\times$ is the homomorphism induced by h_2. We have $(M_X^{\rm gp}/\mathcal{O}_X^\times)_x/(P')^\times \xrightarrow{\sim} (M_Y^{\rm gp}/\mathcal{O}_Y^\times)_y$. Hence (x, V, h) is the image of $(y, V_1, h_1) \in Y_{\rm val}$, where $V_1 = V/(P')^\times$ and where $h_1 : (\tilde{V}_1)^\times \to \mathbf{C}^\times$ is the homomorphism induced by h and $\mathcal{O}_{Y,y}^\times \to \mathbf{C}^\times$, $f \mapsto f(y)$, by the fact that $(\tilde{V}_1)^\times$ is the push-out of $(\tilde{V})^\times \leftarrow (P')^\times \to \mathcal{O}_{Y,y}^\times$. □

Theorem 3.6.28 *Let X be an object of $\mathcal{B}(\log)$, let $(N_{\mathbf{Q}}, s)$ be as in* 3.6.7, *and let Σ be a rational fan in $N_{\mathbf{R}}$. Then the following six conditions* (1)–(6) *are equivalent.*

(1) *$X(\Sigma)$ is a logarithmic modification of X.*
(2) *The map $X(\Sigma) \to X$ is proper and surjective.*

(3) *Locally on X, there is a logarithmic modification Y of X and a morphism $Y \to X(\Sigma)$ over X.*

(4) *For any $x \in X$, $\pi_1^+(x^{\log}) = \bigcup_{\sigma \in \Sigma} \pi_1^+(x^{\log}) \cap s_x^{-1}(\sigma)$, and the set of submonoids $\pi_1^+(x^{\log}) \cap s_x^{-1}(\sigma)$ of $\pi_1^+(x^{\log})$, where σ ranges over Σ, is finite.*

(5) $X(\Sigma)_{\mathrm{val}} = X_{\mathrm{val}}$.

(6) *For any $x \in X$ and any valuative submonoid V of $(M_X^{\mathrm{gp}}/\mathcal{O}_X^\times)_x$ with $(M_X/\mathcal{O}_X^\times)_x \subset V$ and $V^\times \cap (M_X/\mathcal{O}_X^\times)_x = \{1\}$, there is $(x, P, h) \in X(\Sigma) \subset Q(X)$ such that $P \subset V$.*

Proof. The implication (1) $\Rightarrow$ (2) is already shown in 3.6.12.

We prove the implication (2) $\Rightarrow$ (3). Assume that $X(\Sigma) \to X$ is proper and surjective. Since $\{X(\sigma)\}_{\sigma \in \Sigma}$ is an open covering of $X(\Sigma)$ and $X(\Sigma) \to X$ is proper, by working locally on X, we may assume that there is a finite subset Σ_1 of Σ such that $\bigcup_{\sigma \in \Sigma_1} X(\sigma) = X(\Sigma)$. Hence we may assume that there is a finite subfan Σ_2 of Σ such that $X(\Sigma_2) = X(\Sigma)$. Since Σ_2 is finite, locally on X, by 3.6.21 (i), 3.6.6 (i), and 3.6.13, there is a logarithmic modification $Y \to X$ such that the fiber product $X(\Sigma) \times_X Y$ is an open subobject of Y. It is sufficient to prove $X(\Sigma) \times_X Y = Y$. Since $X(\Sigma) \to X$ is proper and surjective by the assumption and since $X(\Sigma) \times_X Y \to X(\Sigma)$ is proper and surjective (3.6.12), the composition $X(\Sigma) \times_X Y \to X$ is proper and surjective. Hence, for each $x \in X$, the fiber F_1 in $X(\Sigma) \times_X Y$ over x is a nonempty compact open subset of the fiber F_2 in Y over x, and F_2 is Hausdorff and connected by 3.6.12. Hence $F_1 = F_2$, and thus we have $X(\Sigma) \times_X Y = Y$, and (3) is proved.

We prove the implication (3) $\Rightarrow$ (4). Let $x \in X$. By 3.6.15 (iii), there are only finitely many submonoids σ of $\pi_1^+(x^{\log})$ such that (x, σ) is in the image of $q'_{Y,1} : Y \to Q'_1(X)$. Let $\sigma_1, \dots, \sigma_n$ be all such submonoids. We have $\pi_1^+(x^{\log}) = \bigcup_{j=1}^n \sigma_j$. Since there is a morphism $Y \to X(\Sigma)$, we have by 3.6.15 (vi) that for each j, there is a submonoid τ_j of $\pi_1^+(x^{\log})$ such that (x, τ_j) is in the image of $q'_{X(\Sigma),1} : X(\Sigma) \to Q'_1(X)$ and $\sigma_j \subset \tau_j$. We have $\pi_1^+(x^{\log}) = \bigcup_{j=1}^n \tau_j$. It remains to show that there are only finitely many submonoids τ of $\pi_1^+(x^{\log})$ such that (x, τ) belongs to the image of $q'_{X(\Sigma),1} : X(\Sigma) \to Q'_1(X)$. If τ is such submonoid, $\tau = \bigcup_{j=1}^n \tau \cap \tau_j$, and by 3.6.15 (iii), $\tau \cap \tau_j$ is a face of τ. Hence $\tau = \tau \cap \tau_j$ for some j. This shows that τ is a face of τ_j. But there are only finitely many faces of τ_j.

We prove the implication (4) $\Rightarrow$ (1). Let $x \in X$. Working locally at x, take a chart $\mathcal{S} \to M_X$ such that $\mathcal{S} \xrightarrow{\sim} (M_X/\mathcal{O}_X^\times)_x$ (2.1.5 (iii)). Take $N_{\mathbf{Z}} \subset N_{\mathbf{Q}}$ as in the proof of 3.6.8. Since the homomorphism of sheaves $\mathcal{S}^{\mathrm{gp}} \to M_X^{\mathrm{gp}}/\mathcal{O}_X^\times$ is surjective, locally at x there is a homomorphism $a : \mathrm{Hom}(N_{\mathbf{Z}}, \mathbf{Z}) \to \mathcal{S}^{\mathrm{gp}}$ such that the homomorphism $s : \mathrm{Hom}(N_{\mathbf{Z}}, \mathbf{Z}) \to M_X^{\mathrm{gp}}/\mathcal{O}_X^\times$ coincides with the composition $\mathrm{Hom}(N_{\mathbf{Z}}, \mathbf{Z}) \xrightarrow{a} \mathcal{S}^{\mathrm{gp}} \to M_X^{\mathrm{gp}}/\mathcal{O}_X^\times$. Let $b : \mathrm{Hom}(\mathcal{S}, \mathbf{R}_{\geq 0}^{\mathrm{add}}) \to N_{\mathbf{R}}$ be the homomorphism induced by a. Since $s_x : \pi_1^+(x^{\log}) \to N_{\mathbf{R}}$ factors as $\pi_1^+(x^{\log}) \xrightarrow{\sim} \mathrm{Hom}(\mathcal{S}, \mathbf{N}) \subset \mathrm{Hom}(\mathcal{S}, \mathbf{R}_{\geq 0}^{\mathrm{add}}) \xrightarrow{b} N_{\mathbf{R}}$, we have $\mathrm{Hom}(\mathcal{S}, \mathbf{R}_{\geq 0}^{\mathrm{add}}) = \bigcup_{\sigma \in \Sigma} b^{-1}(\sigma)$ and that the set of cones $b^{-1}(\sigma)$ ($\sigma \in \Sigma$) is finite. Hence $X(\Sigma)$ coincides with the logarithmic modification defined by the finite subdivision of $\mathrm{Hom}(\mathcal{S}, \mathbf{R}_{\geq 0}^{\mathrm{add}})$ consisting of all faces of $b^{-1}(\sigma)$ for all $\sigma \in \Sigma$.

Thus we have shown that the conditions (1)–(4) are equivalent.

The implication (1) $\Rightarrow$ (5) is clear.

We prove the implication (5) $\Rightarrow$ (2). In general, let $A \to B$ and $B \to C$ be continuous maps of topological spaces, and assume that the composition $A \to C$ is proper, that $A \to B$ is surjective, and that $B \to C$ is separated (see 0.7.5). Then $B \to C$ is proper. We apply this to $A = X_{\mathrm{val}}$, $B = X(\Sigma)$, $C = X$. Note $X_{\mathrm{val}} \to X$ is proper by 3.6.24, $X_{\mathrm{val}} = X(\Sigma)_{\mathrm{val}} \to X(\Sigma)$ is surjective by 3.6.24, and $X(\Sigma) \to X$ is separated by 3.6.11 (ii).

Finally, the equivalence (5) $\Leftrightarrow$ (6) follows from 3.6.27 (ii). □

Chapter Four

Main Results

In this chapter, we state the main results of this book: Theorem A in Section 4.1, and Theorem B in Section 4.2. We extend Griffiths' period maps in Section 4.3 and the infinitesimal period maps in Section 4.4.

4.1 THEOREM A: THE SPACES E_σ, $\Gamma\backslash D_\Sigma$, AND $\Gamma\backslash D_{\Sigma^\sharp}$

4.1.1

We state our first main theorem, whose proof will be given in Chapter 7 below. A subgroup Γ of $G_{\mathbf{Z}}$ is said to be *neat* if, for each $\gamma \in \Gamma$, the subgroup of $\mathbf{C}^\times$ generated by all the eigenvalues of γ is torsion-free. It is known that there exists a neat subgroup of $G_{\mathbf{Z}}$ of finite index.

THEOREM A *Let Σ be a fan in $\mathfrak{g}_{\mathbf{Q}}$ and let Γ be a subgroup of $G_{\mathbf{Z}}$ that is strongly compatible with Σ* (1.3.10). *Then we have:*

(i) *For $\sigma \in \Sigma$, E_σ is open in $\tilde{E}_\sigma$ in the strong topology of $\tilde{E}_\sigma$ in $\check{E}_\sigma$* (3.3.4). *In particular, by* 3.5.10, *E_σ is a logarithmic manifold* (*Definition* 3.5.7).

(ii) *If Γ is neat, then $\Gamma\backslash D_\Sigma$ is also a logarithmic manifold.*

(iii) *Let $\sigma \in \Sigma$ and define the action of $\sigma_{\mathbf{C}}$ on E_σ over $\Gamma(\sigma)^{\mathrm{gp}}\backslash D_\sigma$ by*

$$a \cdot (q, F) := (\mathbf{e}(a)q, \exp(-a)F) \quad (a \in \sigma_{\mathbf{C}},\ (q, F) \in E_\sigma),$$

where $\mathbf{e}(a) \in \mathrm{torus}_\sigma$ is as in 3.3.5 *and $\mathbf{e}(a)q$ is defined by the natural action of* torus_σ *on* toric_σ. *Then, $E_\sigma \to \Gamma(\sigma)^{\mathrm{gp}}\backslash D_\sigma$ is a $\sigma_{\mathbf{C}}$-torsor in the category of logarithmic manifolds. That is, locally on the base $\Gamma(\sigma)^{\mathrm{gp}}\backslash D_\sigma$, E_σ is isomorphic as a logarithmic manifold to the product of $\sigma_{\mathbf{C}}$ and the base endowed with the evident action of $\sigma_{\mathbf{C}}$.*

(iv) *If Γ is neat, then, for any $\sigma \in \Sigma$, the map*

$$\Gamma(\sigma)^{\mathrm{gp}}\backslash D_\sigma \to \Gamma\backslash D_\Sigma$$

is open and locally an isomorphism of logarithmic manifolds.

(v) *The topological space $\Gamma\backslash D_\Sigma$ is Hausdorff.*

(vi) *If Γ is neat, then there is a homeomorphism of topological spaces*

$$(\Gamma\backslash D_\Sigma)^{\log} \simeq \Gamma\backslash D_\Sigma^\sharp,$$

that is compatible with $\tau : (\Gamma\backslash D_\Sigma)^{\log} \to \Gamma\backslash D_\Sigma$ and the projection $\Gamma\backslash D_\Sigma^\sharp \to \Gamma\backslash D_\Sigma$ induced by $D_\Sigma^\sharp \to D_\Sigma$ in 1.3.9.

4.2 THEOREM B: THE FUNCTOR $\underline{\mathrm{PLH}}_{\Phi}$

In this section, we state the second main result of this book, Theorem B, whose proof will be given in Chapter 8 below.

4.2.1

Let Σ be a fan in $\mathfrak{g}_{\mathbf{Q}}$ and let Γ be a subgroup of $G_{\mathbf{Z}}$ which is strongly compatible with Σ (1.3.10). As in 2.5.8, we denote by

$$\Phi = (w, (h^{p,q})_{p,q\in\mathbf{Z}}, H_0, \langle\ ,\ \rangle_0, \Gamma, \Sigma) \tag{1}$$

the 6-tuple consisting of the 4-tuple $(w, (h^{p,q})_{p,q\in\mathbf{Z}}, H_0, \langle\ ,\ \rangle_0)$ as in Section 0.7 and of the above Σ and Γ.

THEOREM B *We assume Γ is neat* (4.1.1). *Define a contravariant functor $\underline{\mathrm{PLH}}_{\Phi}$ from the category $\overline{\mathcal{A}}_2(\log)$* (3.2.4) *to the category of sets by*

$$\underline{\mathrm{PLH}}_{\Phi}(X) := (\text{isomorphism classes of PLH on } X \text{ of type } \Phi) \quad (2.5.8).$$

Then this functor $\underline{\mathrm{PLH}}_{\Phi}$ is represented by $\Gamma\backslash D_{\Sigma}$, i.e., there exists an isomorphism of functors

$$\underline{\mathrm{PLH}}_{\Phi} \simeq \mathrm{Mor}(\ \ ,\Gamma\backslash D_{\Sigma})$$

such that, for any object X of $\overline{\mathcal{A}}_2(\log)$ and any element of $\underline{\mathrm{PLH}}_{\Phi}(X)$, the induced maps $X \to \Gamma\backslash D_{\Sigma}$ and $X^{\log} \to (\Gamma\backslash D_{\Sigma})^{\log} \simeq \Gamma\backslash D_{\Sigma}^{\sharp}$ coincide with the maps in 2.5.10.

4.2.2

Since $\mathcal{A}(\log) \subset \mathcal{B}(\log) \subset \overline{\mathcal{A}}_2(\log)$ by Theorem 3.2.5, Theorem B contains Theorem 0.4.27 (i) in Chapter 0.

Proof of Theorem 0.4.27 (ii). We show that Theorem 0.4.27 (ii) follows from Theorem A, Theorem B, and Theorem 3.2.5. Once we know

$$\underline{\mathrm{PLH}}_{\Phi}\,|_{\mathcal{A}(\log)} \simeq \mathrm{Mor}(\ \ ,\Gamma\backslash D_{\Sigma})|_{\mathcal{A}(\log)} = h_{\mathcal{A}(\log)}^{\Gamma\backslash D_{\Sigma}}$$

by Theorem B, Theorem 0.4.27 (ii) becomes equivalent to

$$\mathrm{Mor}(\Gamma\backslash D_{\Sigma}, Z) \xrightarrow{\sim} \mathrm{Mor}(h_{\mathcal{A}(\log)}^{\Gamma\backslash D_{\Sigma}}, h_{\mathcal{A}(\log)}^{Z})$$

for logarithmic local ringed spaces Z over $\mathbf{C}$, and hence to the statement that $\Gamma\backslash D_{\Sigma}$ belongs to $\overline{\mathcal{A}(\log)}'$. Since $\Gamma\backslash D_{\Sigma}$ is a logarithmic manifold by Theorem A, it is an object of $\mathcal{B}^*(\log)$ and, due to Theorem 3.2.5, it is an object of $\overline{\mathcal{A}(\log)}'$. □

If we restrict our attention to nilpotent cones of rank one, Theorem B shows the following.

COROLLARY 4.2.3 *Let X be an object of $\overline{\mathcal{A}}_2(\log)$ such that* $\mathrm{rank}_{\mathbf{Z}}(M_X^{\mathrm{gp}}/\mathcal{O}_X^{\times})_x \le 1$ *for any $x \in X$. Let Γ be a neat subgroup of $G_{\mathbf{Z}}$ of finite index. Then, isomorphism* *classes of PLH on X endowed with a Γ-level structure correspond bijectively to morphisms from X to $\Gamma\backslash D_{\Xi}$ of logarithmic local ringed spaces over $\mathbf{C}$.*

4.3 EXTENSIONS OF PERIOD MAPS

Assuming the results in Sections 4.1 and 4.2, in this section we extend the period map associated with a variation of polarized Hodge structure over a boundary. We will give other extension theorems in Sections 8.4, 9.4, and 12.6. These extension theorems represent one of the main motivations for Griffiths' hope to enlarge D.

The authors are grateful to Professor Chikara Nakayama for his essential contribution to the proof of the following theorem.

THEOREM 4.3.1 *Let X be a connected, logarithmically smooth, fs logarithmic analytic space and let $U = X_{\mathrm{triv}}$ be the open subspace of X consisting of all points at which the logarithmic structure is trivial. Let $H = (H_{\mathbf{Z}}, \langle\ ,\ \rangle, F)$ be a variation of polarized Hodge structure on U with unipotent local monodromy along $X - U$. Fix a base point $u \in U$ and let $(H_0, \langle\ ,\ \rangle_0) = (H_{\mathbf{Z},u}, \langle\ ,\ \rangle_u)$. Let Γ be a subgroup of $G_{\mathbf{Z}}$ that contains the global monodromy group* $\mathrm{Image}(\pi_1(U,u) \to G_{\mathbf{Z}})$, *and assume that Γ is neat. Let $\varphi : U \to \Gamma\backslash D$ be the associated period map.*

(i) *Assume that $X - U$ is a smooth divisor. Then there exists a fan Σ in $\mathfrak{g}_{\mathbf{Q}}$ which is strongly compatible with Γ such that the period map $\varphi : U \to \Gamma\backslash D$ extends to a morphism $X \to \Gamma\backslash D_{\Sigma}$ of logarithmic manifolds.*

$$\begin{array}{ccc} U & \subset & X \\ \downarrow & & \downarrow \\ \Gamma\backslash D & \subset & \Gamma\backslash D_{\Sigma}. \end{array}$$

(A natural choice of Σ is given in 4.3.2 *below.) In the case where Γ is of finite index in $G_{\mathbf{Z}}$, we can take the fan Ξ in* (0.4.5, 1.3.11) *as Σ.*

(ii) *For any point $x \in X$, there exist an open neighborhood W of x, a logarithmic modification W' of W* (3.6.12), *a commutative subgroup Γ' of Γ, and a fan Σ in $\mathfrak{g}_{\mathbf{Q}}$ that is strongly compatible with Γ' such that the period map $\varphi|_{U\cap W}$ lifts to a morphism $U \cap W \to \Gamma'\backslash D$ which extends to a morphism $W' \to \Gamma'\backslash D_{\Sigma}$ of logarithmic manifolds:*

$$\begin{array}{ccccc} U & \supset & U\cap W & \subset & W' \\ \downarrow & & \downarrow & & \downarrow \\ \Gamma\backslash D & \leftarrow & \Gamma'\backslash D & \subset & \Gamma'\backslash D_{\Sigma}. \end{array}$$

(iii) *Assume Γ is commutative. Then we can take $\Gamma' = \Gamma$ in* (ii). *Assume furthermore that the following condition* (1) *is satisfied.*

(1) *There is a finite family $(X_j)_{1\le j\le n}$ of connected locally closed analytic subspaces of X such that $X = \bigcup_{j=1}^{n} X_j$ as a set and such that, for each j, the inverse image of the sheaf $M_X/\mathcal{O}_X^{\times}$ on X_j is locally constant.*

Then there is a logarithmic modification X' of X and a fan Σ in $\mathfrak{g}_{\mathbf{Q}}$ that is strongly compatible with Γ such that the period map φ extends to a morphism $X' \to \Gamma\backslash D_\Sigma$ of logarithmic manifolds.

$$\begin{array}{ccc} U & \subset & X' \\ \downarrow & & \downarrow \\ \Gamma\backslash D & \subset & \Gamma\backslash D_\Sigma. \end{array}$$

Note that a logarithmic modification $X' \to X$ (or $W' \to W$) is proper and surjective, and the induced morphism $U \times_X X' \to U$ (or $(U \cap W) \times_W W' \to U \cap W$) is an isomorphism.

The assertions in 4.3.1 (iii) may be true without the assumption that Γ is commutative, but the authors need this assumption for the proof. The condition (1) in 4.3.1 (iii) is satisfied, for example, if X is an open set of the toric variety $\mathrm{Spec}(\mathbf{C}[\mathcal{S}])_{\mathrm{an}}$ for some fs monoid $\mathcal{S}$, or if $X - U$ is compact. (If $X - U$ is compact, X has a finite open covering with one member U and with all other members isomorphic to open sets of toric varieties, and hence X satisfies (1) in 4.3.1 (iii).)

In the following, assuming results in Sections 4.1 and 4.2, we prove 4.3.1 (i) in 4.3.2, and prove 4.3.1 (ii) and (iii) in 4.3.9 at the end of this section after preparation.

4.3.2

Proof of 4.3.1 (i). By the nilpotent orbit theorem of Schmid interpreted as Theorem 2.5.14, H extends to a PLH H' on X. Denote by Σ the set of local monodromy cones of H' in $\mathfrak{g}_{\mathbf{R}}$ (2.5.11). Then Γ is strongly compatible with Σ (2.5.12). We show that $\varphi : U \to \Gamma\backslash D$ extends to a morphism $X \to \Gamma\backslash D_\Sigma$ of logarithmic manifolds. Since $\mathrm{rank}_{\mathbf{Z}}(M_X^{\mathrm{gp}}/\mathcal{O}_X^\times)_x \le 1$ for any $x \in X$, H' is of type $(w, (h^{p,q}), H_0, \langle\ ,\ \rangle, \Gamma, \Sigma)$ where w is the weight of H and $(h^{p,q})$ is the Hodge type of H (2.5.12). Hence we have the morphism $X \to \Gamma\backslash D_\Sigma$ associated with H' whose restriction to U is the original period map $U \to \Gamma\backslash D$. If Γ is of finite index in $G_{\mathbf{Z}}$, since Γ is compatible with Ξ and $\Sigma \subset \Xi$, we have the composite morphism $X \to \Gamma\backslash D_\Sigma \to \Gamma\backslash D_\Xi$. □

COROLLARY 4.3.3 *Let H be a variation of polarized Hodge structure on a punctured disc $\Delta^* = \Delta - \{0\}$ with unipotent local monodromy at $0 \in \Delta$. Fix a base point $u \in \Delta^*$ and let $(H_0, \langle\ ,\ \rangle_0) = (H_{\mathbf{Z},u}, \langle\ ,\ \rangle_u)$. Let Γ be the image of $\pi_1(\Delta^*, u) \simeq \mathbf{Z} \to G_{\mathbf{Z}}$, and let σ be the cone of $\mathfrak{g}_{\mathbf{R}}$ generated by the logarithm of the image in $G_{\mathbf{Z}}$ of positive generator* (appendix A1) *of $\pi_1(\Delta^*)$. Then the period map $\varphi : \Delta^* \to \Gamma\backslash D$ extends to a morphism $\Delta \to \Gamma\backslash D_\sigma$ of logarithmic manifolds.*

4.3.4

The reason that we need a logarithmic modification in 4.3.1 (ii) and (iii) is the following. We still have a PLH H' on X that extends H on U. However, if the condition $\mathrm{rank}_{\mathbf{Z}}(M_X^{\mathrm{gp}}/\mathcal{O}_X^\times)_x \le 1$ is not satisfied, it can happen that there is no fan in $\mathfrak{g}_{\mathbf{Q}}$ which contains the set of local monodromy cones of H' in $\mathfrak{g}_{\mathbf{R}}$ (2.5.11) as a subset. We will give such examples (Examples (i) and (ii) below).

Let $X = \Delta^n$ with the logarithmic structure associated to the normal crossing divisor $\Delta^n - (\Delta^*)^n$, let $x = (0, \dots, 0) \in \Delta^n$, let $\rho : \pi_1(x^{\log}) \to G_{\mathbf{Z}}$ be the monodromy action, let $(\gamma_j)_{1\le j\le n}$ be the standard generator of $\pi_1^+(x^{\log}) \simeq \mathbf{N}^n$, and let $N_j = \log(\rho(\gamma_j)) \in \mathfrak{g}_{\mathbf{Q}}$. Then, if $x' = (x'_j)_{1\le j\le n}$ is the point of $X = \Delta^n$, the image $\check{\varphi}(x')$ under the period map $\check{\varphi} : X \to \Gamma\backslash\check{D}_{\mathrm{orb}}$ (2.5.3) has the form $((\sigma(x'), Z(x')) \bmod \Gamma)$ with $\sigma(x') = \sum_{j\,;\,x'_j=0}(\mathbf{R}_{\ge 0})N_j$.

Examples.

(i) It can happen that $\sigma(x)$ is not a sharp cone. In fact, it can happen that $n = 2$ and $N_2 = -N_1 \neq 0$. In this case, $\sigma(x) \simeq \mathbf{Z}$ and $\sigma(x)$ is not sharp.

(ii) It can happen that $n = 3$, N_1 and N_2 are linearly independent over $\mathbf{R}$, and $N_3 = N_1 + N_2$. If $x' = (x'_j)_{1\le j\le 3}$ is a point of Δ^3 such that $x'_3 = 0$ and x'_1 and x'_2 are not zero, then $\sigma(x') = (\mathbf{R}_{\ge 0})N_3$ is not a face of $\sigma(x) = \sum_{j=1}^{3}(\mathbf{R}_{\ge 0})N_j = (\mathbf{R}_{\ge 0})N_1 + (\mathbf{R}_{\ge 0})N_2$. Hence there is no fan in $\mathfrak{g}_{\mathbf{Q}}$ which contains both $\sigma(x)$ and $\sigma(x')$.

Assume that Γ is commutative so that the adjoint action of Γ on $\sigma(x)$ is trivial. In the above examples, our method for extending the period map, which is explained below in detail, is as follows. In Example (i), we subdivide $\sigma(x)$ into sharp cones $(\mathbf{R}_{\ge 0})N_1$ and $(\mathbf{R}_{\ge 0})N_2$. In Example (ii), we subdivide $\sigma(x)$ into the cones $(\mathbf{R}_{\ge 0})N_1 + (\mathbf{R}_{\ge 0})N_3$ and $(\mathbf{R}_{\ge 0})N_2 + (\mathbf{R}_{\ge 0})N_3$. Then we have a fan Σ in $\mathfrak{g}_{\mathbf{Q}}$ that is strongly compatible with Γ and that subdivides the monodromy cones $\sigma(x')$ of any point x' of X. As is shown below, our subdivision yields a logarithmic modification $X(\Sigma)$ of X to which we can extend the period morphism.

4.3.5

Let X be an object of $\mathcal{B}(\log)$, let Γ be a subgroup of $G_{\mathbf{Z}}$, and let H be a pre-PLH on X of weight w and of Hodge type $(h^{p,q})$ endowed with a Γ-level structure μ. Let Σ be a fan in $\mathfrak{g}_{\mathbf{Q}}$ which is compatible with Γ (here Γ and Σ need not be strongly compatible). We define an object $X(\Sigma)$ of $\mathcal{B}(\log)$ over X.

Consider the homomorphism

$$\mathcal{N} : H_{\mathbf{Q}} \to (M_X^{\mathrm{gp}}/\mathcal{O}_X^{\times}) \otimes H_{\mathbf{Q}} \tag{1}$$

defined in 2.3.4. Via the Γ-level structure, $\mathcal{N}$ induces

$$\mathcal{N} \in \Gamma(X, \Gamma\backslash((M_X^{\mathrm{gp}}/\mathcal{O}_X^{\times}) \otimes \mathfrak{g}_{\mathbf{Q}})), \tag{2}$$

a global section of the sheaf of sets $\Gamma\backslash((M_X^{\mathrm{gp}}/\mathcal{O}_X^{\times}) \otimes \mathfrak{g}_{\mathbf{Q}})$ on X, where $\gamma \in \Gamma$ acts on $(M_X^{\mathrm{gp}}/\mathcal{O}_X^{\times}) \otimes \mathfrak{g}_{\mathbf{Q}}$ as $a \otimes b \mapsto a \otimes \mathrm{Ad}(\gamma)b$.

Locally on X, lift $\mathcal{N}$ to an element of $\Gamma(X, M_X^{\mathrm{gp}}/\mathcal{O}_X^{\times}) \otimes \mathfrak{g}_{\mathbf{Q}}$. By 3.6.10, using this lifting of $\mathcal{N}$ as s there, we have an object $X(\Sigma)$ of $\mathcal{B}(\log)$ over X. Since Σ is stable under the adjoint action of Γ, these local constructions are patched up to yield a global object $X(\Sigma)$ of $\mathcal{B}(\log)$ over X.

Let $x \in X$ and let y be a point of $X^{\log}$ lying over x. Take a representation $\tilde{\mu}_y : (H_{\mathbf{Z},y}, \langle\ ,\ \rangle_y) \xrightarrow{\sim} (H_0, \langle\ ,\ \rangle_0)$ of the stalk μ_y of the Γ-level structure μ. Then we have a lifting $\mathcal{N} \in (M_X^{\mathrm{gp}}/\mathcal{O}_X^\times)_x \otimes \mathfrak{g}_\mathbf{Q}$ associated to $\tilde{\mu}_y$. The homomorphism $s_x : \pi_1(x^{\log}) \to \mathfrak{g}_\mathbf{Q}$ in 3.6.7 (we take this lifting of $\mathcal{N}$ as s in 3.6.7) is nothing but the logarithm of the local monodromy action of $\pi_1(x^{\log})$ on H_0 via $\tilde{\mu}_y$.

In the following, for an object X' of $\mathcal{B}(\log)$ over X and for $x' \in X'$, let $\sigma(x')$ be the cone in $\mathfrak{g}_\mathbf{R}$ generated by the image of $\pi_1^+((x')^{\log})$ under the logarithm $\pi_1((x')^{\log}) \to \mathfrak{g}_\mathbf{Q}$ of the local monodromy action on H_0 given by the inverse image of H on X' and by the Γ-level structure of H. Then $\sigma(x')$ is determined modulo the adjoint action of Γ.

The object $X(\Sigma)$ over X is characterized as follows (3.6.10). For any object X' of $\mathcal{B}(\log)$ over X, $\mathrm{Mor}_X(X', X(\Sigma))$ is either empty or a one point set, and it is nonempty if and only if, for any $x' \in X'$, there exists $\tau \in \Sigma$ such that $\sigma(x') \subset \tau$.

Proposition 4.3.6 *Let X be an object of $\mathcal{B}(\log)$, let Γ be a subgroup of $G_\mathbf{Z}$, and let H be a PLH on X of weight w and of Hodge type $(h^{p,q})$ endowed with a Γ-level structure. Let C be the set of local monodromy cones of H in $\mathfrak{g}_\mathbf{Q}$* (2.5.11). *Assume that we are given a fan Σ in $\mathfrak{g}_\mathbf{Q}$ satisfying the following conditions* (1)–(3).

(1) $\bigcup_{\sigma\in\Sigma} \sigma = \bigcup_{\sigma\in C} \sigma$.
(2) *Σ is compatible with Γ.*
(3) *For any $\sigma \in C$, $\sigma = \bigcup_{j=1}^n \tau_j$ for some $n \geq 1$ and for some $\tau_j \in \Sigma$.*

Then

(i) *Σ is strongly compatible with Γ.*
(ii) *The inverse image of H on $X(\Sigma)$ is of type $\Phi = (w, (h^{p,q}), H_0, \langle\ ,\ \rangle_0, \Gamma, \Sigma)$. Consequently, when Γ is neat, we have the corresponding period morphism $X(\Sigma) \to \Gamma\backslash D_\Sigma$.*
(iii) *$X(\Sigma)$ is a logarithmic modification of X.*

Proof. The assertion (i) follows from the condition (1) and (2).

We prove (ii). Let $x' \in X(\Sigma)$ and let $x \in X$ be the image of x' in X. By the characterizing property of $X(\Sigma)$ (3.6.10 restated at the end of 4.3.5), there is $\tau \in \Sigma$ such that $\sigma(x') \subset \tau$. Take the smallest such τ. Our task is to prove that the pullback of H' to x' induces a τ-nilpotent orbit, not only a $\sigma(x')$-nilpotent orbit. Since H induces a $\sigma(x)$-nilpotent orbit, for the proof of the fact that the pullback of H' to x' induces a τ-nilpotent orbit, it is sufficient to prove $\tau \subset \sigma(x)$. Let p be an element in the interior (0.7.7) of $\sigma(x')$. By the minimality of τ, p belongs to the interior of τ. On the other hand, $\sigma(x') \subset \sigma(x)$ and $\sigma(x)$ is the union of some elements τ_j of Σ such that $\tau_j \subset \sigma(x)$ by the condition (3). Hence we have $p \in \tau_j$ for some j. Since $\tau \cap \tau_j$ is a face of τ and contains a point in the interior of τ, we have $\tau \cap \tau_j = \tau$, that is, $\tau \subset \tau_j$. Hence $\tau \subset \sigma(x)$.

We prove (iii). It is sufficient to prove that Σ satisfies the condition (4) in 3.6.28. Let $x \in X$. By the condition (1), we have $\pi_1^+(x^{\log}) = \bigcup_{\sigma\in\Sigma} \pi_1^+(x^{\log}) \cap s_x^{-1}(\sigma)$. It remains to prove that the set of submonoids $\pi_1^+(x^{\log}) \cap s_x^{-1}(\sigma)$ of $\pi_1^+(x^{\log})$, where σ ranges over Σ, is finite. For this it is sufficient to prove that the set of cones $\sigma(x) \cap \tau$, where τ ranges over Σ, is finite. By the condition (3), $\sigma(x) = \bigcup_{j=1}^n \tau_j$ for some $\tau_j \in \Sigma$. If $\tau \in \Sigma$, $\sigma(x) \cap \tau = \bigcup_{j=1}^n \tau_j \cap \tau$. Since $\tau_j \cap \tau$ is a face of τ_j and since the number of faces of each τ_j is finite, the set of cones $\sigma(x) \cap \tau$, where τ ranges over Σ, is finite as desired. $\square$

Lemma 4.3.7 *Let X be an object of $\mathcal{B}(\log)$, Γ a subgroup of $G_{\mathbf{Z}}$, H a pre-PLH of weight w and of Hodge type $(h^{p,q})$ endowed with a Γ-level structure, and let C be the set of local monodromy cones of H in $\mathfrak{g}_{\mathbf{R}}$ (2.5.11).*

(i) *Let $x \in X$. Fix a cone $\sigma(x)$ which is a representative of the class ($\sigma(x)$ mod Γ). Then, if we replace X by a sufficiently small open neighborhood of x, the Γ-level structure of H comes from a $\Gamma(\sigma(x))^{\mathrm{gp}}$-level structure of H, $\sigma(x') \subset \sigma(x)$ for any $x' \in X$ for some choice of a representative $\sigma(x')$ of ($\sigma(x')$ mod Γ), and the quotient set $\Gamma\backslash C$ of C by the adjoint action of Γ is finite.*

(ii) *If $M_X/\mathcal{O}_X^\times$ is a locally constant sheaf, the map $x' \mapsto$ ($\sigma(x')$ mod Γ) is a locally constant function on X.*

(iii) *Assume that the condition* (1) *in* 4.3.1 (iii) *is satisfied (here Γ need not be commutative). Then the quotient set $\Gamma\backslash C$ is finite.*

(iv) *Assume that Γ is commutative. Then $\mathrm{Ad}(\gamma)v = v$ for any $\sigma \in C$, any $v \in \sigma$, and any $\gamma \in \Gamma$.*

Proof.

(i) We first prove the assertion concerning the Γ-level structure. Take $y \in X^{\log}$ lying over x, and let $\tilde{\mu}_y : (H_{\mathbf{Z},y}, \langle\ ,\ \rangle_y) \xrightarrow{\sim} (H_0, \langle\ ,\ \rangle_0)$ be a representative of the germ μ_y of the Γ-level structure μ of H. Via $\tilde{\mu}_y$, the local monodromy action of $\pi_1(x^{\log})$ on $(H_{\mathbf{Z},y}, \langle\ ,\ \rangle_y)$ induces a homomorphism $\pi_1(x^{\log}) \to \Gamma$. If we replace $\tilde{\mu}_y$ by $\gamma\tilde{\mu}_y$ for some suitable $\gamma \in \Gamma$, the image of $\pi_1(x^{\log})$ in Γ is contained in $\Gamma(\sigma(x))^{\mathrm{gp}}$. For this new $\tilde{\mu}_y$, the inverse image $H|_x$ of H on the fs logarithmic point x has a unique $\Gamma(\sigma(x))^{\mathrm{gp}}$-level structure μ' whose germ at y is the class of $\tilde{\mu}_y$. The Γ-level structure μ of $H|_x$ is induced from μ'. By the proper base change theorem (Appendix A2) applied to the proper map $X^{\log} \to X$ and to the sheaves of sets $\Gamma\backslash\underline{\mathrm{Isom}}((H_{\mathbf{Z}}, \langle\ ,\ \rangle), (H_0, \langle\ ,\ \rangle_0))$ and $\Gamma(\sigma(x))^{\mathrm{gp}}\backslash\underline{\mathrm{Isom}}((H_{\mathbf{Z}}, \langle\ ,\ \rangle), (H_0, \langle\ ,\ \rangle_0))$ on $X^{\log}$, for some open neighborhood W of x, μ' extends to a $\Gamma(\sigma(x))^{\mathrm{gp}}$-level structure of $H|_W$ which induces μ of $H|_W$.

We prove the rest of (i). Locally at x, take a chart $\mathcal{S} \to M_X$ such that $\mathcal{S} \xrightarrow{\sim} (M_X/\mathcal{O}_X^\times)_x$. Locally at x, $\mathcal{N} \in \Gamma(X, \Gamma\backslash((M_X^{\mathrm{gp}}/\mathcal{O}_X^\times) \otimes \mathfrak{g}_{\mathbf{Q}}))$ comes from an element $\tilde{\mathcal{N}} \in \mathcal{S}^{\mathrm{gp}} \otimes \mathfrak{g}_{\mathbf{Q}}$. In this situation, for $x' \in X$, ($\sigma(x')$ mod Γ) is described as follows. Let $\sigma_1(x')$ be the image of $\pi_1^+((x')^{\log}) \simeq \mathrm{Hom}((M_X/\mathcal{O}_X^\times)_{x'}, \mathbf{N}) \to \mathrm{Hom}(\mathcal{S}, \mathbf{N})$, where the first isomorphism is by the duality 2.2.9 and the next arrow is induced by $\mathcal{S} \to (M_X/\mathcal{O}_X^\times)_{x'}$. Let $\sigma(x')$ be the cone in $\mathfrak{g}_{\mathbf{R}}$ generated by the images of $\tilde{\mathcal{N}}$ under the maps $\mathcal{S}^{\mathrm{gp}} \otimes \mathfrak{g}_{\mathbf{Q}} \to \mathfrak{g}_{\mathbf{Q}}$ induced by all elements of $\sigma_1(x')$. Then $\sigma(x')$ is a representative of ($\sigma(x')$ mod Γ). Since $\sigma_1(x) = \mathrm{Hom}(\mathcal{S}, \mathbf{N})$, we have $\sigma_1(x') \subset \sigma_1(x)$ and hence $\sigma(x') \subset \sigma(x)$. Furthermore, $\sigma_1(x')$ is a face of $\mathrm{Hom}(\mathcal{S}, \mathbf{N})$. This is shown as follows. Let $\mathcal{S}(x') \subset \mathcal{S}$ be the kernel of $\mathcal{S} \to (M_X/\mathcal{O}_X^\times)_{x'}$. Then $\mathcal{S}(x')$ is a face of $\mathcal{S}$ and we have an isomorphism $(\mathcal{S}(x')^{\mathrm{gp}}\mathcal{S})/\mathcal{S}(x')^{\mathrm{gp}} \xrightarrow{\sim} (M_X/\mathcal{O}_X^\times)_{x'}$. Hence $\sigma_1(x')$ coincides with the set of all homomorphisms $\mathcal{S} \to \mathbf{N}$ which kill $\mathcal{S}(x')$. Hence $\sigma_1(x')$ is a face of $\mathrm{Hom}(\mathcal{S}, \mathbf{N})$. Since there are only finitely many faces of $\mathrm{Hom}(\mathcal{S}, \mathbf{N})$, there are only finitely many ($\sigma(x')$ mod Γ).

(ii) Consider the homomorphism $\mathcal{N} : H_{\mathbf{Q}} \to (M_X^{\mathrm{gp}}/\mathcal{O}_X^\times) \otimes H_{\mathbf{Q}}$ (2.3.4). Since $M_X/\mathcal{O}_X^\times$ is locally constant, the map $\mathcal{N}$ is locally constant. Hence $(\sigma(x') \bmod \Gamma)$ is locally constant when x' varies.

The assertion (iii) follows from (ii).

The assertion (iv) follows from the fact that each $\sigma \in C$ is generated as a cone by logarithms of some unipotent elements of Γ. □

Proposition 4.3.8 *Let X be an object of $\mathcal{B}(\log)$, let Γ be a subgroup of $G_{\mathbf{Z}}$, and let H be a PLH on X of weight w and of Hodge type $(h^{p,q})$ endowed with a Γ-level structure.*

(i) *For any point $x \in X$, there exist an open neighborhood W of x, a logarithmic modification W' of W, a commutative subgroup Γ' of Γ, and a fan Σ in $\mathfrak{g}_{\mathbf{Q}}$ which is strongly compatible with Γ' such that the Γ-level structure of H comes from a Γ'-level structure of H, with which and with Σ the inverse image of H on X' is of type $(w, (h^{p,q}), H_0, \langle\ ,\ \rangle_0, \Gamma', \Sigma)$.*

(ii) *Assume that Γ is commutative and assume that the set C of local monodromy cones of H in $\mathfrak{g}_{\mathbf{R}}$ (2.5.11) is finite. Then there exist a logarithmic modification X' of X and a fan Σ in $\mathfrak{g}_{\mathbf{Q}}$ which is strongly compatible with Γ such that the inverse image of H on X' is of type $(w, (h^{p,q}), H_0, \langle\ ,\ \rangle_0, \Gamma, \Sigma)$.*

Proof. By 4.3.7, it is sufficient to prove (ii). By 4.3.6, for the proof of 4.3.8, it is sufficient to prove that there exists a fan Σ in $\mathfrak{g}_{\mathbf{Q}}$ satisfying the conditions (1)–(3) in 4.3.6.

For each $\tau \in C$, subdivide τ into sharp cones. Let B be the set of all cones that appear as a result. For each $\tau \in B$, take a finite fan Σ_τ in $\mathfrak{g}_{\mathbf{Q}}$ such that $\bigcup_{\tau' \in \Sigma_\tau} \tau' = \mathfrak{g}_{\mathbf{Q}}$ and $\tau \in \Sigma_\tau$. Let Σ' be the set of all cones of the form $\bigcap_{\tau \in B} c(\tau)$ where $c(\tau)$ is an element of Σ_τ for each $\tau \in B$. Then Σ' is a fan. Let Σ be the subset of Σ' consisting of all $\sigma \in \Sigma'$ such that $\sigma \subset \tau$ for some $\tau \in C$. Then Σ is a fan satisfying the conditions (1)–(3) in 4.3.6. In fact, (1) and (3) are checked easily, and (2) follows from the fact that, by the commutativity of Γ, the adjoint action of Γ on $\bigcup_{\sigma \in \Sigma} \sigma$ is trivial. □

4.3.9

Proofs of 4.3.1 (ii) *and* (iii). By the nilpotent orbit theorem of Schmid interpreted as Theorem 2.5.14, we have a PLH on X which extends H on U. Hence 4.3.1 (ii) is reduced to 4.3.8 (i), and 4.3.1 (iii) is reduced to 4.3.7 (iii) and (iv) and 4.3.8 (ii). □

4.4 INFINITESIMAL PERIOD MAPS

The purpose of this section is to generalize Griffiths' formulation of infinitesimal period maps to our extended period maps.

Let the notation be as in 4.2.1.

4.4.1

Let X be an object of $\overline{\mathcal{A}}_1(\log)$ (2.2.1) and let $(H_{\mathbf{Z}}, \langle\ ,\ \rangle, F)$ be a PLH on X (2.4.8). Then, as in 2.4.1 and 2.3.3, $\mathcal{M} = \tau_*(\mathcal{O}_X^{\log} \otimes_{\mathbf{Z}} H_{\mathbf{Z}})$ is a locally free $\mathcal{O}_X$-module of rank $= \operatorname{rank}_{\mathbf{Z}} H_{\mathbf{Z}}$ and $\mathcal{M}^p = \tau_*(F^p)$ is locally a direct summand of $\mathcal{M}$. Let

$$\begin{aligned}\mathcal{E}nd_{\langle\, ,\,\rangle}(\mathcal{M}) := \{\nu \in \mathcal{H}om_{\mathcal{O}_X}(\mathcal{M}, \mathcal{M}) \mid \\ \langle \nu(x), y\rangle + \langle x, \nu(y)\rangle = 0\ (\forall x, \forall y \in \mathcal{M})\}, \\ F^p\mathcal{E}nd_{\langle\, ,\,\rangle}(\mathcal{M}) := \{\nu \in \mathcal{E}nd_{\langle\, ,\,\rangle}(\mathcal{M}) \mid \nu(\mathcal{M}^q) \subset \mathcal{M}^{p+q}\ (\forall q \in \mathbf{Z})\}.\end{aligned}$$

Then, as an $\mathcal{O}_X$-module, $\mathcal{E}nd_{\langle\, ,\,\rangle}(\mathcal{M})$ is locally free of finite rank and $F^p\mathcal{E}nd_{\langle\, ,\,\rangle}(\mathcal{M})$ is locally a direct summand of it.

4.4.2

Let the notation be as in 4.4.1 and assume that X is a logarithmically smooth object of $\mathcal{B}(\log)$ (3.5.4, 3.2.4). Then, $d \otimes 1_{H_{\mathbf{C}}} : \mathcal{O}_X^{\log} \otimes_{\mathbf{C}} H_{\mathbf{C}} \to \omega_X^{1,\log} \otimes_{\mathbf{C}} H_{\mathbf{C}}$ induces a connection

$$\nabla : \mathcal{M} \to \omega_X^1 \otimes_{\mathcal{O}_X} \mathcal{M}. \tag{1}$$

Let θ_X be the sheaf of logarithmic vector fields on X (3.5.4 (iii)). We define a map

$$\theta_X \to \bigoplus_p \mathcal{H}om_{\mathcal{O}_X}(\mathcal{M}^p, \mathcal{M}/\mathcal{M}^p) \tag{2}$$

by sending $\delta \in \theta_X$ to the element of $\oplus_p \mathcal{H}om_{\mathcal{O}_X}(\mathcal{M}^p, \mathcal{M}/\mathcal{M}^p)$ induced by the map

$$\mathcal{M} \xrightarrow{\nabla} \omega_X^1 \otimes_{\mathcal{O}_X} \mathcal{M} \xrightarrow{\delta \otimes 1_{\mathcal{M}}} \mathcal{M}. \tag{3}$$

Note that the map (3) is not $\mathcal{O}_X$-linear but it induces an $\mathcal{O}_X$-linear map $\mathcal{M}^p \to \mathcal{M}/\mathcal{M}^p$. The image of the map (2) is contained in the image of the injection

$$\mathcal{E}nd_{\langle\, ,\,\rangle}(\mathcal{M})/F^0\mathcal{E}nd_{\langle\, ,\,\rangle}(\mathcal{M}) \hookrightarrow \bigoplus_p \mathcal{H}om_{\mathcal{O}_X}(\mathcal{M}^p, \mathcal{M}/\mathcal{M}^p).$$

Thus, we have a homomorphism of $\mathcal{O}_X$-modules

$$\theta_X \to \mathcal{E}nd_{\langle\, ,\,\rangle}(\mathcal{M})/F^0\mathcal{E}nd_{\langle\, ,\,\rangle}(\mathcal{M}). \tag{4}$$

Proposition 4.4.3 *Let $X = \Gamma\backslash D_\Sigma$ with Γ neat* (4.1.1), *let $(H_{\mathbf{Z}}, \langle\ ,\ \rangle, F, \mu)$ be the universal PLH of type Φ on X (Theorem B), and define $\mathcal{E}nd_{\langle\, ,\,\rangle}(\mathcal{M})$ as in* 4.4.1. *Then the homomorphism* 4.4.2 (4) *is an isomorphism*

$$\theta_X \xrightarrow{\sim} \mathcal{E}nd_{\langle\, ,\,\rangle}(\mathcal{M})/F^0\mathcal{E}nd_{\langle\, ,\,\rangle}(\mathcal{M}).$$

Proof. Let U be an open set of X and let $U' := U[T]/(T^2)$ (3.5.6). Define the sets P, Q, and R as follows. Let P be the set of all morphisms $U' \to X$ that extend the inclusion morphism $U \hookrightarrow X$, let $Q = \operatorname{Hom}_{\mathcal{O}_U}(\omega_U^1, \mathcal{O}_U) = \Gamma(U, \theta_X)$, and let R be the set of all isomorphism classes of PLH $(H'_{\mathbf{Z}}, \langle\ ,\ \rangle', F', \mu')$ on U' of type

Φ whose pullbacks to U coincide with the restriction of $(H_{\mathbf{Z}}, \langle\ ,\ \rangle, F, \mu)$ to U. Since $X = \Gamma\backslash D_\Sigma$ is the fine moduli space of PLH of type Φ, the map $P \to R$, obtained by pulling back the universal object, is bijective. On the other hand, by 3.5.5, the trivial extension $\iota : U' \to U \hookrightarrow X$ of the inclusion morphism $U \hookrightarrow X$ induces a bijection $Q \xrightarrow{\sim} P$, $\delta \mapsto \delta\iota$. Furthermore, we have a bijection between R and $\Gamma\big(U, \mathcal{E}nd_{\langle\,,\,\rangle}(\mathcal{M})/F^0\mathcal{E}nd_{\langle\,,\,\rangle}(\mathcal{M})\big)$ given as follows. For an element ν of the last set, the corresponding $(H'_{\mathbf{Z}}, \langle\ ,\ \rangle', F', \mu')$ is given by $(H'_{\mathbf{Z}}, \langle\ ,\ \rangle', \mu') = (H_{\mathbf{Z}}, \langle\ ,\ \rangle, \mu)|_U$ and by F'^p, whose germ at each point $y \in U'^{\log}$ is the $\mathcal{O}^{\log}_{U',y}$-submodule of $\mathcal{O}^{\log}_{U',y} \otimes_{\mathbf{Z}} H_{\mathbf{Z}}$ generated by $a + T\nu(a)$ $(a \in \tau^{-1}(\mathcal{M}^p_y)$, where y is regarded as a point of $U^{\log}$). It is easily seen that this correspondence gives a bijection from $\Gamma\big(U, \mathcal{E}nd_{\langle\,,\,\rangle}(\mathcal{M})/F^0\mathcal{E}nd_{\langle\,,\,\rangle}(\mathcal{M})\big)$ onto R. It is also easily seen that the composite map $\Gamma(U, \theta_X) = Q \xrightarrow{\sim} P \xrightarrow{\sim} R \xrightarrow{\sim} \Gamma\big(U, \mathcal{E}nd_{\langle\,,\,\rangle}(\mathcal{M})/F^0\mathcal{E}nd_{\langle\,,\,\rangle}(\mathcal{M})\big)$ is nothing but the homomorphism 4.4.2 (4). □

4.4.4

Let $X = \Gamma\backslash D_\Sigma$ with Γ neat. We define the *horizontal logarithmic tangent bundle* T^h_X of X, which is a subbundle of the logarithmic tangent bundle T_X (3.5.6), as follows. Let $\mathcal{M}$ be as in 4.4.1 and let θ^h_X be the $\mathcal{O}_X$-submodule of θ_X whose image in $\mathcal{E}nd_{\langle\,,\,\rangle}(\mathcal{M})/F^0\mathcal{E}nd_{\langle\,,\,\rangle}(\mathcal{M})$ under the isomorphism of 4.4.3 coincides with $\mathrm{gr}_F^{-1}\, \mathcal{E}nd_{\langle\,,\,\rangle}(\mathcal{M})$. Then, θ^h_X is locally a direct summand of θ_X, and is called the *horizontal submodule* of θ_X. We define T^h_X to be the vector bundle associated with the sheaf θ^h_X (cf. 3.5.6). We have

$$\begin{aligned}\theta^h_X &\simeq \mathrm{gr}_F^{-1}\, \mathcal{E}nd_{\langle\,,\,\rangle}(\mathcal{M}) \qquad (1)\\ &\simeq \Big\{(h_p)_p \in \bigoplus_p \mathcal{H}om_{\mathcal{O}_X}(\mathrm{gr}^p_{\mathcal{M}}, \mathrm{gr}^{p-1}_{\mathcal{M}}) \ \Big|\ h_{2w+1-p} = -{}^t h_p \ (\forall p \in \mathbf{Z})\Big\}.\end{aligned}$$

Here ${}^t(\ )$ means the transposed with respect to $\langle\ ,\ \rangle$.

4.4.5

Assume that Γ is neat. Denote $\Gamma\backslash D_\Sigma$ by Z. Let X be a logarithmically smooth object of $\mathcal{B}(\log)$ (3.5.4), let $(H_{\mathbf{Z}}, \langle\ ,\ \rangle, F, \mu)$ be a PLH on X of type Φ and let $\varphi : X \to Z$ be the corresponding period map. Then $\mathcal{M}$, $\mathcal{M}^p$ and $F^p\mathcal{E}nd_{\langle\,,\,\rangle}(\mathcal{M})$ on X (4.4.1) are the pullbacks of $\mathcal{M}$, $\mathcal{M}^p$ and $F^p\mathcal{E}nd_{\langle\,,\,\rangle}(\mathcal{M})$ of the universal PLH on Z by φ, respectively. If we regard the isomorphism in 4.4.3 as identification, the map $\theta_X \to \mathcal{E}nd_{\langle\,,\,\rangle}(\mathcal{M})/F^0\mathcal{E}nd_{\langle\,,\,\rangle}(\mathcal{M})$ in 4.4.2 (4) is identified with the canonical map $\theta_X \to \varphi^*(\theta_Z)$ and $\mathrm{gr}_F^{-1}\, \mathcal{E}nd_{\langle\,,\,\rangle}(\mathcal{M})$ is identified with $\varphi^*(\theta^h_Z)$. Hence we have

Proposition 4.4.6 *In the notation in* 4.4.5, *the following four conditions are equivalent.*

(1) *The PLH* $(H_{\mathbf{Z}}, \langle\ ,\ \rangle, F)$ *satisfies the big Griffiths transversality*

$$\nabla(\mathcal{M}^p) \subset \omega_X^1 \otimes_{\mathcal{O}_X} \mathcal{M}^{p-1} \quad (\forall p \in \mathbf{Z})$$

$$(\textit{or equivalently,}\ \ (d \otimes 1_{H_{\mathbf{C}}})(F^p) \subset \omega_X^{1,\log} \otimes_{\mathcal{O}_X^{\log}} F^{p-1} \quad (\forall p \in \mathbf{Z})).$$

(2) *The map* 4.4.2 (4) *factors through* $\mathrm{gr}_F^{-1} \mathcal{E}nd_{\langle\, ,\,\rangle}(\mathcal{M})$.
(3) *The map* $\theta_X \to \varphi^*(\theta_Z)$ *factors through* $\varphi^*(\theta_Z^h)$.
(4) *The map* $d\varphi : T_X \to T_Z$ *factors through* T_Z^h.

4.4.7

Relation to the logarithmic Kodaira-Spencer map. Let the notation be as in 4.4.5, and assume that X is a logarithmically smooth fs logarithmic analytic space and that $(H_{\mathbf{Z}}, \langle\ ,\ \rangle, F, \mu)$ is obtained from "geometry" as in 0.2.21. In this case,

$$\mathcal{M} = R^w f_*(\omega_{Y/X}^\bullet), \quad \mathcal{M}^p = R^w f_*(\omega_{Y/X}^{\geq p}), \quad \mathrm{gr}_{\mathcal{M}}^p = R^{w-p} f_*(\omega_{Y/X}^p),$$

and this PLH satisfies the big Griffiths transversality, i.e., the LVPH (2.4.9, 4.4.6). We define $\theta_{Y/X} := \mathrm{Ker}(\theta_Y \to f^*(\theta_X))$.

The following theorem about the *infinitesimal period map* $d\varphi$ is proved in the same way as in the nonlogarithmic case ([G2, 1.23]).

THEOREM 4.4.8 *In the notation in 4.4.7, the following diagram is commutative:*

$$\begin{array}{ccc} \theta_X & \xrightarrow{d\varphi} & \varphi^*\theta^h_{\Gamma\backslash D_\Sigma} = \mathrm{gr}_F^{-1} \mathcal{E}nd_{\langle\, ,\,\rangle}(\mathcal{M}) \\ {\scriptstyle K\text{-}S}\downarrow & & \cap\downarrow \\ R^1 f_*\theta_{Y/X} & \xrightarrow{\text{via coupling}} & \bigoplus_p \mathcal{H}om_{\mathcal{O}_X}(R^{m-p} f_*\omega^p_{Y/X}, R^{m-p+1} f_*\omega^{p-1}_{Y/X}) \end{array}$$

Here K-S means the logarithmic Kodaira-Spencer map (i.e., the connecting map coming from the exact sequence $0 \to \theta_{Y/X} \to \theta_Y \to f^*(\theta_X) \to 0$*) and the bottom horizontal arrow is induced by the pairing* $\theta_{Y/X} \otimes \omega^p_{Y/X} \to \omega^{p-1}_{Y/X}$.

Chapter Five

Fundamental Diagram

The aim of this chapter is to construct the following "fundamental diagram" which gives an illustration of various enlargements of D and of our method to prove Theorems A and B:

$$\begin{array}{ccccccccc}
 & & & & D_{\mathrm{SL}(2),\mathrm{val}} & \hookrightarrow & D_{\mathrm{BS},\mathrm{val}} & & \\
 & & & & \downarrow & & \downarrow & & \\
 & & D^{\sharp}_{\Sigma,\mathrm{val}} & \to & D_{\mathrm{SL}(2)} & & D_{\mathrm{BS}} & \to & \mathcal{X}_{\mathrm{BS}} \\
 & & \downarrow & & & & & & \\
\Gamma\backslash D_{\Sigma} & \leftarrow & D^{\sharp}_{\Sigma} & & & & & &
\end{array} \tag{5.0.1}$$

In this diagram, D_{Σ}, $D^{\sharp}_{\Sigma}$, $D^{\sharp}_{\Sigma,\mathrm{val}}$, $D_{\mathrm{SL}(2)}$, $D_{\mathrm{SL}(2),\mathrm{val}}$, $D_{\mathrm{BS},\mathrm{val}}$, and D_{BS} are enlargements of D. The first two appeared in Section 1.3, the last four were defined in our previous paper [KU2], as reviewed in Sections 5.1 and 5.2 below, and $D^{\sharp}_{\Sigma,\mathrm{val}}$ will be defined in Section 5.3. Here, $\mathcal{X}$ denotes the space of all maximal compact subgroups of $G_{\mathbf{R}}$ and $\mathcal{X}_{\mathrm{BS}}$ denotes the Borel-Serre space ($\overline{\mathcal{X}}$ in their notation) constructed in [BS] as a nice enlargement of $\mathcal{X}$. In (5.0.1), all arrows are continuous, all the vertical arrows are kinds of projective limits of blow-ups and are proper and surjective, and the map $D_{\mathrm{BS}} \to \mathcal{X}_{\mathrm{BS}}$ and the induced map $\Gamma\backslash D^{\sharp}_{\Sigma} \to \Gamma\backslash D_{\Sigma}$ are also proper and surjective.

To prove that $\Gamma\backslash D_{\Sigma}$ (Σ is a fan and $\Gamma \subset G_{\mathbf{Z}}$ is strongly compatible with Σ) has the nice properties stated in Theorems A and B (Chapter 4), our method is to transport the nice properties of $\mathcal{X}_{\mathrm{BS}}$ along this fundamental diagram from the right to the left. For example, we explain roughly how we prove that the quotient space $\Gamma\backslash D_{\Sigma}$ is Hausdorff. It is known that the action of Γ on $\mathcal{X}_{\mathrm{BS}}$ is proper. From this, we deduce that the actions of Γ on D_{BS}, $D_{\mathrm{BS},\mathrm{val}}$, and $D_{\mathrm{SL}(2),\mathrm{val}}$ are proper. Using the fact that $D_{\mathrm{SL}(2),\mathrm{val}} \to D_{\mathrm{SL}(2)}$ is proper and surjective, we obtain that the action of Γ on $D_{\mathrm{SL}(2)}$ is proper. (These results were obtained in [KU2].) From this, we deduce that the action of Γ on $D^{\sharp}_{\Sigma,\mathrm{val}}$ is proper. By using the fact that $D^{\sharp}_{\Sigma,\mathrm{val}} \to D^{\sharp}_{\Sigma}$ is proper and surjective, we obtain that the action of Γ on $D^{\sharp}_{\Sigma}$ is proper. Hence the quotient space $\Gamma\backslash D^{\sharp}_{\Sigma}$ is Hausdorff. Since $\Gamma\backslash D^{\sharp}_{\Sigma} \to \Gamma\backslash D_{\Sigma}$ is proper and surjective, this shows that $\Gamma\backslash D_{\Sigma}$ is Hausdorff. (These arguments will be given in Section 7.4.)

In Section 5.1, we review the spaces $\mathcal{X}_{\mathrm{BS}}$, D_{BS}, and $D_{\mathrm{BS,val}}$. In Section 5.2, we review the spaces $D_{\mathrm{SL(2),val}}$ and $D_{\mathrm{SL(2)}}$. In Section 5.3, we define the space $D^{\sharp}_{\Sigma,\mathrm{val}}$. In Section 5.4, we show how to define the map $D^{\sharp}_{\Sigma,\mathrm{val}} \to D_{\mathrm{SL(2)}}$ in (5.0.1) by using the work of Cattani, Kaplan, and Schmit [CKS]. Theorems 5.4.3 and 5.4.4 state that this map is actually well defined and continuous; the proofs will be given in Chapter 6.

5.1 BOREL-SERRE SPACES (REVIEW)

5.1.1

Summary. Let $\mathcal{X}$ be the set of all maximal compact subgroups of $G_{\mathbf{R}}$. Then $G_{\mathbf{R}}$ acts on $\mathcal{X}$ transitively by inner automorphisms. Since the normalizer of $G_{\mathbf{R}}$ at each $K \in \mathcal{X}$ is K itself, we have a $G_{\mathbf{R}}$-equivariant isomorphism

$$G_{\mathbf{R}}/K \xrightarrow{\sim} \mathcal{X}, \quad g \mapsto \mathrm{Int}(g)K = gKg^{-1},$$

for each fixed $K \in \mathcal{X}$. By using this isomorphism, we introduce a topology of $\mathcal{X}$. This topology does not depend on the choice of K. Borel and Serre constructed in [BS] a topological space $\mathcal{X}_{\mathrm{BS}}$ that contains $\mathcal{X}$ as an open dense subset.

This Section 5.1 is a review of [BS] and [KU2, §2] for our later use, i.e., we review the Borel-Serre space $\mathcal{X}_{\mathrm{BS}}$ associated with $\mathcal{X}$, a similar enlargement D_{BS} associated with D, and the projective limit $D_{\mathrm{BS,val}}$ of the blow-ups of D_{BS}. These spaces are related by continuous proper surjective maps in the following way:

$$\begin{array}{ccc} D_{\mathrm{BS,val}} & & \\ \downarrow & & \\ D_{\mathrm{BS}} & \longrightarrow & \mathcal{X}_{\mathrm{BS}}. \end{array}$$

5.1.2

For $F \in D$, define compact groups K_F, K'_F such that

$$K'_F \subset K_F \subset G_{\mathbf{R}}$$

as follows. Let

$$K'_F := \{g \in G_{\mathbf{R}} \mid gF = F\}.$$

Let K_F be the subgroup of $G_{\mathbf{R}}$ consisting of all elements that preserve the Hermitian inner product $(x, y) \mapsto \langle C_F(x), \overline{y}\rangle$ on $H_{0,\mathbf{C}}$. Here C_F is the Weil operator (1.1.2) associated with F. Then K_F is a maximal compact subgroup of $G_{\mathbf{R}}$, and we have a canonical continuous map

$$D \to \mathcal{X}, \quad F \mapsto K_F,$$

which is proper and surjective.

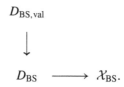

5.1.3

Borel-Serre action. Let P be a parabolic subgroup of $G_{\mathbf{R}}$, P_u its unipotent radical, and C the center of P/P_u. In what follows, we regard P, P_u, and C as the groups of their $\mathbf{R}$-valued points, respectively.

Let $K \in \mathcal{X}$. Then, for each $a \in C$, there exists a unique element a_K of P having the following properties (1) and (2).

(1) The image of a_K in P/P_u coincides with a.
(2) For the Cartan involution $\theta_K : G_{\mathbf{R}} \to G_{\mathbf{R}}$ associated with the maximal compact subgroup K, we have

$$\theta_K(a_K) = a_K^{-1}.$$

Recall that θ_K is the unique automorphism of $G_{\mathbf{R}}$ such that $\theta_K^2 = \mathrm{id}$ and $K = \{g \in G_{\mathbf{R}} \mid \theta_K(g) = g\}$. If $K = K_F$ with $F \in D$, $\theta_K(g) = C_F g C_F^{-1}$ for $g \in G_{\mathbf{R}}$ where C_F is the Weil operator (1.1.2).

We call a_K the *Borel-Serre lifting of a at K*. The map

$$C \to P, \quad a \mapsto a_K,$$

is a homomorphism of algebraic groups over $\mathbf{R}$.

Furthermore, we have an action $\circ$ of the group C on D (resp. $\mathcal{X}$) defined by

$$a \circ F = a_{K_F} F \quad (\text{resp. } a \circ K = \mathrm{Int}(a_K)K).$$

We call this the *Borel-Serre action*.

This action satisfies

$$a \circ pF = p(a \circ F) \quad (\text{resp. } a \circ \mathrm{Int}(p)K = \mathrm{Int}(p)(a \circ K)) \qquad (3)$$

for $a \in C$, $p \in P$, $F \in D$, $K \in \mathcal{X}$ ([KU2, 2.4]).

5.1.4

Now let P be a $\mathbf{Q}$-parabolic subgroup of $G_{\mathbf{R}}$, S_P the maximal $\mathbf{Q}$-split torus in the center C of P/P_u, and A_P the connected component of the group of $\mathbf{R}$-valued points of S_P which contains the unity.

Definition 5.1.5 ([KU2, 2.5]) *The Borel-Serre space D_{BS} (resp. $\mathcal{X}_{\mathrm{BS}}$) is defined by*

$$D_{\mathrm{BS}} \ (\textit{resp. } \mathcal{X}_{\mathrm{BS}}) := \left\{ (P, Z) \;\middle|\; \begin{array}{l} P \textit{ is a } \mathbf{Q}\textit{-parabolic subgroup of } G_{\mathbf{R}}, \\ Z \textit{ is an } (A_P\circ)\textit{-orbit in } D \ (\textit{resp. } \mathcal{X}) \end{array} \right\}.$$

Definition 5.1.6 ([KU2, 2.6]) *We define the space $D_{\mathrm{BS,val}}$ by*

$$D_{\mathrm{BS,val}} := \left\{ (T, Z, V) \;\middle|\; \begin{array}{l} T \textit{ is an } \mathbf{R}\textit{-split torus of } G_{\mathbf{R}}, \\ Z \textit{ is a } (T_{>0})\textit{-orbit in } D, \\ V \textit{ is a valuative submonoid of } X(T), \\ \textit{which satisfy the following conditions } (1)\text{–}(3) \end{array} \right\}.$$

Here, $T_{>0}$ is the connected component of the group of $\mathbf{R}$-valued points of T containing the unity, and $X(T)$ denotes the character group of T. A submonoid V of $X(T)$ is said to be valuative if $X(T) = V \cup V^{-1}$ (cf. 3.6.16).

(1) $\theta_{K_F}(t) = t^{-1} \quad (\forall F \in Z, \forall t \in T)$.
(2) $V^\times = \{1\}$.
(3) *Let* $H_{0,\mathbf{R}} = \bigoplus_{\chi \in X(T)} H_{0,\mathbf{R}}(\chi)$ *be the decomposition into eigenspaces* $H_{0,\mathbf{R}}(\chi) := \{v \in H_{0,\mathbf{R}} \mid tv = \chi(t)v \ (\forall t \in T)\}$. *Then, for* $\alpha \in X(T)$, *the* $\mathbf{R}$*-subspace*

$$W_\alpha := \bigoplus_{\chi \in V} H_{0,\mathbf{R}}(\alpha\chi^{-1})$$

of $H_{0,\mathbf{R}}$ *is* $\mathbf{Q}$*-rational.*

Note that, since the set $\{W_\alpha \mid \alpha \in X(T)\}$ of $\mathbf{R}$-subspaces of $H_{0,\mathbf{R}}$ is totally ordered with respect to the inclusion, we have a $\mathbf{Q}$-parabolic subgroup

$$P_{T,V} := \{g \in (G^\circ)_\mathbf{R} \mid g \text{ preserves } (W_\alpha)_{\alpha \in X(T)}\}.$$

(For G°, see Section 0.7.)

5.1.7

We have morphisms

$$\begin{array}{ccc} D_{\mathrm{BS,val}} & & \\ \downarrow{\scriptstyle\alpha} & & \\ D_{\mathrm{BS}} & \xrightarrow{\ \beta\ } & \mathcal{X}_{\mathrm{BS}}, \end{array}$$

$$\alpha : (T, Z, V) \mapsto (P_{T,V}, A_{P_{T,V}} \circ Z),$$
$$\beta : \text{the map induced by } D \to \mathcal{X}, \ \ F \mapsto K_F.$$

5.1.8

For a $\mathbf{Q}$-parabolic subgroup P of $G_\mathbf{R}$, we define

$$D_{\mathrm{BS}}(P) := \{(Q, Z) \in D_{\mathrm{BS}} \mid Q \supset P\},$$
$$\mathcal{X}_{\mathrm{BS}}(P) := \{(Q, Z) \in \mathcal{X}_{\mathrm{BS}} \mid Q \supset P\},$$
$$D_{\mathrm{BS,val}}(P) := \{(T, Z, V) \in D_{\mathrm{BS,val}} \mid P_{T,V} \supset P\}.$$

In 5.1.9–5.1.12 below, we give preliminaries needed to define the topologies on the spaces D_{BS}, $\mathcal{X}_{BS}$, and $D_{\mathrm{BS,val}}$.

5.1.9

Let P be a $\mathbf{Q}$-parabolic subgroup of $G_\mathbf{R}$. A subset Δ_P of $X(S_P)$ is defined as follows. Let $\tilde{S}_P \subset P$ be any torus such that the projection $P \to P/P_u$ induces an isomorphism $\tilde{S}_P \xrightarrow{\sim} S_P$. Let

$$\Delta'_P := \{\chi \in X(\tilde{S}_P) \mid 0 \neq \exists v \in \mathrm{Lie}(P_u) \text{ such that } \mathrm{Ad}(a)v = \chi(a)^{-1}v \ (\forall a \in \tilde{S}_P)\}.$$

Identify $X(\tilde{S}_P)$ with $X(S_P)$ via the canonical isomorphism $\tilde{S}_P \xrightarrow{\sim} S_P$. Then the subset Δ'_P of $X(S_P)$ is independent of the choice of the liftings $\tilde{S}_P$, and it is a finite subset of $X(S_P)$ which generates $\mathbf{Q} \otimes X(S_P)$ over $\mathbf{Q}$. There exists a unique subset Δ_P of Δ'_P satisfying the following two conditions:

(1) The number $\sharp(\Delta_P)$ of the elements of Δ_P coincides with rank S_P.
(2) Δ'_P is contained in the submonoid of $X(S_P)$ generated by Δ_P.

Let P be as above. Let Q be a $\mathbf{Q}$-parabolic subgroup of $G_{\mathbf{R}}$ containing P. Then there are injective maps

$$S_Q \hookrightarrow S_P, \tag{3}$$

$$\Delta_Q \hookrightarrow \Delta_P, \tag{4}$$

obtained as follows, which are regarded as inclusion maps. Note that $Q \supset P \supset P_u \supset Q_u$. We have that $S_Q \subset P/Q_u$ in Q/Q_u, that the canonical map $S_Q \to P/P_u$ is injective, and that the image of this map is contained in S_P. This gives the injection (3). Let $I := \{\chi \in \Delta_P \mid \chi \text{ annihilates } S_Q\}$, and let $J \subset \Delta_P$ be the complement of I in Δ_P. Then the restriction to S_Q gives a bijection $J \xrightarrow{\sim} \Delta_Q$. The injection (4) is obtained as the composite $\Delta_Q \xleftarrow{\sim} J \hookrightarrow \Delta_P$. It is known that we have a bijection

$$\{\mathbf{Q}\text{-parabolic subgroup of } G_{\mathbf{R}} \text{ containing } P\} \xrightarrow{\sim} \{\text{subset of } \Delta_P\}, \quad Q \mapsto \Delta_Q, \tag{5}$$

(cf. [BS, §4],[B, §11]).

5.1.10

Identification of $D_{\mathrm{BS}}(P)$ with $D \times^{A_P} \overline{A}_P$. Let P be a $\mathbf{Q}$-parabolic subgroup of $G_{\mathbf{R}}$. Let $X(S_P)_+$ be the submonoid of $X(S_P)$ consisting of elements χ such that, for some $n \geq 1$, χ^n belongs to the submonoid of $X(S_P)$ generated by Δ_P. Then $X(S_P)_+$ is an fs monoid, and $(X(S_P)_+)^{\mathrm{gp}} = X(S_P)$. We define

$$\overline{A}_P := \mathrm{Hom}(X(S_P)_+, \mathbf{R}^{\mathrm{mult}}_{\geq 0}) \supset \mathrm{Hom}(X(S_P)_+, \mathbf{R}_{>0}) = \mathrm{Hom}(X(S_P), \mathbf{R}_{>0}) = A_P.$$

Then,

$$\begin{array}{ccccc} \overline{A}_P & \xrightarrow{\sim} & \mathrm{Map}(\Delta_P, \mathbf{R}_{\geq 0}) & \simeq & \mathbf{R}^r_{\geq 0} \\ \cup & & \cup & & \cup \\ A_P & \xrightarrow{\sim} & \mathrm{Map}(\Delta_P, \mathbf{R}_{>0}) & \simeq & \mathbf{R}^r_{>0}, \end{array}$$

where $r := \sharp(\Delta_P) = \mathrm{rank}\, S_P$. The group A_P acts on $\overline{A}_P$ in the natural way. Denote $D \times^{A_P} \overline{A}_P := (D \times \overline{A}_P)/A_P$ under the action

$$a \cdot (F, b) = (a \circ F, a^{-1}b) \quad (a \in A_P,\ (F, b) \in D \times \overline{A}_P).$$

Then we have a bijection

$$D_{\mathrm{BS}}(P) \simeq D \times^{A_P} \overline{A}_P, \quad (Q, Z) \longleftrightarrow (F, b),$$

defined as follows. For $(Q, Z) \in D_{\mathrm{BS}}(P)$, F is any element of Z, and $b \in \overline{A}_P$ is defined by $b(\chi) = 0$ if $\chi \in \Delta_Q$ and $b(\chi) = 1$ if $\chi \in \Delta_P - \Delta_Q$. Conversely, for $(F, b) \in D \times^{A_P} \overline{A}_P$, Q is the $\mathbf{Q}$-parabolic subgroup of $G_{\mathbf{R}}$ containing P such that $\Delta_Q = \{\chi \in \Delta_P \mid b(\chi) = 0\}$, and $Z := \{a \circ F \mid a \in A_P,\ \chi(a) = b(\chi) \text{ for any } \chi \in \Delta_P - \Delta_Q\}$.

5.1.11

Valuative spaces associated with toric varieties. Let $\mathcal{S}$ be an fs monoid. We denote the space $\mathrm{Spec}(\mathbf{C}[\mathcal{S}])_{\mathrm{an,val}}$ (the space X_{val} in 3.6.18–3.6.24 for $X = \mathrm{Spec}(\mathbf{C}[\mathcal{S}])_{\mathrm{an}} = \mathrm{Hom}(\mathcal{S}, \mathbf{C}^{\mathrm{mult}})$) also by $\mathrm{Hom}(\mathcal{S}, \mathbf{C}^{\mathrm{mult}})_{\mathrm{val}}$. We regard the open subset $\mathrm{Hom}(\mathcal{S}^{\mathrm{gp}}, \mathbf{C}^\times)$ of $\mathrm{Hom}(\mathcal{S}, \mathbf{C}^{\mathrm{mult}})$ as an open subset of $\mathrm{Hom}(\mathcal{S}, \mathbf{C}^{\mathrm{mult}})_{\mathrm{val}}$ via the homeomorohism

$$\mathrm{Hom}(\mathcal{S}^{\mathrm{gp}}, \mathbf{C}^\times) \times_{\mathrm{Hom}(\mathcal{S}, \mathbf{C}^{\mathrm{mult}})} \mathrm{Hom}(\mathcal{S}, \mathbf{C}^{\mathrm{mult}})_{\mathrm{val}} \xrightarrow{\sim} \mathrm{Hom}(\mathcal{S}^{\mathrm{gp}}, \mathbf{C}^\times).$$

Let

$$\mathrm{Hom}(\mathcal{S}, \mathbf{R}^{\mathrm{mult}}_{\geq 0})_{\mathrm{val}} \subset \mathrm{Hom}(\mathcal{S}, \mathbf{C}^{\mathrm{mult}})_{\mathrm{val}}$$

be the closure of $\mathrm{Hom}(\mathcal{S}, \mathbf{R}_{>0})$. When $\mathcal{S} = \mathbf{N}^n$, $\mathrm{Hom}(\mathcal{S}, \mathbf{C}^{\mathrm{mult}})_{\mathrm{val}}$ and $\mathrm{Hom}(\mathcal{S}, \mathbf{R}^{\mathrm{mult}}_{\geq 0})_{\mathrm{val}}$ are also denoted by $(\mathbf{C}^n)_{\mathrm{val}}$ and $(\mathbf{R}^n_{\geq 0})_{\mathrm{val}}$, respectively.

We identify, as sets,

$$\mathrm{Spec}(\mathbf{C}[\mathcal{S}])_{\mathrm{an,val}} = \left\{ (V, h) \;\middle|\; \begin{array}{l} V \text{ is a valuative submonoid of } \mathcal{S}^{\mathrm{gp}} \\ \text{with } V \supset \mathcal{S}, \\ h : V \to \mathbf{C}^{\mathrm{mult}} \text{ is a homomorphism} \\ \text{such that } h^{-1}(\mathbf{C}^\times) = V^\times \end{array} \right\}, \tag{1}$$

$$\mathrm{Hom}(\mathcal{S}, \mathbf{R}^{\mathrm{mult}}_{\geq 0})_{\mathrm{val}} = \left\{ (V, h) \;\middle|\; \begin{array}{l} V \text{ is a valuative submonoid of } \mathcal{S}^{\mathrm{gp}} \\ \text{with } V \supset \mathcal{S}, \\ h : V \to \mathbf{R}^{\mathrm{mult}}_{\geq 0} \text{ is a homomorphism} \\ \text{such that } h^{-1}(\mathbf{R}_{>0}) = V^\times \end{array} \right\}. \tag{2}$$

Here (V, h) as in (1) corresponding to $(x, V', h') \in \mathrm{Spec}(\mathbf{C}[\mathcal{S}])_{\mathrm{an,val}}$ is as follows: V is the inverse image of V' under the canonical surjective homomorphism $\mathcal{S}^{\mathrm{gp}} \to (M_X^{\mathrm{gp}}/\mathcal{O}_X^\times)_x$, and h is the homomorphism induced by h'. The identification (2) is the one induced from (1).

In the identification (1) (resp. (2)), a directed family $((V_\lambda, h_\lambda))_\lambda$ converges to (V, h) in $\mathrm{Spec}(\mathbf{C}[\mathcal{S}])_{\mathrm{an,val}}$ (resp. $\mathrm{Hom}(\mathcal{S}, \mathbf{R}^{\mathrm{mult}}_{\geq 0})_{\mathrm{val}}$) if and only if the following condition (3) is satisfied.

(3) For each $\chi \in V$, $\chi \in V_\lambda$ for sufficiently large λ and $h_\lambda(\chi)$ converges to $h(\chi)$.

Since the logarithmic structure of the open set $\mathrm{Spec}(\mathbf{C}[\mathcal{S}^{\mathrm{gp}}])_{\mathrm{an}} \subset \mathrm{Spec}(\mathbf{C}[\mathcal{S}])_{\mathrm{an}}$ is trivial, we have dense open immersions

$$\mathrm{Spec}(\mathbf{C}[\mathcal{S}^{\mathrm{gp}}])_{\mathrm{an}} \subset \mathrm{Spec}(\mathbf{C}[\mathcal{S}])_{\mathrm{an,val}},$$

$$\mathrm{Hom}(\mathcal{S}^{\mathrm{gp}}, \mathbf{R}_{>0}) \subset \mathrm{Hom}(\mathcal{S}, \mathbf{R}^{\mathrm{mult}}_{\geq 0})_{\mathrm{val}}.$$

Furthermore, the natural action of the group $\mathrm{Spec}(\mathbf{C}[\mathcal{S}^{\mathrm{gp}}])_{\mathrm{an}}$ (resp. $\mathrm{Hom}(\mathcal{S}^{\mathrm{gp}}, \mathbf{R}_{>0})$) on $\mathrm{Spec}(\mathbf{C}[\mathcal{S}])_{\mathrm{an}}$ (resp. $\mathrm{Hom}(\mathcal{S}, \mathbf{R}^{\mathrm{mult}}_{\geq 0})$) extends uniquely to a continuous action of this group on $\mathrm{Spec}(\mathbf{C}[\mathcal{S}])_{\mathrm{an,val}}$ (resp. $\mathrm{Hom}(\mathcal{S}, \mathbf{R}^{\mathrm{mult}}_{\geq 0})_{\mathrm{val}}$). In the identification (1) (resp. (2)), an element a of this group sends (V, h) to (V, ah), where ah is the homomorphism sending $x \in V$ to $a(x)h(x)$.

5.1.12

Identification of $D_{\mathrm{BS,val}}(P)$ with $D \times^{A_P} (\overline{A}_P)_{\mathrm{val}}$. Let P be a $\mathbf{Q}$-parabolic subgroup of $G_{\mathbf{R}}$. Define

$$(\overline{A}_P)_{\mathrm{val}} := \mathrm{Hom}(X(S_P)_+, \mathbf{R}^{\mathrm{mult}}_{\geq 0})_{\mathrm{val}},$$

where $X(S_P)_+$ is the submonoid of $X(S_P)$ introduced in 5.1.10. If we identify the submonoid of $X(S_P)_+$ generated by Δ_P with $\mathbf{N}^r$ ($r := \mathrm{rank}(S_P)$), the inclusion $\mathbf{N}^r \subset X(S_P)_+$ induces

$$(\overline{A}_P)_{\mathrm{val}} \xrightarrow{\sim} (\mathbf{R}^r_{\geq 0})_{\mathrm{val}}.$$

For the set $D_{\mathrm{BS,val}}(P)$ in 5.1.8, we have a bijection

$$D_{\mathrm{BS,val}}(P) \to D \times^{A_P} (\overline{A}_P)_{\mathrm{val}}, \quad (T, Z, V') \mapsto (F, (V, h)), \tag{1}$$

which is given by

$$\begin{cases} F \in Z \ \text{(any element)}, \\ V := (\text{the inverse image of } V' \text{ under } X(S_P) \to X(T)), \\ h : V \to \mathbf{R}^{\mathrm{mult}}_{\geq 0}, \quad h(\chi) = 1 \text{ (resp. 0) if } \chi \in V^{\times} \text{ (resp. } \chi \notin V^{\times}). \end{cases} \tag{2}$$

Here, in the definition of V, we regard T as a subtorus of S_P via the composite of the embeddings $T \hookrightarrow S_{P_{T,V'}} \hookrightarrow S_P$ (5.1.6 and 5.1.9).

The inverse map $D \times^{A_P} \overline{A}_P \to D_{\mathrm{BS,val}}(P)$, $(F, (V, h)) \mapsto (T, Z, V')$, is given by

$$\begin{cases} T := \begin{pmatrix} \text{the image of the annihilator of } V^{\times} \text{ in } S_P \\ \text{under the Borel-Serre lifting } S_P \hookrightarrow G_{\mathbf{R}} \text{ at } K_F \end{pmatrix}, \\ Z := \{a \circ F \mid a \in A_P,\ \chi(a) = h(\chi) \ (\forall \chi \in V^{\times})\}, \\ V' := (\text{the image of } V \text{ under } X(S_P) \to X(T)) \ \text{(so that } V' \simeq V/V^{\times}). \end{cases}$$

5.1.13

Topologies of D_{BS}, $\mathcal{X}_{\mathrm{BS}}$ and $D_{\mathrm{BS,val}}$. Let P be a $\mathbf{Q}$-parabolic subgroup of $G_{\mathbf{R}}$. We have

$$\begin{aligned} D_{\mathrm{BS}}(P) &\simeq D \times^{A_P} \overline{A}_P \quad \text{(see 5.1.10)}, \\ \mathcal{X}_{\mathrm{BS}}(P) &\simeq \mathcal{X} \times^{A_P} \overline{A}_P \quad \text{(analogously to the above)}, \\ D_{\mathrm{BS,val}}(P) &\simeq D \times^{A_P} (\overline{A}_P)_{\mathrm{val}} \quad \text{(see 5.1.12)}. \end{aligned}$$

Via these isomorphisms, we introduce topologies on $D_{\mathrm{BS}}(P)$, $\mathcal{X}_{\mathrm{BS}}(P)$, and $D_{\mathrm{BS,val}}(P)$, respectively. We introduce the strongest topology on D_{BS} (resp. $\mathcal{X}_{\mathrm{BS}}$,

$D_{\mathrm{BS,val}}$) for which the map $D_{\mathrm{BS}}(P) \hookrightarrow D_{\mathrm{BS}}$ (resp. $\mathcal{X}_{\mathrm{BS}}(P) \hookrightarrow \mathcal{X}_{\mathrm{BS}}$, $D_{\mathrm{BS,val}}(P) \hookrightarrow D_{\mathrm{BS,val}}$) is continuous for every $\mathbf{Q}$-parabolic subgroup P of $G_{\mathbf{R}}$. Then, it can be shown, as in [BS], that all these maps $D_{\mathrm{BS}}(P) \hookrightarrow D_{\mathrm{BS}}$, $\mathcal{X}_{\mathrm{BS}}(P) \hookrightarrow \mathcal{X}_{\mathrm{BS}}$ and $D_{\mathrm{BS,val}}(P) \hookrightarrow D_{\mathrm{BS,val}}$ are open embeddings.

We have the following results on D_{BS} and $D_{\mathrm{BS,val}}$ which are analogous to the similar results on $\mathcal{X}_{\mathrm{BS}}$ obtained in [BS].

Theorem 5.1.14 ([KU2, 2.17, 5.2, 5.8])

(i) *The maps $\alpha : D_{\mathrm{BS,val}} \to D_{\mathrm{BS}}$, $\beta : D_{\mathrm{BS}} \to \mathcal{X}_{\mathrm{BS}}$ in* 5.1.7 *are proper and surjective.*

(ii) *The spaces D_{BS}, $D_{\mathrm{BS,val}}$ are Hausdorff and moreover locally compact.*

(iii) *For a subgroup Γ of $G_{\mathbf{Z}}$, the actions of Γ on D_{BS} and on $D_{\mathrm{BS,val}}$ are proper, and the quotient spaces $\Gamma\backslash D_{\mathrm{BS}}$ and $\Gamma\backslash D_{\mathrm{BS,val}}$ are Hausdorff. If Γ is neat, the projections $D_{\mathrm{BS}} \to \Gamma\backslash D_{\mathrm{BS}}$ and $D_{\mathrm{BS,val}} \to \Gamma\backslash D_{\mathrm{BS,val}}$ are local homeomorphisms. If Γ is of finite index, $\Gamma\backslash D_{\mathrm{BS}}$ and $\Gamma\backslash D_{\mathrm{BS,val}}$ are compact spaces.*

See Section 7.2 below for the definition of "proper action" and the basic facts concerning this notion.

5.1.15

Relationship to Iwasawa decomposition. The theory of Borel-Serre spaces is closely related to the theory of Iwasawa decomposition.

Let P be a minimal parabolic subgroup of $G_{\mathbf{R}}$ and K a maximal compact subgroup of $G_{\mathbf{R}}$ so that we have the Iwasawa decomposition associated with (P, K),

$$G_{\mathbf{R}} \simeq P_u \times \mathbf{R}^r_{>0} \times K \quad \text{(homeomorphism)}, \tag{1}$$

where r is the dimension of a maximal $\mathbf{R}$-split torus of $G_{\mathbf{R}}$. This homeomorphism is defined in the following way. Let $S_P^{\mathbf{R}}$ be the maximal $\mathbf{R}$-split torus of the center of P/P_u, let $A_P^{\mathbf{R}}$ be the connected component of the group of $\mathbf{R}$-valued points of $S_P^{\mathbf{R}}$ containing the unity, and identify $A_P^{\mathbf{R}}$ with $\mathbf{R}^r_{>0}$ just as in 5.1.10. Then the homeomorphism (1) is defined by $pt_K k \mapsto (p, t, k)$.

Take $F \in D$. Then the Iwasawa decomposition (1) associated with (P, K_F) induces

$$\begin{aligned} &\mathcal{X} \simeq P_u \times \mathbf{R}^r_{>0}, \quad \mathrm{Int}(pt_{K_F})K_F \leftrightarrow (p, t), \\ &D \simeq P_u \times \mathbf{R}^r_{>0} \times (K_F/K'_F), \quad pt_{K_F}kF \leftrightarrow (p,\ t,\ k \bmod K'_F). \end{aligned} \tag{2}$$

Now let Q be a $\mathbf{Q}$-rational parabolic subgroup of $G_{\mathbf{R}}$, and take a minimal parabolic subgroup P of $G_{\mathbf{R}}$ such that $P \subset Q$. Then $S_Q \subset P/Q_u \to P/P_u$ induces injections $S_Q \hookrightarrow S_P^{\mathbf{R}}$, $A_Q \hookrightarrow A_P^{\mathbf{R}} = \mathbf{R}^r_{>0}$. Since $a \circ \mathrm{Int}(pt_{K_F})K_F = \mathrm{Int}(p(at)_{K_F})K_F$ and $a \circ pt_{K_F}kF = p(at)_{K_F}kF$ for $a \in A_Q$, (2) induces

$$\begin{aligned} \mathcal{X}_{\mathrm{BS}}(Q) &\simeq P_u \times (\mathbf{R}^r_{>0} \times^{A_Q} \overline{A}_Q), \\ D_{\mathrm{BS}}(Q) &\simeq P_u \times (\mathbf{R}^r_{>0} \times^{A_Q} \overline{A}_Q) \times (K_F/K'_F), \\ D_{\mathrm{BS,val}}(Q) &\simeq P_u \times (\mathbf{R}^r_{>0} \times^{A_Q} \overline{A}_{Q,\mathrm{val}}) \times (K_F/K'_F). \end{aligned} \tag{3}$$

In the case $P = Q$ and $S_P^{\mathbf{R}} = S_Q$, the homeomorphisms in (3) have simpler forms

$$\begin{aligned} \mathcal{X}_{\mathrm{BS}}(P) &\simeq P_u \times \mathbf{R}^r_{\geq 0}, \\ D_{\mathrm{BS}}(P) &\simeq P_u \times \mathbf{R}^r_{\geq 0} \times (K_F/K'_F), \\ D_{\mathrm{BS,val}}(P) &\simeq P_u \times (\mathbf{R}^r_{\geq 0})_{\mathrm{val}} \times (K_F/K'_F). \end{aligned} \tag{4}$$

5.2 SPACES OF SL(2)-ORBITS (REVIEW)

This Section 5.2 is a review of [KU2, §3, §4] for our later use, i.e., we review spaces of SL(2)-orbits $D_{\mathrm{SL}(2)}$ and the projective limit $D_{\mathrm{SL}(2),\mathrm{val}}$ of the blow-ups of $D_{\mathrm{SL}(2)}$. These spaces, together with the spaces in the previous section, will form the following diagram:

$$\begin{array}{ccccc} D_{\mathrm{SL}(2),\mathrm{val}} & \hookrightarrow & D_{\mathrm{BS,val}} & & \\ \downarrow & & \downarrow & & \\ D_{\mathrm{SL}(2)} & & D_{\mathrm{BS}} & \rightarrow & \mathcal{X}_{\mathrm{BS}}. \end{array}$$

In general, there is no direct relation between $D_{\mathrm{SL}(2)}$ and D_{BS} (see [KU2, §6]), which is the reason that we introduce $D_{\mathrm{SL}(2),\mathrm{val}}$ and $D_{\mathrm{BS,val}}$.

5.2.1

SL(2)-*orbits in several variables.* We review the definition of SL(2)-orbits in several variables ([CKS, 4], [KU2, §3]).

Let (ρ, φ) be a pair of a homomorphism $\rho : \mathrm{SL}(2, \mathbf{C})^n \to G_{\mathbf{C}}$ of algebraic groups, which is defined over $\mathbf{R}$ and a map $\varphi : \mathbf{P}^1(\mathbf{C})^n \to \check{D}$. Throughout this book, such a pair (ρ, φ) is called an SL(2)-*orbit in n variables* if it satisfies the following three conditions (1)–(3):

(1) $\varphi(gz) = \rho(g)\varphi(z)$ for all $g \in \mathrm{SL}(2, \mathbf{C})^n$ and all $z \in \mathbf{P}^1(\mathbf{C})^n$.

(2) The Lie algebra homomorphism $\rho_* : \mathfrak{sl}(2, \mathbf{C})^{\oplus n} \to \mathfrak{g}_{\mathbf{C}}$, associated with ρ, satisfies

$$\rho_*(F^p(z)(\mathfrak{sl}(2, \mathbf{C})^{\oplus n})) \subset F^p(\varphi(z))(\mathfrak{g}_{\mathbf{C}}) \quad (\forall z \in \mathbf{P}^1(\mathbf{C})^n, \forall p \in \mathbf{Z}).$$

Here

$$\begin{aligned} F^p(z)(\mathfrak{sl}(2, \mathbf{C})^{\oplus n}) &:= \{(X_j)_j \in \mathfrak{sl}(2, \mathbf{C})^{\oplus n} \mid X_j F^q(z_j) \\ &\qquad \subset F^{q+p}(z_j)\ (\forall q \in \mathbf{Z}, 1 \leq j \leq n)\}, \\ F^p(\varphi(z))(\mathfrak{g}_{\mathbf{C}}) &:= \{X \in \mathfrak{g}_{\mathbf{C}} \mid XF^q(\varphi(z)) \subset F^{q+p}(\varphi(z))\ (\forall q \in \mathbf{Z})\}, \end{aligned}$$

where $F(z_j)$ (resp. $F(\varphi(z))$) denotes the filtration on $\mathbf{C}^2$ (resp. $H_{0,\mathbf{C}}$) corresponding to $z_j \in \mathbf{P}^1(\mathbf{C})$ (1.2.3) (resp. $\varphi(z) \in \check{D}$ (1.2.2)).

(3) Let $\mathfrak{h}$ be the upper-half plane. Then $\varphi(\mathfrak{h}^n) \subset D$.

We can replace the conditions (2) and (3) by

(4) Let $\mathbf{i} := (i, \ldots, i) \in \mathfrak{h}^n$. Then $\varphi(\mathbf{i}) \in D$, and

$$\rho_*(F^p(\mathbf{i})(\mathfrak{sl}(2, \mathbf{C})^{\oplus n})) \subset F^p(\varphi(\mathbf{i}))(\mathfrak{g}_{\mathbf{C}}) \quad (\forall\, p \in \mathbf{Z}).$$

5.2.2

Let (ρ, φ) be an SL(2)-orbit in n variables.

For $1 \leq j \leq n$, let

$$N_j := \rho_{*j}\begin{pmatrix} 0 & 1 \\ 0 & 0 \end{pmatrix} \in \mathfrak{g}_{\mathbf{R}},$$

where $\rho_{*j} : \mathfrak{sl}(2, \mathbf{C}) \to \mathfrak{g}_{\mathbf{C}}$ denotes the restriction of ρ_* to the jth factor of $\mathfrak{sl}(2, \mathbf{C})^{\oplus n}$.

Let $\Delta : \mathbf{G}^n_{\mathbf{m},\mathbf{R}} \to \mathrm{SL}(2, \mathbf{R})^n$ be the homomorphism defined by

$$(t_1, \ldots, t_n) \mapsto \left(\begin{pmatrix} t_1^{-1} & 0 \\ 0 & t_1 \end{pmatrix}, \ldots, \begin{pmatrix} t_n^{-1} & 0 \\ 0 & t_n \end{pmatrix}\right).$$

Define a homomorphism

$$\tilde{\rho} : \mathbf{G}^n_{\mathbf{m},\mathbf{R}} \to G_{\mathbf{R}}, \quad \tilde{\rho}(t_1, \ldots, t_n) := \rho(\Delta(t_1 \cdots t_n,\ t_2 \cdots t_n,\ \ldots,\ t_{n-1}t_n,\ t_n)). \tag{1}$$

For $1 \leq j \leq n$, let

$$\tilde{\rho}_j : \mathbf{G}_{\mathbf{m},\mathbf{R}} \to G_{\mathbf{R}} \tag{2}$$

be the restriction of $\tilde{\rho}$ to the j-th factor of $\mathbf{G}^n_{\mathbf{m},\mathbf{R}}$. Then, $\tilde{\rho}_j(t) = \rho(\Delta(\underbrace{t, \ldots, t}_{j}, 1 \ldots, 1))$.

5.2.3

Rank of an SL(2)*-orbit.* Let (ρ, φ) be an SL(2)-orbit in n variables.

Let J be the set of all j $(1 \leq j \leq n)$ such that the associated homomorphism ρ_* of Lie algebras is not the zero map on the jth factor of $\mathfrak{sl}(2, \mathbf{C})^{\oplus n}$, and let r be the number of elements of the set J. Then there exists a unique SL(2)-orbit (ρ', φ') in r variables such that $(\rho, \varphi) = (\rho', \varphi') \circ \pi_J$ where $\pi_J : (\mathrm{SL}(2, \mathbf{C}) \times \mathbf{P}^1(\mathbf{C}))^n \to (\mathrm{SL}(2, \mathbf{C}) \times \mathbf{P}^1(\mathbf{C}))^r$ is the projection to the J-factor. We call r the *rank of* (ρ, φ), and (ρ', φ') the SL(2)*-orbit of rank r associated with* (ρ, φ).

5.2.4

For a nilpotent element $N \in \mathfrak{g}_{\mathbf{R}}$, the *weight filtration associated with N* is the increasing filtration $W = W(N)$ of $H_{0,\mathbf{R}}$ characterized by the following two conditions ([D5]):

(1) $NW_k \subset W_{k-2}$ for all $k \in \mathbf{Z}$.

(2) $N^k : \mathrm{gr}^W_k \xrightarrow{\sim} \mathrm{gr}^W_{-k}$ for all $k \in \mathbf{Z}_{\geq 0}$.

The following is proved in [CK2].

THEOREM ([CK2]) *Let (σ, Z) be either a nilpotent orbit or a nilpotent i-orbit* (1.3.6). *Then, for any elements N, N' of the interior* (0.7.7) *of σ, the filtrations $W(N)$ and $W(N')$ of $H_{0,\mathbf{R}}$ coincide.*

This common filtration is denoted by $W(\sigma)$.

5.2.5

Family of weight filtrations associated with an SL(2)*-orbit.* As in [CKS, §4], an SL(2)-orbit (ρ, φ) in n variables defines an ordered family of nilpotent i-orbits $(\sigma_j, Z_j)_{1\le j\le n}$ consisting of

$$\sigma_j := (\mathbf{R}_{\ge 0})N_1 + \cdots + (\mathbf{R}_{\ge 0})N_j \quad (N_k \text{ is as in 5.2.2}), \tag{1}$$
$$Z_j := \exp(i\sigma_{j,\mathbf{R}})\varphi(\underbrace{0, \ldots, 0}_{j}, i, \ldots, i) = \varphi(\underbrace{i\mathbf{R}, \ldots, i\mathbf{R}}_{j}, i, \ldots, i).$$

We call the family $W = (W(\sigma_j))_{1\le j\le n}$ the *family of weight filtrations associated with* (ρ, φ).

We say the weight filtrations of (ρ, φ) are *rational* if the filtrations $W(\sigma_j)$ are defined over $\mathbf{Q}$ for $1 \le j \le n$.

The set J in 5.2.3 coincides with $\{j \mid 1 \le j \le n,\ W(\sigma_j) \ne W(\sigma_{j-1})\}$, where we understand $\sigma_0 = \{0\}$. We have $W(\sigma_j) \ne W(\sigma_k)$ if $j, k \in J$ and $j \ne k$.

DEFINITION 5.2.6 ([KU2, 3.6]) *We define the set $D_{\mathrm{SL}(2)}$ as follows.*

For $j = 1, 2$, let (ρ_j, φ_j) be an SL(2)*-orbit in n_j variables. We say (ρ_1, φ_1) and (ρ_2, φ_2) are equivalent if the following condition* (1) *is satisfied. Let r_j be the rank of (ρ_j, φ_j), and let (ρ'_j, φ'_j) be the* SL(2)*-orbit of rank r_j associated to (ρ_j, φ_j) as in* 5.2.3.

(1) $r_1 = r_2$ *(denoted $r := r_1 = r_2$), and there exists $t \in \mathbf{R}^r_{>0}$ such that $\rho'_2 = \mathrm{Int}(\rho'_1(\Delta(t))) \circ \rho'_1$ and $\varphi'_2 = \rho'_1(\Delta(t)) \cdot \varphi'_1$.*

We define $D_{\mathrm{SL}(2)}$ to be the set of all equivalence classes of SL(2)*-orbits (ρ, φ) of n variables for all $n \ge 0$ whose weight filtrations are rational.*

We denote by $[\rho, \varphi]$ the point of $D_{\mathrm{SL}(2)}$ represented by (ρ, φ).

We denote by $D_{\mathrm{SL}(2),n} \subset D_{\mathrm{SL}(2)}$ the part of $D_{\mathrm{SL}(2)}$ consisting of classes of SL(2)*-orbits of rank n. Then, we have*

$$D_{\mathrm{SL}(2)} = \bigsqcup_{n\ge 0} D_{\mathrm{SL}(2),n}, \quad D = D_{\mathrm{SL}(2),0}.$$

Let $p \in D_{\mathrm{SL}(2),n}$ and let (ρ, φ) be a representative of $[\rho, \varphi]$ in n variables. Then, the family $W = (W(\sigma_j))_{1\le j\le n}$ of weight filtrations of $H_{0,\mathbf{R}}$ associated with (ρ, φ) (cf. 5.2.5) depends only on p and is independent of the choice of (ρ, φ). We call W the family of weight filtrations associated to p.

In our previous paper [KU2, §3], we considered only SL(2)-orbits in n variables of rank n, but the space $D_{\mathrm{SL}(2)}$ in this book coincides with the one in that paper.

Definition 5.2.7 ([KU2, 3.7]) *For a non-negative integer n, we define*

$$D_{\mathrm{SL}(2),\mathrm{val},n} := \left\{ (p, Z, V) \;\middle|\; \begin{array}{l} p \in D_{\mathrm{SL}(2),n},\ Z \subset D,\ V \subset X(\mathbf{G}^n_{\mathbf{m},\mathbf{R}}) \\ \text{which satisfy the following conditions (1) and (2)} \end{array} \right\}.$$

Let $X(\mathbf{G}^n_{\mathbf{m},\mathbf{R}})_+$ be the inverse image of $\mathbf{N}^n \subset \mathbf{Z}^n$ under the canonical isomorphism $X(\mathbf{G}^n_{\mathbf{m},\mathbf{R}}) \simeq \mathbf{Z}^n$.

(1) *V is a valuative submonoid of $X(\mathbf{G}^n_{\mathbf{m},\mathbf{R}})$ such that $X(\mathbf{G}^n_{\mathbf{m},\mathbf{R}})_+ \subset V$ and $X(\mathbf{G}^n_{\mathbf{m},\mathbf{R}})_+ \cap V^\times = \{1\}$.*

(2) *Let*

$$T := \{t \in \mathbf{G}^n_{\mathbf{m},\mathbf{R}} \mid \chi(t) = 1\ (\forall \chi \in V^\times)\},$$

and let (ρ, φ) be a representative of p in n variables. Then Z is a $\tilde{\rho}(T_{>0})$-orbit in $\varphi(i\mathbf{R}^n_{>0})$. Here we denote by $T_{>0}$ the connected component of the group T containing the unity.

We define

$$D_{\mathrm{SL}(2),\mathrm{val}} := \bigsqcup_{n \geq 0} D_{\mathrm{SL}(2),\mathrm{val},n}.$$

We have the canonical surjection $D_{\mathrm{SL}(2),\mathrm{val}} \to D_{\mathrm{SL}(2)}$, $(p, Z, V) \mapsto p$.

Note that $\tilde{\rho}(T_{>0})$ and $\varphi(i\mathbf{R}^n_{>0})$ in 5.2.7 (2) depend only on p and are independent of the choice of a representative (ρ, φ).

Lemma 5.2.8 ([KU2, 3.8]) *Let (ρ, φ) be an* SL(2)*-orbit and put $\mathbf{r} = \varphi(\mathbf{i})$. Then*

$$\theta_{K_\mathbf{r}}(\tilde{\rho}(t)) = \tilde{\rho}(t)^{-1} \quad (\forall t \in \mathbf{G}^n_{\mathbf{m},\mathbf{R}}).$$

Lemma 5.2.9 ([KU2, 3.9]) *Let (ρ, φ) and $\mathbf{r}$ be as in 5.2.8. For $1 \leq j \leq n$, let $W^{(j)} = W(\sigma_j)$ be as in 5.2.5 and let P_j be the $\mathbf{Q}$-parabolic subgroup of $G_\mathbf{R}$ defined by $W^{(j)}$. Then, the Borel-Serre lifting at $K_\mathbf{r}$ of the jth weight map*

$$\mathbf{G}_{\mathbf{m},\mathbf{R}} \to P_j/P_{j,u}, \quad t_j \mapsto (t_j^k \text{ on } \mathrm{gr}_k^{W^{(j)}})_k$$

coincides with $\tilde{\rho}_j$ in 5.2.2 (2) (i.e., the restriction of $\tilde{\rho}$ to the jth factor of $\mathbf{G}^n_{\mathbf{m},\mathbf{R}}$).

Lemma 5.2.10 ([KU2, 3.10]) *Let (ρ, φ), $\mathbf{r} = \varphi(\mathbf{i})$ and W be as in 5.2.8 and 5.2.5. Then the* SL(2)*-orbit (ρ, φ) is determined by $(W, \mathbf{r})$.*

Theorem 5.2.11 ([KU2, 3.11]) *There is an injective map*

$$D_{\mathrm{SL}(2),\mathrm{val}} \to D_{\mathrm{BS},\mathrm{val}}, \quad ([\rho, \varphi], Z, V) \mapsto (\tilde{\rho}(T), Z, V'). \tag{1}$$

Here T is the subtorus of $\mathbf{G}^n_{\mathbf{m},\mathbf{R}}$ in 5.2.7 (2), and $V' \simeq V/V^\times$, which is regarded as a subset of the character group of $\tilde{\rho}(T)$.

5.2.12

A family $(W^{(j)})_{1 \leq j \leq n}$ of increasing filtrations $W^{(j)}$ of $H_{0,\mathbf{R}}$ is called a *compatible family* if there exists a direct sum decomposition $H_{0,\mathbf{R}} = \bigoplus_{m \in \mathbf{Z}^n} H(m)$ such that

$W_k^{(j)} = \bigoplus_{m \in \mathbf{Z}^n,\, m_j \le k} H(m)$ for any j and k. (We used this terminology "compatible family" in our previous paper [KU2] and also use it in this book. But, after writing this book, we realized that this notion is equivalent to the notion "distributive family" of Kashiwara [K].)

Note that, for $[\rho, \varphi] \in D_{\mathrm{SL}(2),n}$, the family of weight filtrations $(W(\sigma_j))_{1 \le j \le n}$ associated with $[\rho, \varphi]$ in 5.2.5 is a compatible family.

Let $W = (W^{(j)})_{1 \le j \le n}$ be a compatible family of $\mathbf{Q}$-rational increasing filtrations $W^{(j)}$ of $H_{0,\mathbf{R}}$. We denote

$$G_{W,\mathbf{R}} = \{g \in G_{\mathbf{R}} \mid gW^{(j)} = W^{(j)}\ (1 \le j \le n)\}.$$

For $W = (W^{(j)})_{1 \le j \le n}$ as above, we define the subset $D_{\mathrm{SL}(2)}(W)$ of $D_{\mathrm{SL}(2)}$ by

$$D_{\mathrm{SL}(2)}(W) := \bigcup_{0 \le m \le n} \left\{ p \in D_{\mathrm{SL}(2),m} \;\middle|\; \begin{array}{l} \exists s_j \in \mathbf{Z}\ (1 \le j \le m) \text{ such that} \\ 1 \le s_1 < \cdots < s_m \le n \text{ and} \\ W(p)^{(j)} = W^{(s_j)}\ (\forall\, j) \end{array} \right\}.$$

Here $(W(p)^{(j)})_{1 \le j \le m}$ is the family of weight filtrations associated with $p \in D_{\mathrm{SL}(2),m}$ (5.2.6).

We define the subset $D_{\mathrm{SL}(2),\mathrm{val}}(W)$ of $D_{\mathrm{SL}(2),\mathrm{val}}$ by the pullback of $D_{\mathrm{SL}(2)}(W)$.

Definition 5.2.13 ([KU2, 3.13]) *We define the* topology *of $D_{\mathrm{SL}(2),\mathrm{val}}$ as the weakest one in which the following two families of subsets are open:*

(1) *The pullbacks on $D_{\mathrm{SL}(2),\mathrm{val}}$ of open subsets of $D_{\mathrm{BS},\mathrm{val}}$.*
(2) *The subset $D_{\mathrm{SL}(2),\mathrm{val}}(W)$ for any n and any compatible family of $\mathbf{Q}$-rational increasing filtrations $W = (W^{(j)})_{1 \le j \le n}$.*

We induce the quotient topology on $D_{\mathrm{SL}(2)}$ of the above one under the projection $D_{\mathrm{SL}(2),\mathrm{val}} \to D_{\mathrm{SL}(2)}$.

5.2.14

The topology of $D_{\mathrm{SL}(2)}$ defined in 5.2.13 has the following property (see [KU2, 4.19]). For an SL(2)-orbit (ρ, φ) of rank n, $[\rho, \varphi] \in D_{\mathrm{SL}(2)}$ is the limit of

$$\varphi(iy_1, \ldots, iy_n) \in D, \quad \text{as } y_j > 0 \text{ and } \frac{y_j}{y_{j+1}} \to \infty \text{ for } 1 \le \forall\, j \le n\ (y_{n+1} \text{ denotes } 1).$$

Note that the space $D_{\mathrm{SL}(2),\mathrm{val}}$ is Hausdorff from 5.1.14 and 5.2.13. We have, moreover,

Theorem 5.2.15 ([KU2, 3.14, 4.18, 5.2, 5.8])

(i) *The canonical map $D_{\mathrm{SL}(2),\mathrm{val}} \to D_{\mathrm{SL}(2)}$ is proper and surjective.*
(ii) *The spaces $D_{\mathrm{SL}(2)}$ and $D_{\mathrm{SL}(2),\mathrm{val}}$ are Hausdorff.*
(iii) *The spaces $D_{\mathrm{SL}(2)}(W)$ and $D_{\mathrm{SL}(2),\mathrm{val}}(W)$ are regular spaces (cf. 6.4.6 below).*
(iv) *For a subgroup Γ of $G_{\mathbf{Z}}$, the actions of Γ on $D_{\mathrm{SL}(2)}$ and on $D_{\mathrm{SL}(2),\mathrm{val}}$ are proper, and the quotient spaces $\Gamma\backslash D_{\mathrm{SL}(2)}$ and $\Gamma\backslash D_{\mathrm{SL}(2),\mathrm{val}}$ are Hausdorff. If Γ is neat, the projections $D_{\mathrm{SL}(2)} \to \Gamma\backslash D_{\mathrm{SL}(2)}$ and $D_{\mathrm{SL}(2),\mathrm{val}} \to \Gamma\backslash D_{\mathrm{SL}(2),\mathrm{val}}$ are local homeomorphisms.*

Proposition 5.2.16 ([KU2, 4.15]) *Let (ρ, φ) be an* SL(2)*-orbit in n variables and of rank n whose associated weight filtrations are rational, and let $\mathbf{r} = \varphi(\mathbf{i})$. Then the sets*

$$B(U, U', U'') := \{\tilde{\rho}(t)gk \cdot \mathbf{r} \mid k \in U,\ t \in \mathbf{R}^n_{>0} \cap U', \\ \mathrm{Int}(\tilde{\rho}(t))^{\nu}(g) \in U''\ (\nu = 0, \pm 1)\}$$

form a basis of the filter $\{D \cap V \mid V$ is a neighborhood of $[\rho, \varphi]$ in $D_{\mathrm{SL}(2)}\}$, when U (resp. U', U'') ranges over all neighborhoods of 1 in $K_{\mathbf{r}}$ (resp. 0 in $\mathbf{R}^n_{\geq 0}$, 1 in $G_{\mathbf{R}}$).

In Chapter 10, we will give new proofs of the results 5.2.15 (i)–(iii) and 5.2.16.

5.2.17

Correction. In [KU2, Lemma 4.7], the definition of $B(U, U', U'')$ is written wrongly as $\{g\tilde{\rho}(t)k\mathbf{r} \mid \cdots\}$. It should be corrected to $\{\tilde{\rho}(t)gk\mathbf{r} \mid \cdots\}$, that is, g and $\tilde{\rho}(t)$ in [KU2, Lemma 4.7] should be interchanged, as in Proposition 5.2.16 above. The proof of [KU2, Lemma 4.7] and the rest of the paper [KU2] concerning this lemma are correct without any change.

5.3 SPACES OF VALUATIVE NILPOTENT ORBITS

In this section, we introduce the spaces $D^\sharp_{\mathrm{val}}$ and D_{val}. $D^\sharp_{\mathrm{val}}$ contains, as a subspace, the project limit $D^\sharp_{\Sigma,\mathrm{val}}$ of the blow-ups of the space of nilpotent i-orbits $D^\sharp_\Sigma$, with which we can describe the relationship between the space of SL(2)-orbits $D_{\mathrm{SL}(2)}$ and the space $D^\sharp_\Sigma$.

Definition 5.3.1 *We define*

$$\mathcal{V} := \left\{ (A, V) \;\middle|\; \begin{array}{l} A \text{ is a } \mathbf{Q}\text{-linear subspace of } \mathfrak{g}_{\mathbf{Q}} \text{ consisting of} \\ \text{mutually commutative nilpotent elements,} \\ V \text{ is a valuative submonoid of } A^* := \mathrm{Hom}_{\mathbf{Q}}(A, \mathbf{Q}) \\ \text{with } V \cap (-V) = \{0\} \end{array} \right\}.$$

Note that a valuative submonoid V of A^* as above is stable under the multiplication by elements of $\mathbf{Q}_{\geq 0}$. In fact, assume $x \in A^*$ and $nx \in V$ for some $n \in \mathbf{Z}_{>0}$. We prove $x \in V$. Assume $x \notin V$. Then $-x \in V$ and hence $-nx \in V \cap (-V) = \{0\}$. Therefore $x = 0$, which contradicts $x \notin V$.

5.3.2

For $(A, V) \in \mathcal{V}$, let $\mathfrak{F}(A, V)$ be the set of all rational nilpotent cones $\sigma \subset \mathfrak{g}_{\mathbf{R}}$ satisfying the following conditions (1) and (2).

(1) $\sigma_{\mathbf{R}} = A_{\mathbf{R}}$.
(2) Let $(\sigma \cap A)^\vee := \{h \in A^* \mid h(\sigma \cap A) \subset \mathbf{Q}_{\geq 0}\}$. Then $(\sigma \cap A)^\vee \subset V$.

Note that $(\sigma \cap A)^\vee = \{h \in A^* \mid h(\sigma) \subset \mathbf{R}_{\geq 0}\}$. This set $\mathfrak{F}(A, V)$ has the following properties (3)–(5).

(3) If $\sigma, \tau \in \mathfrak{F}(A, V)$, then $\sigma \cap \tau \in \mathfrak{F}(A, V)$.

(4) If $\sigma \subset A_{\mathbf{R}}$ is a rational nilpotent cone containing some $\tau \in \mathfrak{F}(A, V)$, then $\sigma \in \mathfrak{F}(A, V)$.

(5) $V = \bigcup_{\sigma \in \mathfrak{F}(A,V)} (\sigma \cap A)^{\vee}$.

(4) is evident. We prove (3) and (5).

For a finitely generated ($\mathbf{Q}_{\geq 0}$)-cone C in A (resp. A^*) such that $C_{\mathbf{Q}} = A$ (resp. A^*) and $C \cap (-C) = \{0\}$, we have $C^{\vee\vee} = C$, where $C^{\vee} = \mathrm{Hom}(C, \mathbf{Q}_{\geq 0}^{\mathrm{add}})$.

For the proof of (3), it is sufficient to show that $(\sigma \cap \tau \cap A)^{\vee} = (\sigma \cap A)^{\vee} + (\tau \cap A)^{\vee}$. But $(\ )^{\vee}$ of both sides coincide with $\sigma \cap \tau \cap A$.

Next we prove (5). When C ranges over all finitely generated ($\mathbf{Q}_{\geq 0}$)-subcones of V such that $C_{\mathbf{Q}} = A^*$, then $V = \bigcup C$. If σ_C denotes $\{a \in A_{\mathbf{R}} \mid C(a) \subset \mathbf{R}_{\geq 0}\}$, then $C^{\vee} = \sigma_C \cap A$. We have $\sigma_C \in \mathfrak{F}(A, V)$ since $C^{\vee\vee} = C \subset V$. These prove (5).

DEFINITION 5.3.3

(i) *We define*

$$\check{D}_{\mathrm{val}} \ (\textit{resp. } \check{D}^{\sharp}_{\mathrm{val}}) := \left\{ (A, V, Z) \;\middle|\; \begin{array}{l} (A, V) \in \mathcal{V},\ Z \textit{ is an } \exp(A_{\mathbf{C}}) \\ (\textit{resp. } \exp(iA_{\mathbf{R}}))\textit{-orbit in } \check{D} \end{array} \right\}.$$

(ii) *We define*

$$D_{\mathrm{val}} \ (\textit{resp. } D^{\sharp}_{\mathrm{val}}) := \left\{ (A, V, Z) \;\middle|\; \begin{array}{l} (A, V, Z) \in \check{D}_{\mathrm{val}} \ (\textit{resp. } \check{D}^{\sharp}_{\mathrm{val}}), \\ \exists \sigma \in \mathfrak{F}(A, V) \textit{ such that } Z \textit{ is a} \\ \sigma\textit{-nilpotent orbit } (\textit{resp. } i\textit{-orbit}) \end{array} \right\}$$

(*cf.* 1.3.7). *There is a natural map*

$$D^{\sharp}_{\mathrm{val}} \to D_{\mathrm{val}}, \quad (A, V, Z) \mapsto (A, V, \exp(A_{\mathbf{C}})Z).$$

LEMMA 5.3.4 *Let Σ be a fan in $\mathfrak{g}_{\mathbf{Q}}$. Let $(A, V, Z) \in \check{D}_{\mathrm{val}}$ (resp. $\check{D}^{\sharp}_{\mathrm{val}}$). Assume that there is $\sigma \in \Sigma$ such that $\sigma \cap A_{\mathbf{R}} \in \mathfrak{F}(A, V)$, and let σ_0 be the smallest one among such σ. (Note that the smallest one exists by* 5.3.2 (3).*) Assume that $\exp(\sigma_{0,\mathbf{C}})Z$ (resp. $\exp(i\sigma_{0,\mathbf{R}})Z$) is a σ_0-nilpotent orbit (resp. i-orbit). Then Z is a $(\sigma_0 \cap A_{\mathbf{R}})$-nilpotent orbit (resp. i-orbit).*

Proof. It is sufficient to prove that $\sigma_0 \cap A_{\mathbf{R}}$ contains a point of the interior (0.7.7) of σ_0. Assume $\sigma_0 \cap A_{\mathbf{R}} = \cup_{\tau} \tau \cap A_{\mathbf{R}}$, where τ ranges over all faces of σ_0 that are different from σ_0. Then, since $\tau \cap A_{\mathbf{R}}$ is a face of $\sigma_0 \cap A_{\mathbf{R}}$ and its dimension is strictly smaller than the dimension of $\sigma_0 \cap A_{\mathbf{R}}$ unless $\tau \cap A_{\mathbf{R}} = \sigma_0 \cap A_{\mathbf{R}}$, we should have $\tau \cap A_{\mathbf{R}} = \sigma_0 \cap A_{\mathbf{R}}$ for some face $\tau \neq \sigma_0$ of σ_0. But this contradicts the minimality of σ_0. □

DEFINITION 5.3.5 *For a fan Σ in $\mathfrak{g}_{\mathbf{Q}}$, we define $D_{\Sigma,\mathrm{val}} \subset \check{D}_{\mathrm{val}}$ (resp. $D^{\sharp}_{\Sigma,\mathrm{val}} \subset \check{D}^{\sharp}_{\mathrm{val}}$) as*

$$D_{\Sigma,\mathrm{val}} \ (\textit{resp. } D^{\sharp}_{\Sigma,\mathrm{val}}) := \left\{ (A, V, Z) \;\middle|\; \begin{array}{l} (A, V, Z) \in \check{D}_{\mathrm{val}} \ (\textit{resp. } \check{D}^{\sharp}_{\mathrm{val}}), \\ \sigma \cap A_{\mathbf{R}} \in \mathfrak{F}(A, V) \textit{ for some } \sigma \in \Sigma, \\ \exp(\sigma_{0,\mathbf{C}})Z \ (\textit{resp. } \exp(i\sigma_{0,\mathbf{R}})Z) \textit{ is} \\ \textit{a } \sigma_0\textit{-nilpotent orbit } (\textit{resp. } i\textit{-orbit}) \end{array} \right\}.$$

Here σ_0 *is as in Lemma* 5.3.4. *Then*

$$D_{\Sigma,\mathrm{val}} \subset D_{\mathrm{val}}, \quad D^\sharp_{\Sigma,\mathrm{val}} \subset D^\sharp_{\mathrm{val}}.$$

For a sharp rational nilpotent cone σ, *we define*

$$D_{\sigma,\mathrm{val}} := D_{\{\text{face of } \sigma\},\mathrm{val}}, \quad D^\sharp_{\sigma,\mathrm{val}} := D^\sharp_{\{\text{face of } \sigma\},\mathrm{val}}.$$

We have

$$D_{\Sigma,\mathrm{val}} = \bigcup_{\sigma\in\Sigma} D_{\sigma,\mathrm{val}}, \quad D^\sharp_{\Sigma,\mathrm{val}} = \bigcup_{\sigma\in\Sigma} D^\sharp_{\sigma,\mathrm{val}}.$$

There is a natural map

$$D^\sharp_{\Sigma,\mathrm{val}} \to D_{\Sigma,\mathrm{val}}, \quad (A, V, Z) \mapsto (A, V, \exp(A_{\mathbf{C}})Z).$$

We define maps

$$D_{\Sigma,\mathrm{val}} \to D_{\Sigma}, \quad (A, V, Z) \mapsto (\sigma_0, \exp(\sigma_{0,\mathbf{C}})Z),$$
$$D^\sharp_{\Sigma,\mathrm{val}} \to D^\sharp_{\Sigma}, \quad (A, V, Z) \mapsto (\sigma_0, \exp(i\sigma_{0,\mathbf{R}})Z),$$

where $\sigma_0 \in \Sigma$ *is as in Lemma* 5.3.4.

5.3.6

Let (Γ, Σ) be a strongly compatible pair of a subgroup Γ of $G_{\mathbf{Z}}$ and a fan Σ in $\mathfrak{g}_{\mathbf{Q}}$.

For $\sigma \in \Sigma$, we denote

$$\begin{aligned} \mathrm{toric}_{\sigma,\mathrm{val}} &:= \mathrm{Spec}(\mathbf{C}[\Gamma(\sigma)^\vee])_{\mathrm{an,val}}, \\ |\mathrm{toric}|_{\sigma,\mathrm{val}} &:= \mathrm{Hom}(\Gamma(\sigma)^\vee, \mathbf{R}^{\mathrm{mult}}_{\geq 0})_{\mathrm{val}} \subset \mathrm{toric}_{\sigma,\mathrm{val}}, \end{aligned} \tag{1}$$

(5.1.11).

There is a canonical bijection between $\mathrm{toric}_{\sigma,\mathrm{val}}$ (resp. $|\mathrm{toric}|_{\sigma,\mathrm{val}}$) and the set of all triples (A, V, z) (resp. (A, V, y)) where (A, V) is an element of $\mathcal{V}$ such that $A \subset \sigma_{\mathbf{R}}$ and $\sigma \cap A \in \mathfrak{F}(A, V)$, and $z \in \sigma_{\mathbf{C}}/(A_{\mathbf{C}} + \log(\Gamma(\sigma)^{\mathrm{gp}})$ (resp. $y \in \sigma_{\mathbf{R}}/A_{\mathbf{R}}$).

The bijection is given as follows. For (A, V, z) (resp. (A, V, y)), the corresponding element of $\mathrm{toric}_{\sigma,\mathrm{val}}$ (resp. $|\mathrm{toric}|_{\sigma,\mathrm{val}}$) is the pair (V', h) (5.1.11) where V' is the set of all elements χ of $\Gamma(\sigma)^{\vee\,\mathrm{gp}}$ such that the induced homomorphism

$$A \hookrightarrow \sigma_{\mathbf{R}} \cap \mathfrak{g}_{\mathbf{Q}} \xrightarrow[\sim]{\exp} \mathbf{Q} \otimes \Gamma(\sigma)^{\mathrm{gp}} \xrightarrow{\chi} \mathbf{Q}$$

belongs to V, and h is the homomorphism $V' \to \mathbf{C}^{\mathrm{mult}}$ (resp. $V' \to \mathbf{R}^{\mathrm{mult}}_{\geq 0}$) that sends $V' - (V')^\times$ to 0 and an element χ of $(V')^\times$ to the image of $\mathbf{e}(z)$ (resp. $\mathbf{e}(iy)$) (3.3.5) under $\chi : \mathbf{C}^\times \otimes \Gamma(\sigma)^{\mathrm{gp}} \to \mathbf{C}^\times$ (resp. $(\mathbf{R}_{>0}) \otimes \Gamma(\sigma)^{\mathrm{gp}} \to \mathbf{R}_{>0}$).

5.3.7

Let (Γ, Σ) be as in 5.3.6. For $\sigma \in \Sigma$, we have

$$E_{\sigma,\mathrm{val}} = \mathrm{toric}_{\sigma,\mathrm{val}} \times_{\mathrm{toric}_\sigma} E_\sigma$$

as a topological space. Define the topological space $E^{\sharp}_{\sigma,\mathrm{val}}$ by

$$E^{\sharp}_{\sigma,\mathrm{val}} := |\mathrm{toric}|_{\sigma,\mathrm{val}} \times_{|\mathrm{toric}|_{\sigma}} E^{\sharp}_{\sigma}. \tag{1}$$

Then we have commutative diagrams with surjective horizontal arrows

$$\begin{array}{ccccccc} E_{\sigma,\mathrm{val}} & \longrightarrow & \Gamma(\sigma)^{\mathrm{gp}}\backslash D_{\sigma,\mathrm{val}} & \qquad & E^{\sharp}_{\sigma,\mathrm{val}} & \longrightarrow & D^{\sharp}_{\sigma,\mathrm{val}} \\ \downarrow & & \downarrow & & \downarrow & & \downarrow \\ E_{\sigma} & \longrightarrow & \Gamma(\sigma)^{\mathrm{gp}}\backslash D_{\sigma}, & & E^{\sharp}_{\sigma} & \longrightarrow & D^{\sharp}_{\sigma}, \end{array} \tag{2}$$

where the upper horizontal maps are defined, respectively, as follows:

$$(V, h, F) \mapsto (A, V', \exp(A_{\mathbf{C}})\exp(z)F),$$
$$(V, h, F) \mapsto (A, V', \exp(iA_{\mathbf{R}})\exp(iy)F),$$

where $F \in \check{D}$, and (A, V', z) (resp. (A, V', y)) corresponds to the element (V, h) of $\mathrm{toric}_{\sigma,\mathrm{val}}$ (resp. $|\mathrm{toric}|_{\sigma,\mathrm{val}}$) (5.1.11) as in 5.3.6.

Definition 5.3.8 *We introduce the topology on $\Gamma(\sigma)^{\mathrm{gp}}\backslash D_{\sigma,\mathrm{val}}$ (resp. $D^{\sharp}_{\sigma,\mathrm{val}}$) as a quotient of $E_{\sigma,\mathrm{val}}$ (resp. $E^{\sharp}_{\sigma,\mathrm{val}}$) by* 5.3.7 (2).

We introduce the strongest topology on $\Gamma\backslash D_{\Sigma,\mathrm{val}}$ (resp. $D^{\sharp}_{\Sigma,\mathrm{val}}$) for which the map $\Gamma(\sigma)^{\mathrm{gp}}\backslash D_{\sigma,\mathrm{val}} \to \Gamma\backslash D_{\Sigma,\mathrm{val}}$ (resp. $D^{\sharp}_{\sigma,\mathrm{val}} \hookrightarrow D^{\sharp}_{\Sigma,\mathrm{val}}$) is continuous for every $\sigma \in \Sigma$.

5.4 VALUATIVE NILPOTENT *i*-ORBITS AND SL(2)-ORBITS

The aim of Section 5.4 is to give the map $D^{\sharp}_{\Sigma,\mathrm{val}} \to D_{\mathrm{SL}(2)}$ in the fundamental diagram (5.0.1), by using the work of Cattani, Kaplan, and Schmid [CKS].

5.4.1

Let $N_1, \ldots, N_n \in \mathfrak{g}_{\mathbf{R}}$ be mutually commutative nilpotent elements. Let $F \in \check{D}$, and assume that $(N_1, \ldots, N_n, F)$ generates a nilpotent orbit. This means that, for $\sigma = \sum_{1\le j\le n}(\mathbf{R}_{\ge 0})N_j$, $\exp(\sigma_{\mathbf{C}})F$ is a σ-nilpotent orbit (1.3.6). In [CKS], Cattani, Kaplan, and Schmid defined an SL(2)-orbit in n variables associated with the family $(N_1, \ldots, N_n, F)$. Here the order of $N_1, \ldots, N_n$ is important. The definition of this SL(2)-orbit will be reviewed in Section 6.1.

Here we state three theorems 5.4.2, 5.4.3, and 5.4.4, related to this SL(2)-orbit. Their proofs will be the main subjects of Chapter 6.

Theorem 5.4.2 *Let $(N_j)_{1\le j\le n}$ and F be as in* 5.4.1. *Put $\sigma_j := \sum_{1\le k\le j}(\mathbf{R}_{\ge 0})N_k$ $(1 \le j \le n)$. Assume that the associated weight filtration $W(\sigma_j)$ is $\mathbf{Q}$-rational for any j. Then the class $[\rho, \varphi]$ in $D_{\mathrm{SL}(2)}$ of the* SL(2)*-orbit associated with*

$(N_1, \dots, N_n, F)$ *by* [CKS] *coincides with the limit*

$$\lim_{\substack{y_j/y_{j+1}\to\infty \\ 1\le j\le n}} \exp\left(\sum_{1\le j\le n} iy_jN_j\right)F$$

in $D_{\mathrm{SL}(2)}$, *where we put* $y_{n+1} = 1$ (*so* $\frac{y_n}{y_{n+1}} \to \infty$ *means* $y_n \to \infty$).

THEOREM 5.4.3 *Let* $p = (A, V, Z) \in D^{\sharp}_{\mathrm{val}}$ (5.3.3).

(i) *There exists a family* $(N_j)_{1\le j\le n}$ *of elements of* $A_{\mathbf{R}}$ *satisfying the following two conditions*:

(1) *If* $F \in Z$, $(N_1, \dots, N_n, F)$ *generates a nilpotent orbit.*
(2) *Via* $(N_j)_{1\le j\le n} : A^* = \mathrm{Hom}_{\mathbf{Q}}(A, \mathbf{Q}) \to \mathbf{R}^n$, V *coincides with the set of all elements of* A^* *whose images in* $\mathbf{R}^n$ *are* ≥ 0 *with respect to the lexicographic order.*

(ii) *Take* $(N_j)_{1\le j\le n}$ *as in* (i), *let* $F \in Z$ *and define* $\psi(p) \in D_{\mathrm{SL}(2)}$ *to be the class of the* SL(2)*-orbit associated with* $(N_1, \dots, N_n, F)$ *by* [CKS]. *Then* $\psi(p)$ *is independent of the choices of* $(N_j)_{1\le j\le n}$ *and* F.

By Theorem 5.4.3, we obtain a map

$$\psi : D^{\sharp}_{\mathrm{val}} \to D_{\mathrm{SL}(2)}, \tag{3}$$

which we call the *CKS map*.

THEOREM 5.4.4 *Let* Σ *be a fan in* $\mathfrak{g}_{\mathbf{Q}}$. *Then,* $\psi : D^{\sharp}_{\Sigma,\mathrm{val}} \to D_{\mathrm{SL}(2)}$ *is continuous.*

Thus $\psi : D^{\sharp}_{\Sigma,\mathrm{val}} \to D_{\mathrm{SL}(2)}$ is the unique continuous extension of the identity map of D.

Chapter Six

The Map $\psi : D^\sharp_{\mathrm{val}} \to D_{\mathrm{SL}(2)}$

This chapter is devoted to the proofs of Theorems 5.4.2, 5.4.3, and 5.4.4. In Section 6.1, we recall the theory of SL(2)-orbits in several variables in [CKS]. We prove Theorem 5.4.2 in Section 6.2, Theorem 5.4.3 (i) in Section 6.3, and Theorem 5.4.3 (ii) and Theorem 5.4.4 in Section 6.4.

6.1 REVIEW OF [CKS] AND SOME RELATED RESULTS

In this section, we review some results in [CKS]. We review in 6.1.1 the theory of the associated SL(2)-orbit in one variable, in 6.1.2 the theory of $\mathbf{R}$-split mixed Hodge stuructures, and in 6.1.3 the theory of the associated SL(2)-orbit in several variables. In the remaining part of this section, we give some supplements.

6.1.1

The associated SL(2)*-orbit in one variable.* Let N be a nonzero nilpotent element of $\mathfrak{g}_{\mathbf{R}}$, let $F \in \check{D}$, and assume that (N, F) generates a nilpotent orbit, that is, $((\mathbf{R}_{\geq 0})N, \exp(\mathbf{C}N)F)$ is a nilpotent orbit. Then, an SL(2)-orbit (ρ, φ) in one variable is associated to (N, F) canonically in [Sc],[CKS, 3]. This SL(2)-orbit has the following two properties (1), (2).

(1) The homomorphism $\rho_* : \mathfrak{sl}(2, \mathbf{C}) \to \mathfrak{g}_{\mathbf{C}}$, associated with ρ, sends $\begin{pmatrix} 0 & 1 \\ 0 & 0 \end{pmatrix}$ to N.

(2) There are $a_n \in \mathfrak{g}_{\mathbf{R}}$ $(n \geq 1)$ such that $\sum_{n\geq 1} a_n t^n$ absolutely converges for $t \in \mathbf{C}$ with $|t|$ sufficiently small, that

$$\exp(iyN)F = \exp\left(\sum_{n\geq 1} a_n y^{-n}\right)\varphi(iy) = \exp\left(\sum_{n\geq 1} a_n y^{-n}\right)\tilde{\rho}(\sqrt{y})^{-1}\varphi(i)$$

for $y \gg 0$, and that the component of a_n of $(\mathrm{Ad} \circ \tilde{\rho})$-weight e $(e \in \mathbf{Z})$ is zero unless $e \leq n - 1$ (see 5.2.2 for the notation $\tilde{\rho}$).

Here the component of $(\mathrm{Ad} \circ \tilde{\rho})$-weight e means the component on which $\mathrm{Ad}(\tilde{\rho}(t))$ acts as the multiplication by t^e for $t \in \mathbf{G}_{\mathbf{m},\mathbf{R}} = \mathbf{R}^\times$.

From (2), we obtain another presentation (3) of $\exp(iyN)F$.

(3) There are $b_n \in \mathfrak{g}_{\mathbf{R}}$ $(n \geq 1)$ such that $\sum_{n\geq 1} b_n t^n$ absolutely converges for $t \in \mathbf{C}$ with $|t|$ sufficiently small,

$$\exp(iyN)F = \tilde{\rho}(\sqrt{y})^{-1}\exp\left(\sum_{n\geq 1} b_n \sqrt{y}^{-n}\right)\varphi(i)$$

for $y \gg 0$, and such that the component of b_n of $(\mathrm{Ad} \circ \tilde{\rho})$-weight e ($e \in \mathbf{Z}$) is zero unless $|e| \leq n - 1$.

In fact, (3) is deduced from (2) as

$$\exp\left(\sum_{n\geq 1} a_n y^{-n}\right)\tilde{\rho}(\sqrt{y})^{-1} = \tilde{\rho}(\sqrt{y})^{-1} \exp\left(\sum_{n\geq 1, e\in\mathbf{Z}} a_n(e)\sqrt{y}^{-(2n-e)}\right)$$
$$= \tilde{\rho}(\sqrt{y})^{-1} \exp\left(\sum_{n\geq 1} b_n \sqrt{y}^{-n}\right),$$

where $a_n(e)$ denotes the component of a_n of $(\mathrm{Ad} \circ \tilde{\rho})$-weight e and $b_n = \sum_k a_k(2k - n)$. Note that $a_k(2k - n) = 0$ unless $|2k - n| < n$, for the conditions $|2k - n| < n$ and $2k - n < k$ are equivalent (since $k > 0$).

The associated SL(2)-orbit (ρ, φ) has the following characterization in terms of the Borel-Serre action.

Let P be the parabolic subgroup $(G^\circ)_{W(N),\mathbf{R}} = (G^\circ)_\mathbf{R} \cap G_{W(N),\mathbf{R}}$ of $G_\mathbf{R}$. Let $w_N : \mathbf{G}_{\mathbf{m},\mathbf{R}} \to P/P_u$ be the weight map characterized by the property that, for $t \in \mathbf{G}_{\mathbf{m},\mathbf{R}}$, $w_N(t)$ acts on $\mathrm{gr}_k^{W(N)}$ as the multiplication by t^k for any k. Then (ρ, φ) is characterized by the following (4) and (5).

(4) The weight filtration associated with (ρ, φ) is $W(N)$.
(5) $\varphi(i) = \lim_{y\to\infty} w_N(\sqrt{y}) \circ \exp(iyN)F$, where $\circ$ is the Borel-Serre action with respect to P (5.1.3).

This characterization is essentially obtained in [Sc], but not so explicitly. We give a proof of this characterization in 6.1.12 below.

We will use the following three facts concerning the associated SL(2)-orbit in one variable.

(6) For $y \in \mathbf{R}$, $(N,\ \exp(iyN)F)$ also generates a nilpotent orbit, and the SL(2)-orbit associated with this pair is also (ρ, φ).
(7) For $g \in G_\mathbf{R}$, $(\mathrm{Ad}(g)N, gF)$ also generates a nilpotent orbit, and the SL(2)-orbit associated with this pair is $(\mathrm{Int}(g) \circ \rho,\ g\varphi)$.
(8) Let (ρ, φ) be an SL(2)-orbit in one variable of rank 1 and let $N = \rho_* \left(\begin{smallmatrix} 0 & 1 \\ 0 & 0 \end{smallmatrix}\right)$. Then, the SL(2)-orbit associated with $(N, \varphi(0))$ is (ρ, φ) itself.

6.1.2

R-*split mixed Hodge structure.* The point $\varphi(0) \in \check{D}$ of the SL(2)-orbit (ρ, φ) associated with (N, F) as above depends, in fact, only on the pair $(W(N), F)$. As is explained in [CKS], $\varphi(0)$ is the **R**-split mixed Hodge structure associated with the mixed Hodge structure $(W(N)[-w], F)$. Here, for an increasing filtration W and $l \in \mathbf{Z}$, we define the increasing filtration $W[l]$ by $W[l]_k := W_{k+l}$. We review the theory of the associated **R**-split mixed Hodge structure in [CKS, 2, 3].

Let V be a finite-dimensional **R**-vector space. Let (W, F) be a mixed Hodge structure on V, i.e., W is the weight filtration, which is an increasing filtration on V, and F is the Hodge filtration, which is a decreasing filtration on $V_\mathbf{C} := \mathbf{C} \otimes_\mathbf{R} V$ such that, for each integer k, F induces on gr_k^W an **R**-Hodge structure of weight k.

A *splitting* of (W, F) is a bigrading $V_{\mathbf{C}} = \bigoplus J^{p,q}$ such that

$$W_{k,\mathbf{C}} = \bigoplus_{p+q \le k} J^{p,q}, \quad F^p = \bigoplus_{r \ge p,\, q} J^{r,q}. \tag{1}$$

We say (W, F) *splits over* $\mathbf{R}$ if it admits a splitting $(J^{p,q})$ satisfying

$$\overline{J^{p,q}} = J^{q,p}, \tag{2}$$

(called an $\mathbf{R}$-*splitting*). This is equivalent to the existence of a decomposition $V = \bigoplus_{k \in \mathbf{Z}} S_k$, such that

$$W_k = \bigoplus_{l \le k} S_l \ \ (\forall k), \quad F^p = \bigoplus_k F^p \cap S_{k,\mathbf{C}} \ \ (\forall p). \tag{3}$$

If (W, F) splits over $\mathbf{R}$, an $\mathbf{R}$-splitting of (W, F) and (S_k) as in (3) are unique and are given by

$$J^{p,q} = F^p \cap \overline{F}^q \cap W_{p+q,\mathbf{C}}, \quad S_k = \Big(\bigoplus_{p+q=k} J^{p,q}\Big) \cap V,$$

respectively.

In [CKS, 3], with a mixed Hodge structure (W, F), the authors canonically associate a pair $(\hat{F}, \varepsilon)$ of a decreasing filtration $\hat{F} = (W, F)^{\wedge}$ on $V_{\mathbf{C}}$, such that $(W, \hat{F})$ is an $\mathbf{R}$-split mixed Hodge structure, and a nilpotent endomorphism $\varepsilon = \varepsilon(W, F)$ of $V_{\mathbf{C}}$, such that $F = \exp(\varepsilon)\hat{F}$. (In the notation of [CKS], the above $\hat{F}$ and $\exp(\varepsilon)$ are written as $\tilde{F}_0$ and $\exp(i\delta)\exp(-\zeta)$, respectively.) The following hold.

(4) Consider the case $V = H_{0,\mathbf{R}}$. Let (σ, Z) be a nilpotent orbit, let $W = W(\sigma)[-w]$, and let $F \in Z$. Then (W, F) is a mixed Hodge structure ([CKS]). Let $\hat{F} = (W, F)^{\wedge}$, let $H_{0,\mathbf{R}} = \bigoplus_k S_k$ be the decomposition given by the $\mathbf{R}$-splitting of $(W, \hat{F})$, and let ν be the action of $\mathbf{G}_{\mathbf{m},\mathbf{R}}$ on $H_{0,\mathbf{R}}$ given by $\nu(t)v = t^{k-w}v$ for $v \in S_k$. Then, for any element N of the interior (0.7.7) of σ, the SL(2)-orbit (ρ, φ) in one variable associated to the pair (N, F) satisfies

$$\varphi(0) = \hat{F}, \quad \tilde{\rho} = \nu \quad \text{(see 5.2.2)}.$$

From this we have

$$\mathrm{Ad}(\nu(t))N = t^{-2}N \quad (\forall\, N \in \sigma_{\mathbf{R}}).$$

This is because $N \in \sigma_{\mathbf{R}}$ is written as an $\mathbf{R}$-linear combination of a finite number of elements of the interior of σ.

(5) If $V = H_{0,\mathbf{R}}$ and W satisfies $\{x \in H_{0,\mathbf{R}} \mid \langle x, W_k\rangle_0 = 0\} = W_{2w-k-1}$ for all k, then $\varepsilon(W, F)$ belongs to $\mathfrak{g}_{\mathbf{C}}$.

(6) Fix W and let $\mathfrak{F}^W$ be the set of all decreasing filtrations F on $V_{\mathbf{C}}$ such that (W, F) is a mixed Hodge structure, and regard $\mathfrak{F}^W$ as a real manifold in the natural way. Then the map $F \mapsto \big((W, F)^{\wedge}, \varepsilon(W, F)\big)$ is a real analytic function on $\mathfrak{F}^W$.

The constructions of δ and ζ are given below (we will review the definition of δ precisely, but will refer to [CKS, 3, (6.60)] for the details of the construction of ζ). Then $(W, F)^\wedge$ is defined to be $\exp(-\varepsilon)F$, where $\varepsilon \in \mathfrak{g}_{\mathbf{C}}$ is the unique nilpotent element satisfying $\exp(\varepsilon) = \exp(i\delta)\exp(-\zeta)$. ε can also be written using the Hausdorff formula $\varepsilon = H(i\delta, -\zeta)$, where $H(x, y) = x + y + (1/2)[x, y] + \cdots$.

There is a unique splitting $(I^{p,q}) = (I^{p,q}(W, F))$ of (W, F) such that $\overline{I^{p,q}} \equiv I^{q,p} \mod \left(\bigoplus_{r<p, s<q} I^{r,s}\right)$ for any $p, q \in \mathbf{Z}$, which we call *Deligne splitting* (see [CKS, 2.13]). It is defined explicitly by

$$I^{p,q} := (F^p \cap W_{p+q,\mathbf{C}}) \cap \left(\overline{F}^q \cap W_{p+q,\mathbf{C}} + \sum_{j\geq 0} \overline{F}^{q-1-j} \cap W_{p+q-2-j,\mathbf{C}}\right).$$

Let

$$L^{-1,-1} = L^{-1,-1}(W, F) := \left\{h \in \mathrm{End}_{\mathbf{C}}(V_{\mathbf{C}}) \;\middle|\; h(I^{p,q}) \subset \bigoplus_{r<p, s<q} I^{r,s}\right\}.$$

Since $\overline{L^{-1,-1}} = L^{-1,-1}$,

$$L_{\mathbf{R}}^{-1,-1} := L^{-1,-1} \cap \mathrm{End}_{\mathbf{R}}(V)$$

is a real form of $L^{-1,-1}$.

It can be shown [CKS (2.20)] that there exists a unique $\delta \in L_{\mathbf{R}}^{-1,-1}$ such that $(W, \exp(-i\delta)F)$ is an $\mathbf{R}$-split mixed Hodge structure. This is the definition of δ.

Let $\tilde{I}^{p,q} = I^{p,q}(W, \exp(-i\delta)F)$ (= the $\mathbf{R}$-splitting of $\exp(-i\delta)F$), and let $(\delta_{p,q})$ be (p, q) components of δ with respect to $(\tilde{I}^{p,q})$, i.e.,

$$\delta = \sum_{p,q\in\mathbf{Z}} \delta_{p,q}, \quad \delta_{p,q}(\tilde{I}^{r,s}) \subset \tilde{I}^{r+p,s+q} \tag{7}$$

for any r, s. Then, ζ is defined as a Lie polynomial in the $\delta_{p,q}$ as in [CKS, (6.60)]. By [CKS, (3.25), (3.26)],

$$\zeta \in L^{-1,-1}(W, F).$$

For small p, q, the (p, q) components of ζ with respect to $\tilde{I}^{p,q}$ are calculated as follows. (Note that $\delta_{p,q} = \zeta_{p,q} = 0$ unless $p, q < 0$.)

$$\begin{aligned}
\zeta_{-1,-1} &= 0, \\
\zeta_{-2,-1} &= \frac{i}{2}\delta_{-2,-1}, \\
\zeta_{-1,-2} &= -\frac{i}{2}\delta_{-1,-2}, \\
\zeta_{-3,-1} &= \frac{3i}{4}\delta_{-3,-1}, \\
\zeta_{-2,-2} &= 0, \\
\zeta_{-1,-3} &= -\frac{3i}{4}\delta_{-1,-3}.
\end{aligned} \tag{8}$$

The following (9) is shown in [CKS, 2.20, 3]:

(9) $\delta h = h\delta$, $\zeta h = h\zeta$ and $\varepsilon h = h\varepsilon$ for any $\mathbf{C}$-homomorphism $h : V_{\mathbf{C}} \to V_{\mathbf{C}}$ defined over $\mathbf{R}$, such that for some $r \in \mathbf{Z}$, $h(F^p) \subset F^{p+r}$ for any p and $h(W_k) \subset W_{k+2r}$ for any k.

We have

(10) If (W, F) is an $\mathbf{R}$-split mixed Hodge structure, then $\varepsilon(W, F) = 0$.

In fact, we have $\delta = 0$ and this shows that $\zeta = 0$ since ζ is a Lie polynomial in $\delta_{p,q}$.

From (10) and 6.1.1 (8), we have

(11) The map $(\rho, \varphi) \mapsto \left(\rho_* \left(\begin{smallmatrix} 0 & 1 \\ 0 & 0 \end{smallmatrix}\right), \varphi(0)\right)$ gives a bijection from the set of all SL(2)-orbits in one variable to the set of all pairs (N, F) of a nilpotent element N of $\mathfrak{g}_{\mathbf{R}}$ and an element F of $\check{D}$ which generate a nilpotent orbit such that F is $\mathbf{R}$-split with respect to the weight filtration $W(N)[-w]$. The inverse map is to take the associated SL(2)-orbit.

6.1.3

The associated SL(2)*-orbit in several variables.* Let $N_1, \ldots, N_n \in \mathfrak{g}_{\mathbf{R}}$ be mutually commutative nilpotent elements, let $F \in \check{D}$, and assume that $(N_1, \ldots, N_n, F)$ generates a nilpotent orbit (5.4.1). We review the definition of the SL(2)-orbit in n variables associated with $(N_1, \ldots, N_n, F)$ in [CKS, 4].

For $1 \leq j \leq n$, put $\sigma_j = (\mathbf{R}_{\geq 0})N_1 + \cdots + (\mathbf{R}_{\geq 0})N_j$ and let $W(\sigma_j)$ be the associated weight filtration.

Define an $\mathbf{R}$-split mixed Hodge structure $(W(\sigma_j)[-w], \hat{F}_{(j)})$ for $j = 1, \ldots, n$ as follows. Since $(N_1, \ldots, N_n, F)$ generates a nilpotent orbit, $(W(\sigma_n)[-w], F)$ is a mixed Hodge structure. Let

$$\hat{F}_{(n)} := (W(\sigma_n)[-w], F)^{\wedge} \tag{1}$$

be the associated $\mathbf{R}$-split mixed Hodge structure. Then $\left(N_1, \ldots, N_{n-1}, \exp(iN_n)\hat{F}_{(n)}\right)$ generates a nilpotent orbit. Hence $\left(W(\sigma_{n-1})[-w], \exp(iN_n)\hat{F}_{(n)}\right)$ is a mixed Hodge structure. Let

$$\hat{F}_{(n-1)} := \left(W(\sigma_{n-1})[-w], \exp(iN_n)\hat{F}_{(n)}\right)^{\wedge} \tag{2}$$

be the associated $\mathbf{R}$-split mixed Hodge structure. $\left(N_1, \ldots, N_{n-2}, \exp(iN_{n-1})\hat{F}_{(n-1)}\right)$ generates a nilpotent orbit. Continuing this process, we obtain $\hat{F}_{(n)}, \ldots, \hat{F}_{(1)} \in \check{D}$.

It is proved in [CKS, (4.20)] that there is a unique SL(2)-orbit (ρ, φ) in n variables satisfying the following (3) and (4):

$$\varphi(\mathbf{0}_j, \mathbf{i}_{n-j}) = \hat{F}_{(j)} \quad \text{for } 1 \leq j \leq n. \tag{3}$$

Here $\mathbf{0}_j := (0, \ldots, 0) \in \mathbf{C}^j$ and $\mathbf{i}_k := (i, \ldots, i) \in \mathfrak{h}^k \subset \mathbf{C}^k$.

(4) Let $1 \leq j \leq n$, and let $\hat{N}_j$ be the image of the element $\left(\begin{smallmatrix} 0 & 1 \\ 0 & 0 \end{smallmatrix}\right)$ of the jth factor of $\mathfrak{sl}(2, \mathbf{C})^{\oplus n}$ under the homomorphism $\rho_* : \mathfrak{sl}(2, \mathbf{C})^{\oplus n} \to \mathfrak{g}_{\mathbf{C}}$ associated with ρ. Consider the decomposition of $\mathfrak{g}_{\mathbf{R}}$ by the action of $\mathbf{G}_{\mathbf{m},\mathbf{R}}^{j-1}$ under $(t_1, \ldots, t_{j-1}) \mapsto \mathrm{Ad}(\rho(\Delta(t_1, \ldots, t_{j-1}, 1, \ldots, 1)))$ (see 5.2.2). Then $\hat{N}_j$ is the component of N_j on which $\mathbf{G}_{\mathbf{m},\mathbf{R}}^{j-1}$ acts trivially. In particular, $\hat{N}_1 = N_1$.

We call (ρ, φ) the SL(2)-orbit in n variables associated with $(N_1, \ldots, N_n, F)$. Concerning (ρ, φ), we have

(5) For $1 \le j \le n$, $\tilde{\rho}_j$ (5.2.2 (2)) is determined in the following way: If $H_{0,\mathbf{R}} = \bigoplus_k S_k$ denotes the decomposition given by the $\mathbf{R}$-splitting of $(W(\sigma_j)[-w], \hat{F}_{(j)})$ (6.1.2 (3)), then $\tilde{\rho}_j(t)$ acts on S_k as the multiplication by t^{k-w}.

(6) The family of weight filtrations of (ρ, φ) in 5.2.5 is $(W(\sigma_j))_{1 \le j \le n}$.

Let $\{s_1, \ldots, s_m\}$ $(1 \le s_1 < \cdots < s_m \le n)$ be the set of all j such that $1 \le j \le n$ and $W(\sigma_j) \ne W(\sigma_{j-1})$, where $\sigma_0 = \{0\}$. By 5.2.3 and 5.2.5, we have the following:

(7) (ρ, φ) is of rank m, and the SL(2)-orbit in m variables associated to (ρ, φ) is the $(s_1, \ldots, s_m)$ component of (ρ, φ).

6.1.4

Example. We consider the case $D = \mathfrak{h}_2$, the Siegel upper half space of genus 2. We assume $h^{1,0} = h^{0,1} = 2$, $h^{p,q} = 0$ for $(p, q) \ne (1, 0), (0, 1)$, and there is a $\mathbf{Z}$-basis $(e_j)_{1 \le j \le 4}$ of H_0 such that $\langle e_j, e_k \rangle_0 = 1$ (resp. -1) if $j - k = 2$ (resp. $k - j = 2$), and $\langle e_j, e_k \rangle_0 = 0$ otherwise.

For a symmetric complex matrix

$$A = \begin{pmatrix} a & c \\ c & b \end{pmatrix} \quad (a, b, c \in \mathbf{C}),$$

there corresponds a point $F \in \check{D}$ defined by $F^0 = H_{0,\mathbf{C}}$, $F^2 = 0$, and

$$F^1 := \{ae_1 + ce_2 + e_3,\ ce_1 + be_2 + e_4\}_{\mathbf{C}}.$$

Here $\{x, \ldots, y\}_{\mathbf{C}}$ is the $\mathbf{C}$-vector subspace of $H_{0,\mathbf{C}}$ generated by $x, \ldots, y$. It is well known that $F \in D$ if and only if the imaginary part $\mathrm{Im}(A)$ is positive definite.

Let $N_1, N_2 \in \mathfrak{g}_{\mathbf{Q}}$ be

$$N_1(e_3) = e_1,\ N_1(e_j) = 0\ (j \ne 3); \quad N_2(e_4) = e_2,\ N_2(e_j) = 0\ (j \ne 4).$$

Then, for any symmetric complex matrix A, (N_1, N_2, F) generates a nilpotent orbit. This is because $\exp(z_1 N_1 + z_2 N_2)F$ $(z_1, z_2 \in \mathbf{C})$ corresponds to the matrix $A + \left(\begin{smallmatrix} z_1 & 0 \\ 0 & z_2 \end{smallmatrix}\right)$ and the imaginary part of this matrix becomes positive definite if $\mathrm{Im}(z_1)$ and $\mathrm{Im}(z_2)$ are sufficiently large, and because the small Griffiths transversality is trivially satisfied.

The weight filtration $W(\sigma_2) = W(N_1 + N_2)$ is given by

$$W(\sigma_2)_{-2} = 0, \quad W(\sigma_2)_1 = H_{0,\mathbf{R}},$$
$$W(\sigma_2)_{-1} = W(\sigma_2)_0 = \mathrm{Image}(N_1 + N_2) = \mathrm{Ker}(N_1 + N_2) = \{e_1, e_2\}_{\mathbf{R}}.$$

Deligne splitting $I_{(2)}^{p,q} := I^{p,q}(W(\sigma_2)[-1], F)$ becomes

$$I_{(2)}^{1,1} = F^1 \cap (\overline{F}^1 + W(\sigma_2)_{-1,\mathbf{C}}) = F^1, \quad I_{(2)}^{0,0} = W(\sigma_2)_{-1,\mathbf{C}},$$
$$I_{(2)}^{p,q} = 0 \quad \text{for } (p, q) \ne (1, 1), (0, 0).$$

Hence the mixed Hodge structure $(W(\sigma_2), F)$ is $\mathbf{R}$-split if and only if $a, b, c \in \mathbf{R}$.

Put $a = a_1 + ia_2$, $b = b_1 + ib_2$, and $c = c_1 + ic_2$, where $a_j, b_j, c_j \in \mathbf{R}$. The nilpotent element δ in 6.1.2 is given by

$$\delta(e_1) = \delta(e_2) = 0, \quad \delta(e_3) = a_2e_1 + c_2e_2, \quad \delta(e_4) = c_2e_1 + b_2e_2.$$

The nilpotent element ζ in 6.1.2 (8) is in the present case $\zeta = \zeta_{-1,-1} = 0$. Hence, the $\mathbf{R}$-split mixed Hodge structure $(W(\sigma_2)[-1], \hat{F}_{(2)})$ associated with $(W(\sigma_2)[-1], F)$ is given by

$$\hat{F}^1_{(2)} = \exp(-i\delta)F^1 = \{a_1e_1 + c_1e_2 + e_3,\ c_1e_1 + b_1e_2 + e_4\}_{\mathbf{C}}.$$

Deligne splitting $\hat{I}^{p,q}_{(2)} := I^{p,q}(W(\sigma_2)[-1], \hat{F}_{(2)}) = \exp(-i\delta)I^{p,q}(W(\sigma_2)[-1], F)$ (see 6.1.8 (ii) below for the last equality) is

$$\hat{I}^{1,1}_{(2)} = \hat{F}^1_{(2)}, \quad \hat{I}^{0,0}_{(2)} = W(\sigma_2)_{0,\mathbf{C}},$$
$$\hat{I}^{p,q}_{(2)} = 0 \quad \text{for } (p,q) \neq (1,1), (0,0).$$

The weight filtration $W(\sigma_1) = W(N_1)$ is given by

$$W(\sigma_1)_{-2} = 0, \quad W(\sigma_1)_{-1} = \mathrm{Image}(N_1) = \{e_1\}_{\mathbf{R}},$$
$$W(\sigma_1)_0 = \mathrm{Ker}(N_1) = \{e_1, e_2, e_4\}_{\mathbf{R}}, \quad W(\sigma_1)_1 = H_{0,\mathbf{R}}.$$

Put $F_{(1)} := \exp(iN_2)\hat{F}_{(2)}$. Deligne splitting $I^{p,q}_{(1)} := I^{p,q}(W(\sigma_1)[-1], F_{(1)})$ becomes

$$I^{1,1}_{(1)} = F^1_{(1)} \cap (\overline{F}^1_{(1)} + W(\sigma_1)_{-1,\mathbf{C}}) = \{a_1e_1 + c_1e_2 + e_3\}_{\mathbf{C}},$$
$$I^{1,0}_{(1)} = F^1_{(1)} \cap W(\sigma_1)_{0,\mathbf{C}} = \{c_1e_1 + (b_1 + i)e_2 + e_4\}_{\mathbf{C}},$$
$$I^{0,1}_{(1)} = \overline{F}^1_{(1)} \cap W(\sigma_1)_{0,\mathbf{C}} = \{c_1e_1 + (b_1 - i)e_2 + e_4\}_{\mathbf{C}},$$
$$I^{0,0}_{(1)} = W(\sigma_1)_{-1,\mathbf{C}},$$
$$I^{p,q}_{(1)} = 0 \quad \text{otherwise}.$$

Hence the mixed Hodge structure $(W(\sigma_1)[-1], F_{(1)})$ splits over $\mathbf{R}$. Therefore

$$\hat{F}_{(1)} = F_{(1)} = \exp(iN_2)\hat{F}_{(2)}.$$

From the above descriptions of $I^{p,q}(W(\sigma_j)[-1], \hat{F}_{(j)})$ $(j = 1, 2)$ and 6.1.3 (5), we see that $\rho(\Delta(t_1, t_2))$ for $t_1, t_2 \in \mathbf{G}_{\mathbf{m},\mathbf{R}} = \mathbf{R}^\times$ is given by

$$\rho(\Delta(t_1, t_2))(a_1e_1 + c_1e_2 + e_3) = t_1(a_1e_1 + c_1e_2 + e_3),$$
$$\rho(\Delta(t_1, t_2))(c_1e_1 + b_1e_2 + e_4) = t_2(c_1e_1 + b_1e_2 + e_4)$$
$$\rho(\Delta(t_1, t_2))(e_j) = t_j^{-1}e_j \quad \text{for } j = 1, 2.$$

Since $\mathrm{Ad}(\rho(\Delta(t_1, 1)))(N_2) = N_2$, we have $\hat{N}_2 = N_2$ by 6.1.3 (4). In conclusion, the SL(2)-orbit (ρ, φ) in two variables associated with (N_1, N_2, F) is described as follows.

(1) Let $e'_3 = a_1e_1 + c_1e_2 + e_3$, $e'_4 = c_1e_1 + b_1e_2 + e_4$. Then, for $g_1, g_2 \in \mathrm{SL}(2, \mathbf{C})$, $\rho(g_1, g_2)$ acts on $\{e_1, e'_3\}_{\mathbf{C}}$ as g_1 with respect to the basis (e_1, e'_3), and acts on $\{e_2, e'_4\}_{\mathbf{C}}$ as g_2 with respect to the basis (e_2, e'_4).

(2) For $z_1, z_2 \in \mathbf{C}$, $\varphi(z_1, z_2)$ is the element of $\check{D}$ corresponding to the symmetric matrix

$$\begin{pmatrix} a_1 + z_1 & c_1 \\ c_1 & b_1 + z_2 \end{pmatrix}.$$

6.1.5

Let $(N_1, \ldots, N_n, F)$ be as in 6.1.3, and let (ρ, φ) be the associated SL(2)-orbit in n variables. In [CKS, (4.20)], it is proved that there are $a_h, b_h \in \mathfrak{g}_{\mathbf{R}}$ $(h \in \mathbf{N}^n)$ such that $\sum_h a_h \prod_{1\le j\le n} t_j^{h_j}$ and $\sum_h b_h \prod_{1\le j\le n} t_j^{h_j}$ $(t_j \in \mathbf{C})$ absolutely converge when the $|t_j|$ are sufficiently small, such that $a_0 = b_0 = 0$, and such that

$$\begin{aligned}
&\exp\Big(\sum_{1\le j\le n} iy_jN_j\Big)F \\
&= \exp\Big(\sum_{h\in\mathbf{N}^n} a_h \prod_{1\le j\le n}\Big(\frac{y_j}{y_{j+1}}\Big)^{-h_j}\Big)\varphi(iy_1, \ldots, iy_n) \\
&= \exp\Big(\sum_{h\in\mathbf{N}^n} a_h \prod_{1\le j\le n}\Big(\frac{y_j}{y_{j+1}}\Big)^{-h_j}\Big)\rho(\Delta(\sqrt{y_1}, \ldots, \sqrt{y_n}))^{-1}\cdot \mathbf{r} \\
&= \rho(\Delta(\sqrt{y_1}, \ldots, \sqrt{y_n}))^{-1}\exp\Big(\sum_{h\in\mathbf{N}^n} b_h \prod_{1\le j\le n}\sqrt{\frac{y_j}{y_{j+1}}}^{\,-h_j}\Big)\cdot \mathbf{r}
\end{aligned}$$

when $\frac{y_j}{y_{j+1}} \gg 0$ $(1 \le j \le n)$, where $y_{n+1} := 1$, $\mathbf{r} := \varphi(\mathbf{i})$.

6.1.6

It is shown in [CKS] that a_h, b_h, and $\mathbf{r}$ in 6.1.5 are real analytic functions in $(N_1, \ldots, N_n, F)$ if $W(N_1 + \cdots + N_j)$ $(1 \le j \le n)$ are fixed. More precisely, fix a family of increasing filtrations $W^{(j)}$ $(1 \le j \le n)$ on $H_{0,\mathbf{R}}$, and let S be the subset of $\mathfrak{g}_{\mathbf{R}}^n \times \check{D}$ consisting of all elements $(N_1, \ldots, N_n, F)$ which generate nilpotent orbits and which satisfy $W(N_1 + \cdots + N_j) = W^{(j)}$ for $1 \le j \le n$. Then, for each $s \in S$ there exist an open neighborhood U of s in $\mathfrak{g}_{\mathbf{R}}^n \times \check{D}$ and real analytic functions $a_h, b_h : U \to \mathfrak{g}_{\mathbf{R}}$ $(h \in \mathbf{N}^n)$ and $\mathbf{r} : U \to D$ which give a_h, b_h, and $\mathbf{r}$ for any $(N_1, \ldots, N_n, F) \in S \cap U$, respectively.

In the rest of Section 6.1, we prove results concerning 6.1.2 and 6.1.3 that will be used later, and give the proof of the characterization of the associated SL(2)-orbit in one variable by (4) and (5) in 6.1.1.

Proposition 6.1.7 *Let (N, F) be as in* 6.1.1, *and let $W = W(N)[-w]$. Then, for $z \in \mathbf{C}$, we have*

(1) $(W, \exp(zN)F)^\wedge = \exp(\mathrm{Re}(z)N)(W, F)^\wedge$,
(2) $\varepsilon(W, \exp(zN)F) = \varepsilon(W, F) + i\,\mathrm{Im}(z)N$.

We use the following lemma for the proof of 6.1.7.

Lemma 6.1.8 *Let (W, F) be as in 6.1.2.*

(i) *If $h \in L^{-1,-1}(W, F)$, then $(W, \exp(h)F)$ is a mixed Hodge structure.*
(ii) *If $h \in L^{-1,-1}(W, F)$, we have*

$$I^{p,q}(W, \exp(h)F) = \exp(h)I^{p,q}(W, F), \quad L^{-1,-1}(W, \exp(h)F) = L^{-1,-1}(W, F).$$

(iii) *If $h_1, h_2 \in L^{-1,-1}(W, F)$ and $\exp(h_1)F = \exp(h_2)F$, then $h_1 = h_2$.*
(iv) *If $h \in \mathrm{End}_{\mathbf{C}}(V_{\mathbf{C}})$ and h satisfies*

$$hF^p \subset F^{p-1}, \quad h\overline{F}^p \subset \overline{F}^{p-1} \quad (\forall\, p \in \mathbf{Z}), \quad hW_{k,\mathbf{C}} \subset W_{k-2,\mathbf{C}} \ \ (\forall\, k \in \mathbf{Z}),$$

then $h \in L^{-1,-1}(W, F)$.

Proof.

(i) is easy. Put $I^{p,q} := I^{p,q}(W, F)$, $A := I^{p,q}$, $B := \overline{I^{q,p}}$, $C := \sum_{r<p,\, s<q} I^{r,s}$, and $D := \sum_{r<p,\, s<q} I^{r,s}(W, \exp(h)F)$.

We prove the first equation of (ii). By the characterizing property of Deligne splitting $I^{p,q}$ in 6.1.2, we have $A + C = B + C$, and it is enough to show that

$$\exp(h)A + D = \exp(\overline{h})B + D. \tag{1}$$

Since $h, \overline{h} \in L^{-1,-1}(W, F)$, we have

$$\exp(h)C = \exp(\overline{h})C = C = D, \tag{2}$$

$$\exp(h)(A + C) = A + C, \quad \exp(\overline{h})(A + C) = A + C. \tag{3}$$

By (2) and the second equation of (3), we have

$$\exp(\overline{h})B + D = \exp(\overline{h})(B + C) = \exp(\overline{h})(A + C) = A + C. \tag{4}$$

(1) follows from (4), the first equation of (3) and (2).

The second equation of (ii) follows from the first one of (ii) and (2).

Under the assumption of (iii), $\exp(-h_2)\exp(h_1)$ preserves $I^{p,q}$ by (ii). This and $(\mathrm{id} - \exp(-h_2)\exp(h_1))(I^{p,q}) \subset \bigoplus_{r<p,\, s<q} I^{r,s}$ show that $\exp(h_1) = \exp(h_2)$, and this shows that $h_1 = h_2$ because h_1, h_2 are nilpotent.

Finally (iv) follows from the definition of $I^{p,q}$ in 6.1.2. □

6.1.9

Proof of Proposition 6.1.7. We prove 6.1.7 (1). Let (ρ, φ) be the SL(2)-orbit associated with (N, F). Then, by 6.1.1 (6) and 6.1.1 (7), the SL(2)-orbit (ρ', φ') associated with $(N, \exp(zN)F)$ is given by $\rho' = \mathrm{Int}(\exp(\mathrm{Re}(z)N)) \circ \rho$, $\varphi' = \exp(\mathrm{Re}(z)N)\varphi$. Hence $(W, \exp(zN)F)^\wedge = \varphi'(0) = \exp(\mathrm{Re}(z)N)\varphi(0) = \exp(\mathrm{Re}(z)N)(W, F)^\wedge$.

We prove 6.1.7 (2). By 6.1.7 (1) and 6.1.2 (9), we have

$$\begin{aligned} \exp(-\varepsilon(W, \exp(zN)F) + zN)F &= (W, \exp(zN)F)^\wedge \\ &= \exp(\mathrm{Re}(z)N)(W, F)^\wedge \\ &= \exp(-\varepsilon(W, F) + \mathrm{Re}(z)N)F. \end{aligned} \tag{1}$$

We have $N \in L^{-1,-1}(W, F)$ by 6.1.8 (iv), and, by 6.1.8 (ii),

$$\varepsilon(W, F), \varepsilon(W, \exp(zN)F) \in L^{-1,-1}(W, F).$$

Hence (1) shows $-\varepsilon(W, \exp(zN)F) + zN = -\varepsilon(W, F) + \mathrm{Re}(z)N$ by 6.1.8 (iii), and 6.1.7 (2) follows. □

Lemma 6.1.10 *Let* $(N_1, \dots, N_n, F)$ *be as in* 6.1.3, *and let* (ρ, φ) *be the associated* SL(2)*-orbit in n variables. Then*

$$\mathrm{Ad}(\tilde{\rho}_j(t))(N_k) = t^{-2}N_k \quad \textit{for } 1 \le k \le j \le n.$$

Proof. This follows from 6.1.2 (4) applied to the nilpotent orbit associated with $(W(\sigma_j)[-w], \hat{F}_{(j)})$ and from 6.1.3 (5). □

Lemma 6.1.11 *Let V be as in* 6.1.2, *W an increasing filtration on V, and F and* F_λ $(\lambda \in \Lambda, \Lambda$ *a directed ordered set) decreasing filtrations on* $V_{\mathbf{C}}$ *such that* (W, F) *and* (W, F_λ) *are mixed Hodge structures. Assume* $F_\lambda \to F$ *in the topological set of decreasing filtrations on* $V_{\mathbf{C}}$.

(i) $F_\lambda(W_{k,\mathbf{C}}) \to F(W_{k,\mathbf{C}})$, $F_\lambda(\mathrm{gr}^W_{k,\mathbf{C}}) \to F(\mathrm{gr}^W_{k,\mathbf{C}})$ *for each* $k \in \mathbf{Z}$.

(ii) *Let* $\hat{F} = (W, F)^\wedge$, *let* $V = \bigoplus_k S_k$ *be the decomposition given by the* $\mathbf{R}$*-splitting of* $\hat{F}$ (6.1.2 (3)) *and let* $\nu : \mathbf{G}_{\mathbf{m},\mathbf{R}} \to \mathrm{Aut}(V)$ *be a homomorphism such that there is an integer r for which*

$$\nu(t)v = t^{k+r}v \quad (\forall k \in \mathbf{Z}, \forall v \in S_k).$$

Then, if $t_\lambda \in \mathbf{R}_{>0}$ *and* $t_\lambda \to \infty$, *we have* $\nu(t_\lambda)F_\lambda \to \hat{F}$.

Here $F_\lambda \to F$ means that, for each $p \in \mathbf{Z}$, there is a basis $(e_j)_{1 \le j \le m}$ of F^p and bases $(e_{\lambda,j})_{1 \le j \le m}$ of F^p_λ given for sufficiently large λ such that $e_{\lambda,j} \to e_j$ for any j.

Proof of Lemma 6.1.11. We prove (i) by downward induction on k. By

$$F^p_\lambda(W_{k,\mathbf{C}}) = \mathrm{Ker}\left(F^p_\lambda(W_{k+1,\mathbf{C}}) \to F^p_\lambda(\mathrm{gr}^W_{k+1,\mathbf{C}})\right),$$
$$F^p(W_{k,\mathbf{C}}) = \mathrm{Ker}\left(F^p(W_{k+1,\mathbf{C}}) \to F^p(\mathrm{gr}^W_{k+1,\mathbf{C}})\right),$$

and by the hypothesis of induction, we have $F^p_\lambda(W_{k,\mathbf{C}}) \to F^p(W_{k,\mathbf{C}})$. It remains to prove that $F^p_\lambda(\mathrm{gr}^W_{k,\mathbf{C}}) \to F^p(\mathrm{gr}^W_{k,\mathbf{C}})$. Since $F_\lambda(\mathrm{gr}^W_k)$ is an $\mathbf{R}$-Hodge structure of weight k,

$$F^p_\lambda(\mathrm{gr}^W_{k,\mathbf{C}}) \oplus \overline{F}^{k+1-p}_\lambda(\mathrm{gr}^W_{k,\mathbf{C}}) = \mathrm{gr}^W_{k,\mathbf{C}} \quad (\forall\, \lambda). \tag{1}$$

Let $(e_j)_{j \in J}$ (resp. $(e'_j)_{j \in J'}$) be a family of elements of $F^p(W_{k,\mathbf{C}})$ (resp. $F^{k+1-p}(W_{k,\mathbf{C}})$) whose image via

$$\pi : F^p(W_{k,\mathbf{C}}) \to F^p(\mathrm{gr}^W_{k,\mathbf{C}}) \quad (\text{resp. } \pi' : F^{k+1-p}(W_{k,\mathbf{C}}) \to F^{k+1-p}(\mathrm{gr}^W_{k,\mathbf{C}}))$$

is a basis of the target. Then, since $F_\lambda(W_{k,\mathbf{C}})$ converges to $F(W_{k,\mathbf{C}})$, there exist $e_{\lambda,j} \in F^p_\lambda(W_{k,\mathbf{C}})$ $(j \in J)$ (resp. $e'_{\lambda,j} \in F^{k+1-p}_\lambda(W_{k,\mathbf{C}})$ $(j \in J')$) such that $e_{\lambda,j}$ (resp.

$e'_{\lambda,j}$) converges to e_j $(j \in J)$ (resp. e'_j $(j \in J')$). It follows by (1) that the image $(\pi(e_{\lambda,j}))_{j\in J}$ (resp. $(\pi'(e'_{\lambda,j}))_{j\in J'}$) is a basis of $F^p_\lambda(\mathrm{gr}^W_{k,\mathbf{C}})$ (resp. $F^{k+1-p}_\lambda(\mathrm{gr}^W_k)$) if λ is sufficiently large. This proves (i).

We prove (ii). Fix $p \in \mathbf{Z}$. By (i), there exists a basis $(e_{k,j})_{k\in\mathbf{Z},j\in J_k}$ of F^p and bases $(e_{\lambda,k,j})_{k\in\mathbf{Z},j\in J_k}$ of F^p_λ given for sufficiently large λ such that $e_{k,j}, e_{\lambda,k,j} \in W_{k,\mathbf{C}}$, the image of $(e_{k,j})_{j\in J_k}$ (resp. $(e_{\lambda,k,j})_{j\in J_k}$) in $\mathrm{gr}^W_{k,\mathbf{C}}$ is a basis of $F^p(\mathrm{gr}^W_{k,\mathbf{C}})$ (resp. $F^p_\lambda(\mathrm{gr}^W_{k,\mathbf{C}})$), and $e_{\lambda,k,j}$ converges to $e_{k,j}$ for each k, j. For an element $v \in W_{k,\mathbf{C}}$, write $v = v' + v''$ with $v' \in S_{k,\mathbf{C}}$, $v'' \in W_{k-1,\mathbf{C}}$. Then, $(e'_{k,j})_{k\in\mathbf{Z},j\in J_k}$ is a basis of $\hat{F}^p$. We have

$$t_\lambda^{-k-r}\nu(t_\lambda)e_{\lambda,k,j} = e'_{\lambda,k,j} + t_\lambda^{-k-r}\nu(t_\lambda)e''_{\lambda,k,j} \quad \text{with} \quad t_\lambda^{-k-r}\nu(t_\lambda)e''_{\lambda,k,j} \to 0. \qquad (2)$$

Since $e_{\lambda,k,j} \to e_{k,j}$, we have $e'_{\lambda,k,j} \to e'_{k,j}$ and hence (2) shows $t_\lambda^{-k-r}\nu(t_\lambda)e_{\lambda,k,j} \to e'_{k,j}$. This proves $\nu(t_\lambda)F_\lambda \to \hat{F}$. □

6.1.12

Here we prove that the SL(2)-orbit (ρ, φ) in one variable associated with the pair (N, F) in 6.1.1 is characterized by the properties 6.1.1 (4) and 6.1.1 (5).

If we prove that (ρ, φ) satisfies 6.1.1 (4), 6.1.1 (5), the uniqueness of such (ρ, φ) follows from 5.2.10. By 6.1.1 (1), (ρ, φ) satisfies 6.1.1 (4).

We prove that it satisfies 6.1.1 (5). Let $(b_n)_{n\geq 1}$ be as in 6.1.1 (3). Let $P = (G^\circ)_{W(N),\mathbf{R}}$ be the parabolic subgroup of $G_\mathbf{R}$ in 6.1.1. Put $\mathbf{r} = \varphi(i)$.

For an element v of $\mathfrak{g}_\mathbf{R}$ of $(\mathrm{Ad}\circ\tilde{\rho})$-weight e $(e \in \mathbf{Z})$, $v \in \mathrm{Lie}(P)$ if $e \leq 0$, and, in the case $e \geq 0$, v is written as $-\theta_{K_\mathbf{r}}(v) + (\theta_{K_\mathbf{r}}(v) + v)$, where $-\theta_{K_\mathbf{r}}(v)$ is an element of $\mathrm{Lie}(P)$ of $(\mathrm{Ad}\circ\tilde{\rho})$-weight $-e$ and $v + \theta_{K_\mathbf{r}}(v) \in \mathrm{Lie}(K_\mathbf{r})$. From this, by the method of [KU2, Lemma 4.4], we obtain

$$\exp\left(\sum_{n\geq 1} b_n\sqrt{y}^{-n}\right) = \exp\left(\sum_{n\geq 1} p_n\sqrt{y}^{-n}\right)\exp\left(\sum_{n\geq 1} k_n\sqrt{y}^{-n}\right),$$

where $p_n \in \mathrm{Lie}(P)$, $k_n \in \mathrm{Lie}(K_\mathbf{r})$, $\sum_{n\geq 1} p_n t^n$ and $\sum_{n\geq 1} k_n t^n$ converge to elements of P and of $K_\mathbf{r}$, respectively, when $|t|$ $(t \in \mathbf{C})$ is sufficiently small, and the component $p_n(-e)$ of p_n of $(\mathrm{Ad}\circ\tilde{\rho})$-weight $-e$ $(e \geq 0)$ is zero unless $e < n$.

Hence, from 6.1.1 (3), we have

$$\begin{aligned}
&w_N(\sqrt{y})\circ\exp(iyN)F\\
&= w_N(\sqrt{y})\circ\tilde{\rho}(\sqrt{y})^{-1}\exp\left(\sum_{n\geq 1} p_n\sqrt{y}^{-n}\right)\exp\left(\sum_{n\geq 1} k_n\sqrt{y}^{-n}\right)\cdot\mathbf{r}\\
&= \tilde{\rho}(\sqrt{y})^{-1}\exp\left(\sum_{n\geq 1} p_n\sqrt{y}^{-n}\right)\left(w_N(\sqrt{y})\circ\exp\left(\sum_{n\geq 1} k_n\sqrt{y}^{-n}\right)\cdot\mathbf{r}\right)\\
&= \tilde{\rho}(\sqrt{y})^{-1}\exp\left(\sum_{n\geq 1} p_n\sqrt{y}^{-n}\right)\tilde{\rho}(\sqrt{y})\exp\left(\sum_{n\geq 1} k_n\sqrt{y}^{-n}\right)\cdot\mathbf{r}
\end{aligned}$$

$$= \exp\Big(\sum_{n\ge 1,\, e\ge 0} p_n(-e)\sqrt{y}^{\,e-n}\Big)\exp\Big(\sum_{n\ge 1} k_n\sqrt{y}^{\,-n}\Big)\cdot \mathbf{r}$$

$$\to \mathbf{r} \quad \text{when } y\to\infty,$$

where the second equation follows from 5.1.3 (3), the third from 5.2.9. □

6.2 PROOF OF THEOREM 5.4.2

In this section, we will deduce Theorem 5.4.2 from Proposition 6.2.2 below, which is a result for the weights of the coefficients in the Taylor expansion appearing in SL(2)-orbit theorem. The proof of Proposition 6.2.2 will be given in 6.2.4–6.2.6.

6.2.1

Let $N_j \in \mathfrak{g}_{\mathbf{R}}$ $(1\le j\le n)$ be mutually commuting nilpotent elements, let $F\in\check{D}$, and assume that $(N_1,\dots,N_n,F)$ generates a nilpotent orbit. Let (ρ,φ) be the associated SL(2)-orbit in n variables and let $\mathbf{r}:=\varphi(\mathbf{i})$. Then, from 6.1.5, we see that there are $c_h\in\mathfrak{g}_{\mathbf{R}}^-$, $k_h\in\mathfrak{g}_{\mathbf{R}}^+$ $(h\in\mathbf{N}^n)$, where $\mathfrak{g}_{\mathbf{R}}^{\pm}=\mathfrak{g}_{\mathbf{R}}^{\pm,\mathbf{r}}$ are the (±1)-eigensubspaces of $\mathfrak{g}_{\mathbf{R}}$ under the Cartan involution associated with $K_{\mathbf{r}}$, respectively, such that $\sum_{h\in\mathbf{N}^n} c_h\prod_{1\le j\le n} t_j^{h_j}$ and $\sum_{h\in\mathbf{N}^n} k_h\prod_{1\le j\le n} t_j^{h_j}$ $(t_j\in\mathbf{C})$ converge when the $|t_j|$ are sufficiently small, that $c_0=k_0=0$ and that, for $\frac{y_j}{y_{j+1}}\gg 0$ $(1\le j\le n$, $y_{n+1}=1)$,

$$\exp\Big(\sum_{1\le j\le n} iy_jN_j\Big)F=\rho(\Delta(\sqrt{y_1},\dots,\sqrt{y_n}))^{-1}\exp\Big(\sum_{h\in\mathbf{N}^n} c_h\prod_{1\le j\le n}\sqrt{\frac{y_j}{y_{j+1}}}^{\,-h_j}\Big)$$
$$\cdot\exp\Big(\sum_{h\in\mathbf{N}^n} k_h\prod_{1\le j\le n}\sqrt{\frac{y_j}{y_{j+1}}}^{\,-h_j}\Big)\cdot\mathbf{r}. \qquad (1)$$

Note that these c_h $(h\in\mathbf{N}^n)$ are uniquely determined (whereas the k_h are not).

PROPOSITION 6.2.2 *We use the notation in* 6.2.1. *For* $v\in\mathfrak{g}_{\mathbf{R}}$ *and* $e\in\mathbf{Z}^n$, *let* $v(e)$ *be the component of* v *on which* $\mathrm{Ad}(\tilde\rho(t))$ $(t=(t_j)_{1\le j\le n}\in\mathbf{G}^n_{m,\mathbf{R}})$ *acts as* $\prod_{1\le j\le n} t_j^{e_j}$. *We denote* $|e|=(|e_j|)_{1\le j\le n}$. *Let* $h\in\mathbf{N}^n-\{0\}$, $e\in\mathbf{Z}^n$.

(i) $c_h(e)=0$ *unless* $|e|<h$ *for the product order in* $\mathbf{N}^n$, *i.e.,* $|e_j|\le h_j$ $(1\le \forall\, j\le n)$ *and* $|e|\ne h$.

(ii) *Let* l *be the largest integer satisfying* $h_l\ne 0$. *Then* $c_h(e)=0$ *unless* $|e_l|<h_l$.

6.2.3

Deduction of Theorem 5.4.2 *from Proposition* 6.2.2. Assuming Proposition 6.2.2, whose proof will be given in 6.2.4–6.2.6 below, we prove Theorem 5.4.2. Let $(N_1,\dots,N_n,F)$ be as in the hypothesis of 5.4.2, let (ρ,φ) be the associated SL(2)-orbit in n variables, let m be the rank of (ρ,φ), and let (ρ',φ') be the associated SL(2)-orbit in m variables (5.2.3). Consider the presentation 6.2.1 (1) of

$\exp(\sum_{1\le j\le n} iy_jN_j)F$. By 5.2.2 and 5.2.3, we have

$$\begin{aligned}\tilde{\rho}'\left(\sqrt{\frac{y_{s_1}}{y_{s_2}}},\dots,\sqrt{\frac{y_{s_m}}{y_{s_{m+1}}}}\right) &= \rho'(\Delta'(\sqrt{y_{s_1}},\dots,\sqrt{y_{s_m}}))\\ &= \rho(\Delta(\sqrt{y_1},\dots,\sqrt{y_n})) = \prod_{1\le j\le n}\tilde{\rho}_j\left(\sqrt{\frac{y_j}{y_{j+1}}}\right). \quad (1)\end{aligned}$$

Here $\Delta' : \mathbf{G}^m_{\mathbf{m},\mathbf{R}} \to \mathrm{SL}(2,\mathbf{R})^m$ is as Δ in 5.2.2, and $y_{s_{m+1}} := 1$. When $\frac{y_j}{y_{j+1}} \to \infty$ for $1 \le j \le n$, we have, by (1) and Proposition 6.2.2,

$$\begin{aligned}&\mathrm{Ad}\left(\tilde{\rho}'\left(\sqrt{\frac{y_{s_1}}{y_{s_2}}},\dots,\sqrt{\frac{y_{s_m}}{y_{s_{m+1}}}}\right)\right)^{\nu}\left(\sum_{h\in\mathbf{N}^n} c_h \prod_{1\le j\le n}\sqrt{\frac{y_j}{y_{j+1}}}^{\,-h_j}\right)\\ &= \sum_{h\in\mathbf{N}^n}\sum_{e\in\mathbf{Z}^n} c_h(e)\prod_{1\le j\le n}\sqrt{\frac{y_j}{y_{j+1}}}^{\,\nu e_j-h_j} \to 0 \quad \text{for } \nu = 0, \pm 1.\end{aligned}$$

By Proposition 5.2.16, this shows that $\exp(\sum_{1\le j\le n} iy_jN_j)F$ converges to $[\rho,\varphi] = [\rho',\varphi']$ in $D_{\mathrm{SL}(2)}$. □

6.2.4

We prove the following weaker version of Proposition 6.2.2 (i): $c_h(e) = 0$ unless $|e| \le h$.

Fix l such that $1 \le l \le n$, and fix $y_j > 0$ for $1 \le j \le n$ such that $\frac{y_j}{y_{j+1}} \gg 0$ ($1 \le j \le n$, $y_{n+1} = 1$). Then the pair $(\sum_{1\le j\le l} y_jN_j,\ \exp(\sum_{l<j\le n} iy_jN_j)F)$ generates a nilpotent orbit. Let (ρ',φ') be the SL(2)-orbit in one variable associated to this pair. By 6.1.1 (3), we have $f'_s \in \mathfrak{g}_{\mathbf{R}}$ ($s \ge 1$) such that $\sum_{s\ge1} f'_s t^s$ absolutely converges for $t \in \mathbf{C}$ with $|t|$ sufficiently small, that

$$\exp\left(iv\left(\sum_{1\le j\le l} y_jN_j\right) + \sum_{l<j\le n} iy_jN_j\right)F = \tilde{\rho}'(\sqrt{v})^{-1}\exp\left(\sum_{s\ge1} f'_s\sqrt{v}^{\,-s}\right)\cdot\mathbf{r}' \quad (1)$$

for $v \gg 0$, where $\mathbf{r}' := \varphi'(i)$, and that the part of f'_s of $(\mathrm{Ad}\circ\tilde{\rho}')$-weight e' ($e' \in \mathbf{Z}$) is zero unless $|e'| < s$.

Since $P := (G^\circ)_{W(\sigma_l),\mathbf{R}}$ is parabolic, there exist $g \in P$ and $k \in K_{\mathbf{r}}$ such that $\mathbf{r}' = gk\mathbf{r}$. Since $\tilde{\rho}_l$ (resp. $\tilde{\rho}'$) is the Borel-Serre lifting of the weight map $\mathbf{G}_{\mathbf{m},\mathbf{R}} \to P/P_u$ at the maximal compact subgroup $K_{\mathbf{r}}$ (resp. $K_{\mathbf{r}'} = gK_{\mathbf{r}}g^{-1}$) of $G_{\mathbf{R}}$, we have $\tilde{\rho}' = \mathrm{Int}(g)\circ\tilde{\rho}_l$. Let $f_s := \mathrm{Ad}(g)^{-1}(f'_s)$. Then clearly

(2) The component of f_s of $(\mathrm{Ad}\circ\tilde{\rho}_l)$-weight e' ($e' \in \mathbf{Z}$) is zero unless $|e'| < s$.

The right-hand side of (1) is equal to

$$\tilde{\rho}_l(\sqrt{v})^{-1}\,\mathrm{Int}(\tilde{\rho}_l(\sqrt{v}))(g)\exp\left(\sum_{s\ge1} f_s\sqrt{v}^{\,-s}\right)k\cdot\mathbf{r}.$$

On the other hand, by 6.2.1 (1), the left-hand side of (1) belongs to

$$\tilde{\rho}_l(\sqrt{v})^{-1}\rho(\Delta(\sqrt{y_1},\ldots,\sqrt{y_n}))^{-1}\exp\left(\sum_{h\in\mathbf{N}^n} c_h \prod_{1\le j\le n}\sqrt{\frac{y_j}{y_{j+1}}}^{-h_j}\sqrt{v}^{-h_l}\right)K_{\mathbf{r}}\cdot\mathbf{r}.$$

Hence

$$\exp\left(\sum_{h\in\mathbf{N}^n} c_h \prod_{1\le j\le n}\sqrt{\frac{y_j}{y_{j+1}}}^{-h_j}\sqrt{v}^{-h_l}\right)K_{\mathbf{r}}\cdot\mathbf{r}$$

$$= \rho(\Delta(\sqrt{y_1},\ldots,\sqrt{y_n}))\,\mathrm{Int}(\tilde{\rho}_l(\sqrt{v}))(g)\exp\left(\sum_{s\ge 1} f_s\sqrt{v}^{-s}\right)K_{\mathbf{r}}\cdot\mathbf{r}. \quad (3)$$

Write $g \in P$ as $g = g_0\exp(\sum_{s\ge1} g_{-s})$ where g_0 is an element of $G_{\mathbf{R}}$ which commutes with $\tilde{\rho}_l(t)$ for any t and g_{-s} is an element of $\mathfrak{g}_{\mathbf{R}}$ of $(\mathrm{Ad}\circ\tilde{\rho}_l)$-weight $-s$. Then

$$\mathrm{Int}(\tilde{\rho}_l(\sqrt{v}))(g) = g_0\exp\left(\sum_{s\ge1} g_{-s}\sqrt{v}^{-s}\right). \quad (4)$$

By (2) and (4), the right-hand side of (3) has the form

$$\exp\left(\sum_{s\ge 0} d_s\sqrt{v}^{-s}\right)K_{\mathbf{r}}\cdot\mathbf{r},$$

such that the component of d_s of $(\mathrm{Ad}\circ\tilde{\rho}_l)$-weight e_l is zero unless $|e_l| \le s$. This proves that the component of c_h of $(\mathrm{Ad}\circ\tilde{\rho}_l)$-weight e_l is zero unless $|e_l| \le h_l$. □

For the proof of Proposition 6.2.2, it remains to prove (ii) of 6.2.2. We prove the following preliminary lemma.

Lemma 6.2.5 *Let the notation be as in* 6.2.1 *and, for* $\frac{y_j}{y_{j+1}} \gg 0$ $(1 \le j \le n,$ $y_{n+1} = 1)$, *write*

$$\exp\left(\sum_{1\le j\le n} iy_jN_j\right)F$$

$$= \rho(\Delta(\sqrt{y_1},\ldots,\sqrt{y_n}))^{-1}\exp\left(\sum_{h\in\mathbf{N}^n} b_h \prod_{1\le j\le n}\sqrt{\frac{y_j}{y_{j+1}}}^{-h_j}\right)\cdot\mathbf{r} \quad (1)$$

$(b_h \in \mathfrak{g}_{\mathbf{R}})$. *Let* $1 \le l \le n$. *Let* $A = \rho(\Delta(\sqrt{y_1},\ldots,\sqrt{y_l},1,\ldots,1))$. *Then, for* $y_j > 0$ *with* $\frac{y_j}{y_{j+1}} \gg 0$ $(1 \le j \le n,\ y_{n+1} = 1)$, *we have the following* (P_l) *and* (Q_l):

$$\exp(iy_1N_1 + \cdots + iy_lN_l + iN_{l+1})\varphi(\mathbf{0}_{l+1},\mathbf{i}_{n-l-1})$$

$$= A^{-1}\exp\left(\sum_{h\in\mathbf{N}^l} b_{h,\mathbf{0}_{n-l}}\left(\prod_{1\le j<l}\sqrt{\frac{y_j}{y_{j+1}}}^{-h_j}\right)\sqrt{y_l}^{-h_l}\right)\cdot\mathbf{r}. \quad (\mathrm{P}_l)$$

Here, in the case $l = n$, we define $N_{n+1} = 0$ and $\varphi(\mathbf{0}_{l+1}, \mathbf{i}_{n-l-1})$ denotes F.

$$\exp(iy_1N_1 + \cdots + iy_lN_l)\varphi(\mathbf{0}_l, \mathbf{i}_{n-l})$$
$$= A^{-1}\exp\left(\sum_{h\in\mathbf{N}^{l-1}} b_{h,\mathbf{0}_{n-l+1}}\left(\prod_{1\le j<l-1}\sqrt{\frac{y_j}{y_{j+1}}}^{-h_j}\right)\sqrt{y_{l-1}}^{-h_{l-1}}\right)\cdot\mathbf{r}. \qquad (\mathrm{Q}_l)$$

Proof. We prove (P_l) and (Q_l) simultaneously by downward induction on l.

In the case $l = n$, (P_n) is nothing but (1).

For each l, we can deduce (Q_l) from (P_l) as follows. Consider the element

$$\exp(iy_1N_1 + \cdots + iy_lN_l)\tilde{\rho}_l(\sqrt{v})\exp(iN_{l+1})\varphi(\mathbf{0}_{l+1}, \mathbf{i}_{n-l-1}) \quad (v > 0). \qquad (2)$$

Since the **R**-split mixed Hodge structure associated with the mixed Hodge structure $\big(W(\sigma_l)[-w], \exp(iN_{l+1})\varphi(\mathbf{0}_{l+1}, \mathbf{i}_{n-l-1})\big)$ is $\varphi(\mathbf{0}_l, \mathbf{i}_{n-l})$ (6.1.3 (3)), and since the **R**-splitting 6.1.3 (5) of this **R**-split mixed Hodge structure is given by $\tilde{\rho}_l$, the limit of (2) for $v \to \infty$ is equal to the left-hand side of (Q_l). Here we use

$$\lim_{v\to\infty} \tilde{\rho}_l(\sqrt{v})\cdot\exp(iN_{l+1})\varphi(\mathbf{0}_{l+1}, \mathbf{i}_{n-l-1}) = \varphi(\mathbf{0}_l, \mathbf{i}_{n-l}).$$

On the other hand, by 6.1.10, the element (2) is equal to $\tilde{\rho}_l(\sqrt{v})\exp(ivy_1N_1 + \cdots + ivy_lN_l + iN_{l+1})\varphi(\mathbf{0}_{l+1}, \mathbf{i}_{n-l-1})$ and, by (P_l), this element is equal to

$$A^{-1}\exp\left(\sum_{h\in\mathbf{N}^l} b_{h,\mathbf{0}_{n-l}}\left(\prod_{1\le j<l}\sqrt{\frac{y_j}{y_{j+1}}}^{-h_j}\right)\sqrt{y_l}^{-h_l}\sqrt{v}^{-h_l}\right)\cdot\mathbf{r}.$$

The limit of this element for $v \to \infty$ is the right-hand side of (Q_l).

Finally, for $l > 1$, we deduce (P_{l-1}) from (Q_l), just by applying $\tilde{\rho}_l(\sqrt{y_l})$ to (Q_l) and by using 6.1.10 and the fact $\tilde{\rho}_l(t)\varphi(\mathbf{0}_l, \mathbf{i}_{n-l}) = \varphi(\mathbf{0}_l, \mathbf{i}_{n-l})$ for any t. □

6.2.6

Proof of Proposition 6.2.2 (ii). Take $y_j > 0$ such that $\frac{y_j}{y_{j+1}} \gg 0$ $(1 \le j \le n)$.

The pair $(\sum_{1\le j\le l} y_jN_j, \exp(iN_{l+1})\varphi(\mathbf{0}_{l+1}, \mathbf{i}_{n-l-1}))$ generates a nilpotent orbit. Let (ρ', φ') be the SL(2)-orbit in one variable associated to this pair. Using 6.1.1 (3), write

$$\exp\left(iv\left(\sum_{1\le j\le l} y_jN_j\right) + iN_{l+1}\right)\varphi(\mathbf{0}_{l+1}, \mathbf{i}_{n-l-1})$$
$$= \tilde{\rho}'(\sqrt{v})^{-1}\exp\left(\sum_{s\ge1} f'_s\sqrt{v}^{-s}\right)\cdot\mathbf{r}', \qquad (1)$$

where

$$\mathbf{r}' := \varphi'(i) = \exp\left(\sum_{1\le j\le l} iy_jN_j\right)\varphi(\mathbf{0}_l, \mathbf{i}_{n-l})$$

and the component of f'_s of $(\mathrm{Ad}\circ\tilde{\rho}')$-weight e' $(e' \in \mathbf{Z})$ is 0 unless $|e'| < s$. By the assumption of 6.2.2 (ii) and 6.2.5 (Q_l), we have $\mathbf{r}' \in gK_{\mathbf{r}}\cdot\mathbf{r}$ with $g \in G_{\mathbf{R}}$ such that

$g\tilde{\rho}_l(t) = \tilde{\rho}_l(t)g$ for any t. Let $P := (G^\circ)_{W(\sigma_l),\mathbf{R}}$. Since $\tilde{\rho}'$ (resp. $\tilde{\rho}_l$) is the Borel-Serre lifting of the weight map $\mathbf{G}_{\mathbf{m},\mathbf{R}} \to P/P_u$ at $K_{\mathbf{r}'} = gK_{\mathbf{r}}g^{-1}$ (resp. $K_{\mathbf{r}}$), we have $\tilde{\rho}' = \mathrm{Int}(g) \circ \tilde{\rho}_l = \tilde{\rho}_l$. Hence

$$\tilde{\rho}'(\sqrt{v})^{-1} \exp\Big(\sum_{s\geq 1} f'_s \sqrt{v}^{-s}\Big) \cdot \mathbf{r}' = \tilde{\rho}_l(\sqrt{v})^{-1} g \exp\Big(\sum_{s\geq 1} f_s \sqrt{v}^{-s}\Big) \cdot \mathbf{r}, \qquad (2)$$

where $f_s = \mathrm{Ad}(g)^{-1}(f'_s)$. The component of f_s of $(\mathrm{Ad} \circ \tilde{\rho}_l)$-weight e' ($e' \in \mathbf{Z}$) is zero unless $|e'| < s$. On the other hand, by 6.2.5 (P_l), the left-hand side of (1) is equal to

$$\tilde{\rho}_l(\sqrt{v})^{-1} A^{-1} \exp\Big(\sum_{h\in \mathbf{N}^l} b_{h,\mathbf{0}_{n-l}} \Big(\prod_{1\leq j<l} \sqrt{\frac{y_j}{y_{j+1}}}^{-h_j}\Big) \sqrt{y_l}^{-h_l} \sqrt{v}^{-h_l}\Big) \cdot \mathbf{r}, \qquad (3)$$

with $A = \rho(\Delta(\sqrt{y_1}, \ldots, \sqrt{y_l}, 1, \ldots, 1))$. Hence, by (1), (2), and (3), we have

$$\exp\Big(\sum_{h\in \mathbf{N}^l} c_{h,\mathbf{0}_{n-l}} \Big(\prod_{1\leq j<l} \sqrt{\frac{y_j}{y_{j+1}}}^{-h_j}\Big) \sqrt{y_l}^{-h_l} \sqrt{v}^{-h_l}\Big) K_{\mathbf{r}} \cdot \mathbf{r} = AgBK_{\mathbf{r}} \cdot \mathbf{r},$$

where $B := \exp(\sum_{s\geq 1} f_s \sqrt{v}^{-s})$ and the component of f_s of $(\mathrm{Ad} \circ \tilde{\rho}_l)$-weight e' ($e' \in \mathbf{Z}$) is zero unless $|e'| < s$, and where Ag satisfy $\tilde{\rho}_l(t)Ag = Ag\tilde{\rho}_l(t)$ for any t (see the notation 6.2.1). This proves that the component of $c_{h,\mathbf{0}_{n-l}}$ of $(\mathrm{Ad} \circ \tilde{\rho}_l)$-weight e_l ($e_l \in \mathbf{Z}$) is zero unless $|e_l| < h_l$. □

6.3 PROOF OF THEOREM 5.4.3 (i)

In this section, we prove Theorem 5.4.3 (i).

Let A be a finite-dimensional $\mathbf{Q}$-vector space and let V be a valuative submonoid of $A^* = \mathrm{Hom}_{\mathbf{Q}}(A, \mathbf{Q})$ such that $V^\times = \{0\}$.

6.3.1

By a face of V, we mean a submonoid V' of V such that if $a, b \in V$ and $a + b \in V'$, then $a, b \in V'$. A face is stable under the multiplication by $\mathbf{Q}_{\geq 0}$. By [R, chapter A], we have the following.

The set of all faces of V is finite and is totally ordered with respect to the inclusion. Let

$$V = V^0 \supsetneq V^1 \supsetneq \cdots \supsetneq V^n = \{0\}$$

be all the faces of V. For each $1 \leq j \leq n$, there exists an injective $\mathbf{Q}$-homomorphism

$$\nu_j : V^{j-1}_{\mathbf{Q}} / V^j_{\mathbf{Q}} \hookrightarrow \mathbf{R}$$

such that the image of V^{j-1} in $V^{j-1}_{\mathbf{Q}}/V^j_{\mathbf{Q}}$ coincides with $\nu_j^{-1}(\mathbf{R}_{\geq 0})$. This ν_j is unique up to multiplication by an element of $\mathbf{R}_{>0}$. We have the following description of V:

$$V = \{\varphi \in A^* \mid \text{if } 1 \leq j \leq n, \varphi \in V^{j-1}_{\mathbf{Q}} \text{ and } \varphi \notin V^j_{\mathbf{Q}}, \text{ then } \nu_j(\varphi \bmod V^j_{\mathbf{Q}}) > 0\}.$$

Let $A_j \subset A$ $(0 \leq j \leq n)$ be the annihilator of $V_{\mathbf{Q}}^j$. Then we have

$$\{0\} = A_0 \subsetneq A_1 \subsetneq \cdots \subsetneq A_n = A.$$

For $1 \leq j \leq n$, since A_j/A_{j-1} is the dual $\mathbf{Q}$-vector space of $V_{\mathbf{Q}}^{j-1}/V_{\mathbf{Q}}^j$, ν_j is regarded as an element of $\mathbf{R} \otimes_{\mathbf{Q}} (A_j/A_{j-1})$. The injectivity of ν_j in (1) implies that, if $(e_s)_s$ is a $\mathbf{Q}$-basis of A_j/A_{j-1} and $\nu_j = \sum_s a_s e_s$ $(a_s \in \mathbf{R})$, then $(a_s)_s$ is linearly independent over $\mathbf{Q}$.

6.3.2

Example. Assume $\dim_{\mathbf{Q}}(A) = 2$. Then one of the following two cases occurs.

(1) There exists a $\mathbf{Q}$-basis (e_1, e_2) of A for which V is the set of $xe_1^* + ye_2^*$ $(x, y \in \mathbf{Q})$ such that (x, y) is ≥ 0 with respect to the lexicographic order. Here (e_1^*, e_2^*) denotes the dual basis in A^* of (e_1, e_2).

(2) There exists a $\mathbf{Q}$-basis (e_1, e_2) of A and an irrational number $a > 0$ for which $V = \{xe_1^* + ye_2^* \mid x, y \in \mathbf{Q},\ x + ay \geq 0\}$.

In case (1), $n = 2$, $A_1 = \mathbf{Q}e_1$, $A_2 = A$. In case (2), $n = 1$, $A_1 = A$, and ν_1 is a homomorphism of the form $A^* \to \mathbf{R}$, $xe_1^* + ye_2^* \mapsto a_1 x + a_2 y$, with $a_1, a_2 \in \mathbf{R}_{>0}$, $(a_1 : a_2) = (1 : a)$, and ν_1 is identified with $a_1 e_1 + a_2 e_2 \in \mathbf{R} \otimes_{\mathbf{Q}} A$.

Definition 6.3.3 *By a* good basis *for (A, V), we mean a $\mathbf{Q}$-basis $(N_s)_{s \in S}$ of A satisfying the following condition: There exist $a_s \in \mathbf{R}_{>0}$ $(s \in S)$ and a division of the index set S into nonempty subsets S_j $(1 \leq j \leq n)$, with n as in* 6.3.1, *such that, under the map*

$$\left(\sum_{s \in S_j} a_s N_s\right)_{1 \leq j \leq n} : A^* \to \mathbf{R}^n,$$

V coincides with the set of all elements of A^* whose images in $\mathbf{R}^n$ are ≥ 0 with respect to the lexicographic order.

6.3.4

Example. In Example 6.3.2, in both cases (1) and (2), (e_1, e_2) is a good basis for (A, V), but $(-e_1, e_2)$, $(e_1, -e_2)$, $(-e_1, -e_2)$ are not good bases for (A, V). In case (1), for the good basis (e_1, e_2), we have $S_1 = \{1\}$, $S_2 = \{2\}$, and a_1, a_2 are any elements of $\mathbf{R}_{>0}$. In case (2), for the good basis (e_1, e_2), (a_1, a_2) is any pair of elements of $\mathbf{R}_{>0}$ such that $(a_1 : a_2) = (1 : a)$.

Proposition 6.3.5

(i) *A good basis for (A, V) exists.*

(ii) *Let $(N_s)_{s \in S}$ be a good basis for (A, V). Let $(a_s)_{s \in S}$ and $(S_j)_{1 \leq j \leq n}$ be as in* 6.3.3, *and let $(a'_s)_s$ and $(S'_j)_{1 \leq j \leq n'}$ be also families satisfying the same condition. Then $S'_j = S_j$ for any j, and there exists $c_j \in \mathbf{R}_{>0}$ $(1 \leq j \leq n)$ such that $a'_s = c_j a_s$ for any $s \in S_j$.*

The proof of 6.3.5 will be given in 6.3.7 below after a preliminary Lemma 6.3.6, which can be proved easily.

LEMMA 6.3.6 *Let v_j $(1 \le j \le n)$ and A_j $(0 \le j \le n)$ be as in* 6.3.1. *Let $(N_s)_{s\in S}$ be a **Q**-basis of A. Assume we are given $a_s \in \mathbf{R}_{>0}$ $(s \in S)$ and a division of the index set S into nonempty subsets S_j $(1 \le j \le n)$. Then $(N_s)_{s\in S}$ is a good basis for (A, V) and $(a_s)_s$ and $(S_j)_j$ have the properties described in* 6.3.3, *if and only if the following three conditions are satisfied for each j $(1 \le j \le n)$.*

(1) $S_j = \{s \in S \mid N_s \in A_j, N_s \notin A_{j-1}\}$.
(2) $(N_s \bmod A_{j-1})_{s\in S_j}$ *is a* **Q**-*basis of* A_j/A_{j-1}.
(3) $v_j = \left(\sum_{s\in S_j} ca_sN_s \bmod A_{j-1,\mathbf{R}}\right)$ *for some* $c \in \mathbf{R}_{>0}$.

6.3.7

Proof of Proposition 6.3.5. We can find easily a **Q**-basis $(N_s)_{s\in S}$ of A with a division of S into nonempty subsets S_j $(1 \le j \le n)$ which satisfy the conditions 6.3.6 (1), 6.3.6 (2). Write $v_j = \left(\sum_{s\in S_j} a_sN_s \bmod A_{j-1,\mathbf{R}}\right)$ with $a_s \in \mathbf{R}$. Replacing N_s by $-N_s$ in the case $a_s < 0$, we have $a_s > 0$ for any $s \in S$. Then $(N_s)_{s\in S}$ is a good basis for (A, V) by Lemma 6.3.6. This proves Proposition 6.3.5 (i). Proposition 6.3.5 (ii) follows directly from Lemma 6.3.6. □

DEFINITION 6.3.8 *Let $p = (A, V, Z) \in D^\sharp_{\mathrm{val}}$. By an* excellent basis *for p, we mean a good basis $(N_s)_{s\in S}$ for (A, V) such that for any $F \in Z$, $((N_s)_{s\in S}, F)$ generates a nilpotent orbit.*

PROPOSITION 6.3.9 *Let $p = (A, V, Z) \in D^\sharp_{\mathrm{val}}$. Then an excellent basis for p exists. Furthermore, for any $\sigma \in \mathfrak{F}(A, V)$* (5.3.2), *there exists an excellent basis $(N_s)_{s\in S}$ for p such that $N_s \in \sigma$ for any $s \in S$.*

The proof of 6.3.9 will be given in 6.3.14 below after preliminaries 6.3.11–6.3.13.

6.3.10

Deduction of Theorem 5.4.3 (i) *from Proposition* 6.3.9. Theorem 5.4.3 (i) follows from Proposition 6.3.9 if we take $N_j := \sum_{s\in S_j} a_sN_s$ $(1 \le j \le n)$. □

6.3.11

Fix a good basis $(N_s)_{s\in S}$ for (A, V), and let $(a_s)_s$, $(S_j)_j$ be as in Definition 6.3.3. Denote $S_{\le j} := \bigsqcup_{k\le j} S_k$ and $S_{\ge j} := \bigsqcup_{k\ge j} S_k$.

Let $c \in \mathbf{Q}_{>0}$, and let $\alpha = (\alpha_{j,s,t})$ and $\beta = (\beta_{j,s,t})$ $(1 \le j \le n, s, t \in S_j)$ be families of elements of $\mathbf{Q}_{>0}$ such that $\alpha_{j,s,t} > \frac{a_s}{a_t} > \beta_{j,s,t}$ for any j, s, t. We define a rational nilpotent cone $\sigma_{c,\alpha,\beta}$ in $\mathfrak{g}_\mathbf{R}$ as the set of $\sum_{s\in S} y_sN_s \in A$ with $y_s \in \mathbf{R}_{\ge 0}$ satisfying the following two conditions:

(1) If $1 \le j < n$ and $s \in S_{\le j}$, $t \in S_{\ge j+1}$, then $y_s \ge cy_t$.
(2) If $1 \le j \le n$ and $s, t \in S_j$, then $\alpha_{j,s,t}y_t \ge y_s \ge \beta_{j,s,t}y_t$.

Note that $\sigma_{c,\alpha,\beta,\mathbf{R}} = A_\mathbf{R}$.

PROPOSITION 6.3.12 *Let $\mathfrak{F}(A, V)$ be as in 5.3.2, and fix a good basis $(N_s)_{s\in S}$ for (A, V). Then, for a rational nilpotent cone σ such that $\sigma_{\mathbf{R}} = A_{\mathbf{R}}$, the following (1) and (2) are equivalent.*

(1) $\sigma \in \mathfrak{F}(A, V)$.
(2) $\sigma_{c,\alpha,\beta} \subset \sigma$ *for some* c, α, β *as in* 6.3.11.

Proof. We prove that (1) implies (2). Assume (1). Take a finite subset I of $(\sigma \cap A)^\vee$ such that $(\sigma \cap A)^\vee = \sum_{\varphi\in I}(\mathbf{Q}_{\geq 0})\varphi$ and such that $0 \notin I$. Then $\sigma = \{a \in A_{\mathbf{R}} \mid \varphi(a) \in \mathbf{R}_{\geq 0}\ (\forall \varphi \in I)\}$. Here we denote the $\mathbf{R}$-linear map $A_{\mathbf{R}} \to \mathbf{R}$ induced by φ by the same letter φ. It is sufficient to prove that there exist c, α, β such that $\varphi(\sigma_{c,\alpha,\beta}) \subset \mathbf{R}_{\geq 0}$ for any $\varphi \in I$. Let $\varphi \in I$, and assume $\varphi(A_j) \neq 0$, $\varphi(A_{j-1}) = 0$. It is sufficient to prove that if c is sufficiently large and if $\alpha_{j,s,t}$ and $\beta_{j,s,t}$ are sufficiently near to $\frac{a_s}{a_t}$, then, for any $\sum_{s\in S} y_{\lambda,s}N_s \in \sigma_{c,\alpha,\beta}$ $(y_{\lambda,s} \in \mathbf{R}_{\geq 0})$, $\varphi(\sum_{s\in S} y_{\lambda,s}N_s) = \sum_{s\in S_j} y_{\lambda,s}\varphi(N_s) + \sum_{t\in S_{\geq j+1}} y_{\lambda,t}\varphi(N_t)$ belongs to $\mathbf{R}_{\geq 0}$. If $y_{\lambda,s} = 0$ for any $s \in S_j$, then $y_{\lambda,t} = 0$ for any $t \in S_{\geq j+1}$ because $y_{\lambda,s} \geq c y_{\lambda,t}$. Hence we may assume that $y_{\lambda,s} > 0$ for some $s \in S_j$. Then, since $\frac{y_{\lambda,t}}{y_{\lambda,s}}$ is sufficiently near to 0 for $s \in S_j$ and $t \in S_{\geq j+1}$ and since the ratio of $(y_{\lambda,s})_{s\in S_j}$ is sufficiently near to the ratio of $(a_s)_{s\in S_j}$ for which $\sum_{s\in S_j} a_s\varphi(N_s) > 0$, we have $\sum_{s\in S_j} y_{\lambda,s}\varphi(N_s) + \sum_{t\in S_{\geq j+1}} y_{\lambda,t}\varphi(N_t) > 0$.

Next we prove that (2) implies (1). It is sufficient to prove that $\sigma_{c,\alpha,\beta} \in \mathfrak{F}(A, V)$. Let $\varphi \in A^*$ and assume $\varphi(\sigma_{c,\alpha,\beta}) \subset \mathbf{R}_{\geq 0}$. It is sufficient to prove $\varphi \in V$. Take elements $y_{\lambda,s}$ $(s \in S, \lambda \in \Lambda)$ of $\mathbf{R}_{>0}$ with a directed set Λ such that $\frac{y_{\lambda,s}}{y_{\lambda,t}} \to \infty$ for any $s \in S_{\leq j}$, $t \in S_{\geq j+1}$ with $1 \leq j < n$ and that the ratio of $(y_{\lambda,s})_{s\in S_j}$ converges to the ratio of $(a_s)_{s\in S_j}$ for $1 \leq j \leq n$. Then, for a sufficiently large $\lambda \in \Lambda$, $\sum_{s\in S} y_{\lambda,s}N_s$ belongs to $\sigma_{c,\alpha,\beta}$ and hence $\varphi(\sum_{s\in S} y_{\lambda,s}N_s) \geq 0$. Let $1 \leq j \leq n$ and assume $\varphi \in V_{\mathbf{Q}}^{j-1}$, $\varphi \notin V_{\mathbf{Q}}^{j}$. Take $\sum_{s\in S_j} a_sN_s$ as ν_j. Then

$$0 \leq \varphi\Bigl(\sum_{s\in S} y_{\lambda,s}N_s\Bigr) = \sum_{s\in S_j} y_{\lambda,s}\varphi(N_s) + \sum_{t\in S_{\geq j+1}} y_{\lambda,t}\varphi(N_t).$$

Since $\frac{y_{\lambda,s}}{y_{\lambda,t}} \to \infty$ for any $s \in S_j$ and $t \in S_{\geq j+1}$ and since the ratio of $(y_{\lambda,s})_{s\in S_j}$ converges to the ratio of $(a_s)_{s\in S_j}$, we obtain $\nu_j(\varphi) = \sum_{s\in S} a_s\varphi(N_s) \geq 0$. This proves $\varphi \in V$ by the description of V in 6.3.1. □

LEMMA 6.3.13 *Let $p = (A, V, Z) \in D^\sharp_{\mathrm{val}}$ and let $(N_s)_{s\in S}$ be a good basis for (A, V). Let c, α, β and $\sigma_{c,\alpha,\beta}$ be as in 6.3.11 (defined with respect to $(N_s)_{s\in S}$). Then there exists a good basis $(N'_s)_{s\in S}$ for (A, V) such that $N'_s \in \sigma_{c,\alpha,\beta}$ for any $s \in S$.*

Proof. If $C_j > 0$ $(1 \leq j \leq n)$ and if C_j/C_{j+1} $(1 \leq j \leq n$, put $C_{n+1} = 1)$ are sufficiently large, then the following conditions (1) and (2) hold.

(1) If $1 \leq j < k \leq n$ and $s \in S_j$ and $t \in S_k$, then

$$C_ja_s > (1 + C_ka_t)c.$$

(2) If $1 \le j \le n$ and $s, t \in S_j$ and $s \ne t$, then

$$\alpha_{j,s,t} > \frac{1+C_j a_s}{C_j a_t}, \quad \frac{C_j a_s}{1+C_j a_t} > \beta_{j,s,t}.$$

We fix such C_j.

For $s \in S$, take a rational number b_s which is sufficiently near to a_s. Then we have the following conditions (3)–(5).

(3) If $1 \le j < k \le n$ and $s \in S_j$ and $t \in S_k$, then

$$C_j b_s \ge (1 + C_k b_t)c.$$

(4) If $1 \le j \le n$ and $s, t \in S_j$ and $s \ne t$, then

$$\alpha_{j,s,t} \ge \frac{1+C_j b_s}{C_j b_t}, \quad \frac{C_j b_s}{1+C_j b_t} \ge \beta_{j,s,t}.$$

(5) If $1 \le j \le n$ and $s \in S_j$, then

$$a_s - \frac{C_j b_s \sum_{t \in S_j} a_t}{1 + C_j \sum_{t \in S_j} b_t} > 0.$$

For $s \in S$, define N'_s as follows. Let j be the integer such that $s \in S_j$. Let

$$N'_s = N_s + \sum_{k=1}^{j} C_k \sum_{t \in S_k} b_t N_t.$$

From (3) and (4), we see easily that

$$N'_s \in \sigma_{c,\alpha,\beta}.$$

Furthermore, if we denote the left-hand side of (5) by a'_s, then

$$\sum_{s \in S_j} a_s N_s \equiv \sum_{s \in S_j} a'_s N'_s \mod A_{j-1,\mathbf{R}}.$$

Since $a'_s > 0$ for all $s \in S$, $(N'_s)_s$ is a good basis for (A, V) by Definition 6.3.3 and the description of V in 6.3.1. □

6.3.14

Proof of Proposition 6.3.9. Take $\tau \in \mathfrak{F}(A, V)$ such that (τ, Z) is a nilpotent i-orbit. Replacing σ by $\sigma \cap \tau \in \mathfrak{F}(A, V)$, we may assume that (σ, Z) is a nilpotent i-orbit. Then any good basis $(N_s)_{s \in S}$ for (A, V) such that $N_s \in \sigma$ for all $s \in S$ is an excellent basis for p. To construct such a good basis, take first any good basis $(N'_s)_{s \in S}$ for (A, V) (6.3.5). Then, by 6.3.12, we may assume $\sigma = \sigma_{c,\alpha,\beta}$ for some c, α, β as in 6.3.11, where $\sigma_{c,\alpha,\beta}$ is defined with respect to this good basis $(N'_s)_{s \in S}$. Hence we are reduced to Lemma 6.3.13. □

6.4 PROOFS OF THEOREM 5.4.3 (ii) AND THEOREM 5.4.4

In this section, we prove Theorem 5.4.3 (ii) and Theorem 5.4.4.

PROPOSITION 6.4.1 *Let $(N_s)_{s\in S}$ be a finite family of mutually commuting nilpotent elements of $\mathfrak{g}_{\mathbf{R}}$, let $F \in \check{D}$, and assume that $((N_s)_{s\in S}, F)$ generates a nilpotent orbit. Let $a_s \in \mathbf{R}_{>0}$ for $s \in S$. Assume that S is the disjoint union of nonempty subsets S_j $(1 \le j \le n)$. Let $S_{\le j}$, $S_{\ge j}$ be as in* 6.3.11. *For $1 \le j \le n$, let $\check{D}_j$ be the subset of $\check{D}$ consisting of all $F' \in \check{D}$ such that $((N_s)_{s\in S_{\le j}}, F')$ generates a nilpotent orbit. Let Λ be a directed ordered set, let $F_\lambda \in \check{D}$ $(\lambda \in \Lambda)$, $y_{\lambda,s} \in \mathbf{R}_{>0}$ $(\lambda \in \Lambda, s \in S)$, and assume that the following five conditions are satisfied.*

(1) *F_λ converges to F.*
(2) *$y_{\lambda,s} \to \infty$ for any $s \in S$.*
(3) *If $1 \le j < n$, $s \in S_{\le j}$ and $t \in S_{\ge j+1}$, then $\frac{y_{\lambda,s}}{y_{\lambda,t}} \to \infty$.*
(4) *If $1 \le j \le n$ and $s, t \in S_j$, then $\frac{y_{\lambda,s}}{y_{\lambda,t}} \to \frac{a_s}{a_t}$.*
(5) *For $1 \le j \le n$ and $e \ge 0$, there exist $F^*_\lambda \in \check{D}$ $(\lambda \in \Lambda)$ and $y^*_{\lambda,t} \in \mathbf{R}_{>0}$ $(\lambda \in \Lambda, t \in S_{\ge j+1})$ such that*

$$\exp\Bigl(\sum_{t\in S_{\ge j+1}} iy^*_{\lambda,t}N_t\Bigr)F^*_\lambda \in \check{D}_j \quad (\text{for } \lambda\text{: for sufficiently large}),$$

$$y^e_{\lambda,s}d(F_\lambda, F^*_\lambda) \to 0 \quad (\forall\, s \in S_j),$$

$$y^e_{\lambda,s}|y_{\lambda,t} - y^*_{\lambda,t}| \to 0 \quad (\forall s \in S_j, \forall t \in S_{\ge j+1}).$$

Here d is a metric on a neighborhood of F in $\check{D}$ that is compatible with the analytic structure (3.1.4).

For each $1 \le j \le n$, take $c_j \in S_j$ and denote $N_j := \sum_{s\in S_j} \frac{a_s}{a_{c_j}} N_s$. Let (ρ, φ) be the SL(2)*-orbit in n variables associated to $(N_1, \dots, N_n, F)$* (6.1.3)*, and let $\mathbf{r} = \varphi(\mathbf{i})$.*

Then, for sufficiently large λ, we have

$$\rho(\Delta(\sqrt{y_{\lambda,c_1}}, \dots, \sqrt{y_{\lambda,c_n}}))\exp\Bigl(\sum_{s\in S} iy_{\lambda,s}N_s\Bigr)F_\lambda = g_\lambda k_\lambda \cdot \mathbf{r}$$

for some $g_\lambda \in G_{\mathbf{R}}$ and $k_\lambda \in K_{\mathbf{r}}$ such that

$$k_\lambda \to 1,$$

$$\mathrm{Int}(\rho(\Delta(\sqrt{y_{\lambda,c_1}}, \dots, \sqrt{y_{\lambda,c_n}})))^\nu(g_\lambda) \to 1 \quad (\nu = 0, \pm 1).$$

We give preliminaries 6.4.2 and 6.4.3 for the proof of Proposition 6.4.1.

LEMMA 6.4.2 *Let the notation be as in* 6.4.1. *Let $1 \le j \le n$. Then*

$$\Bigl(\prod_{j\le k\le n} \tilde{\rho}_k\Bigl(\sqrt{\frac{y_{\lambda,c_k}}{y_{\lambda,c_{k+1}}}}\Bigr)\Bigr)\exp\Bigl(\sum_{s\in S_{\ge j}} iy_{\lambda,s}N_s\Bigr)F_\lambda \to \exp(iN_j)\varphi(\mathbf{0}_j, \mathbf{i}_{n-j}) \quad \text{in } \check{D},$$

where $y_{\lambda,c_{n+1}} := 1$. (Recall that $N_j := \sum_{s\in S_j} \frac{a_s}{a_{c_j}} N_s$.)

Proof. We prove 6.4.2 by downward induction on j.

First we show that we may assume

$$\exp\Big(\sum_{s\in S_{\geq j+1}} iy_{\lambda,s}N_s\Big)F_\lambda \in \check{D}_j \quad (\forall\,\lambda). \tag{1}$$

Let $u_\lambda = \Big(\prod_{j\leq k\leq n}\tilde{\rho}_k\Big(\sqrt{\frac{y_{\lambda,c_k}}{y_{\lambda,c_{k+1}}}}\Big)\Big)\exp\big(\sum_{s\in S_{\geq j}} iy_{\lambda,s}N_s\big)$. Since the N_s are nilpotent and they mutually commute, $\exp(\sum_{s\in S_{\geq j}} iy_{\lambda,s}N_s)$ is a polynomial in $y_{\lambda,s}$ with coefficients in $\mathfrak{g}_{\mathbf{C}}$. From this, we see that, if $e \geq 0$ is a sufficiently large integer, then the following holds: If $x_\lambda \in \mathfrak{g}_{\mathbf{C}}$ converges to 0 satisfying $y^e_{\lambda,c_j}x_\lambda \to 0$, then $\mathrm{Ad}(u_\lambda)(x_\lambda) \to 0$. Take such e and take F^*_λ $(\lambda \in \Lambda)$ and $y^*_{\lambda,t}$ $(\lambda\in\Lambda, t\in S_{\geq j+1})$ as in 6.4.1 (5). Then (1) is satisfied if we replace $y_{\lambda,s}$ by $y^*_{\lambda,s}$ and F_λ by F^*_λ. Let $y^*_{\lambda,s} = y_{\lambda,s}$ $(\lambda\in\Lambda, s\in S_j)$ and define u^*_λ in the same way as u_λ replacing $y_{\lambda,s}$ by $y^*_{\lambda,s}$. Then $F_\lambda = \exp(x_\lambda)F^*_\lambda$ for some $x_\lambda \in \mathfrak{g}_{\mathbf{C}}$ such that $y^e_{\lambda,c_j}x_\lambda \to 0$. We have $u_\lambda F_\lambda = u_\lambda\exp(x_\lambda)F^*_\lambda = \exp(\mathrm{Ad}(u_\lambda)(x_\lambda))(u_\lambda(u^*_\lambda)^{-1})u^*_\lambda F^*_\lambda$, and $\mathrm{Ad}(u_\lambda)(x_\lambda)\to 0$, $u_\lambda(u^*_\lambda)^{-1} \to 1$. Hence we can assume (1).

We assume (1). By Lemma 6.1.10, we have

$$\begin{aligned}&\Big(\prod_{k\geq j}\tilde{\rho}_k\Big(\sqrt{\frac{y_{\lambda,c_k}}{y_{\lambda,c_{k+1}}}}\Big)\Big)\exp\Big(\sum_{s\in S_{\geq j}} iy_{\lambda,s}N_s\Big)F_\lambda\\ &\quad= \exp\Big(\sum_{s\in S_j} i\frac{y_{\lambda,s}}{y_{\lambda,c_j}}N_s\Big)\Big(\prod_{k\geq j}\tilde{\rho}_k\Big(\sqrt{\frac{y_{\lambda,c_k}}{y_{\lambda,c_{k+1}}}}\Big)\Big)\exp\Big(\sum_{t\in S_{\geq j+1}} iy_{\lambda,t}N_t\Big)F_\lambda.\end{aligned}$$

If $j < n$, the hypothesis of induction is

$$\begin{aligned}&\Big(\prod_{k\geq j+1}\tilde{\rho}_k\Big(\sqrt{\frac{y_{\lambda,c_k}}{y_{\lambda,c_{k+1}}}}\Big)\Big)\exp\Big(\sum_{t\in S_{\geq j+1}} iy_{\lambda,t}N_t\Big)F_\lambda\\ &\quad\to \exp(iN_{j+1})\varphi(\mathbf{0}_{j+1},\mathbf{i}_{n-j-1}).\end{aligned} \tag{2}$$

On the other hand, in the case $j = n$, consider the following convergence:

$$F_\lambda \to F. \tag{3}$$

Let $W^{(j)} = W(\sigma_j)$ where $\sigma_j = (\mathbf{R}_{\geq 0})N_1 + \cdots + (\mathbf{R}_{\geq 0})N_j$. By (1), in the case $j < n$ (resp. $j = n$), (2) (resp. (3)) is a convergence of mixed Hodge structures for the weight filtration $W^{(j)}[-w]$. The **R**-split mixed Hodge structure associated to the right-hand side of (2) (resp. (3)) is $\varphi(\mathbf{0}_j, \mathbf{i}_{n-j})$, and the **R**-splitting 6.1.2 (3) of this **R**-split mixed Hodge structure is given by $\tilde{\rho}_j$ (6.1.3 (5)). By 6.1.11 (ii), it follows that

$$\Big(\prod_{k\geq j}\tilde{\rho}_k\Big(\sqrt{\frac{y_{\lambda,c_k}}{y_{\lambda,c_{k+1}}}}\Big)\Big)\exp\Big(\sum_{t\in S_{\geq j+1}} iy_{\lambda,t}N_t\Big)F_\lambda \to \varphi(\mathbf{0}_j,\mathbf{i}_{n-j}). \tag{4}$$

On the other hand, by 6.4.1 (4), we have

$$\sum_{s\in S_j}\frac{y_{\lambda,s}}{y_{\lambda,c_j}}N_s \to N_j. \tag{5}$$

Applying the exponential of i-times of (5) to (4), we obtain Lemma 6.4.2. □

Lemma 6.4.3 *Let the notation be as in* 6.4.1. *Fix j such that $1 \le j \le n$, and assume that* $\exp\big(\sum_{t\in S_{\ge j+1}} iy_{\lambda,t}N_t\big)F_\lambda \in \check{D}_j$ *for any λ. Put*

$$N_{\lambda,k} := \sum_{s\in S_k} \frac{y_{\lambda,s}}{y_{\lambda,c_k}} N_s \quad (1 \le k \le j),$$

$$U_\lambda := \left(\prod_{l\ge j+1} \tilde{\rho}_l\left(\sqrt{\frac{y_{\lambda,c_l}}{y_{\lambda,c_{l+1}}}}\right)\right)\exp\left(\sum_{t\in S_{\ge j+1}} iy_{\lambda,t}N_t\right)F_\lambda.$$

For each λ, let $(\rho_\lambda, \varphi_\lambda)$ be the SL(2)*-orbit in j variables associated with*

$$((N_{\lambda,k})_{1\le k\le j}, U_\lambda), \tag{1}$$

and let $\mathbf{r}_\lambda := \varphi_\lambda(\mathbf{i})$. Then $\mathbf{r}_\lambda$ converges to $\mathbf{r}$.

Note that (1) generates a nilpotent orbit by 6.1.10.

Proof. By Lemma 6.4.2,

$$U_\lambda \to \exp(iN_{j+1})\varphi(\mathbf{0}_{j+1}, \mathbf{i}_{n-j-1}), \tag{2}$$

where we understand $\exp(iN_{n+1})\varphi(\mathbf{0}_{n+1}, \mathbf{i}_{-1}) = F$. By the condition 6.4.1 (4),

$$N_{\lambda,k} \to N_k \quad (1 \le k \le j). \tag{3}$$

As in 6.1.3, $((N_k)_{1\le k\le j}, \exp(iN_{j+1})\varphi(\mathbf{0}_{j+1}, \mathbf{i}_{n-j-1}))$ generates a nilpotent orbit and the associated SL(2)-orbit (ρ', φ') in j variables has the reference point $\varphi'(\mathbf{i}) = \mathbf{r}$. Hence, by 6.1.6, $\mathbf{r}_\lambda$ converges to $\mathbf{r}$ as $\lambda \to \infty$. □

6.4.4

Proof of Proposition 6.4.1. Let the notation be as in 6.4.1. We prove the following assertion (C_j) by a downward induction on j. Let $0 \le j \le n$, and let $\check{D}_0 := \check{D}$.

(C_j) *Assume that* $\exp\big(\sum_{t\in S_{\ge j+1}} iy_{\lambda,t}N_t\big)F_\lambda \in \check{D}_j$ *for any λ. Then, for a sufficiently large λ, we have*

$$\rho(\Delta(\sqrt{y_{\lambda,c_1}}, \ldots, \sqrt{y_{\lambda,c_n}}))\exp\left(\sum_{s\in S} iy_{\lambda,s}N_s\right)F_\lambda$$
$$= \exp\left(\sum_{h\in\mathbf{N}^j} b_{\lambda,h}\prod_{1\le k\le j}\sqrt{\frac{y_{\lambda,c_k}}{y_{\lambda,c_{k+1}}}}^{\,-h_k}\right)k_\lambda\cdot\mathbf{r},$$

where $b_{\lambda,h} \in \mathfrak{g}_{\mathbf{R}}^-$ ($\mathfrak{g}_{\mathbf{R}}^-$ denotes the (-1)-eigenspace of $\mathfrak{g}_{\mathbf{R}}$ under the Cartan involution associated with $K_{\mathbf{r}}$), $k_\lambda \in K_{\mathbf{r}}$, and $\sum_{h\in\mathbf{N}^j} b_{\lambda,h}\prod_{1\le k\le j} x_k^{h_k}$ ($x_k \in \mathbf{C}$) absolutely converges when $|x_k|$ ($1 \le k \le j$) are sufficiently small, which satisfy the following three conditions. Here $y_{c_{n+1}} := 1$.

1. *(1)* $k_\lambda \to 1$.
2. *(2)* $|(\mathrm{Ad}\circ\tilde{\rho}_k)_{1\le k\le j}$*-weight of* $b_{\lambda,h}| \le h$ *for the product order in* $\mathbf{N}^j$.
3. *(3) For each* $h \in \mathbf{N}^j$, $\mathrm{Ad}\left(\prod_{k\ge j+1}\tilde{\rho}_k\left(\sqrt{\frac{y_{\lambda,c_k}}{y_{\lambda,c_{k+1}}}}\right)\right)^{\nu}(b_{\lambda,h})$ *converges for* $\nu = 0, \pm 1$. *Moreover, if $h = 0$ then it converges to* 0.

Note that Proposition 6.4.1 is nothing but (C_0).

We prove (C_j). By $\rho(\Delta(\sqrt{y_{\lambda,c_1}},\ldots,\sqrt{y_{\lambda,c_n}})) = \prod_{1\le k\le n}\tilde{\rho}_k\left(\sqrt{\frac{y_{\lambda,c_k}}{y_{\lambda,c_{k+1}}}}\right)$ and Lemma 6.1.10, we have

$$\begin{aligned}
&\rho(\Delta(\sqrt{y_{\lambda,c_1}},\ldots,\sqrt{y_{\lambda,c_n}}))\exp\left(\sum_{s\in S} iy_{\lambda,s}N_s\right)F_\lambda\\
&=\left(\prod_{k\le j}\tilde{\rho}_k\left(\sqrt{\frac{y_{\lambda,c_k}}{y_{\lambda,c_{k+1}}}}\right)\right)\exp\left(\sum_{s\in S_{\le j}} i\frac{y_{\lambda,s}}{y_{\lambda,c_{j+1}}}N_s\right)U_\lambda\\
&=\left(\prod_{k\le j}\tilde{\rho}_k\left(\sqrt{\frac{y_{\lambda,c_k}}{y_{\lambda,c_{k+1}}}}\right)\right)\exp\left(\sum_{k\le j} i\frac{y_{\lambda,c_k}}{y_{\lambda,c_{j+1}}}N_{\lambda,k}\right)U_\lambda,
\end{aligned}$$

where $N_{\lambda,k}$ and U_λ are as in Lemma 6.4.3. Let $(\rho_\lambda,\varphi_\lambda)$ be the SL(2)-orbit in j variables associated with $((N_{\lambda,k})_{1\le k\le j}, U_\lambda)$, and put $\mathbf{r}_\lambda := \varphi_\lambda(\mathbf{i})$. Then, by 6.2.1,

$$\begin{aligned}
&\rho(\Delta(\sqrt{y_{\lambda,c_1}},\ldots,\sqrt{y_{\lambda,c_n}}))\exp\left(\sum_{s\in S} iy_{\lambda,s}N_s\right)F_\lambda\\
&=\left(\prod_{k\le j}\tilde{\rho}_k\left(\sqrt{\frac{y_{\lambda,c_k}}{y_{\lambda,c_{k+1}}}}\right)\right)\left(\prod_{k\le j}\tilde{\rho}_{\lambda,k}\left(\sqrt{\frac{y_{\lambda,c_k}}{y_{\lambda,c_{k+1}}}}\right)\right)^{-1} f_\lambda k_{1,\lambda}\cdot\mathbf{r}_\lambda, \quad \text{where}
\end{aligned}$$

$$f_\lambda := \exp\left(\sum_{h\in\mathbf{N}^j} a_{\lambda,h}\prod_{k\le j}\sqrt{\frac{y_{\lambda,c_k}}{y_{\lambda,c_{k+1}}}}^{\,-h_k}\right),\quad a_{\lambda,h}\in\mathfrak{g}_{\mathbf{R}}^{-,\mathbf{r}_\lambda},$$

$$k_{1,\lambda}\in K_{\mathbf{r}_\lambda},\quad k_{1,\lambda}\to 1.$$

Here $\mathfrak{g}_{\mathbf{R}}^{-,\mathbf{r}_\lambda}$ denotes the (-1)-eigenspace of $\mathfrak{g}_{\mathbf{R}}$ under the Cartan involution associated to the maximal compact sugroup $K_{\mathbf{r}_\lambda}$ of $G_{\mathbf{R}}$.

Claim 1 *We can write*

$$\mathbf{r}_\lambda = g_\lambda k_{2,\lambda}\cdot\mathbf{r},\quad \tilde{\rho}_{\lambda,k} = \operatorname{Int}(g_\lambda)\circ\tilde{\rho}_k,$$
$$g_\lambda\in(G^\circ)_{W^{(1)},\ldots,W^{(j)},\mathbf{R}},\quad k_{2,\lambda}\in K_{\mathbf{r}},\quad g_\lambda\to 1,\quad k_{2,\lambda}\to 1.$$

We prove this claim. Take a $\mathbf{Q}$-parabolic subgroup P of $G_{\mathbf{R}}$ such that

$$(G^\circ)_{W^{(1)},\ldots,W^{(j)},\mathbf{R}}\subset P,\quad G_{W^{(1)},\ldots,W^{(j)},\mathbf{R},u}\subset P_u,$$
$$\operatorname{Image}\left(\rho(\Delta(\mathbf{G}^j_{\mathbf{m},\mathbf{R}}\times\mathbf{1}_{n-j}))\to P/P_u\right)\subset(\text{center of } P/P_u).$$

For the existence of P, see [KU2, Proof of 4.12]. Since $\mathbf{r}_\lambda\to\mathbf{r}$ by Lemma 6.4.3, we can write

$$\mathbf{r}_\lambda = g_\lambda k_{2,\lambda}\cdot\mathbf{r},\quad g_\lambda\in P,\quad k_{2,\lambda}\in K_{\mathbf{r}},\quad g_\lambda\to 1,\quad k_{2,\lambda}\to 1.$$

We show that $g_\lambda\in(G^\circ)_{W^{(1)},\ldots,W^{(j)},\mathbf{R}}$. In fact, since both $(\tilde{\rho}_{\lambda,k})_{1\le k\le j}$ and $(\tilde{\rho}_k)_{1\le k\le j}$ split $(W^{(k)})_{1\le k\le j}$, there exists $u_\lambda\in G_{W^{(1)},\ldots,W^{(j)},\mathbf{R},u}$ such that

$$\tilde{\rho}_{\lambda,k} = \operatorname{Int}(u_\lambda)\circ\tilde{\rho}_k\quad(1\le k\le j).$$

On the other hand, since the homomorphisms $\mathbf{G_{m,R}} \to P/P_u$ induced by $\tilde{\rho}_{\lambda,k}$ and $\tilde{\rho}_k$ coincide and since $\tilde{\rho}_{\lambda,k}$ and $\tilde{\rho}_k$ are the Borel-Serre liftings at $\mathbf{r}_\lambda$ and $\mathbf{r}$, respectively, of this common induced homomorphism, we have

$$\tilde{\rho}_{\lambda,k} = \mathrm{Int}(g_\lambda) \circ \tilde{\rho}_k \quad (1 \le k \le j).$$

Hence

$$\mathrm{Int}(u_\lambda^{-1} g_\lambda) \circ \tilde{\rho}_k = \tilde{\rho}_k \quad (1 \le k \le j).$$

Therefore

$$u_\lambda^{-1} g_\lambda \in (G^\circ)_{W^{(1)},\dots,W^{(j)},\mathbf{R}}, \quad g_\lambda \in (G^\circ)_{W^{(1)},\dots,W^{(j)},\mathbf{R}}.$$

This proves Claim 1.

We go back to the proof of (C_j). By Claim 1, we have

$$\begin{aligned}&\rho(\Delta(\sqrt{y_{\lambda,c_1}}, \dots, \sqrt{y_{\lambda,c_n}})) \exp\left(\sum_{s\in S} i y_{\lambda,s} N_s\right) F_\lambda \\ &= \mathrm{Int}\left(\prod_{k\le j} \tilde{\rho}_k\left(\sqrt{\frac{y_{\lambda,c_k}}{y_{\lambda,c_{k+1}}}}\right)\right)(g_\lambda)\, \mathrm{Int}(g_\lambda)^{-1}(f_\lambda)\, \mathrm{Int}(g_\lambda)^{-1}(k_{1,\lambda}) k_{2,\lambda} \cdot \mathbf{r}. \qquad (4)\end{aligned}$$

Here $\mathrm{Int}(g_\lambda)^{-1}(k_{1,\lambda}) \in K_{\mathbf{r}}$ and, concerning $a_{\lambda,h} \in \mathfrak{g}_{\mathbf{R}}^{-,\mathbf{r}_\lambda}$ in the definition of f_λ, we have $\mathrm{Ad}(g_\lambda)^{-1}(a_{\lambda,h}) \in \mathfrak{g}_{\mathbf{R}}^{-} = \mathfrak{g}_{\mathbf{R}}^{-,\mathbf{r}}$. Furthermore, if we write

$$g_\lambda = \exp\left(\sum_{h\in\mathbf{N}^j} g_{\lambda,-h}\right), \quad g_{\lambda,-h} \in \mathrm{Lie}(G_{W^{(1)},\dots,W^{(j)},\mathbf{R}}), \qquad (5)$$

$$((\mathrm{Ad} \circ \tilde{\rho}_k)_{1\le k\le j}\text{-weight of } g_{\lambda,-h}) = -h, \quad g_{\lambda,-h} \to 0,$$

we have

$$\mathrm{Int}\left(\prod_{k\le j} \tilde{\rho}_k\left(\sqrt{\frac{y_{\lambda,c_k}}{y_{\lambda,c_{k+1}}}}\right)\right)(g_\lambda) = \exp\left(\sum_{h\in\mathbf{N}^j} g_{\lambda,-h} \prod_{k\le j} \sqrt{\frac{y_{\lambda,c_k}}{y_{\lambda,c_{k+1}}}}^{\,-h_k}\right). \qquad (6)$$

By Proposition 6.2.2,

$$|(\mathrm{Ad} \circ \tilde{\rho}_k)_{1\le k\le j}\text{-weight of } \mathrm{Ad}(g_\lambda)^{-1}(a_{\lambda,h})| < h. \qquad (7)$$

From (4)–(7), we obtain

$$\begin{aligned}&\rho(\Delta(\sqrt{y_{\lambda,c_1}}, \dots, \sqrt{y_{\lambda,c_n}})) \exp\left(\sum_{s\in S} i y_{\lambda,s} N_s\right) F_\lambda \\ &= \exp\left(\sum_{h\in\mathbf{N}^j} b_{\lambda,h} \prod_{k\le j} \sqrt{\frac{y_{\lambda,c_k}}{y_{\lambda,c_{k+1}}}}^{\,-h_k}\right) k_\lambda \cdot \mathbf{r}, \quad \text{where}\end{aligned}$$

$$\begin{aligned}&k_\lambda := \mathrm{Int}(g_\lambda)^{-1}(k_{1,\lambda}) k_{2,\lambda} \in K_{\mathbf{r}}, \quad k_\lambda \to 1, \\ &b_{\lambda,h} \in \mathfrak{g}_{\mathbf{R}}^{-}, \quad b_{\lambda,h} \text{ converges for each } h, \quad b_{\lambda,h}(\pm h) \to 0, \\ &|(\mathrm{Ad} \circ \tilde{\rho}_k)_{1\le k\le j}\text{-weight of } b_{\lambda,h}| \le h. \qquad (8)\end{aligned}$$

Here $b_{\lambda,h}(\pm h)$ denotes the parts of $b_{\lambda,h}$ of weight $\pm h$ with respect to $(\mathrm{Ad} \circ \tilde{\rho}_k)_{1 \le k \le j}$.

In the case $j = n$, (8) already completes the proof of (C_j).

If $0 \le j < n$, it remains to prove (3) of (C_j). This is shown by downward induction on j, as we have mentioned. First we show the following.

CLAIM 2 *To prove* (C_j) (3), *we may assume* $\exp\left(\sum_{t \in S_{\ge j+1}} iy_{\lambda,t} N_t\right) F_\lambda \in \check{D}_{j+1}$ *for any* λ.

We prove Claim 2. By the condition (5) of 6.4.1 for $j+1$, there exist $e \ge 0$, $F_\lambda^* \in \check{D}$, $y_{\lambda,t}^* \in \mathbf{R}$ $(t \in S_{\ge j+2})$ satisfying the following conditions (9)–(13).

(9) $\exp\left(\sum_{t \in S_{\ge j+1}} iy_{\lambda,t}^* N_t\right) F_\lambda^* \in \check{D}_{j+1}$. Here $y_{\lambda,t}^* := y_{\lambda,t}$ $(t \in S_{j+1})$.

(10) $y_{\lambda,c_{j+1}}^{2e} d(F_\lambda, F_\lambda^*) \to 0$.

(11) $y_{\lambda,t} - y_{\lambda,t}^* \to 0$ for any $t \in S_{\ge j+1}$.

(12) Let $\alpha_\lambda = \left(\prod_{k \ge j+1} \tilde{\rho}_k\left(\sqrt{\frac{y_{\lambda,c_k}}{y_{\lambda,c_{k+1}}}}\right)\right) \exp\left(\sum_{t \in S_{\ge j+1}} iy_{\lambda,t} N_t\right)$ and define α_λ^* analogously by replacing $y_{\lambda,t}$ by $y_{\lambda,t}^*$ $(t \in S_{\ge j+1})$. (Note $U_\lambda = \alpha_\lambda F_\lambda$.) Then,

$$\alpha_\lambda (\alpha_\lambda^*)^{-1} \to 1 \text{ in } G_{\mathbf{C}}, \quad \text{and } y_{\lambda,c_{j+1}}^e \log(\alpha_\lambda (\alpha_\lambda^*)^{-1}) \to 0 \text{ in } \mathfrak{g}_{\mathbf{C}}.$$

Furthermore,

$$y_{\lambda,c_{j+1}}^{-e} \mathrm{Ad}(\alpha_\lambda) \to 0$$

in the space of linear endomorphisms of $\mathfrak{g}_{\mathbf{C}}$.

(13) Let $\beta_\lambda = \prod_{k \ge j+1} \tilde{\rho}_k\left(\sqrt{\frac{y_{\lambda,c_k}}{y_{\lambda,c_{k+1}}}}\right)$ and define β_λ^* analogously by replacing y_{λ,c_k} by y_{λ,c_k}^* $(k \ge j+1)$. Then,

$$\mathrm{Ad}(\beta_\lambda)^\nu - \mathrm{Ad}(\beta_\lambda^*)^\nu \to 0 \quad \text{and} \quad y_{\lambda,c_{j+1}}^{-e} \mathrm{Ad}(\beta_\lambda)^\nu \to 0 \quad \text{for } \nu = 0, \pm 1$$

in the space of linear endomorphisms of $\mathfrak{g}_{\mathbf{C}}$.

Define $y_{\lambda,s}^* := y_{\lambda,s}$ for $s \in S_{\le j+1}$. Define $b_{\lambda,h}^*$ $(h \in \mathbf{N}^j)$ just as $b_{\lambda,h}$ by replacing $y_{\lambda,s}$, F_λ by $y_{\lambda,s}^*$, F_λ^*, respectively. To prove Claim 2, it is sufficient to prove that

$$\mathrm{Ad}(\beta_\lambda)^\nu (b_{\lambda,h}) - \mathrm{Ad}(\beta_\lambda^*)^\nu (b_{\lambda,h}^*) \to 0 \quad \text{for } \nu = 0, \pm 1.$$

The left-hand side of this is equal to

$$y_{\lambda,c_{j+1}}^{-e} \mathrm{Ad}(\beta_\lambda)^\nu (y_{\lambda,c_{j+1}}^e (b_{\lambda,h} - b_{\lambda,h}^*)) + (\mathrm{Ad}(\beta_\lambda)^\nu - \mathrm{Ad}(\beta_\lambda^*)^\nu)(b_{\lambda,h}^*).$$

Hence, by (13), it is sufficient to prove that

$$y_{\lambda,c_{j+1}}^e (b_{\lambda,h} - b_{\lambda,h}^*) \to 0.$$

Since $b_{\lambda,h}$ is a real analytic function in $((N_{\lambda,k})_{1 \le k \le j}, U_\lambda)$ (6.1.6), it is sufficient to prove that

$$y_{\lambda,c_{j+1}}^e d(U_\lambda, U_\lambda^*) \to 0 \tag{14}$$

where $U_\lambda^* = \alpha_\lambda^* F_\lambda^*$. By (10), we can write

$$F_\lambda = \exp(x_\lambda) F_\lambda^* \quad \text{with} \quad x_\lambda \in \mathfrak{g}_{\mathbf{C}}, \quad y_{\lambda,c_{j+1}}^{2e} x_\lambda \to 0.$$

We have

$$U_\lambda = \alpha_\lambda F_\lambda = \exp(\mathrm{Ad}(\alpha_\lambda)(x_\lambda))\alpha_\lambda(\alpha_\lambda^*)^{-1}U_\lambda^*.$$

Using (12) and

$$y_{\lambda,c_{j+1}}^{e}\,\mathrm{Ad}(\alpha_\lambda)(x_\lambda) = (y_{\lambda,c_{j+1}}^{-e}\,\mathrm{Ad}(\alpha_\lambda))(y_{\lambda,c_{j+1}}^{2e}x_\lambda) \to 0,$$

we obtain (14). Thus Claim 2 is proved.

Now we assume $\exp\left(\sum_{t\in S_{\geq j+1}} iy_{\lambda,t}N_t\right)F_\lambda \in \check{D}_{j+1}$ for any λ. Then, by 6.2.1, we have elements $b_{\lambda,h'}$ for $h' \in \mathbf{N}^{j+1}$ and we have, for $h \in \mathbf{N}^j$,

$$b_{\lambda,h} = \sum_{k=0}^{\infty} b_{\lambda,(h,k)}\sqrt{\frac{y_{\lambda,c_{j+1}}}{y_{\lambda,c_{j+2}}}}^{\,-k}. \tag{15}$$

Using (8), applied to $b_{\lambda,(h,k)}$ replacing j by $j+1$, we have

(16) $|(\mathrm{Ad}\circ\tilde{\rho}_{j+1})$-weight of $b_{\lambda,(h,k)}| \leq k$. If $h=0$, the parts of $b_{\lambda,(h,k)}$ of $(\mathrm{Ad}\circ\tilde{\rho}_{j+1})$-weight $\pm k$ converge to 0.

By the hypothesis of induction,

(17) $\mathrm{Ad}\left(\prod_{k\geq j+2}\tilde{\rho}_k\left(\sqrt{\frac{y_{\lambda,c_k}}{y_{\lambda,c_{k+1}}}}\right)\right)^{\nu}(b_{\lambda,(h,k)})$ converges for $\nu = 0, \pm 1$.

By (15)–(17), we have (3) of (C_j). □

6.4.5

Proof of Theorem 5.4.3 (ii). Let $\mathfrak{F}$ be the filter on the set $A_{\mathbf{R}}$ for which the sets $\sigma + y$ ($\sigma \in \mathfrak{F}(A, V)$, $y \in \sigma$) (5.3.2) is a base. For $F \in Z$, let $\mathfrak{F}'_F$ be the filter on the set D whose basis is given by the sets $\{\exp(iy)F \mid y \in U\}$ where U ranges over sufficiently small elements in $\mathfrak{F}$. Then $\mathfrak{F}'_F$ is independent of the choice of $F \in Z$. So we denote $\mathfrak{F}'_F$ by $\mathfrak{F}'$. By Proposition 6.4.1, for any $(N_j)_{1\leq j\leq n}$ and $F \in Z$ as in the hypothesis of 5.4.3 (ii), the class $[\rho, \varphi] \in D_{\mathrm{SL}(2)}$ of the SL(2)-orbit (ρ, φ) associated with $(N_1, \dots, N_n, F)$ (6.1.3) is the limit of $\mathfrak{F}'$ in $D_{\mathrm{SL}(2)}$ (5.2.16) and hence it is independent of the choice of $(N_1, \dots, N_n, F)$. □

We use the notion of "regular space," which we review here.

Definition 6.4.6 [Bn, Ch. 1, §8, no. 4, Definition 2] *A topological space is called* regular *if it is Hausdorff and satisfies the following axiom: Given any closed subset F of X and any point $x \notin F$, there is a neighborhood of x and a neighborhood of F that are disjoint.*

Lemma 6.4.7 [Bn, Ch. 1, §8, no. 5, Theorem 1] *Let X be a topological space, A a dense subset of X, $f : A \to Y$ a map from A into a regular space Y. A necessary and sufficient condition for f to extend to a continuous map $\overline{f} : X \to Y$ is that, for each $x \in X$, $f(y)$ tends to a limit in Y when y tends to x while remaining in A. The continuous extension $\overline{f}$ of f to X is then unique.*

6.4.8

Our method for the proof of Theorem 5.4.4 is as follows. It is sufficient to prove that, for $\sigma \in \Sigma$, the composite map $\tilde{\psi} : E^{\sharp}_{\sigma,\mathrm{val}} \to D_{\sigma,\mathrm{val}} \xrightarrow{\psi} D_{\mathrm{SL}(2)}$ is continuous. Let $p \in E^{\sharp}_{\sigma,\mathrm{val}}$, and let W be the family of weight filtrations associated with $\tilde{\psi}(p) \in D_{\mathrm{SL}(2)}$. We show in 6.4.10 that there is an open neighborhood U of p such that $\tilde{\psi}(U) \subset D_{\mathrm{SL}(2)}(W)$. Now since $D_{\mathrm{SL}(2)}(W)$ is a regular space (5.2.15 (iii)), to prove the continuity of ψ, it is sufficient to prove the following (1).

(1) Let $(x_\lambda)_\lambda$ be a directed family in $E^{\sharp}_{\sigma,\mathrm{val}}$ such that the image of x_λ in $|\mathrm{toric}|_{\sigma,\mathrm{val}}$ is contained in $|\mathrm{torus}|_\sigma$ for any λ, and assume x_λ converges to p in $E^{\sharp}_{\sigma,\mathrm{val}}$. Then $\psi(x_\lambda)$ converges to $\psi(p)$ in $D_{\mathrm{SL}(2)}$.

We will prove this by using Proposition 6.4.1.

Lemma 6.4.9 *Let $p = (A, V, Z) \in D^{\sharp}_{\mathrm{val}}$, and let W be the family of weight filtrations associated to $\psi(p)$. Let $(N_s)_{s \in S}$ be an excellent basis for p, and let S_j $(1 \le j \le n)$ be as in 6.3.3. Let $\tau \in \mathfrak{F}(A, V)$ and assume that τ satisfies the following condition* (1).

(1) *Let $y_s \in \mathbf{R}$ and assume $\sum_{s \in S} y_s N_s \in \tau$. Then $y_s \ge 0$ for all s. Furthermore, if $s \in S_j$ and $y_s \ne 0$, then $y_t \ne 0$ for any $k \le j$ and any $t \in S_k$.*

Then for any $p' = (A', V', Z') \in D^{\sharp}_{\mathrm{val}}$ such that $\tau \cap A' \in \mathfrak{F}(A', V')$, we have $\psi(p') \in D_{\mathrm{SL}(2)}(W)$.

Proof. Let W' be the family of weight filtrations associated to $\psi(p')$. By Proposition 6.3.9, there exists an excellent basis $(N'_s)_{s \in S'}$ for p' such that $N'_s \in \tau$ for any $s \in S'$. Let $S'_j \subset S'$ $(1 \le j \le n')$ be as in 6.3.3. Recall that W and W' are described as follows (6.1.3 (6)). Let J (resp. J') be the subset of $\{1, \dots, n\}$ (resp. $\{1, \dots, n'\}$) consisting of elements k such that

$$W\Big(\sum_{s \in S_{\le k-1}} (\mathbf{R}_{\ge 0}) N_s\Big) \ne W\Big(\sum_{s \in S_{\le k}} (\mathbf{R}_{\ge 0}) N_s\Big),$$

$$W\Big(\sum_{s \in S'_{\le k-1}} (\mathbf{R}_{\ge 0}) N'_s\Big) \ne W\Big(\sum_{s \in S'_{\le k}} (\mathbf{R}_{\ge 0}) N'_s\Big).$$

Write

$$J = \{b(1), \dots, b(m)\} \text{ with } b(1) < \cdots < b(m),$$
$$J' = \{b'(1), \dots, b'(m')\} \text{ with } b'(1) < \cdots < b'(m').$$

Then

$$W = (W^{(j)})_{1 \le j \le m} \text{ where } W^{(j)} = W\Big(\sum_{s \in S_{\le b(j)}} (\mathbf{R}_{\ge 0}) N_s\Big),$$

$$W' = (W'^{(j)})_{1 \le j \le m'} \text{ where } W'^{(j)} = W\Big(\sum_{s \in S'_{\le b'(j)}} (\mathbf{R}_{\ge 0}) N'_s\Big).$$

Let $c : \{1, \ldots, n\} \to \{1, \ldots, m\}$ and $d : \{1, \ldots, m'\} \to \{1, \ldots, n\}$ be the maps defined as follows. Let $c(k) = \max\{j \mid 1 \le j \le m, b(j) \le k\}$. Then $W\big(\sum_{s \in S_{\le k}} (\mathbf{R}_{\ge 0})N_s\big) = W^{(c(k))}$ for any k. Let $1 \le j \le m'$, write $\sum_{s \in S_{\le b'(j)}} N'_s = \sum_{s \in S} y_s N_s$, and let $d(j) = \max\{k \mid 1 \le k \le n,\ y_s \ne 0 \text{ for some } s \in S_k\}$. Then, by the assumption (1) on τ, $y_s \ne 0$ for any $s \in S_{\le d(j)}$. Hence by Cattani and Kaplan (5.2.4), $W'^{(j)} = W\big(\sum_{s \in S_{\le b'(j)}} N'_s\big) = W^{(c(d(j)))}$. Since $c(d(j)) \le c(d(k))$ if $1 \le j \le k \le m'$, this shows $\psi(p') \in D_{\mathrm{SL}(2)}(W)$. □

Lemma 6.4.10 *Let $p \in E^{\sharp}_{\sigma,\mathrm{val}}$, and let W be the family of weight filtrations associated to $\tilde{\psi}(p) \in D_{\mathrm{SL}(2)}$. Then there exists an open neighborhood U of p such that $\tilde{\psi}(U) \subset D_{\mathrm{SL}(2)}(W)$.*

Proof. Let $\overline{p} = (A, V, Z)$ be the image of p in $D^{\sharp}_{\sigma,\mathrm{val}}$. Take an excellent basis $(N_s)_{s \in S}$ for $\overline{p}$ such that $N_s \in \sigma$ for any $s \in S$ (Proposition 6.3.9). Then, $\tau = \sigma_{c,\alpha,\beta}$ for any c, α, β as in 6.3.11 ($\sigma_{c,\alpha,\beta}$ is defined here with respect to $(N_s)_{s \in S}$) satisfies the condition (1) in 6.4.9. Take τ satisfying (1) in 6.4.9, and let U be the subset of $E^{\sharp}_{\sigma,\mathrm{val}}$ consisting of all elements whose images (A', V', Z') in $D^{\sharp}_{\sigma,\mathrm{val}}$ satisfy $\tau \cap A' \in \mathfrak{F}(A', V')$. It is sufficient to prove the following (1)–(3).

(1) $p \in U$.
(2) $\tilde{\psi}(U) \subset D_{\mathrm{SL}(2)}(W)$.
(3) U is open.

(1) is clear. (2) follows from Lemma 6.4.9. We prove (3). The inclusion $\Gamma(\tau) \subset \Gamma(\sigma)$ induces $\Gamma(\sigma)^{\vee} \subset \Gamma(\tau)^{\vee}$. Recall that $|\mathrm{toric}|_{\sigma,\mathrm{val}}$ is identified with the set of all pairs (V'', h) where V'' a valuative submonoid of $\Gamma(\sigma)^{\vee\,\mathrm{gp}} = \Gamma(\tau)^{\vee\,\mathrm{gp}}$ contaning $\Gamma(\sigma)^{\vee}$ and h is a homomorphism $V'' \to \mathbf{R}^{\mathrm{mult}}_{\ge 0}$ such that $h^{-1}(\mathbf{R}_{>0}) = (V'')^{\times}$ (5.1.11). It is easy to see that the set U coincides with the inverse image of the open set of $|\mathrm{toric}|_{\sigma,\mathrm{val}}$ consisting of all $(V'', h) \in |\mathrm{toric}|_{\sigma,\mathrm{val}}$ such that $V'' \supset \Gamma(\tau)^{\vee}$, and hence is open. □

Lemma 6.4.11 *Let $x \in |\mathrm{toric}|_{\sigma,\mathrm{val}}$ and let (A, V, d) be the triple corresponding to x as in 5.3.6. Take a good basis $(N_j)_{j \in S}$ for (A, V), and let $(a_s)_s$ and $(S_j)_j$ be as in 6.3.3. Let B be any $\mathbf{R}$-vector subspace of $\sigma_{\mathbf{R}}$ such that $\sigma_{\mathbf{R}} = A_{\mathbf{R}} \oplus B$, and let $\tilde{d} \in B$ be the unique representative of $d \in \sigma_{\mathbf{R}}/A_{\mathbf{R}}$ in $\sigma_{\mathbf{R}}$ contained in B. Then $\mathbf{e}(iy)$ $(y \in \sigma_{\mathbf{R}})$ converges to x if and only if the following (1)–(4) are satisfied. Write $y = \sum_{s \in S} y_s N_s + b \in \sigma_{\mathbf{R}}$ $(y_s \in \mathbf{R},\ b \in B)$.*

(1) $y_s \to \infty$ for any $s \in S$.
(2) If $1 \le j < k \le n$ and $s \in S_j, t \in S_k$, then $\frac{y_s}{y_t} \to \infty$.
(3) If $1 \le j \le n$ and $s, t \in S_j$, $\frac{y_s}{y_t}$ converges to $\frac{a_s}{a_t}$.
(4) b converges to $\tilde{d}$.

Proof. Write $x = (V', h)$ where V' is a valuative submonoid of $\sigma^{\vee\,\mathrm{gp}}$ containing $\sigma^{\vee}$ and h is a homomorphism $V' \to \mathbf{R}^{\mathrm{mult}}_{\ge 0}$ such that $h^{-1}(\mathbf{R}_{>0}) = (V')^{\times}$ (5.1.11).

Assume $\mathbf{e}(iy)$ converges to x. We prove that (1)–(4) are satisfied.

Let v be an element of V which sends N_s to 1 and N_t $(t \neq s)$ to 0. Then there exists an element $v' \in V'$ such that the composition $A \xrightarrow{\exp} \Gamma(\sigma)^{\mathrm{gp}} \otimes \mathbf{Q} \xrightarrow{v'} \mathbf{Q}$ coincides with a positive multiple of v. Then $v' \notin (V')^{\times}$. Hence $h(v') = 0$. Since $\mathbf{e}(iy)$ converges to x, $v(y)$ converges to ∞ (see 5.1.11 (3)). This shows that (1) is satisfied.

We prove that (2) is satisfied. Let $c > 0$, and let $v \in V$ be the element that sends N_s to 1, N_t to $-c$, and N_u to 0 $(u \in S - \{s, t\})$. As above, we have $v(y) = y_s - cy_t \to \infty$. Since c is arbitrary, this shows that (2) is satisfied.

We prove that (3) is satisfied. Take $\varepsilon > 0$. Let v be the element of V which sends N_s to $-a_t$, N_t to $a_s + \varepsilon$, and N_u to 0 $(u \in S - \{s, t\})$. As above, we have

$$v(y) = -a_t y_s + (a_s + \varepsilon) y_t \to \infty. \tag{5}$$

Next let v be the element of V which sends N_s to a_t, N_t to $-a_s + \varepsilon$, and N_u to 0 $(u \in S - \{s, t\})$. As above, we have

$$v(y) = a_t y_s + (-a_s + \varepsilon) y_t \to \infty. \tag{6}$$

Since ε is arbitrary, these results (5) and (6) prove that (3) is satisfied.

We prove that (4) is satisfied. For any $v' \in (V')^{\times}$, $v'(\mathbf{e}(ib))$ converges to $v'(\mathbf{e}(i\tilde{d}))$ in $\mathbf{R}_{>0}$. This shows that b converges to $\tilde{d}$.

On the other hand, it is easy to see that $e(iy)$ converges to x if (1)–(4) are satisfied. □

6.4.12

Proof of Theorem 5.4.4. We prove Theorem 5.4.4. It is sufficient to prove that $\tilde{\psi} : E^{\sharp}_{\sigma,\mathrm{val}} \to D_{\mathrm{SL}(2)}$ is continuous. Let $p \in E^{\sharp}_{\sigma,\mathrm{val}}$ and let W be the family of weight filtrations associated with $\tilde{\psi}(p)$. By the fact that $D_{\mathrm{SL}(2)}(W)$ is a regular space, and by Lemmas 6.4.7 and 6.4.10, it is sufficient to prove (1) in 6.4.8.

Take an $\mathbf{R}$-subspace B of $\sigma_{\mathbf{R}}$ such that $\sigma_{\mathbf{R}} = A_{\mathbf{R}} \oplus B$. Write $x_\lambda = (\mathbf{e}(y_\lambda + b_\lambda), F'_\lambda)$ with $y_\lambda \in A_{\mathbf{R}}$, $b_\lambda \in B$, $F'_\lambda \in \check{D}$. Let $F_\lambda = \exp(b_\lambda) F'_\lambda$. Fix an excellent basis $(N_s)_{s \in S}$ for (A, V), and let S_j $(1 \le j \le n)$ and a_s $(s \in S)$ be as in 6.3.3. Write $y_\lambda = \sum_{s \in S} y_{\lambda,s} N_s$. Then from Lemma 6.4.11, we see that the assumption of Proposition 6.4.1 is satisfied. (The condition (5) in 6.4.1 is satisfied by Proposition 3.1.6.) Hence, by Propositions 6.4.1 and 5.2.16, $\tilde{\psi}(x_\lambda) = \exp(y_\lambda) F_\lambda$ converges to $\tilde{\psi}(p)$. □

Chapter Seven

Proof of Theorem A

In this chapter we prove Theorem A. In Section 7.1, we prove Theorem A (i). In Section 7.2, we study the actions of $\sigma_{\mathbf{C}}$ (resp. $i\sigma_{\mathbf{R}}$) on E_σ, $E_{\sigma,\mathrm{val}}$ (resp. $E_\sigma^\sharp$, $E_{\sigma,\mathrm{val}}^\sharp$), and then, using these results, we prove Theorem A for $\Gamma(\sigma)^{\mathrm{gp}}\backslash D_\sigma$ in Section 7.3. In Section 7.4, we complete the proof of Theorem A for $\Gamma\backslash D_\Sigma$.

In Chapter 7, let Σ be a fan in $\mathfrak{g}_{\mathbf{Q}}$ and let Γ be a subgroup of $G_{\mathbf{Z}}$ which is strongly compatible with Σ.

7.1 PROOF OF THEOREM A (i)

In this section, we prove Theorem A (i).

We first prove

Proposition 7.1.1 *Let $\sigma \in \Sigma$, let $\tilde{A}(\sigma)$ be the closed analytic subset of $\check{D}$ defined by*

$$\tilde{A}(\sigma) = \{F \in \check{D} \mid N(F^p) \subset F^{p-1} \ (\forall N \in \sigma, \forall p \in \mathbf{Z})\},$$

and let $A(\sigma)$ be the subset of $\check{D}$ consisting of all elements F such that $(\sigma, \exp(\sigma_{\mathbf{C}})F)$ is a nilpotent orbit. Then $A(\sigma)$ is an open set of $\tilde{A}(\sigma)$.

Proof. Let $F_{(0)} \in A(\sigma)$. It is enough to prove that, if $F \in \tilde{A}(\sigma)$ is sufficiently close to $F_{(0)}$, then $F \in A(\sigma)$.

In the rest of the proof of Proposition 7.1.1, F always denotes an element of $\tilde{A}(\sigma)$.

We prove the following (A_k) and (B_k) by downward induction on k. Let $W = W(\sigma)$.

(A_k) If F converges to $F_{(0)}$, then $F(W_k)$ converges to $F_{(0)}(W_k)$.
(B_k) If F converges to $F_{(0)}$, then $F(\mathrm{gr}_k^W)$ converges to $F_{(0)}(\mathrm{gr}_k^W)$.

If k is sufficiently large, (A_k) and (B_k) obviously hold. Furthermore, we can deduce (A_{k-1}) from (A_k) and (B_k), since $F^p(W_{k-1}) = \mathrm{Ker}(F^p(W_k) \to F^p(\mathrm{gr}_k^W))$ and $F_{(0)}^p(W_{k-1}) = \mathrm{Ker}(F_{(0)}^p(W_k) \to F_{(0)}^p(\mathrm{gr}_k^W))$. Hence it is sufficient to prove (B_k), assuming (A_l) for all $l \geq k$. Note that, since $F_{(0)}(\mathrm{gr}_k^W)$ is an $\mathbf{R}$-Hodge structure of weight $w+k$, we have

$$\mathrm{gr}_{k,\mathbf{C}}^W = F_{(0)}^p(\mathrm{gr}_k^W) \oplus \overline{F}_{(0)}^q(\mathrm{gr}_k^W) \quad \text{if } p+q = w+k+1. \tag{1}$$

Take a basis $e_{(0)}'^p$ of $F_{(0)}^p(\mathrm{gr}_k^W)$ and a basis $e_{(0)}^p$ of $F_{(0)}^p(W_k)$ whose image in gr_k^W contains $e_{(0)}'^p$. If F converges to $F_{(0)}$, $F^p(W_k)$ converges to $F_{(0)}^p(W_k)$ by our hypothesis

(A_k), and hence some basis e^p of $F^p(W_k)$ converges to $e^p_{(0)}$. Applying this also replacing p by $q = w+k+1-p$, we have

$$\mathrm{gr}^W_{k,\mathbf{C}} = F^p(\mathrm{gr}^W_k) + \overline{F}^q(\mathrm{gr}^W_k) \tag{2}$$

if F is sufficiently close to $F_{(0)}$. We show that

$$\mathrm{gr}^W_{k,\mathbf{C}} = F^p(\mathrm{gr}^W_k) \oplus \overline{F}^q(\mathrm{gr}^W_k) \tag{3}$$

if F is sufficiently close to $F_{(0)}$. Assume first that $k \geq 0$. If (3) is not true, we should have by (2) that

$$F^p(\mathrm{gr}^W_k) \cap \overline{F}^q(\mathrm{gr}^W_k) \neq 0. \tag{4}$$

Take an element N in the interior (0.7.7) of σ. Then N^k induces $\mathrm{gr}^W_k \xrightarrow{\sim} \mathrm{gr}^W_{-k}$. Since $F \in \tilde{A}(\sigma)$, we have $N^k F^p \subset F^{p-k}$ and $N^k \overline{F}^q \subset \overline{F}^{q-k}$, and hence (4) shows that

$$F^{p-k}(\mathrm{gr}^W_{-k}) \cap \overline{F}^{q-k}(\mathrm{gr}^W_{-k}) \neq 0. \tag{5}$$

On the other hand, $\langle\ ,\ \rangle_0$ induces a nondegenerate pairing

$$\mathrm{gr}^W_{k,\mathbf{C}} \times \mathrm{gr}^W_{-k,\mathbf{C}} \to \mathbf{C}, \tag{6}$$

and $\mathrm{gr}^W_{k,\mathbf{C}} = F^q(\mathrm{gr}^W_k) + \overline{F}^p(\mathrm{gr}^W_k)$ annihilates $F^{p-k}(\mathrm{gr}^W_{-k}) \cap \overline{F}^{q-k}(\mathrm{gr}^W_{-k})$ in this pairing since $p+q-k > w$. Hence $F^{p-k}(\mathrm{gr}^W_{-k}) \cap \overline{F}^{q-k}(\mathrm{gr}^W_{-k}) = 0$, contradicting (5). This proves (3) in the case $k \geq 0$. Assume next that $k \leq 0$. If (3) is not satisfied, then we should have (4). Since $-k \geq k$, (A_{-k}) is assumed and hence $\mathrm{gr}^W_{-k,\mathbf{C}} = F^{q-k}(\mathrm{gr}^W_{-k}) + \overline{F}^{p-k}(\mathrm{gr}^W_{-k})$ if F is sufficiently close to $F_{(0)}$. But this shows that $\mathrm{gr}^W_{-k,\mathbf{C}}$ annihilates $F^p(\mathrm{gr}^W_k) \cap \overline{F}^q(\mathrm{gr}^W_k)$ in the pairing (6), for $p+q-k > w$, contradicting (4). Thus (3) is proved. Now (3) and the fact e^p converges to $e^p_{(0)}$ prove (B_k).

By (B_k), we have

(7) $(W[-w], F)$ is a mixed Hodge structure if F is sufficiently close to $F_{(0)}$.

Now we prove that $F \in A(\sigma)$ if F is sufficiently close to $F_{(0)}$. By [CKS 4.66], it is enough to prove that, if F is sufficiently close to $F_{(0)}$, then

(8) $(W[-w], F, \langle\ ,\ \rangle_0, N)$ is a polarized mixed Hodge structure for any element N of the interior (0.7.7)of σ.

For $k \geq 0$, let $P_{k,N}$ be the primitive part of gr^W_k with respect to N, let $b_{k,N}$ be the nondegenerate bilinear form on gr^W_k defined by $(x, y) \mapsto \langle x, N^k(y)\rangle_0$, and let $h_{k,N,F}$ be the nondegenerate Hermitian form on $P_{k,N,\mathbf{C}}$ defined by $(x, y) \mapsto b_{k,N}(i^{p-q}x, \overline{y})$ for x in the (p, q)-component of the Hodge structure $F(P_{k,N,\mathbf{C}})$. Then (8) is satisfied if and only if the following conditions (9) and (10) are satisfied for any $k \geq 0$.

(9) $b_{k,N}(F^p(P_{k,N,\mathbf{C}}), F^q(P_{k,N,\mathbf{C}})) = 0$ if $p+q > k+w$.
(10) $h_{k,N,F}$ is positive definite.

We prove (9). Since $F \in \tilde{A}(\sigma)$, $N^k(F^p) \subset F^{p-k}$. Since $(p-k)+q > w$, $\langle F^{p-k}, F^q\rangle_0 = 0$. This proves (9).

We show

(11) If (10) is satisfied by one element N of the interior (0.7.7) of σ, it is satisfied for any N in the interior of σ.

Let N, N' be elements of the interior of σ and assume that (10) is true. For $a \in \mathbf{R}$, $0 \le a \le 1$, consider the elements $N_a := aN + (1-a)N'$ of the interior of σ. Take bases of $P_{k,N_a,\mathbf{C}}$ continuously in a. Then $h_{k,N_a,F}$ are regarded as nondegenerate Hermitian matrices. If $s(a)$ denotes the minimum of all eigenvalues of the matrix $h_{k,N_a,F}$, $s(a)$ is a continuous function from the interval [0, 1] to $\mathbf{R}$ which does not have value 0. Since $h_{k,N,F}$ is positive definite, we have $s(1) > 0$. Hence $s(a) > 0$ for any $a \in [0, 1]$ by the continuity. Hence $s(0) > 0$ and this shows that $h_{k,N',F}$ is also positive definite.

For a fixed element N in the interior of σ, since (9) is satisfied, the condition (10) is an open condition on $F(P_{k,N,\mathbf{C}})$, satisfied by $F_{(0)}(P_{k,N,\mathbf{C}})$. Hence, by ($\mathrm{B}_k$), it is satisfied if F is sufficiently close to $F_{(0)}$. □

7.1.2

Proof of Theorem A (i). We have

$$\tilde{E}_{\sigma,\mathrm{val}} = \mathrm{toric}_{\sigma,\mathrm{val}} \times_{\mathrm{toric}_\sigma} \tilde{E}_\sigma$$

as a topological space. Since $\tilde{E}_{\sigma,\mathrm{val}} \to \tilde{E}_\sigma$ is a proper surjective map, the topology of $\tilde{E}_\sigma$ is the quotient topology of the one of $\tilde{E}_{\sigma,\mathrm{val}}$. Hence it is sufficient to prove that $E_{\sigma,\mathrm{val}}$ is open in $\tilde{E}_{\sigma,\mathrm{val}}$.

Assume $(x_\lambda)_\lambda$ is a directed family of elements of $\tilde{E}_{\sigma,\mathrm{val}}$ which converges to $x \in E_{\sigma,\mathrm{val}}$. We prove

(1) $x_\lambda \in E_{\sigma,\mathrm{val}}$ if λ is sufficiently large.

Write $x_\lambda = (q_\lambda, F'_\lambda)$, $x = (q, F')$ $(q_\lambda, q \in \mathrm{toric}_{\sigma,\mathrm{val}},\ F'_\lambda, F' \in \check{D})$. Let $\overline{x} = (A, V, Z) \in D^\sharp_{\sigma,\mathrm{val}}$ be the image of x and take an excellent basis $(N_s)_{s\in S}$ for $\overline{x}$ such that $N_s \in \sigma(q)$ for any s (6.3.9). Let S_j $(1 \le j \le n)$ be as in 6.3.3. For $1 \le j \le n$, let $\sigma_j = \sum_{s\in S_{\le j}} (\mathbf{R}_{\ge 0})N_s$. Take an $\mathbf{R}$-subspace B of $\sigma_{\mathbf{R}}$ such that $\sigma_{\mathbf{R}} = A_{\mathbf{R}} \oplus B$.

We have a unique injective open continuous map

$$(\mathbf{R}^S_{\ge 0})_{\mathrm{val}} \times B \to |\mathrm{toric}|_{\sigma,\mathrm{val}}$$

which sends $((e^{-2\pi y_s})_{s\in S}, b)$ $(y_s \in \mathbf{R}, b \in B)$ to $\mathbf{e}((\sum_{s\in S} iy_s N_s) + ib)$ (cf. 3.3.5). Let U be the image of this map. Define the maps $t_s : U \to \mathbf{R}_{\ge 0}$ $(s \in S)$ and $b : U \to B$ by

$$(t, b) = ((t_s)_{s\in S}, b) : U \simeq (\mathbf{R}^S_{\ge 0})_{\mathrm{val}} \times B \to \mathbf{R}^S_{\ge 0} \times B.$$

Let $|\ | : \mathrm{toric}_{\sigma,\mathrm{val}} \to |\mathrm{toric}|_{\sigma,\mathrm{val}}$ be the canonical projection induced by $\mathbf{C} \to \mathbf{R}$, $z \mapsto |z|$. Then, $|q| \in U$ and $t(|q|) := (t_s(|q|))_{a\in S} = 0$. Since $|q_\lambda| \to |q|$, we may assume $|q_\lambda| \in U$. By Lemma 6.4.11, for each sufficiently large λ, there exists j with $0 \le j \le n$ such that $t_s(|q_\lambda|) = 0$ if $s \in S_{\le j}$ and $t_s(|q_\lambda|) \ne 0$ if $s \in S_{\ge j+1}$. Dividing $(x_\lambda)_\lambda$ into finite subfamilies, we may assume this j is common for any λ. For $s \in S_{\ge j+1}$, define $y_{\lambda,s} \in \mathbf{R}$ by $t_s(|q_\lambda|) = e^{-2\pi y_{\lambda,s}}$. Define $F_\lambda := \exp(b(|q_\lambda|))F'_\lambda$ and $F := \exp(b(|q|))F'$. Note that, since $x_\lambda \in \tilde{E}_{\sigma,\mathrm{val}}$, we have F_λ, $\exp(\sum_{s \in S_{\ge j+1}} iy_{\lambda,s}N_s)F_\lambda \in \tilde{A}(\sigma_j)$ (notation in 7.1.1).

We prove (1) by downward induction on this j.

If $j = n$, $F_\lambda \in \tilde{A}(\sigma_n)$ converges to F and (σ_n, F) generates a nilpotent orbit and hence, by Proposition 7.1.1, if λ is sufficiently large then (σ_n, F_λ) generates a nilpotent orbit, that is, $x_\lambda \in E_{\sigma,\mathrm{val}}$.

Assume $j < n$. Let d be a metric on a neighborhood of F in $\check{D}$ which is compatible with the analytic structure (3.1.4). Take any l such that $j < l \le n$ and take any $e \ge 0$. Then, since $\tilde{E}_\sigma$ is an analytically constructible subset of $\check{E}_\sigma$ (3.5.9, 3.5.10), we can deduce the following fact from Propositions 3.1.5 and 3.1.6: When λ is sufficiently large, there exist $x^*_\lambda = (q^*_\lambda, F^{*\prime}_\lambda) \in \tilde{E}_{\sigma,\mathrm{val}}$ $(\lambda \in \Lambda)$ satisfying the following conditions (2)–(4). Let $F^*_\lambda = \exp(b(|q^*_\lambda|))F^{*\prime}_\lambda$.

(2) For $s \in S$, $t_s(|q^*_\lambda|) = 0$ if $s \in S_{\le l}$, and $t_s(|q^*_\lambda|) \ne 0$ if $s \in S_{\ge l+1}$.

(3) Let $s \in S_{\ge l+1}$ and define $y^*_{\lambda,s} \in \mathbf{R}$ by $t_s(|q_\lambda|) = e^{-2\pi y^*_{\lambda,s}}$. Then

$$y^e_{\lambda,s} d(F_\lambda, F^*_\lambda) \to 0 \quad (\forall s \in S_l),$$
$$y^e_{\lambda,s} |y_{\lambda,t} - y^*_{\lambda,t}| \to 0 \quad (\forall s \in S_l, \ \forall t \in S_{\ge l+1}).$$

(4) x^*_λ converges to x in $\tilde{E}_{\sigma,\mathrm{val}}$.

By the hypothesis of downward induction on j, x^*_λ belongs to $E_{\sigma,\mathrm{val}}$ if λ is sufficiently large.

Fix N in the interior (0.7.7) of σ_j. Let $k \ge 0$ and let $P_{k,N}$ be the primitive part of $\mathrm{gr}^{W(\sigma_j)}_k$ with respect to N. It is enough to prove that, when λ is sufficiently large, $F_\lambda(P_{k,N,\mathbf{C}})$ is a polarized Hodge structure with respect to the intersection form $b_{k,N} : (x, y) \mapsto \langle x, N^k(y)\rangle_0$ (as in the proof of 7.1.1). (Though we have to prove this for any N in the interior of σ_j, it is enough to consider one fixed N by 7.1.1 (11).) By replacing our $H_{0,\mathbf{R}}$ with $\langle\ ,\ \rangle_0$ by $P_{k,N}$ with $b_{k,N}$, we are reduced to the case $j = 0$. Then the assumptions of Proposition 6.4.1 are satisfied, and hence we have $\exp(\sum_{s \in S} iy_{\lambda,s}N_s)F_\lambda \in D$ if λ is sufficiently large. This shows that $(q_\lambda, F_\lambda) \in E_{\sigma,\mathrm{val}}$ if λ is sufficiently large (3.3.7). Thus we have proved Theorem A (i). □

The following result will be used in §7.2.

Corollary 7.1.3 *The map*

$$E_{\sigma,\mathrm{val}} \to E^\sharp_{\sigma,\mathrm{val}}, \quad (q, F) \mapsto (|q|, F),$$

is continuous.

Proof. Since $\tilde{E}_\sigma$ is an analytically constructible subset of $\check{E}_\sigma$ and E_σ is an open set of $\tilde{E}_\sigma$ in the strong topology, we can use Proposition 3.1.5 to check the continuity of $E_\sigma \to E^\sharp_\sigma$, $(q, F) \mapsto (|q|, F)$. The assertion follows from this. □

7.2 ACTION OF $\sigma_{\mathbf{C}}$ ON E_σ

7.2.1

Let $\sigma \in \Sigma$. Consider the action of $\sigma_{\mathbf{C}}$ on E_σ for which $a \in \sigma_{\mathbf{C}}$ sends (q, F) to $(\mathbf{e}(a)q, \exp(-a)F)$ where $q \in \text{toric}_\sigma$, $F \in \check{D}$, $(q, F) \in E_\sigma$. This action is continuous. In fact, by a formal argument, we see that this action $\sigma_{\mathbf{C}} \times E_\sigma \to E_\sigma$ is continuous in the strong topology of $\sigma_{\mathbf{C}} \times E_\sigma$ in $\sigma_{\mathbf{C}} \times \check{E}_\sigma$. By 3.1.8 (2), this proves the continuity of the action $\sigma_{\mathbf{C}} \times E_\sigma \to E_\sigma$ in the product topology on $\sigma_{\mathbf{C}} \times E_\sigma$.

From this continuity, we see that the action of $\sigma_{\mathbf{C}}$ on $E_{\sigma,\text{val}}$, for which $a \in \sigma_{\mathbf{C}}$ sends (q, F) to $(\mathbf{e}(a)q, \exp(-a)F)$ where $q \in \text{toric}_{\sigma,\text{val}}$, $F \in \check{D}$, $(q, F) \in E_{\sigma,\text{val}}$, and $\mathbf{e}$ in 3.3.5, is also continuous.

Consider also the induced action of $i\sigma_{\mathbf{R}}$ on $E_\sigma^\sharp$ (resp. $E_{\sigma,\text{val}}^\sharp$).

The quotient spaces for these actions are identified as

$$\sigma_{\mathbf{C}}\backslash E_\sigma = \Gamma(\sigma)^{\text{gp}}\backslash D_\sigma, \quad \sigma_{\mathbf{C}}\backslash E_{\sigma,\text{val}} = \Gamma(\sigma)^{\text{gp}}\backslash D_{\sigma,\text{val}},$$
$$i\sigma_{\mathbf{R}}\backslash E_\sigma^\sharp = D_\sigma^\sharp, \quad i\sigma_{\mathbf{R}}\backslash E_{\sigma,\text{val}}^\sharp = D_{\sigma,\text{val}}^\sharp.$$

In this section, we prove the following theorem.

Theorem 7.2.2

(i) *The action of $\sigma_{\mathbf{C}}$ on E_σ (resp. $E_{\sigma,\text{val}}$) is proper.*
(ii) *The action of $i\sigma_{\mathbf{R}}$ on $E_\sigma^\sharp$ (resp. $E_{\sigma,\text{val}}^\sharp$) is proper.*

(In fact (ii) is a corollary of (i), because $E_\sigma^\sharp$ is a closed topological subspace of E_σ and $i\sigma_{\mathbf{R}}$ is a closed topological subgroup of $\sigma_{\mathbf{C}}$.) Before proving this theorem, we recall the notion of "proper action" and some related results for our later use. The proof of theorem 7.2.2 will be given in 7.2.13.

Definition 7.2.3 [Bn, Ch. 3, §4, no. 1, Definition 1] *Let H be a Hausdorff topological group acting continuously on a topological space X. H is said to act* properly *on X if the map*

$$H \times X \to X \times X, \quad (h, x) \mapsto (x, hx),$$

is proper.

Recall that, in the convention in this book (0.7.5), the meaning of properness of a continuous map is slightly different from that in [Bn]. However, for a continuous action of a Hausdorff topological group H on a topological space X, the map $H \times X \to X \times X, ; (h, x) \mapsto (x, hx)$ is always separated, and hence this map is proper in our sense if and only if it is proper in the sense of [Bn].

Lemma 7.2.4 (cf. [Bn, Ch. 3, §4, no. 2, Proposition 3]) *If a Hausdorff topological group H acts properly on a topological space X, then the quotient space $H\backslash X$ is Hausdorff.*

Lemma 7.2.5 (cf. [Bn, Ch. 3, §4, no. 4, Corollary]) *If a discrete group H acts properly and freely on a Hausdorff space X, then the projection $X \to H\backslash X$ is a local homeomorphism.*

Here "free" means that any element of $H - \{1\}$ has no fixed point.

Lemma 7.2.6 (cf. [Bn, Ch. 3, §2, no. 2, Proposition 5]) *Let H be a Hausdorff topological group acting continuously on topological spaces X and X'. Let $\psi : X \to X'$ be an equivariant continuous map.*

(i) *If ψ is proper and surjective and if H acts properly on X, then H acts properly on X'.*

(ii) *If H acts properly on X' and if X is Hausdorff, then H acts properly on X.*

Lemma 7.2.7 *Assume that a Hausdorff topological group H acts on a topological space X continuously and freely. Let X' be a dense subset of X. Then, the following two conditions* (1) *and* (2) *are equivalent.*

(1) *The action of H on X is proper.*

(2) *Let $x, y \in X$, Λ be a directed set, $(x_\lambda)_{\lambda\in\Lambda}$ be a family of elements of X' and $(h_\lambda)_{\lambda\in\Lambda}$ be a family of elements of H, such that $(x_\lambda)_\lambda$ (resp. $(h_\lambda x_\lambda)_\lambda$) converges to x (resp. y). Then $(h_\lambda)_\lambda$ converges to an element h of H and $y = hx$.*

If X is Hausdorff, these equivalent conditions are also equivalent to the following condition (3).

(3) *Let Λ be a directed set, $(x_\lambda)_{\lambda\in\Lambda}$ be a family of elements of X' and $(h_\lambda)_{\lambda\in\Lambda}$ be a family of elements of H, such that $(x_\lambda)_\lambda$ and $(h_\lambda x_\lambda)_\lambda$ converge in X. Then $(h_\lambda)_\lambda$ converges in H.*

Proof. Since the action of H on X is free, the action of H on X is proper if and only if the injection $H \times X \to X \times X$, $(h, x) \mapsto (x, hx)$, is closed. From this, it is easily seen that (1) implies (2).

Assume (2). We prove (1). It is sufficient to prove that, for a closed subset W of H and a closed subset Z of X, the image I of $W \times Z$ in $X \times X$ is closed. Let (x, y) be an element of the closure of I in $X \times X$. For a neighborhood U of x and a neighborhood V of y, there exist elements $h_{U,V} \in W$ and $z_{U,V} \in Z$ such that $z_{U,V} \in U$ and $h_{U,V}z_{U,V} \in V$. For an element $x_{U,V}$ of X' which is sufficiently near to $z_{U,V}$, we have $x_{U,V} \in U$ and $h_{U,V}x_{U,V} \in V$. Let Λ be the set of all pairs (U, V). Since $(x_\lambda)_\lambda$ converges to x and $(h_\lambda x_\lambda)_\lambda$ converges to y, there exists $h \in H$ such that $y = hx$ and such that $(h_\lambda)_\lambda$ converges to h. The last property of $(h_\lambda)_\lambda$ shows $h \in W$. Since $(z_\lambda)_\lambda$ converges to x, x belongs to Z. Hence $y = hx$ shows $(x, y) \in I$.

Assume finally X is Hausdorff. It is clear that (2) implies (3). Conversely, assume (3) and let $x, y \in X$, Λ, $(x_\lambda)_{\lambda\in\Lambda}$, $(h_\lambda)_{\lambda\in\Lambda}$ be as in the hypothesis in (2). Then $(h_\lambda)_\lambda$ converges to some element h of H. Since $(h_\lambda x_\lambda)_\lambda$ converges both to y and to hx and since X is Hausdorff, we have $y = hx$. □

By Theorem 7.2.2 and Lemma 7.2.4, we have

Corollary 7.2.8 *The spaces $\Gamma(\sigma)^{\mathrm{gp}}\backslash D_\sigma$, $\Gamma(\sigma)^{\mathrm{gp}}\backslash D_{\sigma,\mathrm{val}}$, $D_\sigma^\sharp$, $D_{\sigma,\mathrm{val}}^\sharp$ are Hausdorff.*

This Corollary will be generalized in Sections 7.3 and 7.4 by replacing σ and $\Gamma(\sigma)^{\mathrm{gp}}$ by Σ and Γ.

We first prove

PROPOSITION 7.2.9

(i) *The action of* $\sigma_{\mathbf{C}}$ *on* E_σ *(resp.* $E_{\sigma,\mathrm{val}}$*) is free.*
(ii) *The action of* $i\sigma_{\mathbf{R}}$ *on* $E_\sigma^\sharp$ *(resp.* $E_{\sigma,\mathrm{val}}^\sharp$*) is free.*

Here and in the following, free action always means set-theoretic free action as in 7.2.5.

Proof. We show here that the action of $\sigma_{\mathbf{C}}$ on E_σ is free. The rest is proved in the similar way.

If $a\cdot(q,F)=(q,F)$ $(a\in\sigma_{\mathbf{C}},\,(q,F)\in E_\sigma)$ then, by $\mathbf{e}(a)q=q$, we have $a=b+c\,(b\in\sigma(q)_{\mathbf{C}}, c\in\log\Gamma(\sigma)^{\mathrm{gp}})$. On the other hand, taking $\varepsilon(\ )=\varepsilon(W(\sigma(q)),\ )$ of $\exp(-a)F=F$, we have $\varepsilon(F)=\varepsilon(\exp(-a)F)=-i\,\mathrm{Im}(b)+\varepsilon(F)$ by 6.1.7 (2). It follows $\mathrm{Im}(b)=0$ and hence $\exp(\mathrm{Re}(b)+c)F=F$. Take an element y of the interior (0.7.7) of $\sigma(q)$ such that $F':=\exp(iy)F\in D$. Then $\exp(\mathrm{Re}(b)+c)F'=F'$, that is, $\exp(\mathrm{Re}(b)+c)$ belongs to the compact group $K'_{F'}$. Since $\mathrm{Re}(b)+c$ is nilpotent, we have $\mathrm{Re}(b)+c=0$ and hence $a=0$. □

The following proposition is a key for the proof of Theorem 7.2.2.

PROPOSITION 7.2.10 *Let* σ *and* σ' *be rational nilpotent cones. Let* $\alpha\in E_{\sigma,\mathrm{val}}^\sharp$ *and* $\alpha'\in E_{\sigma',\mathrm{val}}^\sharp$*. Assume that* $(\mathbf{e}(iy_\lambda),F_\lambda)\in E_{\sigma,\mathrm{val,triv}}^\sharp$ $(y_\lambda\in\sigma_{\mathbf{R}},\ F_\lambda\in\check{D})$ *(resp.* $(\mathbf{e}(iy'_\lambda),F'_\lambda)\in E_{\sigma',\mathrm{val,triv}}^\sharp$ $(y'_\lambda\in\sigma'_{\mathbf{R}},\ F'_\lambda\in\check{D})$*) converges to* α *(resp.* α'*) in the strong topology, and that*

$$\exp(iy_\lambda)F_\lambda=\exp(iy'_\lambda)F'_\lambda \quad \text{in } D.$$

Then we have

(i) *The images of* α *and* α' *in* $D_{\mathrm{val}}^\sharp$ *coincide.*
(ii) $y_\lambda-y'_\lambda$ *converges in* $\mathfrak{g}_{\mathbf{R}}$.

Proof. Since the composite map $E_{\sigma,\mathrm{val}}^\sharp$ (resp. $E_{\sigma',\mathrm{val}}^\sharp$) $\to D_{\mathrm{val}}^\sharp \xrightarrow{\psi} D_{\mathrm{SL}(2)}$ is continuous by Definition 5.3.8 and Theorem 5.4.4, the image of α (resp. α') under this composite map is the limit of $\exp(iy_\lambda)F_\lambda$ (resp. $\exp(iy'_\lambda)F'_\lambda$) in $D_{\mathrm{SL}(2)}$. Since $\exp(iy_\lambda)F_\lambda=\exp(iy'_\lambda)F'_\lambda$ and since $D_{\mathrm{SL}(2)}$ is Hausdorff ([KU2, 3.14 (ii)]), these images coincide, say $p\in D_{\mathrm{SL}(2)}$. Let m be the rank of p, let (ρ,φ) be an SL(2)-orbit in m variables representing p and let $(W^{(k)})_{1\le k\le m}$ be the associated family of weight filtrations. Let (A,V,Z) (resp. (A',V',Z')) be the image of α (resp. α') in $D_{\mathrm{val}}^\sharp$. Take an excellent basis $(N_s)_{s\in S}$ (resp. $(N'_s)_{s\in S'}$) for (A,V,Z) (resp. (A',V',Z')) (6.3.8) and let $(a_s)_{s\in S}$ and $(S_j)_{1\le j\le n}$ (resp. $(a'_s)_{s\in S'}$ and $(S'_j)_{1\le j\le n'}$) be as in 6.3.3. For each l with $1\le l\le m$, let $f(l)$ (resp. $f'(l)$) be the unique integer such that $1\le f(l)\le n$ (resp. $1\le f'(l)\le n'$), that $W^{(l)}=W(\sum_{s\in S_{\le f(l)}}(\mathbf{R}_{\ge0})N_s)$ (resp. $=W(\sum_{s\in S'_{\le f'(l)}}(\mathbf{R}_{\ge0})N'_s)$) and that $W^{(l)}\ne W(\sum_{s\in S_{\le f(l)-1}}(\mathbf{R}_{\ge0})N_s)$ (resp. $\ne W(\sum_{s\in S'_{\le f'(l)-1}}(\mathbf{R}_{\ge0})N'_s)$). Let $g(l)$ (resp. $g'(l)$) be any element of $S_{f(l)}$ (resp. $S'_{f'(l)}$).

Take an $\mathbf{R}$-subspace B of $\sigma_{\mathbf{R}}$ (resp. B' of $\sigma'_{\mathbf{R}}$) such that $\sigma_{\mathbf{R}} = A_{\mathbf{R}} \oplus B$ (resp. $\sigma'_{\mathbf{R}} = A'_{\mathbf{R}} \oplus B'$). Write

$$y_\lambda = \sum_{s \in S} y_{\lambda,s} N_s + b_\lambda, \quad y'_\lambda = \sum_{s \in S'} y'_{\lambda,s} N'_s + b'_\lambda$$

with $y_{\lambda,s}$, $y'_{\lambda,s} \in \mathbf{R}$, $b_\lambda \in B$, $b'_\lambda \in B'$. We may assume $y_{\lambda,s}$, $y'_{\lambda,s} > 0$. Then, by Lemma 6.4.11, $y_{\lambda,s}$ $(s \in S)$, $y'_{\lambda,s}$ $(s \in S')$, $\frac{y_{\lambda,s}}{y_{\lambda,t}}$ $(s \in S_{\le j},\ t \in S_{\ge j+1})$ and $\frac{y'_{\lambda,s}}{y'_{\lambda,t}}$ $(s \in S'_{\le j}, t \in S'_{\ge j+1})$ tend to ∞, $\frac{y_{\lambda,s}}{y_{\lambda,t}}$ $(s, t \in S_j)$ tends to $\frac{a_s}{a_t}$, $\frac{y'_{\lambda,s}}{y'_{\lambda,t}}$ $(s, t \in S'_j)$ tends to $\frac{a'_s}{a'_t}$, and b_λ and b'_λ converge.

Claim A $\frac{y'_{\lambda,g'(l)}}{y_{\lambda,g(l)}}$ *converges to an element of* $\mathbf{R}_{>0}$ *for* $1 \le l \le m$.

We prove this claim. The assumption of Proposition 6.4.1 is satisfied if we take $(N_s)_{s \in S}$, a_s, S_j, $y_{\lambda,s}$ as above the same ones in 6.4.1, the above $\exp(b_\lambda)F_\lambda$ as F_λ in 6.4.1, the limit of the above $\exp(b_\lambda)F_\lambda$ as F in 6.4.1. By 6.4.1, we have

$$\tilde{\rho}\left(\sqrt{\frac{y_{\lambda,g(1)}}{y_{\lambda,g(2)}}}, \ldots, \sqrt{\frac{y_{\lambda,g(m)}}{y_{\lambda,g(m+1)}}}\right) \exp(iy_\lambda)F_\lambda \to \mathbf{r}.$$

Similarly we have

$$\tilde{\rho}\left(\sqrt{\frac{y'_{\lambda,g'(1)}}{y'_{\lambda,g'(2)}}}, \ldots, \sqrt{\frac{y'_{\lambda,g'(m)}}{y'_{\lambda,g'(m+1)}}}\right) \exp(iy'_\lambda)F'_\lambda \to \tilde{\rho}(t) \cdot \mathbf{r}$$

for some $t = (t_1, \ldots, t_m) \in \mathbf{R}^m_{>0}$. Here $y_{\lambda,g(m+1)} = y'_{\lambda,g'(m+1)} = 1$. As in [KU2, 4.12], take a continuous map

$$\beta : D \to \mathbf{R}^m_{>0} \ \text{ such that } \ \beta(\tilde{\rho}(t) \cdot x) = t\beta(x) \ \ (x \in D,\ t \in \mathbf{R}^m_{>0}).$$

Applying β to the above convergences, taking their ratios and using $\exp(iy_\lambda)F_\lambda = \exp(iy'_\lambda)F'_\lambda$, we have

$$\frac{y'_{\lambda,g'(l)}}{y_{\lambda,g(l)}} \to t_l^2 \quad (1 \le l \le m).$$

Claim A is proved.

Next, we prove the following Claims B_l and C_l $(1 \le l \le m+1)$ by induction on l.

Claim B_l $y_{\lambda,g(l)}^{-1}(\sum_{s \in S_{\le f(l)-1}} y_{\lambda,s}N_s - \sum_{s \in S'_{\le f'(l)-1}} y'_{\lambda,s}N'_s)$ *converges.*

Claim C_l $\sum_{s \in S_{\le f(l)-1}} \mathbf{Q}N_s = \sum_{s \in S'_{\le f'(l)-1}} \mathbf{Q}N'_s$.

Here $y_{\lambda,g(m+1)} := 1$, $S_{\le f(m+1)-1} := S$, $S'_{\le f'(m+1)-1} := S'$.

Note that Proposition 7.2.10 follows from Claims B_{m+1} and C_{m+1}. In fact, $A = A'$ follows from C_{m+1}, $V = V'$ follows from B_{m+1}, and $Z = Z'$ follows from the facts that the limit of $\exp(b_\lambda)F_\lambda$ (resp. $\exp(b'_\lambda)F'_\lambda$) is an element of Z (resp. Z') and that these limits coincide by B_{m+1}, C_{m+1} and the assumption $\exp(iy_\lambda)F_\lambda = \exp(iy'_\lambda)F'_\lambda$.

We prove these claims. First, Claims B_1 and C_1 are trivial. Assume $l > 1$. By the hypothesis Claim C_{l-1} of induction, N_s ($s \in S_{\le f(l-1)-1}$) and N'_s ($s \in S'$) are commutative. Hence, by the formula $\exp(x_1 + x_2) = \exp(x_1)\exp(x_2)$ if $x_1x_2 = x_2x_1$ and by the assumption $\exp(iy_\lambda)F_\lambda = \exp(iy'_\lambda)F'_\lambda$, we have

$$\exp\left(iy_\lambda - \sum_{s\in S_{\le f(l-1)-1}} iy_{\lambda,s}N_s\right)F_\lambda = \exp\left(iy'_\lambda - \sum_{s\in S_{\le f(l-1)-1}} iy_{\lambda,s}N_s\right)F'_\lambda.$$

Applying $\prod_{k=l}^{m} \tilde{\rho}_k\left(\sqrt{\frac{y_{\lambda,g(k)}}{y_{\lambda,g(k+1)}}}\right)$ to this and using 6.1.10, we obtain

$$\begin{aligned}
&\exp\left(\sum_{f(l-1)\le j<f(l)}\sum_{s\in S_j} i\frac{y_{\lambda,s}}{y_{\lambda,g(l)}}N_s\right)\\
&\cdot\prod_{k\ge l}\tilde{\rho}_k\left(\sqrt{\frac{y_{\lambda,g(k)}}{y_{\lambda,g(k+1)}}}\right)\exp\left(\sum_{s\in S_{\ge f(l)}} iy_{\lambda,s}N_s + ib_\lambda\right)F_\lambda\\
&= \exp\left(\sum_{s\in S'_{\le f'(l)-1}} i\frac{y'_{\lambda,s}}{y_{\lambda,g(l)}}N'_s - \sum_{s\in S_{\le f(l-1)-1}} i\frac{y_{\lambda,s}}{y_{\lambda,g(l)}}N_s\right)\\
&\cdot\prod_{k\ge l}\tilde{\rho}_k\left(\sqrt{\frac{y_{\lambda,g(k)}}{y_{\lambda,g(k+1)}}}\right)\exp\left(\sum_{s\in S'_{\ge f'(l)}} iy'_{\lambda,s}N'_s + ib'_\lambda\right)F'_\lambda.
\end{aligned}$$

By Lemma 6.4.2,

$$\prod_{k\ge l}\tilde{\rho}_k\left(\sqrt{\frac{y_{\lambda,g(k)}}{y_{\lambda,g(k+1)}}}\right)\exp\left(\sum_{s\in S_{\ge f(l)}} iy_{\lambda,s}N_s + ib_\lambda\right)F_\lambda \quad\text{and}$$

$$\prod_{k\ge l}\tilde{\rho}_k\left(\sqrt{\frac{y_{\lambda,g(k)}}{y_{\lambda,g(k+1)}}}\right)\exp\left(\sum_{s\in S'_{\ge f'(l)}} iy'_{\lambda,s}N'_s + ib'_\lambda\right)F'_\lambda \quad\text{converge.}$$

Let d (resp. d') be a metric on a neighborhood of the limit of F_λ (resp. F'_λ) which is compatible with the analytic structure. Let $e \ge 1$ be an integer. Then, since $(\mathbf{e}(iy_\lambda), F_\lambda)$ (resp. $(\mathbf{e}(iy'_\lambda), F'_\lambda)$) converges to α (resp. α'), there exist, by Proposition 3.1.6, $y^*_\lambda = \sum_{s\in S} y^*_{\lambda,s}N_s + b^*_\lambda \in \sigma_{\mathbf{R}}$ ($b^*_\lambda \in B_{\mathbf{R}}$), $y'^*_\lambda = \sum_{s\in S'} y'^*_{\lambda,s}N'_s + b'^*_\lambda \in \sigma'_{\mathbf{R}}$ ($b'^*_\lambda \in B'_{\mathbf{R}}$) and $F^*_\lambda, F'^*_\lambda \in \check{D}$ having the following three properties:

(1) $y^e_{\lambda,g(l-1)}(y_\lambda - y^*_\lambda)$, $y^e_{\lambda,g(l-1)}(y'_\lambda, -y'^*_\lambda)$, $y^e_{\lambda,g(l-1)}d(F_\lambda, F^*_\lambda)$ and $y^e_{\lambda,g(l-1)}d(F'_\lambda, F'^*_\lambda)$ converge to 0.

(2) $y_{\lambda,s} = y^*_{\lambda,s}$ ($s \in S_{\le f(l-1)}$) and $y'_{\lambda,s} = y'^*_{\lambda,s}$ ($s \in S'_{\le f'(l-1)}$).

(3) $((N_s)_{s\in S_{\le f(l)-1}}, \exp(\sum_{s\in S_{\ge f(l)}} iy^*_{s,\lambda}N_s + b^*_\lambda)F^*_\lambda)$ and $((N'_s)_{s\in S'_{\le f'(l)-1}}, \exp(\sum_{s\in S'_{\ge f'(l)}} iy'^*_{\lambda,s}N'_s + b'^*_\lambda)F'^*_\lambda)$ generate nilpotent orbits.

Take e sufficiently large. Then, by the hypothesis Claim B_{l-1} of induction and by Proposition 6.1.7 (2), the difference between the values by $\varepsilon(W^{(l-1)},\ \ )$ of

$$\exp\Bigl(\sum_{f(l-1)\le j<f(l)}\sum_{s\in S_j} i\frac{y_{\lambda,s}}{y_{\lambda,g(l)}}N_s\Bigr)$$
$$\cdot\prod_{k\ge l}\tilde{\rho}_k\Bigl(\sqrt{\frac{y_{\lambda,g(k)}}{y_{\lambda,g(k+1)}}}\Bigr)\exp\Bigl(\sum_{s\in S_{\ge f(l)}} iy^*_{\lambda,s}N_s+ib^*_\lambda\Bigr)F^*_\lambda$$

and of

$$\exp\Bigl(\sum_{s\in S'_{\le f'(l)-1}} i\frac{y'_{\lambda,s}}{y_{\lambda,g(l)}}N'_s-\sum_{s\in S_{\le f(l-1)-1}} i\frac{y_{\lambda,s}}{y_{\lambda,g(l)}}N_s\Bigr)$$
$$\cdot\prod_{k\ge l}\tilde{\rho}_k\Bigl(\sqrt{\frac{y_{\lambda,g(k)}}{y_{\lambda,g(k+1)}}}\Bigr)\exp\Bigl(\sum_{s\in S'_{\ge f'(l)}} iy'^*_{\lambda,s}N'_s+ib'^*_\lambda\Bigr)F'^*_\lambda$$

converges to 0. The former is equal to

$$\sum_{f(l-1)\le j<f(l)}\sum_{s\in S_j} i\frac{y_{\lambda,s}}{y_{\lambda,g(l)}}N_s+(\text{a term which converges}),$$

and the latter is equal to

$$\sum_{s\in S'_{\le f'(l)-1}} i\frac{y'_{\lambda,s}}{y_{\lambda,g(l)}}N'_s-\sum_{s\in S_{\le f(l-1)-1}} i\frac{y_{\lambda,s}}{y_{\lambda,g(l)}}N_s+(\text{a term which converges}).$$

This proves Claim B_l. Claim C_l follows from Claim B_l easily. □

7.2.11

Proof of Theorem 7.2.2 (ii). Since the canonical map $E^\sharp_{\sigma,\mathrm{val}}\to E^\sharp_\sigma$ is proper surjective, it is sufficient to prove, by Lemma 7.2.6 (i), that the action of $i\sigma_{\mathbf{R}}$ on $E^\sharp_{\sigma,\mathrm{val}}$ is proper. To prove this, we use Lemma 7.2.7 by taking $H=i\sigma_{\mathbf{R}}$, $X=E^\sharp_{\sigma,\mathrm{val}}$, $X'=E^\sharp_{\sigma,\mathrm{val,triv}}$ (= the inverse image of $|\mathrm{torus}|_\sigma$ under $E^\sharp_{\sigma,\mathrm{val}}\to|\mathrm{toric}|_{\sigma,\mathrm{val}}$). Let $(h_\lambda)_\lambda$ be a directed family in H and $(x_\lambda)_\lambda$ be a directed family in X' such that $(x_\lambda)_\lambda$ and $(h_\lambda x_\lambda)_\lambda$ converge in X. By Lemma 7.2.7, it is sufficient to prove that $(h_\lambda)_\lambda$ converges in H. Write $x_\lambda=(\mathbf{e}(iy_\lambda),F_\lambda)$ and $h_\lambda x_\lambda=(\mathbf{e}(iy'_\lambda),F'_\lambda)$ with $y_\lambda,y'_\lambda\in\sigma_{\mathbf{R}}$ and $F_\lambda,F'_\lambda\in\check{D}$. We have

$$\exp(iy_\lambda)F_\lambda=\exp(iy'_\lambda)F'_\lambda\ \text{ in }\ D.$$

By Proposition 7.2.10 (ii), $h_\lambda=iy'_\lambda-iy_\lambda$ converges in H. □

Lemma 7.2.12 *Let $\sigma\in\Sigma$, let I be the set of all* $\mathbf{Q}$*-parabolic subgroups P of $G_{\mathbf{R}}$ such that* $\exp(\sigma_{\mathbf{R}})\subset P$ *and let J be the set of all compatible families W of* $\mathbf{Q}$*-rational increasing filtrations on $H_{0,\mathbf{R}}$ such that* $\exp(\sigma_{\mathbf{R}})\subset G_W$ (5.2.12). *Then,* $\exp(\sigma_{\mathbf{R}})$ *acts properly on the spaces* $\bigcup_{P\in I}\mathcal{X}_{\mathrm{BS}}(P)$, $\bigcup_{P\in I}D_{\mathrm{BS}}(P)$, $\bigcup_{P\in I}D_{\mathrm{BS,val}}(P)$, $\bigcup_{W\in J}D_{\mathrm{SL}(2)}(W)$ *and* $\bigcup_{W\in J}D_{\mathrm{SL}(2),\mathrm{val}}(W)$.

Proof. The result for $\mathcal{X}_{\mathrm{BS}}$ follows from the strong approximation [BS,10.4]. In fact, let $p, q \in \mathcal{X}_{\mathrm{BS}}$ and assume that the $\mathbf{Q}$-parabolic subgroups P, Q associated with p, q, respectively, belong to I. Assume $x_\lambda \in \mathcal{X}$ converges to p and $h_\lambda x_\lambda$ ($h_\lambda \in \exp(\sigma_{\mathbf{R}})$) converges to q in $\mathcal{X}_{\mathrm{BS}}$. It is sufficient to prove, by Lemma 7.2.7, that h_λ converges in $\exp(\sigma_{\mathbf{R}})$. By [BS, 10.4], we see that $P \cap Q$ is a parabolic subgroup of $G_{\mathbf{R}}$. Hence we are reduced to proving that $\exp(\sigma_{\mathbf{R}})$ acts on $\mathcal{X}_{\mathrm{BS}}(P \cap Q)$ properly. Since $P \cap Q \in I$, we are reduced to proving that $\exp(\sigma_{\mathbf{R}})$ acts on $\mathcal{X}_{\mathrm{BS}}(P)$ properly for $P \in I$. Since $\exp(\sigma_{\mathbf{R}}) \subset P_u$, it is sufficient to prove the fact that P_u acts on $\mathcal{X}_{\mathrm{BS}}(P)$ properly. This fact is deduced from the Iwasawa decomposition 5.1.15 (3) (we take P here as Q there).

Now from the result for $\mathcal{X}_{\mathrm{BS}}$, we obtain the results for D_{BS} and for $D_{\mathrm{BS,val}}$ by Lemma 7.2.6 (ii). Since $\bigcup_{W \in J} D_{\mathrm{SL}(2),\mathrm{val}}(W) \subset \bigcup_{P \in I} D_{\mathrm{BS,val}}(P)$, we have the result for $D_{\mathrm{SL}(2),\mathrm{val}}$. Since $D_{\mathrm{SL}(2),\mathrm{val}} \to D_{\mathrm{SL}(2)}$ is proper and surjective, we have the result for $D_{\mathrm{SL}(2)}$ by Lemma 7.2.6 (i). □

7.2.13

Proof of Theorem 7.2.2 (i). Since $E_{\sigma,\mathrm{val}} \to E_\sigma$ is proper and surjective, it is sufficient to consider $E_{\sigma,\mathrm{val}}$.

We apply Lemma 7.2.7 by taking $H = \sigma_{\mathbf{C}}$, $X = E_{\sigma,\mathrm{val}}$, $X' = E_{\sigma,\mathrm{val,triv}}$. Let $(x_\lambda)_\lambda$ be a directed family of elements of X', let $h_\lambda \in \sigma_{\mathbf{C}}$ and assume that x_λ converges to $\alpha \in E_{\sigma,\mathrm{val}}$ and $h_\lambda x_\lambda$ converges to $\beta \in E_{\sigma,\mathrm{val}}$. By Lemma 7.2.7, it is sufficient to prove that $(h_\lambda)_\lambda$ converges in H.

Let $|\ | : E_{\sigma,\mathrm{val}} \to E^\sharp_{\sigma,\mathrm{val}}$ be the continuous map $(q, F) \mapsto (|q|, F)$ in 7.1.3 and let $\tilde{\psi} : E^\sharp_{\sigma,\mathrm{val}} \to D_{\mathrm{SL}(2)}$ be the composite map $E^\sharp_{\sigma,\mathrm{val}} \to D^\sharp_{\sigma,\mathrm{val}} \xrightarrow{\psi} D_{\mathrm{SL}(2)}$. We have $\tilde{\psi}(|h_\lambda x_\lambda|) = \exp(-\mathrm{Re}(h_\lambda))\tilde{\psi}(|x_\lambda|)$. Hence, in $D_{\mathrm{SL}(2)}$, $\tilde{\psi}(|x_\lambda|)$ converges to $\tilde{\psi}(|\alpha|)$ and $\exp(-\mathrm{Re}(h_\lambda))\tilde{\psi}(|x_\lambda|)$ converges to $\tilde{\psi}(|\beta|)$. Since $\alpha, \beta \in \cup_{W \in J} D_{\mathrm{SL}(2)}(W)$ with J as in Lemma 7.2.12, that lemma for $D_{\mathrm{SL}(2)}$ together with Lemma 7.2.7 shows that $\mathrm{Re}(h_\lambda)$ converges in $\sigma_{\mathbf{R}}$.

Let $x'_\lambda = |\mathrm{Re}(h_\lambda)x_\lambda| \in E^\sharp_{\sigma,\mathrm{val}}$. Then $|h_\lambda x_\lambda| = (i\,\mathrm{Im}(h_\lambda))x'_\lambda$ and hence x'_λ and $(i\,\mathrm{Im}(h_\lambda))x'_\lambda$ converge in $E^\sharp_{\sigma,\mathrm{val}}$. Hence, by Theorem 7.2.2 (ii), $\mathrm{Im}(h_\lambda)$ converges in $\sigma_{\mathbf{R}}$. Thus $\mathrm{Re}(h_\lambda)$ and $\mathrm{Im}(h_\lambda)$ converge in $\sigma_{\mathbf{R}}$ and hence h_λ converges in $H = \sigma_{\mathbf{C}}$. □

7.3 PROOF OF THEOREM A FOR $\Gamma(\sigma)^{\mathrm{gp}}\backslash D_\sigma$

7.3.1

Let $\sigma \in \Sigma$. In this section, we prove Theorem A for $\Gamma(\sigma)^{\mathrm{gp}}\backslash D_\sigma$. That is, we prove the following results.

Theorem A (ii) for $\Gamma(\sigma)^{\mathrm{gp}}\backslash D_\sigma$. *$\Gamma(\sigma)^{\mathrm{gp}}\backslash D_\sigma$ is a logarithmic manifold.*

Theorem A (iii) *$E_\sigma \to \sigma_{\mathbf{C}}\backslash E_\sigma$ $(= \Gamma(\sigma)^{\mathrm{gp}}\backslash D_\sigma)$ is a $\sigma_{\mathbf{C}}$-torsor in the category of logarithmic manifolds.*

Theorem A (v) for $\Gamma(\sigma)^{\mathrm{gp}}\backslash D_\sigma$ *$\Gamma(\sigma)^{\mathrm{gp}}\backslash D_\sigma$ is Hausdorff.*

Theorem A (vi) for $\Gamma(\sigma)^{\mathrm{gp}}\backslash D_\sigma$ $(\Gamma(\sigma)^{\mathrm{gp}}\backslash D_\sigma)^{\log} = \Gamma(\sigma)^{\mathrm{gp}}\backslash D_\sigma^\sharp$.

In fact Theorem A (v) for $\Gamma(\sigma)^{\mathrm{gp}}\backslash D_\sigma$ is already proved in Corollary 7.2.8. We also prove the following related results on $\sharp$-spaces.

Theorem 7.3.2

(i) *$E_\sigma^\sharp \to i\sigma_{\mathbf{R}}\backslash E_\sigma^\sharp = D_\sigma^\sharp$ (resp. $E_{\sigma,\mathrm{val}}^\sharp \to i\sigma_{\mathbf{R}}\backslash E_{\sigma,\mathrm{val}}^\sharp = D_{\sigma,\mathrm{val}}^\sharp$) is an $i\sigma_{\mathbf{R}}$-torsor in the category of topological spaces.*

(ii) *For each $\sigma \in \Sigma$, the inclusion maps $D_\sigma^\sharp \hookrightarrow D_\Sigma^\sharp$ and $D_{\sigma,\mathrm{val}}^\sharp \hookrightarrow D_{\Sigma,\mathrm{val}}^\sharp$ are open maps.*

(iii) *The canonical map $D_{\Sigma,\mathrm{val}}^\sharp \to D_\Sigma^\sharp$ is proper.*

(iv) *$D_\Sigma^\sharp$ and $D_{\Sigma,\mathrm{val}}^\sharp$ are Hausdorff spaces.*

We first give a lemma for torsors.

Lemma 7.3.3 *Let $\mathcal{C}$ be either one of the following two categories:*

(a) *the category of topological spaces.*
(b) *the category of logarithmic manifolds.*

In case (a) (resp. (b)), let H be a topological group (resp. an analytic group), X an object of $\mathcal{C}$ and assume we have an action $H \times X \to X$ in $\mathcal{C}$. (In case (b), we regard H as having the trivial logarithmic structure.) Assume this action is proper topologically and is free set-theoretically. Assume moreover the following condition (1) is satisfied.

(1) *For any point $x \in X$, there exist an object S of $\mathcal{C}$, a morphism $\iota : S \to X$ of $\mathcal{C}$ whose image contains x and an open neighborhood U of 1 in H such that $U \times S \to X$, $(h, s) \mapsto h\iota(s)$, induces an isomorphism in $\mathcal{C}$ from $U \times S$ onto an open set of X.*

Then

(i) *In case (b), the quotient topological space $H\backslash X$ has a unique structure of an object of $\mathcal{C}$ such that, for an open set V of $H\backslash X$, $\mathcal{O}_{H\backslash X}(V)$ (resp. $M_{H\backslash X}(V)$) is the set of all functions on V whose pullbacks to the inverse image V' of V in X belong to $\mathcal{O}_X(V')$ (resp. $M_X(V')$).*

(ii) *$X \to H\backslash X$ is an H-torsor in the category $\mathcal{C}$.*

Proof. Let $x \in X$ and S be as in (1). We prove the following Claim.

Claim *There exists an open neighborhood S' of $\iota^{-1}(x)$ in S such that*

$$H \times S' \to X, \quad (h, s) \mapsto h\iota(s)$$

induces an isomorphism in $\mathcal{C}$ from $H \times S'$ onto an open set of X.

We prove Claim. Let S, ι, U be as in (1). For each $g \in H$, the morphism $Ug \times S \to X$, $(h, s) \mapsto h\iota(s)$, induces an isomorphism in $\mathcal{C}$ from $Ug \times S$ onto an open set of X. Hence, for the proof of Claim, it is sufficient to prove that there exists an open neighborhood S' of $\iota^{-1}(x)$ in S such that $H \times S' \to X$, $(h, s) \mapsto h\iota(s)$, is injective. If this is not true, then for each open neighborhood S' of $\iota^{-1}(x)$ in S, there exist $h_{S'} \in H$ and

$s_{S'} \in S'$ such that $h_{S'} \neq 1$ and $h_{S'}\iota(s_{S'}) \in \iota(S')$. When S' ranges over all open neighborhoods of $\iota^{-1}(x)$ in S, the directed family $(\iota(s_{S'}), h_{S'}\iota(s_{S'}))_{S'}$ converges to (x, x) in $X \times X$. Since the action of H on X is proper and free, $(h_{S'})_{S'}$ converges to 1 in H by Lemma 7.2.7. It follows that, for some S', $(h_{S'}, \iota(s_{S'}))$, $(1, h_{S'}\iota(s_{S'})) \in U \times \iota(S)$ with $(h_{S'}, \iota(s_{S'})) \neq (1, h_{S'}\iota(s_{S'}))$ but they are sent to the same point $h_{S'}\iota(s_{S'})$ under the multiplication map $U \times \iota(S) \hookrightarrow X$, contradicting its injectivity. Claim is proved.

Now, replacing S' in Claim by S, we have an isomorphism $H \times S \xrightarrow{\sim} Y$, $(h, s) \mapsto h\iota(s)$, in $\mathcal{C}$ for some open set Y of X. We have a homeomorphism $S \to H\backslash Y$. Hence, in the case (b), $H\backslash Y$ has a unique structure of an object of $\mathcal{C}$ satisfying the condition in 7.3.3 (i) (with X there replaced by Y). Furthermore, in both cases (a), (b), $Y \to H\backslash Y$ is clearly an H-torsor in $\mathcal{C}$. This proves Lemma 7.3.3. □

7.3.4

We give some basic facts on toric varieties that we will use in this section. Let $\sigma \in \Sigma$ and let τ be a face of σ.

By the surjective homomorphism $\Gamma(\sigma)^\vee \to \Gamma(\tau)^\vee$ induced by the inclusion map $\tau \hookrightarrow \sigma$, toric_τ is regarded as a closed analytic subspace of toric_σ and $|\text{toric}|_\tau$ is regarded as a closed subspace of $|\text{toric}|_\sigma$. We have $|\text{toric}|_\tau = \text{toric}_\tau \cap |\text{toric}|_\sigma$ in toric_σ.

We define an open set $U(\tau)$ (resp. $|U|(\tau)$) of toric_σ (resp. $|\text{toric}|_\sigma$) such that

$$\text{toric}_\tau \subset U(\tau) \subset \text{toric}_\sigma \quad (\text{resp. } |\text{toric}|_\tau \subset |U|(\tau) \subset |\text{toric}|_\sigma).$$

Let $\mathcal{S} \subset \text{Hom}(\Gamma(\sigma)^{\text{gp}}, \mathbf{Z})$ be the inverse image of $\Gamma(\tau)^\vee \subset \text{Hom}(\Gamma(\tau)^{\text{gp}}, \mathbf{Z})$ via the canonical projection $\text{Hom}(\Gamma(\sigma)^{\text{gp}}, \mathbf{Z}) \to \text{Hom}(\Gamma(\tau)^{\text{gp}}, \mathbf{Z})$ and define $U(\tau) := \text{Spec}(\mathbf{C}[\mathcal{S}])_{\text{an}}$. Since $\mathcal{S}$ is the submonoid of $\text{Hom}(\Gamma(\sigma)^{\text{gp}}, \mathbf{Z})$ generated by $\Gamma(\sigma)^\vee$ and the inverses of the elements in the kernel of $\Gamma(\sigma)^\vee \to \Gamma(\tau)^\vee$, we see

$$U(\tau) = \{x \in \text{toric}_\sigma \mid \text{elements of } \text{Ker}(\Gamma(\sigma)^\vee \to \Gamma(\tau)^\vee) \text{ are invertible at } x\}$$

and hence $U(\tau)$ is an open set of $\text{toric}_\sigma = \text{Spec}(\mathbf{C}[\Gamma(\sigma)^\vee])_{\text{an}}$. Let $|U|(\tau) := U(\tau) \cap |\text{toric}|_\sigma$ in toric_σ.

7.3.5

Proofs of Theorem A (ii) *for* $\Gamma(\sigma)^{\text{gp}}\backslash D_\sigma$ *and Theorem* A (iii). We apply Lemma 7.3.3 by taking $H = \sigma_{\mathbf{C}}$, $X = E_\sigma$ and $\mathcal{C}$ to be the category of logarithmic manifolds. By Theorem 7.2.2 (i), Proposition 7.2.9, and Lemma 7.3.3, it is sufficient to prove that the condition 7.3.3 (1) is satisfied.

Let $x = (q, F) \in E_\sigma$. Let $T_{\check{D}}(F)$ be the tangent space of $\check{D}$ at F. Then the morphism $G_{\mathbf{C}} \to \check{D}$, $g \mapsto gF$, induces a surjective homomorphism $\mathfrak{g}_{\mathbf{C}} \to T_{\check{D}}(F)$ whose kernel consists of all elements N of $\mathfrak{g}_{\mathbf{C}}$ such that $NF^p \subset F^p$ for all $p \in \mathbf{Z}$. We have

(1) The homomorphism $\sigma(q)_{\mathbf{C}} \to T_{\check{D}}(F)$ is injective.

In fact, if $N \in \sigma(q)_{\mathbf{C}}$ is in the kernel, then $N \in L^{-1,-1}(W(\sigma(q))[-w], F)$ and $\exp(N)F = F$, and hence $N = 0$ by Lemma 6.1.8 (iii). (1) is proved.

Now take a $\mathbf{C}$-subspace B_1 of $\mathfrak{g}_{\mathbf{C}}$ such that $\sigma(q)_{\mathbf{C}} \oplus B_1 \to T_{\check{D}}(F)$ is an isomorphism. Take also a $\mathbf{C}$-subspace B_2 of $\sigma_{\mathbf{C}}$ such that $\sigma_{\mathbf{C}} = \sigma(q)_{\mathbf{C}} \oplus B_2$. Then the morphism

$$\text{toric}_{\sigma(q)} \times B_2 \to \text{toric}_\sigma, \quad (a, b) \mapsto \mathbf{e}(b)a,$$

is a local homeomorphism as is easily seen. Here $\text{toric}_{\sigma(q)}$ is embedded in toric_σ as in 7.3.4. Write $q = \mathbf{e}(b)1_{\sigma(q)}$ with $b \in B_2$. Then there are an open neighborhood U_0 (resp. U_1, U_2, U_3) of the origin 0 in $\text{toric}_{\sigma(q)}$ (resp. $B_1, B_2, \sigma(q)_{\mathbf{C}}$) such that

$$U_0 \times U_1 \times U_2 \times U_3 \to \check{E}_\sigma, \quad (a_0, a_1, a_2, a_3) \mapsto (\mathbf{e}(a_2 + b)a_0, \exp(a_3)\exp(a_1)F),$$

is a continuous injective open map which sends $(0, 0, 0, 0)$ to x. From this, we see that there are an open neighborhood U_0 (resp. U_1, U) of 0 in $\text{toric}_{\sigma(q)}$ (resp. B_1, $\sigma_{\mathbf{C}}$) such that

$$U \times U_0 \times U_1 \to \check{E}_\sigma,$$
$$(h, a_0, a_1) \mapsto (\mathbf{e}(h + b)a_0, \exp(-h)\exp(a_1)F) = h \cdot (\mathbf{e}(b)a_0, \exp(a_1)F),$$

is a continuous injective open map which sends $(0, 0, 0)$ to x. Define

$$A := \{(\mathbf{e}(b)a_0, \exp(a_1)F) \mid a_0 \in U_0,\ a_1 \in U_1\} \subset \check{E}_\sigma,$$
$$S := A \cap E_\sigma.$$

Endow S with the strong topology in $\check{E}_\sigma$ (this is the same as the strong topology in the analytic space A), and regard S as an fs logarithmic local ringed space over $\mathbf{C}$ endowed with $\mathcal{O}_S := \mathcal{O}_A|_S$ and $M_S := M_A|_S$. Here M_A is the inverse image of the logarithmic structure of $\check{E}_\sigma$. Then the action of $\sigma_{\mathbf{C}}$ on E_σ induces a continuous injective open map

$$U \times S \to E_\sigma, \quad (h, s) \mapsto hs.$$

From the fact that E_σ is a logarithmic manifold, we see that S is also a logarithmic manifold and that this map induces an isomorphism of logarithmic manifolds from $U \times S$ onto an open set of E_σ. The condition 7.3.3 (1) is satisfied. □

7.3.6

Proof of Theorem 7.3.2 (i). This can be deduced from Theorem A (iii) and can also be proved by arguments like those in 7.3.5, by using Lemma 7.3.3 and working in the category of topological spaces. □

7.3.7

Proof of Theorem A (vi) *for* $\Gamma(\sigma)^{\text{gp}} \backslash D_\sigma$. We give a canonical homeomorphism $(\Gamma(\sigma)^{\text{gp}} \backslash D_\sigma)^{\log} \simeq \Gamma(\sigma)^{\text{gp}} \backslash D_\sigma^\sharp$. We have

$$(\text{toric}_\sigma)^{\log} \simeq \text{Hom}(\Gamma(\sigma)^\vee, \mathbf{S}^1 \times \mathbf{R}^{\text{mult}}_{\geq 0}) \simeq (\mathbf{S}^1 \otimes_{\mathbf{Z}} \Gamma(\sigma)^{\text{gp}}) \times |\text{toric}|_\sigma .$$

Hence the homeomorphisms

$$(E_\sigma)^{\log} \simeq (\text{toric}_\sigma)^{\log} \times_{\text{toric}_\sigma} E_\sigma, \quad E_\sigma^\sharp \simeq |\text{toric}|_\sigma \times_{\text{toric}_\sigma} E_\sigma$$

induce a homeomorphism

$$(E_\sigma)^{\log} \simeq (\mathbf{S}^1 \otimes_{\mathbf{Z}} \Gamma(\sigma)^{\mathrm{gp}}) \times E_\sigma^\sharp. \tag{1}$$

This homeomorphism is compatible with the actions of $\sigma_{\mathbf{C}}$. Here $z = x + iy \in \sigma_{\mathbf{C}}$ ($x, y \in \sigma_{\mathbf{R}}$) acts on the left-hand side of (1) as the map $(E_\sigma)^{\log} \to (E_\sigma)^{\log}$ induced by the action of z on E_σ (7.2.1) and on the right-hand side as $(u, q, F) \mapsto (\mathbf{e}(x)u, \mathbf{e}(iy)q, F)$ ($u \in \mathbf{S}^1 \otimes_{\mathbf{Z}} \Gamma(\sigma)^{\mathrm{gp}}$, $q \in |\mathrm{toric}|_\sigma$, $F \in \check{D}$). The homeomorphism between the quotient spaces of these actions is

$$(\Gamma(\sigma)^{\mathrm{gp}} \backslash D_\sigma)^{\log} \simeq \Gamma(\sigma)^{\mathrm{gp}} \backslash D_\sigma^\sharp.$$

□

7.3.8

Proof of Theorem 7.3.2 (ii). For $\sigma, \tau \in \Sigma$, we have $D_\sigma^\sharp \cap D_\tau^\sharp = D_{\sigma\cap\tau}^\sharp$. Hence, by the definition of the topology of $D_\Sigma^\sharp$, it is sufficient to prove the following (1).

(1) If $\sigma \in \Sigma$ and τ is a face of σ, the inclusion $D_\tau^\sharp \to D_\sigma^\sharp$ is a continuous open map.

We prove (1). Let the notation be as in 7.3.4 and let $|\tilde{U}|(\tau) \subset E_\sigma^\sharp$ be the inverse image of $|U|(\tau)$ under $E_\sigma^\sharp \to |\mathrm{toric}|_\sigma$. The inclusion map $|\mathrm{toric}|_\tau \to |\mathrm{toric}|_\sigma$ in 7.3.4 induces the inclusion $E_\tau^\sharp \subset E_\sigma^\sharp$ which is continuous and we have commutative diagrams

$$\begin{array}{ccccc} E_\tau^\sharp & \xrightarrow{\subset} & |\tilde{U}|(\tau) & \xrightarrow{\subset} & E_\sigma^\sharp \\ \downarrow & & \downarrow & & \downarrow \\ |\mathrm{toric}|_\tau & \xrightarrow{\subset} & |U|(\tau) & \xrightarrow{\subset} & |\mathrm{toric}|_\sigma, \end{array} \qquad \begin{array}{ccc} E_\tau^\sharp & \xrightarrow{\subset} & E_\sigma^\sharp \\ \downarrow & & \downarrow \\ D_\tau^\sharp & \xrightarrow{\subset} & D_\sigma^\sharp. \end{array}$$

Furthermore, the inverse image of $D_\tau^\sharp$ under $E_\sigma^\sharp \to D_\sigma^\sharp$ coincides with $|\tilde{U}|(\tau)$. The second diagram shows that $D_\tau^\sharp \to D_\sigma^\sharp$ is continuous. Let B be an $\mathbf{R}$-vector subspace of $\sigma_{\mathbf{R}}$ such that $\tau_{\mathbf{R}} \oplus B = \sigma_{\mathbf{R}}$. Then we have a homeomorphism

$$|\mathrm{toric}|_\tau \times B \xrightarrow{\sim} |U|(\tau), \quad (a, b) \mapsto \mathbf{e}(ib)a.$$

From this, we see that the projection $|\tilde{U}|(\tau) \to D_\tau^\sharp$ factors through $|\tilde{U}|(\tau) \to E_\tau^\sharp$ which sends $(\mathbf{e}(ib)a, F)$ to $(a, \exp(-ib)F)$. This shows that a subset U of $D_\tau^\sharp$ is open in $D_\sigma^\sharp$, that is, its inverse image in $E_\sigma^\sharp$ is open, if and only if it is open in $D_\tau^\sharp$, that is, its inverse image in $E_\tau^\sharp$ is open. This completes the proof of (1). □

7.3.9

Proof of Theorem 7.3.2 (iii). By Theorem 7.3.2 (ii), we are reduced to the case $\Sigma = \{\text{face of } \sigma\}$. In this case, by Theorem 7.3.2 (i), we are reduced to the fact that $E_{\sigma,\mathrm{val}}^\sharp \to E_\sigma^\sharp$ is proper (5.3.8). □

7.3.10

Proof of Theorem 7.3.2 (iv). The statement for $D_\Sigma^\sharp$ follows from that for $D_{\Sigma,\mathrm{val}}^\sharp$, since $D_{\Sigma,\mathrm{val}}^\sharp \to D_\Sigma^\sharp$ is proper by Theorem 7.3.2 (iii) and is surjective.

We prove that $D^\sharp_{\Sigma,\mathrm{val}}$ is Hausdorff. By Theorem 7.3.2 (ii), it is sufficient to prove the following (1).

(1) Let σ and σ' be rational nilpotent cones and let $\beta \in D^\sharp_{\sigma,\mathrm{val}}$ and $\beta' \in D^\sharp_{\sigma',\mathrm{val}}$. Assume $x_\lambda \in D$ converges to β in $D^\sharp_{\sigma,\mathrm{val}}$ and to β' in $D^\sharp_{\sigma',\mathrm{val}}$. Then $\beta = \beta'$ in $D^\sharp_{\mathrm{val}}$.

We prove this. By Theorem 7.3.2 (i), there exist an open neighborhood U of β in $D^\sharp_{\sigma,\mathrm{val}}$ (resp. U' of β' in $D^\sharp_{\sigma',\mathrm{val}}$) and a continuous section $s_\sigma : U \to E^\sharp_{\sigma,\mathrm{val}}$ (resp. $s_{\sigma'} : U' \to E^\sharp_{\sigma',\mathrm{val}}$) of the projection $E^\sharp_{\sigma,\mathrm{val}} \to D^\sharp_{\sigma,\mathrm{val}}$ (resp. $E^\sharp_{\sigma',\mathrm{val}} \to D^\sharp_{\sigma',\mathrm{val}}$). Write

$$s_\sigma(x_\lambda) = (\mathbf{e}(iy_\lambda), F_\lambda), \quad s_{\sigma'}(x_\lambda) = (\mathbf{e}(iy'_\lambda), F'_\lambda).$$

Write $\alpha = s_\sigma(\beta)$, $\alpha' = s_{\sigma'}(\beta')$. Then the assumption of Proposition 7.2.10 is satisfied. Hence we have $\beta = \beta'$ by Proposition 7.2.10 (i). □

7.3.11

Here we explain our method in this chapter, compared to that in [AMRT]. In the classical situation, we have

$$\begin{array}{ccccc}
 & & \mathrm{toric}_\sigma \times \exp(\sigma_{\mathbf{C}})D & \overset{\supset}{\longleftarrow} & E_\sigma \\
 & & \downarrow & & \downarrow \\
\Gamma(\sigma)^{\mathrm{gp}}\backslash \exp(\sigma_{\mathbf{C}})D & \overset{\subset}{\longrightarrow} & \mathrm{toric}_\sigma \times^{\mathrm{torus}_\sigma} (\Gamma(\sigma)^{\mathrm{gp}}\backslash \exp(\sigma_{\mathbf{C}})D) & \overset{\supset}{\longleftarrow} & \Gamma(\sigma)^{\mathrm{gp}}\backslash D_\sigma \\
\downarrow & & \downarrow & & \\
\exp(\sigma_{\mathbf{C}})\backslash \exp(\sigma_{\mathbf{C}})D & = & \exp(\sigma_{\mathbf{C}})\backslash \exp(\sigma_{\mathbf{C}})D & &
\end{array}$$

Note that $\Gamma(\sigma)^{\mathrm{gp}}\backslash \exp(\sigma_{\mathbf{C}}) \simeq \mathrm{torus}_\sigma$. The two upper downarrows are $\sigma_{\mathbf{C}}$-torsors. The upper square is cartesian. The lower square is the torus embedding.

In this book, we do not obtain $\Gamma(\sigma)^{\mathrm{gp}}\backslash D_\sigma$ by the method of torus embedding (instead, we obtain it as a quotient of E_σ as a logarithmic local ringed space over $\mathbf{C}$). This is because we are not sure whether the quotient space $\exp(\sigma_{\mathbf{C}})\backslash \exp(\sigma_{\mathbf{C}})D$, which was the base of torus embedding in the classical situation, is still a nice space in our present general case.

7.4 PROOF OF THEOREM A FOR $\Gamma\backslash D_\Sigma$

7.4.1

In this section, we complete the proof of Theorem A. At the end of this section, we add Theorem 7.4.12 which gives criteria of local compactness for the spaces E_σ, $E^\sharp_\sigma$, $D^\sharp_\Sigma$, and $\Gamma\backslash D_\Sigma$.

Concerning Theorem A, Theorem A (i) is already proved in Section 7.1 and Theorem A (iii) is proved in Section 7.3. The rest follow.

THEOREM A (ii) *Assume Γ is neat. Then $\Gamma\backslash D_\Sigma$ is a logarithmic manifold.*

THEOREM A (iv) *Assume Γ is neat. Then $\Gamma(\sigma)^{\mathrm{gp}}\backslash D_\sigma \to \Gamma\backslash D_\Sigma$ is a local homeomorphism.*

THEOREM A (v) *$\Gamma\backslash D_\Sigma$ is Hausdorff.*

THEOREM A (vi) *Assume Γ is neat. Then $(\Gamma\backslash D_\Sigma)^{\log} = \Gamma\backslash D_\Sigma^\sharp$.*

We also prove the following related result on $\sharp$-spaces.

THEOREM 7.4.2

(i) *The actions of Γ on $D_\Sigma^\sharp$ and $D_{\Sigma,\mathrm{val}}^\sharp$ are proper. In particular, the quotient spaces $\Gamma\backslash D_\Sigma^\sharp$ and $\Gamma\backslash D_{\Sigma,\mathrm{val}}^\sharp$ are Hausdorff.*

(ii) *Assume Γ is neat. Then the canonical maps $D_\Sigma^\sharp \to \Gamma\backslash D_\Sigma^\sharp$ and $D_{\Sigma,\mathrm{val}}^\sharp \to \Gamma\backslash D_{\Sigma,\mathrm{val}}^\sharp$ are local homeomorphisms.*

To prove these results, the following result on the action of Γ plays a key role.

PROPOSITION 7.4.3 *Assume Γ is neat. Let $(\sigma, Z) \in D_\Sigma^\sharp$ (resp. D_Σ), $\gamma \in \Gamma$ and assume $\gamma(\sigma, Z) = (\sigma, Z)$. Then $\gamma = 1$ (resp. $\gamma \in \Gamma(\sigma)^{\mathrm{gp}}$).*

We use the following lemma for the proof of Proposition 7.4.3.

LEMMA 7.4.4 *Let $\mathcal{S}$ be an fs monoid with $\mathcal{S}^\times = \{1\}$. Then the group $\mathrm{Aut}(\mathcal{S})$ of automorphisms of $\mathcal{S}$ is finite.*

Proof. Let B be the set of all irreducible elements of $\mathcal{S}$, that is, elements $x \in \mathcal{S} - \{1\}$ for which there are no elements $y, z \in \mathcal{S} - \{1\}$ such that $x = yz$. Since B is stable under the action of $\mathrm{Aut}(\mathcal{S})$, in order to prove this lemma, it is enough to show the following two assertions.

(1) B generates $\mathcal{S}$.
(2) B is a finite set.

The proof of (1) is essentially the same as the classical proof of the fact that irreducible elements of an Noetherian integral domain A generate $(A - \{0\})/A^\times$ multiplicatively. In fact, since $\mathcal{S}$ is finitely generated, the monoid ring $\mathbf{Z}[\mathcal{S}]$ is finitely generated as a $\mathbf{Z}$-algebra, and hence Noetherian. Let C be the set of elements of $\mathcal{S}$ which are not generated by B, and let $C' := \{\mathbf{Z}[\mathcal{S}]x \mid x \in C\}$ be the set of principal ideals of $\mathbf{Z}[\mathcal{S}]$ which are generated by some element of C. Assume that C is not empty. Then, since $\mathbf{Z}[\mathcal{S}]$ is Noetherian, there is a maximal element $\mathbf{Z}[\mathcal{S}]x$ of C'. Since x is not irreducible, $x = yz$ for some $y, z \in \mathcal{S} - \{1\}$. This means that $\mathbf{Z}[\mathcal{S}]x \subsetneqq \mathbf{Z}[\mathcal{S}]y$, $\mathbf{Z}[\mathcal{S}]x \subsetneqq \mathbf{Z}[\mathcal{S}]z$. By the maximality of $\mathbf{Z}[\mathcal{S}]x$, y and z are generated by B, and hence x is generated by B. This is a contradiction and (1) is proved.

We next prove (2). Since $\mathcal{S}$ is finitely generated and each element of $\mathcal{S}$ is generated by B, there is a finite subset B_0 of B which generates $\mathcal{S}$. On the other hand, each element of B is irreducible and is generated by B_0. This implies $B = B_0$ and (2) is proved. □

7.4.5

Proof of Proposition 7.4.3. Let $(\sigma, Z) \in D_\Sigma$ and $\gamma \in \Gamma$. Assume $\gamma(\sigma, Z) = (\sigma, Z)$, that is, $\mathrm{Ad}(\gamma)(\sigma) = \sigma$, $\gamma Z = Z$. Then $\gamma\Gamma(\sigma)\gamma^{-1} = \Gamma(\sigma)$. Hence we have an automorphism $\mathrm{Int}(\gamma) : y \mapsto \gamma y \gamma^{-1}$ $(y \in \Gamma(\sigma))$ of the fs monoid $\Gamma(\sigma)$. By Lemma 7.4.4, this automorphism is of finite order. Since $\sigma_{\mathbf{C}}$ is generated over $\mathbf{C}$ by $\log(\Gamma(\sigma))$, the $\mathbf{C}$-linear map $\mathrm{Ad}(\gamma) : \sigma_{\mathbf{C}} \to \sigma_{\mathbf{C}}$, $y \mapsto \gamma y \gamma^{-1}$, is of finite order. On the other hand, any eigenvalue a of this $\mathbf{C}$-linear map is equal to bc^{-1} for some eigenvalues b, c of the $\mathbf{C}$-linear map $\gamma : H_{0,\mathbf{C}} \to H_{0,\mathbf{C}}$, and hence the neat property of Γ shows $a = 1$. Thus we have $\gamma y \gamma^{-1} = y$ for any $y \in \sigma_{\mathbf{C}}$. This implies $\gamma W = W$, where $W := W(\sigma)$.

Take $F \in Z$. Since $\gamma Z = Z$, there exists $N \in \sigma_{\mathbf{C}}$ such that

$$\gamma F = \exp(N)F. \tag{1}$$

Since $\exp(N)$ acts on $\mathrm{gr} := (\mathrm{gr}^W)_{\mathbf{C}}$ trivially, we have $\mathrm{gr}(\gamma)F(\mathrm{gr}) = F(\mathrm{gr})$.

CLAIM 1 $\mathrm{gr}(\gamma) = 1$.

In fact, since $F(\mathrm{gr})$ is polarized, the isotropy group of $F(\mathrm{gr})$ is compact, so $\mathrm{gr}(\gamma)$ is contained in the intersection of a discrete subgroup and a compact subgroup of $\mathrm{Aut}(\mathrm{gr})$ and hence is of finite order. Hence, by the neat property, we have $\mathrm{gr}(\gamma) = 1$. Claim 1 is proved.

CLAIM 2 $\log(\gamma)W_k \subset W_{k-2}$ $(\forall k \in \mathbf{Z})$.

In fact, by (1), we have

$$(\log(\gamma) - N)F^p \subset F^p \quad (\forall p \in \mathbf{Z}). \tag{2}$$

Since $N(W_{k,\mathbf{C}}) \subset W_{k-2,\mathbf{C}}$, (2) shows that the map $\mathrm{gr}(\gamma) : \mathrm{gr}_k \to \mathrm{gr}_{k-1}$ satisfies

$$\mathrm{gr}(\log(\gamma))F^p(\mathrm{gr}_k) \subset F^p(\mathrm{gr}_{k-1}) \quad (\forall k, \forall p). \tag{3}$$

By taking the complex conjugation of (3), we have

$$\mathrm{gr}(\log(\gamma))\overline{F}^p(\mathrm{gr}_k) \subset \overline{F}^p(\mathrm{gr}_{k-1}) \quad (\forall k, \forall p). \tag{4}$$

Since

$$\mathrm{gr}_k = \bigoplus_{p+q=w+k} F^p(\mathrm{gr}_k) \cap \overline{F}^q(\mathrm{gr}_k) \quad \text{and}$$

$$F^p(\mathrm{gr}_{k-1}) \cap \overline{F}^q(\mathrm{gr}_{k-1}) = 0 \quad \text{for } p+q = w+k > w+(k-1),$$

(3) and (4) show that the map $\mathrm{gr}(\log(\gamma)) : \mathrm{gr}_k \to \mathrm{gr}_{k-1}$ is the zero map. This proves Claim 2.

By (2), $\log(\gamma)F^p \subset F^{p-1}$, $\log(\gamma)\overline{F}^p \subset \overline{F}^{p-1}$ $(\forall p \in \mathbf{Z})$. These and Claim 2 show, by Lemma 6.1.8 (iv), that

$$\log(\gamma), N \in L^{-1,-1}(W(\sigma)[-w], F).$$

Hence, by Lemma 6.1.8 (iii), $\gamma F = \exp(N)F$ proves $\log(\gamma) = N$. This proves $\gamma \in \Gamma(\sigma)^{\mathrm{gp}}$.

If $\gamma(\sigma, Z') = (\sigma, Z')$ for some $(\sigma, Z') \in D_\Sigma^\sharp$ then, for $Z := \exp(\sigma_{\mathbf{R}})Z'$, we have $(\sigma, Z) \in D_\Sigma$ and $\gamma(\sigma, Z) = (\sigma, Z)$. In the above argument, we take $N \in i\sigma_{\mathbf{R}}$, and have $\log(\gamma) = N$. Since $\log(\gamma)$ is real and N is purely imaginary, this shows that $\log(\gamma) = 0$. Hence $\gamma = 1$. □

7.4.6

Proof of Theorem 7.4.2. We first prove 7.4.2 (i). By Theorem 5.2.15 (iv), the action of Γ on $D_{\mathrm{SL}(2)}$ is proper. By Lemma 7.2.6 (ii), the statement for $D^{\sharp}_{\Sigma,\mathrm{val}}$ follows from this, from the continuity of $\psi : D^{\sharp}_{\Sigma,\mathrm{val}} \to D_{\mathrm{SL}(2)}$ and from Theorem 7.3.2 (iv). By Lemma 7.2.6 (i), the statement for $D^{\sharp}_{\Sigma}$ follows from the above result, since $D^{\sharp}_{\Sigma,\mathrm{val}} \to D^{\sharp}_{\Sigma}$ is proper by Theorem 7.3.2 (iii) and is surjective.

Next, by Lemma 7.2.5, 7.4.2 (ii) follows from 7.4.2 (i) and Proposition 7.4.3. □

Lemma 7.4.7 *Let X be a topological space with a continuous action of a discrete group Γ and let Y be a set with an action of Γ. Let $f : X \to Y$ be a Γ-equivariant surjective map. Let Γ' be a subgroup of Γ. We introduce the quotient topologies of X on $\Gamma'\backslash Y$ and on $\Gamma\backslash Y$. Let V be an open set of $\Gamma'\backslash Y$ and let U be the inverse image of V in $\Gamma'\backslash X$. We assume moreover the three conditions* (1)–(3) *below. Then, $V \to \Gamma\backslash Y$ is a local homeomorphism.*

(1) *$X \to \Gamma\backslash X$ is a local homeomorphism and $\Gamma\backslash X$ is Hausdorff.*
(2) *$U \to V$ is proper.*
(3) *If $x \in X$ and $\gamma \in \Gamma$, and if the images of γx and x in $\Gamma'\backslash Y$ are contained in V and they coincide, then $\gamma \in \Gamma'$.*

Proof. If the conclusion would be wrong, there exists an element $a \in V$ such that for any neighborhood W of a in V, the map from W to $\Gamma\backslash Y$ is not injective. This means that there exist directed families of points $(x_\lambda)_{\lambda\in\Lambda}$ in X and $(\gamma_\lambda)_{\lambda\in\Lambda}$ in Γ such that the images of x_λ and $\gamma_\lambda x_\lambda$ in $\Gamma'\backslash Y$ are contained in V, are different for each λ and converge to a. By (2), replacing Λ by a cofinal directed subset of Λ if necessary, we may assume that the images of x_λ in U converge to some point $b'_1 \in U$ whose image in V coincides with a. Again by (2), replacing Λ by a cofinal directed subset of Λ if necessary, we may assume that the images of $\gamma_\lambda x_\lambda$ in U converge to some point $b'_2 \in U$ whose image in V coincides with a. Take a lifting $b_j \in X$ of b'_j $(j = 1, 2)$. Then, by (1), we can find $\alpha_\lambda, \beta_\lambda \in \Gamma'$ such that $\alpha_\lambda x_\lambda$ and $\beta_\lambda\gamma_\lambda x_\lambda$ converge to b_1 and b_2 in X, respectively. This means that the images of b_1 and b_2 in $\Gamma\backslash X$ coincide because $\Gamma\backslash X$ is Hausdorff by (1). Hence we can find $\delta \in \Gamma$ such that $b_2 = \delta b_1$ in X. Looking at their images in V and using (3), we see that $\delta \in \Gamma'$. On the other hand, since both $\alpha_\lambda x_\lambda$ and $\delta^{-1}\beta_\lambda\gamma_\lambda x_\lambda = (\delta^{-1}\beta_\lambda\gamma_\lambda\alpha_\lambda^{-1})\alpha_\lambda x_\lambda$ converge to b_1, we see, by (1), that $\delta^{-1}\beta_\lambda\gamma_\lambda\alpha_\lambda^{-1} = 1$ for any sufficiently large λ. It follows that $\gamma_\lambda = \beta_\lambda^{-1}\delta\alpha_\lambda \in \Gamma'$ for any sufficiently large λ. This contradicts to the assumption that the images of x_λ and $\gamma_\lambda x_\lambda$ in $\Gamma'\backslash Y$ are different. □

7.4.8

Proof of Theorem A (iv). We use Lemma 7.4.7 for $X = D^{\sharp}_{\Sigma}$, $Y = D_\Sigma$, $\Gamma = \Gamma$, $\Gamma' = \Gamma(\sigma)^{\mathrm{gp}}$, $V = \Gamma(\sigma)^{\mathrm{gp}}\backslash D_\sigma$ and $U = \Gamma(\sigma)^{\mathrm{gp}}\backslash D^{\sharp}_{\sigma}$. Theorem 7.4.2 for $D^{\sharp}_{\Sigma}$ shows that the condition (1) in Lemma 7.4.7 is satisfied. Theorem A (vi) for $\Gamma(\sigma)^{\mathrm{gp}}\backslash D_\sigma$, proved in Section 7.3, shows that the condition (2) in 7.4.7 is satisfied. Proposition 7.4.3 shows that the condition (3) in 7.4.7 is satisfied. □

7.4.9

Proof of Theorem A (ii). Theorem A (ii) follows from Theorem A (ii) for $\Gamma(\sigma)^{\mathrm{gp}}\backslash D_\sigma$ by Theorem A (iv). □

7.4.10

Proof of Theorem A (vi). By using Theorem A (iv), it is easily seen that the canonical homeomorphisms $(\Gamma(\sigma)^{\mathrm{gp}}\backslash D_\sigma)^{\log} \simeq \Gamma(\sigma)^{\mathrm{gp}}\backslash D_\sigma^\sharp$ (Theorem A (vi) for $\Gamma(\sigma)^{\mathrm{gp}}\backslash D_\sigma$, proved in Section 7.3) glue uniquely to a homeomorphism $(\Gamma\backslash D_\Sigma)^{\log} \simeq \Gamma\backslash D_\Sigma^\sharp$ □

7.4.11

Proof of Theorem A (v). Replacing Γ by a subgroup of finite index, we may assume Γ is neat. Then, by Theorem A (vi), the map $\Gamma\backslash D_\Sigma^\sharp \to \Gamma\backslash D_\Sigma$ is proper and surjective. Since $\Gamma\backslash D_\Sigma^\sharp$ is Hausdorff by Theorem 7.4.2 (i), it follows from 7.2.6 (i) and 7.2.4 that $\Gamma\backslash D_\Sigma$ is Hausdorff. □

For criteria of local compactness of the spaces E_σ, $E_\sigma^\sharp$, $D_\Sigma^\sharp$ and $\Gamma\backslash D_\Sigma$, we have

THEOREM 7.4.12

(i) *Let* $\sigma \in \Sigma$ *and let* $(q, F) \in E_\sigma$. *Then the following* (1) *and* (2) *are equivalent.*

(1) *(q, F) has a compact neighborhood in E_σ.*
(2) *If $F' \in \check{D}$ is sufficiently near to F, then $NF'^p \subset F'^{p-1}$ for any $N \in \sigma(q)$ and $p \in \mathbf{Z}$.*

(ii) *Let $X := E_\sigma$ with $\sigma \in \Sigma$ (resp. $:= E_\sigma^\sharp$ with $\sigma \in \Sigma$, resp. $:= D_\Sigma^\sharp$), let $x \in X$, and let $y \in Y := \Gamma\backslash D_\Sigma$ be the image of x. Then x has a compact neighborhood in X if and only if y has a compact neighborhood in Y.*

Proof. (i) follows from Proposition 3.1.10 and Theorem A (i).

For (ii), replacing Γ by a neat subgroup of Γ of finite index, we may assume that Γ itself is neat. Then (ii) follows from the relationship among E_σ, $E_\sigma^\sharp$, $D_\Sigma^\sharp$, $\Gamma\backslash D_\Sigma$ proved in Sections 7.2–7.4 just before this theorem. □

A local description of the logarithmic manifold $\Gamma\backslash D_\Sigma$ is given by the following.

THEOREM 7.4.13 *Let $(\sigma, Z) \in D_\Sigma$ and let p be its image in $\Gamma\backslash D_\Sigma$.*

(i) *Let $F_{(0)} \in Z$. Then there exists a locally closed analytic submanifold Y of $\check{D}$ satisfying the following conditions* (1) and (2).

(1) *$F_{(0)} \in Y$.*
(2) *The canonical map $T_Y(F_{(0)}) \oplus \sigma_{\mathbf{C}} \to T_{\check{D}}(F_{(0)})$ is an isomorphism.*

(ii) *Let* $F_{(0)}$ *and* Y *be as in* (i). *Let* S *be the analytically constructible subset of* $Z := \text{toric}_\sigma \times Y$, *containing* $Z_{\text{triv}} = \text{torus}_\sigma \times Y$, *defined by*

$$S := \{(q, F) \in \check{E}_\sigma \mid F \in Y, N(F^p) \subset F^{p-1} \ (\forall N \in \sigma(q), \forall p \in \mathbf{Z})\},$$

and regard S *as a logarithmic manifold in the natural way. Then, there exist an open neighborhood* U_1 *of* $(0, F_{(0)})$ *in* S *in the strong topology, an open neighborhood* U_2 *of* p *in* $\Gamma\backslash D_\Sigma$, *and an isomorphism* $U_1 \xrightarrow{\sim} U_2$ *of logarithmic manifolds sending* $(0, F_{(0)})$ *to* p.

Proof. This follows from the proof of Theorem A (iii) in 7.3.5 and from Theorem A (iv) which is proved in 7.4.8. □

Chapter Eight

Proof of Theorem B

In this chapter, we prove Theorem B stated in Section 4.2. In the proof of Theorem B, we use the results in Theorem A which are stated in Section 4.1 and proved in Chapter 7.

In Section 8.4, we prove Theorem 0.5.29.

8.1 LOGARITHMIC LOCAL SYSTEMS

We will study moduli of pairs (X, a polarized Hodge structure on X) comparing it with moduli of pairs (X, a local system on $X^{\log}$). The latter is something like to forget the Hodge filtration F in the former. In this section, we give preparations on moduli of local systems on $X^{\log}$.

Let X be an object of the category $\overline{\mathcal{A}}_1(\log)$ (2.2.1).

Definition 8.1.1 *Let $(H_0, \langle\ ,\ \rangle_0)$ be as in §0.7 Notation and let Γ be a subgroup of $G_{\mathbf{Z}}$. By a logarithmic local system on X of type $(H_0, \langle\ ,\ \rangle_0, \Gamma)$, we mean a triple $(H_{\mathbf{Z}}, \langle\ ,\ \rangle, \mu)$, where $H_{\mathbf{Z}}$ is a locally constant sheaf of $\mathbf{Z}$-modules on $X^{\log}$, $\langle\ ,\ \rangle$ is a non-degenerate $\mathbf{Q}$-bilinear form $H_{\mathbf{Q}} \times H_{\mathbf{Q}} \to \mathbf{Q}$ and μ is a Γ-level structure, that is, a global section of the sheaf $\Gamma \backslash \underline{\mathrm{Isom}}((H_{\mathbf{Z}}, \langle\ ,\ \rangle), (H_0, \langle\ ,\ \rangle_0))$ on $X^{\log}$.*

Lemma 8.1.2 *There is an equivalence of categories*

$$\begin{pmatrix}\textit{logarithmic local systems}\\ \textit{on } X \textit{ of type } (H_0, \langle\ ,\ \rangle_0, \Gamma)\end{pmatrix} \xrightarrow{\sim} (\Gamma\textit{-torsors on } X^{\log}),$$

$$(H_{\mathbf{Z}}, \langle\ ,\ \rangle, \mu) \mapsto (\textit{the inverse image of } \mu \textit{ in } \underline{\mathrm{Isom}}((H_{\mathbf{Z}}, \langle\ ,\ \rangle), (H_0, \langle\ ,\ \rangle_0))).$$

This is proved by a formal argument.

8.1.3

For a subgroup Γ of $G_{\mathbf{Z}}$, let $\underline{\mathrm{LS}}_{\Gamma}$ be the sheaf on X obtained as the sheafification of the presheaf

$$U \mapsto \begin{pmatrix}\text{isomorphism classes of logarithmic}\\ \text{local systems on } U \text{ of type } (H_0, \langle\ ,\ \rangle_0, \Gamma)\end{pmatrix}.$$

By Lemma 8.1.2, we have

$$\underline{\mathrm{LS}}_{\Gamma} \simeq R^1\tau_*(\Gamma).$$

LEMMA 8.1.4 *There is a canonical isomorphism*

$$R^1\tau_*(\mathbf{Z}) \simeq M_X^{\mathrm{gp}}/\mathcal{O}_X^\times.$$

Proof. The exact sequence

$$0 \to \mathbf{Z} \xrightarrow{2\pi i} \mathcal{L} \xrightarrow{\exp} \tau^{-1}(M_X^{\mathrm{gp}}) \to 1$$

induces a homomorphsim $M_X^{\mathrm{gp}}/\mathcal{O}_X^\times \to R^1\tau_*(\mathbf{Z})$. We prove that this is an isomorphism. Since τ is proper, by proper base change theorem (Appendix A2), it is sufficient to prove

$$M_{X,x}^{\mathrm{gp}}/\mathcal{O}_{X,x}^\times \xrightarrow{\sim} H^1(\tau^{-1}(x), \mathbf{Z}) = H^1(x^{\log}, \mathbf{Z}). \tag{1}$$

Let

$$H^1(x^{\log}, \mathbf{Z}) \simeq \mathrm{Hom}(\pi_1(x^{\log}), \mathbf{Z})$$

be the canonical isomorphism normalized as Appendix A1.10. Then the map (1) is the $\mathbf{Z}$-dual of the isomorphism

$$\pi_1(x^{\log}) \xrightarrow{\sim} \mathrm{Hom}(M_{X,x}^{\mathrm{gp}}/\mathcal{O}_{X,x}^\times, \mathbf{Z})$$

defined in 2.2.9 (1). □

COROLLARY 8.1.5 *Let Γ be a finitely generated abelian subgroup of $G_{\mathbf{Z}}$.*

(1) *There exists a canonical isomorphism*

$$\underline{\mathrm{LS}}_\Gamma \simeq \Gamma \otimes (M_X^{\mathrm{gp}}/\mathcal{O}_X^\times).$$

(2) *Let $(H_{\mathbf{Z}}, \langle\ ,\ \rangle, \mu)$ be a logarithmic local system on X of type $(H_0, \langle\ ,\ \rangle_0, \Gamma)$ and let s be the associated global section of $\underline{\mathrm{LS}}_\Gamma \simeq \Gamma \otimes (M_X^{\mathrm{gp}}/\mathcal{O}_X^\times)$ on X. Let $x \in X$, $y \in \tau^{-1}(x)$ and let $\tilde{\mu}_y : (H_{\mathbf{Z},y}, \langle\ ,\ \rangle_y) \xrightarrow{\sim} (H_0, \langle\ ,\ \rangle_0)$ be a lifting of the germ of μ at y. Then, the homomorphism*

$$\pi_1(x^{\log}) \to \mathrm{Aut}(H_{\mathbf{Z},y}, \langle\ ,\ \rangle_y) \xrightarrow[\sim]{\text{by } \tilde{\mu}_y} \mathrm{Aut}(H_0, \langle\ ,\ \rangle_0) = G_{\mathbf{Z}},$$

induced via $\tilde{\mu}_y$ from the action of $\pi_1(x^{\log})$ on $H_{\mathbf{Z},y}$, coincides with the composite map

$$\pi_1(x^{\log}) \xrightarrow{\sim} \mathrm{Hom}(M_{X,x}^{\mathrm{gp}}/\mathcal{O}_{X,x}^\times, \mathbf{Z}) \xrightarrow{\text{by } s} \Gamma \subset G_{\mathbf{Z}}.$$

8.1.6

Let Σ be a fan in $\mathfrak{g}_{\mathbf{Q}}$ and let Γ be a subgroup of $G_{\mathbf{Z}}$ which is strongly compatible with Σ. Denote by

$$\Psi := (H_0, \langle\ ,\ \rangle_0, \Gamma, \Sigma) \tag{1}$$

the 4-tuple consisting of the pair $(H_0, \langle\ ,\ \rangle_0)$ as in §0.7 and of the above Σ and Γ.

For $\sigma \in \Sigma$, we denote

$$\Psi_\sigma := (H_0, \langle\ ,\ \rangle_0, \Gamma(\sigma)^{\mathrm{gp}}, \{\text{face of } \sigma\}). \tag{2}$$

DEFINITION 8.1.7 *Let Ψ be as in* 8.1.6 (1).

(i) *We define a* logarithmic local system on X of type Ψ *as a logarithmic local system* $(H_{\mathbf{Z}}, \langle\ ,\ \rangle, \mu)$ *on* X *of type* $(H_0, \langle\ ,\ \rangle_0, \Gamma)$ *which satisfies the following condition* (1).

(1) *For any* $x \in X$ *and any* $y \in x^{\log} = \tau^{-1}(x)$, *if* $\tilde{\mu}_y : (H_{\mathbf{Z},y}, \langle\ ,\ \rangle_y) \xrightarrow{\sim} (H_0, \langle\ ,\ \rangle_0)$ *denotes a representative of the germ of* μ *at* y, *then there exists* $\sigma \in \Sigma$ *such that the image of the composite map*

$$\pi_1^+(x^{\log}) \hookrightarrow \pi_1(x^{\log}) \to \operatorname{Aut}(H_{\mathbf{Z},y}, \langle\ ,\ \rangle_y) \xrightarrow[\sim]{\text{by } \tilde{\mu}_y} G_{\mathbf{Z}}$$

is contained in $\Gamma(\sigma)$.

(ii) *For* $\sigma \in \Sigma$, *we define the canonical logarithmic local system*

$$(H_{\sigma,\mathbf{Z}}, \langle\ ,\ \rangle_\sigma, \mu_\sigma) \text{ on } \operatorname{toric}_\sigma \text{ of type } \Psi_\sigma,$$

as in 3.3.3.

8.1.8

For Ψ in 8.1.6 (1), let $\underline{\mathrm{LS}}_\Psi$ be the sheaf on X associated to the presheaf

$$U \mapsto (\text{isomorphism classes of logarithmic local systems on } U \text{ of type } \Psi).$$

PROPOSITION 8.1.9 *We have*

$$\underline{\mathrm{LS}}_{\Psi_\sigma} \simeq \operatorname{Hom}(\Gamma(\sigma)^\vee, M_X/\mathcal{O}_X^\times) \subset \underline{\mathrm{LS}}_{\Gamma(\sigma)^{\mathrm{gp}}} = \Gamma(\sigma)^{\mathrm{gp}} \otimes (M_X^{\mathrm{gp}}/\mathcal{O}_X^\times).$$

Proof. We have an injective map

$$\underline{\mathrm{LS}}_{\Psi_\sigma}(X) \to H^0(X, R^1\tau_*(\Gamma(\sigma)^{\mathrm{gp}}))$$

whose image consists of all elements t of $H^0(X, R^1\tau_*(\Gamma(\sigma)^{\mathrm{gp}}))$ satisfying the following condition (1).

(1) For any $x \in X$, if $t_x : \pi_1(x^{\log}) \to \Gamma(\sigma)^{\mathrm{gp}}$ denotes the image of t in $H^1(x^{\log}, \Gamma(\sigma)^{\mathrm{gp}}) \simeq \operatorname{Hom}(\pi_1(x^{\log}), \Gamma(\sigma)^{\mathrm{gp}})$ (we normalize this isomorphism as in Appendix A1.10), then t_x sends $\pi_1^+(x^{\log})$ into $\Gamma(\sigma)$.

This (1) is equivalent to the following

(2) The image of $t \in H^0(X, R^1\tau_*(\Gamma(\sigma)^{\mathrm{gp}}))$ under

$$\begin{aligned} H^0(X, R^1\tau_*(\Gamma(\sigma)^{\mathrm{gp}})) &\xrightarrow{\sim} H^0(X, \Gamma(\sigma)^{\mathrm{gp}} \otimes (M_X^{\mathrm{gp}}/\mathcal{O}_X^\times)) \\ &\xrightarrow{\sim} \operatorname{Hom}(\Gamma(\sigma)^\vee, M_X^{\mathrm{gp}}/\mathcal{O}_X^\times) \end{aligned}$$

is contained in $\operatorname{Hom}(\Gamma(\sigma)^\vee, M_X/\mathcal{O}_X^\times)$. □

8.2 PROOF OF THEOREM B

In this section, we prove Theorem B given in Section 4.2.

Let $\overline{\mathcal{A}}_2(\log)$ be the category defined in 3.2.4. Let $\Phi = (w, (h^{p,q})_{p,q\in\mathbf{Z}}, H_0, \langle\ ,\ \rangle_0, \Gamma, \Sigma)$ be as in 2.5.8.

8.2.1

We define contravariant functors

$$C, C_\sigma : \overline{\mathcal{A}}_2(\log) \to (\text{Sets}) \quad (\sigma \in \Sigma) \tag{1}$$

by $C := \underline{\text{PLH}}_\Phi$ and by $C_\sigma := \underline{\text{PLH}}_{\Phi_\sigma}$, where

$$\Phi_\sigma = (w, (h^{p,q})_{p,q\in\mathbf{Z}}, H_0, \langle\ ,\ \rangle_0, \Gamma(\sigma)^{\text{gp}}, \{\text{face of } \sigma\}) \tag{2}$$

for $\sigma \in \Sigma$. Then we have the canonical morphism

$$C_\sigma \to C.$$

We define contravariant functors

$$B_\sigma : \overline{\mathcal{A}}_2(\log) \to (\text{Sets}) \quad (\sigma \in \Sigma) \tag{3}$$

as follows. Let Ψ_σ be as in 8.1.6 (2). Let $X \in \overline{\mathcal{A}}_2(\log)$. For a morphism $\theta : X \to \text{toric}_\sigma$, define the logarithmic local system

$$(H_{\theta,\mathbf{Z}}, \langle\ ,\ \rangle_\theta, \mu_\theta) \ \text{ on } X \text{ of type } \Psi_\sigma$$

to be the inverse image of the canonical logarithmic local system $(H_{\sigma,\mathbf{Z}}, \langle\ ,\ \rangle_\sigma, \mu_\sigma)$ on toric_σ (8.1.7) by $\theta^{\log} : X^{\log} \to \text{toric}_\sigma^{\log}$. We define $B_\sigma(X)$ to be the set of all pairs (θ, F), where θ is a morphism $X \to \text{toric}_\sigma$ and F is a decreasing filtration on the $\mathcal{O}_X^{\log}$-module $\mathcal{O}_X^{\log} \otimes_{\mathbf{Z}} H_{\theta,\mathbf{Z}}$ such that $(H_{\theta,\mathbf{Z}}, \langle\ ,\ \rangle_\theta, F, \mu_\theta)$ is a PLH on X of type Φ_σ.

Then we have a morphism

$$B_\sigma \to C_\sigma, \quad (\theta, F) \mapsto (H_{\theta,\mathbf{Z}}, \langle\ ,\ \rangle_\theta, F, \mu_\theta).$$

8.2.2

Our method is to prove Theorem B in the following order: As functors on $\overline{\mathcal{A}}_2(\log)$,

Step 1 $B_\sigma \simeq \text{Mor}(\ \ , E_\sigma)$,
Step 2 $C_\sigma \simeq \text{Mor}(\ \ , \Gamma(\sigma)^{\text{gp}}\backslash D_\sigma)$,
Step 3 $C \simeq \text{Mor}(\ \ , \Gamma\backslash D_\Sigma)$ (this is Theorem B).

8.2.3

Proof of Step 1. We prove that $B_\sigma \simeq \text{Mor}(\ \ , E_\sigma)$ as functors on $\overline{\mathcal{A}}_2(\log)$.

Let

$$\check{B}_\sigma : \overline{\mathcal{A}}_2(\log) \to (\text{Sets})$$

be the contravariant functor defined as follows. For $X \in \overline{\mathcal{A}}_2(\log)$, $\check{B}_\sigma(X)$ is the set of all pairs (θ, F), where θ is a morphism $X \to \text{toric}_\sigma$ and F is a decreasing filtration on the $\mathcal{O}_X^{\log}$-module $\mathcal{O}_X^{\log} \otimes_{\mathbf{Z}} H_{\theta,\mathbf{Z}}$ such that $(H_{\theta,\mathbf{Z}}, \langle\ ,\ \rangle_\theta, F)$ is a pre-PLH on X of weight w and of Hodge type $(h^{p,q})_{p,q}$ (cf. 8.2.1 (3)). Then we have $B_\sigma \subset \check{B}_\sigma$. We show first that □

Claim 1 $\check{B}_\sigma \simeq \mathrm{Mor}(\ \ , \check{E}_\sigma)$, *where* $\check{E}_\sigma = \mathrm{toric}_\sigma \times \check{D}$ (3.3.2).

Proof. We have $\check{D} = Z_{\mathrm{an}}$, where Z is the scheme of finite type over $\mathbf{C}$ which represents the following functor A from the category of local ringed spaces over $\mathbf{C}$ to the category of sets. For a local ringed space X over $\mathbf{C}$, $A(X)$ is the set of all decreasing filtrations $F = (F^p)_{p\in\mathbf{Z}}$ on the $\mathcal{O}_X$-module $\mathcal{O}_X \otimes_{\mathbf{Z}} H_0$ satisfying the following two conditions (1) and (2).

(1) The $\mathcal{O}_X$-module F^p/F^{p+1} is locally free of rank $h^{p,w-p}$ for any $p \in \mathbf{Z}$.
(2) $\langle F^p, F^q\rangle_0 = 0$ if $p+q > w$.

Let $X \in \overline{\mathcal{A}}_2(\log)$. Then, since X satisfies the condition (A_1') in 2.2.1, we have $\mathrm{Mor}(X, \check{D}) \simeq \mathrm{Mor}(X, Z) = A(X)$. Hence $\mathrm{Mor}(X, \check{E}_\sigma) \simeq \mathrm{Mor}(X, \mathrm{toric}_\sigma) \times A(X)$. By using the identification $\mathcal{O}_X^{\log} \otimes_{\mathbf{Z}} H_{\theta,\mathbf{Z}} = \mathcal{O}_X^{\log} \otimes_{\mathbf{Z}} H_0$ which comes from $\mathcal{O}_{\mathrm{toric}_\sigma}^{\log} \otimes_{\mathbf{Z}} H_{\sigma,\mathbf{Z}} = \mathcal{O}_{\mathrm{toric}_\sigma}^{\log} \otimes_{\mathbf{Z}} H_0$ (3.3.3), we have the canonical bijection

$$\mathrm{Mor}(X, \mathrm{toric}_\sigma) \times A(X) \xrightarrow{\sim} \check{B}_\sigma(X), \quad (\theta, F) \mapsto (\theta, F^{\log}),$$
$$\text{where} \quad F^{\log} := \mathcal{O}_X^{\log} \otimes_{\tau^{-1}(\mathcal{O}_X)} \tau^{-1}(F).$$ □

Claim 2 *The isomorphism in Claim* 1 *induces the isomorphism of subfunctors* $B_\sigma \simeq \mathrm{Mor}(\ \ , E_\sigma)$.

Proof. Let $X \in \overline{\mathcal{A}}_2(\log)$ and let $f : X \to \check{E}_\sigma$ be a morphism. Then by 3.3.8, the following conditions (4) and (5) are equivalent:

(4) The image of $f : X \to \check{E}_\sigma$ is contained in E_σ.
(5) The element of $\check{B}_\sigma(X)$ corresponding to f via the isomorphism in Claim 1 belongs to $B_\sigma(X)$.

By 3.2.8, a morphism $X \to E_\sigma$ is equivalent to a morphism $X \to \check{E}_\sigma$ whose image is contained in E_σ. This proves Claim 2. □

8.2.4

Let $\Phi = (w, (h^{p,q})_{p,q\in\mathbf{Z}}, H_0, \langle\ ,\ \rangle_0, \Gamma, \Sigma)$ and $\Psi = (H_0, \langle\ ,\ \rangle_0, \Gamma, \Sigma)$ be as in 2.5.8 and in 8.1.6 (1), respectively. For $\sigma \in \Sigma$, let Φ_σ and Ψ_σ be as in 8.2.1 (2) and in 8.1.6 (2), respectively. As in 8.1.8, 8.1.9, we have a contravariant functor

$$\overline{C} := \underline{\mathrm{LS}}_\Psi \ (\text{resp. } \overline{C}_\sigma := \underline{\mathrm{LS}}_{\Psi_\sigma}) : \overline{\mathcal{A}}_2(\log) \to (\mathrm{Sets}), \tag{1}$$
$$X \mapsto \begin{pmatrix}\text{isomorphism classes of logarithmic local}\\ \text{systems on } X \text{ of type } \Psi \ (\text{resp. } \Psi_\sigma)\end{pmatrix}.$$

If $(H_{\mathbf{Z}}, \langle\ ,\ \rangle, F, \mu)$ is a PLH on X of type Φ (resp. Φ_σ), then $(H_{\mathbf{Z}}, \langle\ ,\ \rangle, \mu)$ is a logarithmic local system on X of type Ψ (resp. Ψ_σ). By this projection, we have a commutative diagram

$$\begin{array}{ccc} C_\sigma & \longrightarrow & \overline{C}_\sigma \\ \downarrow & & \downarrow \\ C & \longrightarrow & \overline{C}. \end{array} \tag{2}$$

8.2.5

Proof of Step 2. We prove that $C_\sigma \simeq \mathrm{Mor}(\quad, \Gamma(\sigma)^{\mathrm{gp}}\backslash D_\sigma)$ as functors on $\overline{\mathcal{A}}_2(\log)$, using Theorem A.

By Theorem A (iii),

$$\mathrm{Mor}(\quad, \Gamma(\sigma)^{\mathrm{gp}}\backslash D_\sigma) \simeq \mathrm{Mor}(\quad, \sigma_{\mathbf{C}})\backslash \mathrm{Mor}(\quad, E_\sigma)$$

in the category of sheaf-functors on $\overline{\mathcal{A}}_2(\log)$. That is, for each object X of $\overline{\mathcal{A}}_2(\log)$, the restriction of $\mathrm{Mor}(\quad, \Gamma(\sigma)^{\mathrm{gp}}\backslash D_\sigma)$ to the category of open subobjects of X is the quotient of the restriction of $\mathrm{Mor}(\quad, E_\sigma)$ by the action of the restriction of $\mathrm{Mor}(\quad, \sigma_{\mathbf{C}})$ in the category of sheaves on X.

Hence, by the result of Step 1, it is sufficient to prove the following two assertions:

CLAIM 1 *C_σ is a sheaf-functor. (That is, for any object X of $\overline{\mathcal{A}}_2(\log)$, the restriction of C_σ to the category of open subobjects of X is a sheaf on X.)*

CLAIM 2 *$C_\sigma \simeq \mathrm{Mor}(\quad, \sigma_{\mathbf{C}})\backslash B_\sigma$ in the category of sheaf-functors on $\overline{\mathcal{A}}_2(\log)$.*

Here, in Claim 2, the action of $\sigma_{\mathbf{C}}$ on B_σ is the one corresponding to the action of $\sigma_{\mathbf{C}}$ on E_σ. That is, for $X \in \overline{\mathcal{A}}_2(\log)$ and for $a \in \mathrm{Mor}(X, \sigma_{\mathbf{C}})$, a sends

$$B_\sigma(X) \ni (\theta, F) \mapsto (\mathbf{e}(a)\theta, \exp(-a)F),$$

where $\exp(-a)F$ is defined by the action of $\exp(\sigma_{\mathbf{C}})$ on $\mathcal{O}^{\log} \otimes_{\mathbf{Z}} H_{\theta,\mathbf{Z}} = \mathcal{O}^{\log} \otimes_{\mathbf{C}} H_{0,\mathbf{C}}$ which comes from the action of $\exp(\sigma_{\mathbf{C}})$ on $H_{0,\mathbf{C}}$.

For later use, it is convenient to prove a more general assertion than Claim 1:

CLAIM 3 *C is a sheaf-functor.*

In fact, Claim 1 is the special case of Claim 3 for $\Sigma = \{\text{face of } \sigma\}$ and $\Gamma = \Gamma(\sigma)^{\mathrm{gp}}$.

Proof of Claim 3. It is sufficient to prove that, for any $X \in \overline{\mathcal{A}}_2(\log)$ and for any PLH $(H_{\mathbf{Z}}, \langle\ ,\ \rangle, F, \mu)$ on X of type Φ, $\mathrm{Aut}(H_{\mathbf{Z}}, \langle\ ,\ \rangle, F, \mu) = \{1\}$. (In fact, assume we have proved this. Then, if we have an open covering $(U_\lambda)_\lambda$ of X and have PLH a_λ on U_λ of type Φ such that the images of the classes of a_λ and $a_{\lambda'}$ coincide in $C(U_\lambda \cap U_{\lambda'})$ for any λ, λ', we have a unique isomorphism between the restrictions of a_λ and $a_{\lambda'}$ to $U_\lambda \cap U_{\lambda'}$ for any λ, λ', which give a PLH a on X of type Φ. Here the uniqueness of the isomorphism on $U_\lambda \cap U_{\lambda'}$ and also the uniqueness of the class of such a follow from the triviality of the group of automorphisms. Hence C is a sheaf-functor.) Let $\gamma \in \mathrm{Aut}(H_{\mathbf{Z}}, \langle\ ,\ \rangle, F, \mu)$, that is, let γ be an element of $\mathrm{Aut}(H_{\mathbf{Z}})$ which preserves $\langle\ ,\ \rangle$, F and μ. To prove $\gamma = 1$, it is sufficient to show that, for any $x \in X$, the pullback of γ on x is 1. Hence we may assume that $X = x$ whose underlying local ringed space over $\mathbf{C}$ is $\mathrm{Spec}(\mathbf{C})$. By 2.4.8 and 2.4.7, there exist $y \in x^{\log}$ and $s \in \mathrm{sp}(y)$ such that $(H_{\mathbf{Z},y}, \langle\ ,\ \rangle_y, F(s))$ is a polarized Hodge structure in the usual sense. Since γ induces an automorphism of this polarized Hodge structure, γ belongs to the discrete group of automorphisms of $(H_{\mathbf{Z},y}, \langle\ ,\ \rangle_y)$ and also to the compact group of automorphisms of the Hermitian form associated to this polarized Hodge structure. Hence γ is of finite order. On the other hand, take $y \in x^{\log}$ and a representative

$$\tilde{\mu}_y : (H_{\mathbf{Z},y}, \langle\ ,\ \rangle_y) \xrightarrow{\sim} (H_0, \langle\ ,\ \rangle_0)$$

of μ_y. Then $\tilde{\mu}_y \gamma \tilde{\mu}_y^{-1} \in \Gamma$. Since Γ is neat, Γ is torsion free. Hence $\gamma = 1$, as desired. □

Proof of Claim 2. By the exact sequence

$$1 \longrightarrow \Gamma(\sigma)^{\mathrm{gp}} \longrightarrow \sigma_{\mathbf{C}} \longrightarrow \mathrm{torus}_\sigma \longrightarrow 1$$

we have

$$\mathrm{Mor}(\ \ , \sigma_{\mathbf{C}})\backslash B_\sigma \simeq \mathrm{Mor}(\ \ , \mathrm{torus}_\sigma)\backslash(\Gamma(\sigma)^{\mathrm{gp}}\backslash B_\sigma),$$

where the quotients are taken in the category of sheaf-functors on $\overline{\mathcal{A}}_2(\log)$. Hence it is sufficient to prove that C_σ is the quotient of $\Gamma(\sigma)^{\mathrm{gp}}\backslash B_\sigma$ by the action of $\mathrm{Mor}(\ \ , \mathrm{torus}_\sigma)$ in the category of sheaf-functors on $\overline{\mathcal{A}}_2(\log)$. This fact is a consequence of the following (1) and (2).

(1) The following diagram is cartesian:

$$\begin{array}{ccc} \Gamma(\sigma)^{\mathrm{gp}}\backslash B_\sigma & \longrightarrow & \mathrm{Mor}(\ \ , \mathrm{toric}_\sigma) \\ \downarrow & & \downarrow \\ C_\sigma & \longrightarrow & \overline{C}_\sigma. \end{array}$$

Here the upper horizontal arrow is $((\theta, F) \bmod \Gamma(\sigma)^{\mathrm{gp}}) \mapsto \theta$ and the right vertical arrow is $\theta \mapsto (H_{\theta,\mathbf{Z}}, \langle\ ,\ \rangle_\theta, \mu_\theta)$.

(2) $\mathrm{Mor}(\ \ , \mathrm{toric}_\sigma) \to \overline{C}_\sigma$ is a $\mathrm{Mor}(\ \ , \mathrm{torus}_\sigma)$-torsor.

(1) is easy to prove. (2) follows from the commutative diagram

$$\begin{array}{ccc} \mathrm{Mor}(X, \mathrm{toric}_\sigma) & \xrightarrow{\sim} & \mathrm{Hom}(\Gamma(\sigma)^\vee, M_X) \\ \downarrow & & \downarrow \\ \overline{C}_\sigma(X) & \xrightarrow{\sim} & \mathrm{Hom}(\Gamma(\sigma)^\vee, M_X/\mathcal{O}_X^\times) \end{array}$$

for $X \in \overline{\mathcal{A}}_2(\log)$, in which the lower isomorphism is by 8.1.9 and the upper horizontal arrow commutes with the action of $\mathrm{Mor}(X, \mathrm{torus}_\sigma) = \mathrm{Hom}(\Gamma(\sigma)^\vee, \mathcal{O}_X^\times)$.

This completes the proof of Claim 2 and hence Step 2 is proved. □

8.2.6

Proof of Step 3. We prove $C \simeq \mathrm{Mor}(\ \ , \Gamma\backslash D_\Sigma)$ as functors on $\overline{\mathcal{A}}_2(\log)$, using Theorem A.

By Theorem A (iv), we have

$$\left(\bigsqcup_{\sigma\in\Sigma} \mathrm{Mor}(\quad, \Gamma(\sigma)^{\mathrm{gp}}\backslash D_\sigma)\right)\Big/\sim \;\xrightarrow{\sim}\; \mathrm{Mor}(\quad, \Gamma\backslash D_\Sigma), \tag{1}$$

where the left side is the quotient in the category of sheaf-functors by the equivalence relation $\sim$ generated by the following relation: For a triple (σ, τ, γ) with $\sigma, \tau \in \Sigma$ and $\gamma \in \Gamma$ such that $\mathrm{Ad}(\gamma)(\tau) \subset \sigma$, and for $X \in \overline{\mathcal{A}}_2(\log)$, $a \in \mathrm{Mor}(X, \Gamma(\tau)^{\mathrm{gp}}\backslash D_\tau)$ is related by $\sim$ to $\gamma \circ a \in \mathrm{Mor}(X, \Gamma(\sigma)^{\mathrm{gp}}\backslash D_\sigma)$, where the γ denotes the morphism $\Gamma(\tau)^{\mathrm{gp}}\backslash D_\tau \to \Gamma(\sigma)^{\mathrm{gp}}\backslash D_\sigma$ in $\overline{\mathcal{A}}_2(\log)$ induced by γ.

Hence, by the result of Step 2, it is sufficient to show

Claim 1 $(\bigsqcup_{\sigma\in\Sigma} C_\sigma)/\sim\; \xrightarrow{\sim} C$.

Here the equivalence relation $\sim$ is defined in the same way as in (1). That is, the equivalence relation $\sim$ is generated by the following relation: For a triple (σ, τ, γ) with $\sigma, \tau \in \Sigma$ and $\gamma \in \Gamma$ such that $\mathrm{Ad}(\gamma)(\tau) \subset \sigma$ and for $X \in \overline{\mathcal{A}}_2(\log)$, the isomorphism class of PLH $(H_{\mathbf{Z}}, \langle\ ,\ \rangle, F, \mu)$ on X of type Φ_τ is related by $\sim$ to the isomorphism class of PLH $(H_{\mathbf{Z}}, \langle\ ,\ \rangle, F, \gamma\circ\mu)$ on X of type Φ_σ.

We first prove

Claim 2 $(\bigsqcup_{\sigma\in\Sigma} \overline{C}_\sigma)/\sim\; \xrightarrow{\sim} \overline{C}$.

Here the equivalence relation $\sim$ is defined in the same way as in Claim 1.

Proof of Claim 2. For $X \in \overline{\mathcal{A}}_2(\log)$, we have a map

$$\overline{C}(X) \to H^0(X, R^1\tau_*\Gamma) \tag{2}$$

sending the isomorphism class of a logarithmic local system $(H_{\mathbf{Z}}, \langle\ ,\ \rangle, \mu)$ on X of type Ψ to the class of the Γ-torsor on $X^{\log}$ obtained by the inverse image of μ under the map

$$\underline{\mathrm{Isom}}((H_{\mathbf{Z}}, \langle\ ,\ \rangle), (H_0, \langle\ ,\ \rangle_0)) \to \Gamma\backslash\,\underline{\mathrm{Isom}}((H_{\mathbf{Z}}, \langle\ ,\ \rangle), (H_0, \langle\ ,\ \rangle_0)).$$

The map (2) is injective and the image consists of those elements $t \in H^0(X, R^1\tau_*\Gamma)$ which satisfy the following condition: For any point $x \in X$, if t_x denotes the image of t in

$$H^1(x^{\log}, \Gamma) = (\mathrm{Hom}(\pi_1(x^{\log}), \Gamma) \text{ mod conjugation in } \Gamma)$$

and $\tilde{t}_x : \pi_1(x^{\log}) \to \Gamma$ denotes a homomorphism which represents t_x, then there exists $\sigma \in \Sigma$ such that $\tilde{t}_x$ sends $\pi_1^+(x^{\log})$ into $\Gamma(\sigma)$. (This follows from the similar description of $\overline{C}_\sigma(X)$ as a subset of $H^0(X, R^1\tau_*(\Gamma(\sigma)^{\mathrm{gp}}))$ obtained in 8.1.9.) Since $\tau : X^{\log} \to X$ is proper, by the proper base change theorem (Appendix A2), the stalk of $R^1\tau_*\Gamma$ at $x \in X$ is isomorphic to $H^1(\tau^{-1}(x), \Gamma) = H^1(x^{\log}, \Gamma)$ and the stalk of $R^1\tau_*(\Gamma(\sigma)^{\mathrm{gp}})$ at x is isomorphic to $H^1(x^{\log}, \Gamma(\sigma)^{\mathrm{gp}})$. Hence, to prove Claim 2, it is sufficient to prove

$$\left(\bigsqcup_{\sigma\in\Sigma} \overline{C}_\sigma(x)\right)\Big/\sim \;\xrightarrow{\sim}\; \overline{C}(x) \tag{3}$$

for any fs logarithmic analytic space x whose underlying analytic space is $\mathrm{Spec}(\mathbf{C})$.

The surjectivity of (3) is clear by definition. We prove the injectivity. Let $\sigma, \sigma' \in \Sigma$ and let $g : \pi_1^+(x^{\log}) \to \Gamma(\sigma)$, $g' : \pi_1^+(x^{\log}) \to \Gamma(\sigma')$ be homomorphisms. Assume that the homomorphisms $\pi_1^+(x^{\log}) \to \Gamma$ induced by g and g' are conjugate in Γ, that is, assume that there exists $\gamma \in \Gamma$ such that $g'(a) = \gamma g(a)\gamma^{-1}$ for any $a \in \pi_1^+(x^{\log})$. We show that $g \sim g'$ in the sense of (3). Put $\sigma'' := \sigma \cap \mathrm{Ad}(\gamma)^{-1}(\sigma') \in \Sigma$. Then the image of g is contained in $\Gamma(\sigma) \cap \gamma^{-1}\Gamma(\sigma')\gamma = \Gamma(\sigma'')$, and hence g gives a homomorphism $g'' : \pi_1^+(x^{\log}) \to \Gamma(\sigma'')$. We have $\mathrm{Ad}(\gamma)(\sigma'') \subset \sigma'$ and g' is equivalent to g'', but g'' is clearly equivalent to g. (3) and hence Claim 2 is proved. □

Proof of Claim 1. It is easy to see that the canonical morphism from C_σ to the fiber product $C \times_{\overline{C}} \overline{C}_\sigma$ is a surjection in the category of sheaf-functors. Hence, by Claim 2, the morphism in Claim 1 is a surjection in the category of sheaf-functors. We show that it is injective. Let $X \in \overline{\mathcal{A}}_2(\log)$, let $\sigma, \sigma' \in \Sigma$ and let $a \in C_\sigma(X)$, $a' \in C_{\sigma'}(X)$. Assume that the images of a and a' in $C(X)$ coincide. We show that $a \sim a'$ in the sense of Claim 1. By Claim 2, we may assume that $a, a' \in C_\sigma(X)$ for some common $\sigma \in \Sigma$ and that the images of a, a' in $\overline{C}_\sigma(X)$ coincide. Thus we are reduced to the following situation. Let $(H_{\mathbf{Z}}, \langle\ ,\ \rangle, \mu)$ be a logarithmic local system on X of type Ψ_σ, let F and F' be decreasing filtrations on $\mathcal{O}_X^{\log}$-module $\mathcal{O}_X^{\log} \otimes H_{\mathbf{Z}}$ such that $(H_{\mathbf{Z}}, \langle\ ,\ \rangle, F, \mu)$ and $(H_{\mathbf{Z}}, \langle\ ,\ \rangle, F', \mu)$ are PLHs on X of type Φ_σ, and assume that there exists an isomorphism ν from the former to the latter when we regard them as PLH on X of type Φ (not Φ_σ). It is sufficient to prove that the isomorphism classes $a, a' \in C_\sigma(X)$ of $(H_{\mathbf{Z}}, \langle\ ,\ \rangle, F, \mu)$ and $(H_{\mathbf{Z}}, \langle\ ,\ \rangle, F', \mu)$, respectively, satisfy $a \sim a'$ in the sense of Claim 1.

Let $x \in X$, $y \in x^{\log}$ and let $\tilde{\mu}_y : (H_{\mathbf{Z},y}, \langle\ ,\ \rangle_y) \xrightarrow{\sim} (H_0, \langle\ ,\ \rangle_0)$ be a representative of μ_y. Then, since ν is an isomorphism as PLH of type Φ, $\gamma := \tilde{\mu}_y \circ \nu \circ \tilde{\mu}_y^{-1} \in G_{\mathbf{Z}}$ belongs to Γ. Let $\sigma'' = \sigma \cap \mathrm{Ad}(\gamma)^{-1}\sigma \in \Sigma$. We show that, for some open neighborhood U of x in X, the image a_U of a in $C_\sigma(U)$ comes from an element $b \in C_{\sigma''}(U)$ whose image under $\gamma : C_{\sigma''}(U) \to C_\sigma(U)$ coincides with the image a'_U of a' in $C_\sigma(U)$. In fact, since ν commutes with the action of $\pi_1(x^{\log})$ on $H_{\mathbf{Z},y}$, the image of

$$\pi_1^+(x^{\log}) \to \mathrm{Aut}(H_{\mathbf{Z},y}, \langle\ ,\ \rangle_y) \xrightarrow[\sim]{\text{by } \tilde{\mu}_y} G_{\mathbf{Z}}$$

is contained in $\Gamma(\sigma) \cap \gamma^{-1}\Gamma(\sigma)\gamma = \Gamma(\sigma) \cap \Gamma(\gamma^{-1}\sigma\gamma) = \Gamma(\sigma'')$. Since $\tau : X^{\log} \to X$ is proper, the stalk of $R^1\tau_*(\Gamma(\sigma)^{\mathrm{gp}})$ at x is isomorphic to

$$H^1(x^{\log}, \Gamma(\sigma)^{\mathrm{gp}}) = \mathrm{Hom}(\pi_1(x^{\log}), \Gamma(\sigma)^{\mathrm{gp}})$$

and the same holds when σ is replaced by σ''. Hence, on the same open neighborhood U of x in X, the logarithmic local system $(H_{\mathbf{Z}}, \langle\ ,\ \rangle, \mu)$ of type Ψ_σ comes from a logarithmic local system $(H_{\mathbf{Z}}, \langle\ ,\ \rangle, \mu'')$ on U of type $\Psi_{\sigma''}$ such that $\nu : H_{\mathbf{Z}} \xrightarrow{\sim} H_{\mathbf{Z}}$ gives an isomorphism $(H_{\mathbf{Z}}, \langle\ ,\ \rangle, \gamma \circ \mu'') \xrightarrow{\sim} (H_{\mathbf{Z}}, \langle\ ,\ \rangle, \mu)$ of logarithmic local systems on U of type Ψ_σ. Now $(H_{\mathbf{Z}}, \langle\ ,\ \rangle, F, \mu'')$ is a PLH on U of type $\Phi_{\sigma''}$. Let $b \in C_{\sigma''}(U)$ be its isomorphism class. Then the image of b in $C_\sigma(U)$ under the natural map $C_{\sigma''}(U) \to C_\sigma(U)$ is a_U. On the other hand, the image of b under $\gamma : C_{\sigma''}(U) \to C_\sigma(U)$ coincides with a'_U, since ν gives an isomorphism

$(H_{\mathbf{Z}}, \langle\ ,\ \rangle, F, \gamma \circ \mu'') \xrightarrow{\sim} (H_{\mathbf{Z}}, \langle\ ,\ \rangle, F', \mu)$ of PLH on U of type Φ_σ. Claim 1 is proved. □

This completes the proof of Theorem B. □

8.2.7

Remark. As in [KU2, §6] and in 0.4.14, we say that we are in the classical situation if the following conditions (1) and (2) are satisfied.

(1) The horizontal tangent bundle of D coincides with the tangent bundle of D.
(2) Fibers of $D \to \mathcal{X}$ are finite sets.

In the classical situation, $\tilde{E}_\sigma = \check{E}_\sigma$ by (1) and hence E_σ is open in $\check{E}_\sigma$, and the strong topology on $E_\sigma \subset \check{E}_\sigma$ coincides with the topology on E_σ as a subspace of $\check{E}_\sigma$. Thus our space $\Gamma \backslash D_\Sigma$ for a neat Γ becomes an fs logarithmic analytic space whose underlying analytic space is nothing but a toroidal partial compactification of Mumford et al. ([AMRT]). Even in this case, our theory gives the following new things; a new interpretation of a toroidal compactification as the fine moduli space of polarized logarithmic Hodge structures, and a description of the precise relationship between toroidal compactifications and Borel-Serre compactification (0.5.28).

8.2.8

Remark. Concerning the map $\varphi^{\log} : X^{\log} \to \Gamma \backslash D^\sharp_\Sigma$ in 2.5.10(2), in the case where $X = \Gamma \backslash D_\Sigma$ with Γ neat and the PLH is the universal one, then $\varphi^{\log}$ coincides with the homeomorphism $(\Gamma \backslash D_\Sigma)^{\log} \xrightarrow{\sim} \Gamma \backslash D^\sharp_\Sigma$ in Theorem A (vi).

8.3 RELATIONSHIP AMONG CATEGORIES OF GENERALIZED ANALYTIC SPACES

In this section, we prove Theorem 3.2.5. As is explained in 4.2.2, this completes the proof of Theorem 0.4.27.

8.3.1

We prove

Claim $\overline{\mathcal{A}(\log)} \subset \overline{\mathcal{A}}(\log)$.

Proof. Let Z be a local ringed space over $\mathbf{C}$ and endow Z with the trivial logarithmic structure. Let X be an fs logarithmic local ringed space over $\mathbf{C}$. Then, the set $\mathrm{Mor}_{\log}(X, Z)$ of morphisms $X \to Z$ in the category of logarithmic local ringed spaces over $\mathbf{C}$ and the set $\mathrm{Mor}(X, Z)$ of morphisms in the category of local ringed spaces over $\mathbf{C}$ (by forgetting the logarithmic structures of X and Z) are identified naturally.

We show that, for Z as above, the sets $\mathrm{Mor}(h^X_{\mathcal{A}(\log)}, h^Z_{\mathcal{A}(\log)})$ and $\mathrm{Mor}(h^X_{\mathcal{A}}, h^Z_{\mathcal{A}})$ are also identified naturally. Let $f : h^X_{\mathcal{A}(\log)} \to h^Z_{\mathcal{A}(\log)}$ be a morphism. We define the corresponding $g : h^X_{\mathcal{A}} \to h^Z_{\mathcal{A}}$ as follows. Let A be an analytic space and let $h : A \to X$ be a morphism of local ringed spaces over $\mathbf{C}$. Endow A with the inverse image of the logarithmic structure of X. Since the logarithmic structure of X is fs, so is the induced logarithmic structure of A. Let $h(\log) : A \to X$ be the morphism of logarithmic local ringed spaces over $\mathbf{C}$ induced by h, and define $g(h) : A \to Z$ to be the morphism of local ringed spaces over $\mathbf{C}$ underlying $f(h(\log))$. We have defined the map $f \mapsto g$. The inverse map $g \mapsto f$ is defined as follows. For an fs logarithmic analytic space A and for a morphism $h : A \to X$ of logarithmic local ringed spaces over $\mathbf{C}$, we have a morphism of local ringed spaces $g(\overline{h}) : A \to Z$ over $\mathbf{C}$, where $\overline{h}$ is the morphism of local ringed spaces over $\mathbf{C}$ underlying h. We obtain the morphism $f(h) : A \to Z$ of logarithmic local ringed spaces over $\mathbf{C}$ corresponding to $g(\overline{h})$.

Now let X be an object of $\overline{\mathcal{A}(\log)}$. Then, for Z as above, we have

$$\mathrm{Mor}(h^X_{\mathcal{A}}, h^Z_{\mathcal{A}}) = \mathrm{Mor}(h^X_{\mathcal{A}(\log)}, h^Z_{\mathcal{A}(\log)}) = \mathrm{Mor}_{\log}(X, Z) = \mathrm{Mor}(X, Z).$$

This proves that X belongs to $\overline{\mathcal{A}}(\log)$. □

8.3.2

We prove

Claim *For an object X of $\overline{\mathcal{A}}(\log)$, the residue field is $\mathbf{C}$ at any point $x \in X$.*

Proof. Assume, contrarily, that there is a point $x \in X$ at which the residue field is not $\mathbf{C}$. Take any set Z and maps $f, g : X \to Z$ such that $f(x) \neq g(x)$ and $f(y) = g(y)$ for any $y \in X - \{x\}$. Endow Z with the weakest topology, i.e., Z and the empty set are the only open sets, and with the constant sheaf of rings $\mathbf{C}$, and regard Z as a local ringed space over $\mathbf{C}$. Then, for any local ringed space Y over $\mathbf{C}$, $\mathrm{Mor}(Y, Z)$ is identified with the set of maps $Y \to Z$. Hence f and g give morphisms $X \to Z$ of local ringed spaces over $\mathbf{C}$ which are different. Let $\mathcal{A}$ be the category of analytic spaces. We show that f and g induce the same morphism of functors $h^X_{\mathcal{A}} \to h^Z_{\mathcal{A}}$. Let A be an analytic space and $h : A \to X$ be a morphism. Since there is no homomorphism $\mathcal{O}_{X,x}/\mathfrak{m}_{X,x} \to \mathbf{C}$ over $\mathbf{C}$, x is not contained in the image of h. Hence $f \circ h = g \circ h$. Thus the map $\mathrm{Mor}(X, Z) \to \mathrm{Mor}(h^X_{\mathcal{A}}, h^Z_{\mathcal{A}})$ is not injective. □

Proposition 8.3.3 *Let $\mathcal{C} = \overline{\mathcal{A}}$ (resp. $\overline{\mathcal{A}(\log)}$).*

(i) *Let X be an object of $\mathcal{C}$ and U be an open set of X. Then U, with the induced topology and with the restriction of $\mathcal{O}_X$ (resp. the restrictions of $\mathcal{O}_X$ and M_X), is an object of $\mathcal{C}$.*

(ii) *Let X be a local ringed space (resp. a logarithmic local ringed space) over $\mathbf{C}$ and assume that X has an open covering consisting of objects of $\mathcal{C}$. Then X is an object of $\mathcal{C}$.*

Proof. (ii) is easy. We prove (i). Let $\mathcal{D} = \mathcal{A}$ (resp. $\mathcal{D} = \mathcal{A}(\log)$). Let Z be a local ringed space (resp. logarithmic local ringed space) over $\mathbf{C}$. We will give the inverse map of $\mathrm{Mor}(U, Z) \to \mathrm{Mor}(h^U_{\mathcal{D}}, h^Z_{\mathcal{D}})$ (cf. 3.2.2). Let W be the following local ringed space (resp. logarithmic local ringed space) over $\mathbf{C}$. As a set, W is the union of

Z and one point p which does not belong to Z. An open set of W is either W itself or an open set of Z. The structure sheaf $\mathcal{O}_W$ is defined by $\mathcal{O}_W(W) := \mathbf{C}$ and $\mathcal{O}_W(V) := \mathcal{O}_Z(V)$ for any open set V of Z. In the logarithmic case (i.e., the "resp. " case), the logarithmic structure M_W is defined by $M_W(W) := \mathbf{C}^\times$ and $M_W(V) := M_Z(V)$ for any open set V of Z. Let $f : h^U_{\mathcal{D}} \to h^Z_{\mathcal{D}}$ be a morphism. We define the corresponding morphism $g : U \to Z$ as follows. First, let $f' : h^X_{\mathcal{D}} \to h^W_{\mathcal{D}}$ be the following morphism. For an object A of $\mathcal{D}$ and a morphism $h : A \to X$, let $f'(h) : A \to W$ be the unique morphism which coincides on $h^{-1}(U)$ with the morphism $f(h_U)$ where h_U is the restriction $h^{-1}(U) \to U$ of h and which sends all points of $A - h^{-1}(U)$ to p. Since X belongs to $\mathcal{C}$, there exists a unique morphism $g' : X \to W$ which induces f'. Then, the map g' sends U into Z. This follows from the fact that, for any point x of U, there are an object A of $\mathcal{D}$ and a morphism $h_x : A \to U$ whose image coincides with $\{x\}$. (In fact, in the non-logarithmic case, we have by 8.3.2 a unique morphism $h_x : \mathrm{Spec}(\mathbf{C}) \to U$ whose image is $\{x\}$. In the logarithmic case, if we endow $\mathrm{Spec}(\mathbf{C})$ with the inverse image of the logarithmic structure of U under this h_x, $\mathrm{Spec}(\mathbf{C})$ becomes an object of $\mathcal{D}$ and h_x becomes a desired morphism.) We define $g : U \to Z$ to be the morphism induced by g'. It is easily seen that the map $f \mapsto g$ is the inverse of the map $\mathrm{Mor}(U, Z) \to \mathrm{Mor}(h^U_{\mathcal{D}}, h^Z_{\mathcal{D}})$. □

8.3.4

We prove that

CLAIM $\overline{\mathcal{A}} \subset \overline{\mathcal{A}}_2$.

Proof. Let X be an object of $\overline{\mathcal{A}}$.

We first show that X satisfies the condition (A_1') (= (A_1)) in 2.2.1. Let U be an open set of X. By Proposition 8.3.3, U belongs to $\overline{\mathcal{A}}$. Let Z be a scheme locally of finite type over $\mathbf{C}$. Since $h^Z_{\mathcal{A}} = h^{Z_{\mathrm{an}}}_{\mathcal{A}}$, we have

$$\mathrm{Mor}(U, Z) \simeq \mathrm{Mor}(h^U_{\mathcal{A}}, h^Z_{\mathcal{A}}) \simeq \mathrm{Mor}(h^U_{\mathcal{A}}, h^{Z_{\mathrm{an}}}_{\mathcal{A}}) \simeq \mathrm{Mor}(U, Z_{\mathrm{an}}).$$

We next prove that X satisfies 3.2.4 (A_2). Let X' be the following local ringed space over $\mathbf{C}$. The underlying set of X' is that of X, a subset U of X' is defined to be open if and only if, for any analytic space A and any morphism $\lambda : A \to X$, $\lambda^{-1}(U)$ is open in A, and $\mathcal{O}_{X'}$ is defined to be the inverse image of $\mathcal{O}_X$. Then it is trivial to see that the canonical morphism $h^{X'}_{\mathcal{A}} \to h^X_{\mathcal{A}}$ is an isomorphism. Since X is an object of $\overline{\mathcal{A}}$, the inverse $h^X_{\mathcal{A}} \to h^{X'}_{\mathcal{A}}$ of this isomorphism comes from a morphism $X \to X'$ such that the composite map $X \to X' \to X$ is the identity of X. This shows that the topologies of X and X' coincide, that is, X satisfies 3.2.4 (A_2). □

PROPOSITION 8.3.5 *Let X be a local ringed space over $\mathbf{C}$. Then, X belongs to $\overline{\mathcal{A}}$ if and only if X satisfies the following conditions* (A_0) *and* (A_3) *and also the condition* 3.2.4 (A_2).

(A_0) *The residue field is $\mathbf{C}$ at any $x \in X$.*

(A_3) *Let $\mathcal{O}'_X$ be the sheaf of rings on X defined by*

$$\mathcal{O}'_X(U) = \varprojlim_{(A,\lambda)} \mathcal{O}_A(\lambda^{-1}(U)) \quad (U \subset X : \textit{open}),$$

where (A, λ) ranges over the category of pairs consisting of an analytic space A and a morphism $\lambda : A \to X$ of local ringed spaces over **C**. *Then the canonical map $\mathcal{O}_X \to \mathcal{O}'_X$ is an isomorphism.*

Proof. Let X be an object of $\overline{\mathcal{A}}$. Then X belongs to $\overline{\mathcal{A}}_2$ by 8.3.4 and hence satisfies (A_0) (2.2.2) and also 3.2.4 (A_2). We prove that X satisfies (A_3). Let U be an open set of X and let $Z = \mathrm{Spec}(\mathbf{C}[T])$ where T is an indeterminate over **C**. By 8.3.3, U belongs to $\overline{\mathcal{A}}$. Hence we have

$$\mathcal{O}_X(U) \simeq \mathrm{Mor}(U, Z) \simeq \mathrm{Mor}(h^U_{\mathcal{A}}, h^Z_{\mathcal{A}}) \simeq \mathcal{O}'_X(U),$$

and (A_3) is satisfied by X.

Conversely, assume that X is a local ringed space over **C** satisfying (A_0), 3.2.4 (A_2), and (A_3). We prove that X belongs to $\overline{\mathcal{A}}$. Let Z be a local ringed space over **C**. We prove that the map $\mathrm{Mor}(X, Z) \to \mathrm{Mor}(h^X_{\mathcal{A}}, h^Z_{\mathcal{A}})$ is bijective. We give the inverse of this map. Let $f : h^X_{\mathcal{A}} \to h^Z_{\mathcal{A}}$ be a morphism. We define the corresponding morphism $g : X \to Z$ as follows. First, we define the map of sets underlying g. Let $x \in X$. By (A_0), we have a unique morphism $h : \mathrm{Spec}(\mathbf{C}) \to X$ of local ringed spaces over **C** whose image is $\{x\}$. Define $g(x) \in Z$ to be the image of $f(h) : \mathrm{Spec}(\mathbf{C}) \to Z$. By the condition 3.2.4 (A_2), g is continuous. We define $\mathcal{O}_Z \to g_*(\mathcal{O}_X)$ as the composite map $\mathcal{O}_Z \to g_*(\mathcal{O}'_X) \simeq g_*(\mathcal{O}_X)$ where the first arrow is induced by f and the last is the isomorphism by (A_3). This gives a morphism $g : X \to Z$ of local ringed spaces over **C**. It is seen easily that $f \mapsto g$ is the inverse of the map $\mathrm{Mor}(X, Z) \to \mathrm{Mor}(h^X_{\mathcal{A}}, h^Z_{\mathcal{A}})$. □

Proposition 8.3.6 *For an fs logarithmic local ringed space X over* **C**, *the following conditions* (1)–(3) *are equivalent.*

(1) *X belongs to $\overline{\mathcal{A}(\log)}$.*
(2) *X belongs to $\overline{\mathcal{A}(\log)}'_{fs}$.*
(3) *The local ringed space over* **C** *underlying X belongs to $\overline{\mathcal{A}}$, and X satisfies the following condition* (A_4).

(A_4) *Let M'_X be the sheaf on X defined by*

$$M'_X(U) = \varprojlim_{(A,\lambda)} M_A(\lambda^{-1}(U)) \quad (U \subset X : \textit{open}),$$

where (A, λ) ranges over the category of pairs consisting of an analytic space A and a morphism $\lambda : A \to X$ of local ringed spaces over **C** *and where M_A denotes the inverse image on A of the logarithmic structure of X. Then the canonical map $M_X \to M'_X$ is an isomorphism.*

Proof. By using 8.3.1, the proofs of the implications

$$(1) \Rightarrow (3) \Rightarrow (2)$$

go in the same way as the proof of Proposition 8.3.5. (In the place where we used $Z = \operatorname{Spec}(\mathbf{C}[T])$ in the proof of 8.3.5, we use this time $Z = \operatorname{Spec}(\mathbf{C}[\mathbf{N}])$ with the logarithmic structure associated to $\mathbf{N} \to \mathbf{C}[\mathbf{N}]$ ($\simeq \mathbf{C}[T]$).) Since (2) implies (1) clearly, we see that (1), (2), and (3) are equivalent. □

8.3.7

For the proof of Theorem 3.2.5, it remains to prove $\mathcal{B}^* \subset \overline{\mathcal{A}}$ and $\mathcal{B}^*(\log) \subset \overline{\mathcal{A}(\log)}$. We prove $\mathcal{B}^* \subset \overline{\mathcal{A}}$ in 8.3.8–8.3.10 and prove $\mathcal{B}^*(\log) \subset \overline{\mathcal{A}(\log)}$ in 8.3.11.

In 8.3.7–8.3.11, let Z be a reduced analytic space, let B be a closed analytic subspace of Z such that $Z - B$ is dense in Z and let X be a subset of Z containing $Z - B$. Endow X with the strong topology in Z and with the inverse image of $\mathcal{O}_Z$. Let $\mathcal{A}/X$ be the category of pairs (A, λ) where A is an analytic space and λ is a morphism $A \to X$ of local ringed spaces over $\mathbf{C}$.

To prove $\mathcal{B}^* \subset \overline{\mathcal{A}}$, by Proposition 8.3.3, it is sufficient to show that this X belongs to $\overline{\mathcal{A}}$. To show it, by Proposition 8.3.5 and 3.2.10, it is sufficient to show that $\mathcal{O}_X \to \mathcal{O}'_X$ is an isomorphism.

Let $x \in X$ and $h \in \mathcal{O}'_{X,x}$. By taking $(A, \lambda) \in \mathcal{A}/X$ with $A := \operatorname{Spec}(\mathcal{O}_{Z,x}/\mathfrak{m}^n_{Z,x})$ and λ the inclusion morphism $A \hookrightarrow Z$, we have $h_{(A,\lambda)} \in \mathcal{O}_{Z,x}/\mathfrak{m}^n_{Z,x}$ and hence we have $\hat{h} \in \hat{\mathcal{O}}_{Z,x} := \varprojlim \mathcal{O}_{Z,x}/\mathfrak{m}^n_{Z,x}$. This gives a homomorphism $\mathcal{O}'_{X,x} \to \hat{\mathcal{O}}_{Z,x}$ such that the composite map $\mathcal{O}_{Z,x} = \mathcal{O}_{X,x} \to \mathcal{O}'_{X,x} \to \hat{\mathcal{O}}_{Z,x}$ is the inclusion map. Hence we have the injectivity of $\mathcal{O}_{X,x} \to \mathcal{O}'_{X,x}$. For the proof of the surjectivity, it is sufficient to prove the following two claims.

Claim 1 *$\mathcal{O}'_{X,x} \to \hat{\mathcal{O}}_{Z,x}$, $h \mapsto \hat{h}$, is injective.*

Claim 2 *For $h \in \mathcal{O}'_{X,x}$, $\hat{h}$ belongs to $\mathcal{O}_{Z,x} \subset \hat{\mathcal{O}}_{Z,x}$.*

We prove Claim 1 here. The proof of Claim 2 will be given in 8.3.8–8.3.10 below.

Proof of Claim 1. It is sufficient to show that, for an open set U of X (in the strong topology in Z) and for $h_1, h_2 \in \mathcal{O}'_X(U)$, if V denotes the subset of U consisting of all points y such that the images of h_1 and h_2 under $\mathcal{O}'_X(U) \to \mathcal{O}'_{X,y} \to \hat{\mathcal{O}}_{Z,y}$ coincide, then V is open in U (in the strong topology). We prove that, for any $(A, \lambda) \in \mathcal{A}/X$, $\lambda^{-1}(V)$ is open in A. Let U' be the open set $\lambda^{-1}(U)$ of A and let $h_{j,U'} = h_{j,(U',\lambda)} \in \mathcal{O}_A(U')$ $(j = 1, 2)$. Then, $\lambda^{-1}(V)$ coincides with the subset of U' consisting of all points y such that the images of $h_{1,U'}$ and $h_{2,U'}$ in $\mathcal{O}_{A,y}$ coincide, and hence is open. Claim 1 is proved. □

8.3.8

We prove Claim 2 in 8.3.7 in the case when Z is non-singular and B is a divisor with normal crossings on Z.

Since the problem is local at x, we may assume $Z = \mathbf{C}^n$ and $x = 0 \in Z$, and B is defined by $q = \prod_{1 \le j \le r} z_j$ where $(z_j)_{1 \le j \le n}$ is the coordinate of $\mathbf{C}^n$ and $r \le n$. Take an integer $k > r$. Let $\mathcal{S}$ be the submonoid of the group $E := \{\prod_{1 \le j \le n} z_j^{e(j)} \mid e(j) \in \mathbf{Z}\}$ generated by z_j and $z_j^k q^{-1}$ $(1 \le j \le n)$. Then $\mathcal{S}^\times = \{1\}$. This is because the group homomorphism $\mathcal{S}^{\mathrm{gp}} = E \to \mathbf{N}$, $t_j \mapsto 1$ $(\forall j)$, sends $z_j^k q^{-1}$ to $k - r > 0$. Let $A = \mathrm{Spec}(\mathbf{C}[\mathcal{S}])_{\mathrm{an}}$, and let $\lambda : A \to Z$ be the canonical morphism.

We show $\lambda(A) = (Z - B) \cup \{x\}$. This, in particular, shows $(A, \lambda) \in \mathcal{A}/X$.

First we show $\lambda(A) \subset (Z - B) \cup \{x\}$. If $a \in A$ corresponds to a homomorphism $\varphi : \mathcal{S} \to \mathbf{C}$ where $\mathbf{C}$ is regarded as a monoid by multiplication and if the image of a in Z is contained in B, then $\varphi(q) = 0$ and hence $\varphi(z_j)^k = \varphi(z_j^k q^{-1})\varphi(q) = 0$. This shows $\varphi(z_j) = 0$ for all j, that is, $\lambda(a) = x$. Next we prove $x \in \lambda(A)$. In fact, x is the image of the point y of A corresponding to the homomorphism $\mathcal{S} \to \mathbf{C}$ which sends $\mathcal{S} - \{1\}$ to 0. Finally we prove $Z - B \subset \lambda(A)$. Since q is invertible on $Z - B$, an element p of $Z - B$ gives a homomorphism $\mathcal{S} \to \mathbf{C}$, $f \mapsto f(p)$, which defines a unique point of A whose image in Z is p.

Let $y \in A$ be as above. Then the image of h in $\mathcal{O}_{A,y} \subset \hat{\mathcal{O}}_{A,y} = \mathbf{C}[[\mathcal{S}]]$ coincides with the image of $\hat{h}$. This shows that, if we write $\hat{h} = \sum_{m \in \mathbf{N}^n} a_m \prod_j z_j^{m(j)} = \sum_{s \in \mathcal{S}} b_s s$, then, for some homomorphism $\psi : \mathcal{S} \to \mathbf{C}^\times$, $\sum_{s \in \mathcal{S}} b_s \psi(s)$ converges absolutely. From this, we see that $\sum_{m \in \mathbf{N}^n} a_m \prod_j \alpha_j^{m(j)}$ converges absolutely for some $\alpha_j \ne 0$ $(1 \le j \le n)$. Hence $\hat{h}$ belongs to $\mathcal{O}_{Z,x}$. Claim 2 in 8.3.7 in the present case is proved.

8.3.9

We prove Claim 2 in 8.3.7, assuming Z is normal.

By the resolution of singularity of Hironaka [H], [AHV], we may assume that there exists a smooth analytic space Z' and a proper morphism $f : Z' \to Z$ which is an isomorphism over a dense open subset $Z - B$ of Z such that $f^{-1}(B)$ is a divisor with normal crossings on Z'. Since Z is normal, the theory of Stein decomposition shows $f_*(\mathcal{O}_{Z'}) = \mathcal{O}_Z$. Let X' be the inverse image of X in Z', endow X' with the strong topology in Z' and with the inverse image of $\mathcal{O}_{Z'}$ and let $g : X' \to X$ be the induced map. Note that X' belongs to $\overline{\mathcal{A}}$ by 8.3.7 and 8.3.8. By Lemma 3.1.9, the topology of X' coincides with the topology as the fiber product $X \times_Z Z'$. Since f is proper, we have

$$g_*(\mathcal{O}_{X'}) = g_*(\mathcal{O}_{Z'}|_{X'}) = f_*(\mathcal{O}_{Z'})|_X = \mathcal{O}_Z|_X = \mathcal{O}_X. \tag{1}$$

A section of $\mathcal{O}'_X$ induces a section of $g_*(\mathcal{O}'_{X'}) = g_*(\mathcal{O}_{X'}) = \mathcal{O}_X$ by (1). This proves Claim 2 in 8.3.7 in the present case.

8.3.10

We prove Claim 2 in 8.3.7 in general. This will complete the proof of $\mathcal{B}^* \subset \overline{\mathcal{A}}$.

Let Z' be the normalization of Z and let $f : Z' \to Z$ be the canonical morphism. Then

$$\mathcal{O}_{Z,x} = f_*(\mathcal{O}_{Z'})_x \cap \hat{\mathcal{O}}_{Z,x}$$

in the completion of the semi-local ring $f_*(\mathcal{O}_{Z'})_x$. By 8.3.9, this proves Claim 2 in 8.3.7.

8.3.11

We prove $\mathcal{B}^*(\log) \subset \overline{\mathcal{A}(\log)}$.

Let Z, B, X be as in 8.3.7 and assume that Z is endowed with an fs logarithmic structure. We prove that X belongs to $\overline{\mathcal{A}(\log)}$. By Proposition 8.3.6 and 8.3.7–8.3.10, it is sufficient to prove that $M_X \to M'_X$ is an isomorphism.

Let $x \in X$ and $h \in M'_{X,x}$. By taking $(A, \lambda) \in \mathcal{A}/X$ with $A := \mathrm{Spec}(\mathcal{O}_{Z,x}/\mathfrak{m}^n_{Z,x})$ and $\lambda : A \hookrightarrow Z$ the inclusion morphism, we have $h_{(A,\lambda)} \in M_A(A)$ and hence we have $\hat{h} \in \hat{M}_{Z,x} := \varprojlim_{(A,\lambda)} M_A(A)$. This gives a homomorphism $M'_{X,x} \to \hat{M}_{Z,x}$ such that the composite map $M_{Z,x} = M_{X,x} \to M'_{X,x} \to \hat{M}_{Z,x}$ is the inclusion map. Hence we have the injectivity of $M_{X,x} \to M'_{X,x}$. For the proof of the surjectivity, since $M_{Z,x}/\mathcal{O}^\times_{Z,x} = M_{X,x}/\mathcal{O}^\times_{X,x} \to \hat{M}_{Z,x}/\hat{\mathcal{O}}^\times_{Z,x}$ is bijective and since $\mathcal{O}^\times_{X,x} = (\mathcal{O}'_{X,x})^\times$, it is sufficient to prove the following claim.

Claim *$M'_{X,x}/(\mathcal{O}'_{X,x})^\times \to \hat{M}_{Z,x}/\hat{\mathcal{O}}^\times_{Z,x}$ is injective.*

Proof. By a similar argument as in the proof of Claim 1 in 8.3.7, we see the following. Let U be an open set of X, let $h_1, h_2 \in M'_X(U)$, and let V be the subset of U consisting of all points y such that the images of h_1 and h_2 under $M'_X(U) \to M'_{X,y} \to \hat{M}_{Z,y}/\hat{\mathcal{O}}^\times_{Z,y}$ coincide. Then V is open in U. Hence, if $x \in U$ and if the images of h_1 and h_2 in $\hat{M}_{Z,x}/\hat{\mathcal{O}}^\times_{Z,x}$ coincide, then the above V is an open neighborhood of x in X and, on V, $h_1h_2^{-1}$ belongs to $\mathcal{O}'_X(V)$. This proves Claim. □

Thus we have completed the proof of Theorem 3.2.5.

8.4 PROOF OF THEOREM 0.5.29

In this section, we prove Theorem 0.5.29.

8.4.1

Let X be an object of $\mathcal{B}(\log)$. Let Γ be a neat subgroup of $G_{\mathbf{Z}}$, and let H be a polarized Hodge structure on X of weight w and of Hodge type $(h^{p,q})$ with a Γ-level structure. We define period maps

$$X_{\mathrm{val}} \to \Gamma\backslash D_{\mathrm{val}}, \quad X^{\log}_{\mathrm{val}} \to \Gamma\backslash D^\sharp_{\mathrm{val}}, \quad X^{\log}_{\mathrm{val}} \to \Gamma\backslash D_{\mathrm{SL}(2)}, \tag{1}$$

associated with H.

The first map in (1) sends a point $(x, V, h) \in X_{\mathrm{val}}$ (3.6.18) to the following point $((A, V', Z) \bmod \Gamma)$ of $\Gamma\backslash D_{\mathrm{val}}$ (5.3.3). In the duality between $\pi_1(x^{\log})$ and $(M^{\mathrm{gp}}_X/\mathcal{O}^\times_X)_x$, let $A' \subset \pi_1(x^{\log})$ be the annihilator of $V^\times$. Let $\rho : \pi_1(x^{\log}) \to G_{\mathbf{Z}}$ be the local monodromy action on $H_{\mathbf{Z}}$ which is determined up to conjugacy by Γ. Then A is the image of $\mathbf{Q}\otimes A'$ under $\log\circ\rho : \mathbf{Q}\otimes\pi_1(x^{\log}) \to \mathfrak{g}_{\mathbf{Q}}$. Define $V' \subset \mathrm{Hom}_{\mathbf{Q}}(A, \mathbf{Q})$ to be the inverse image of the subset $(\mathbf{Q}_{\geq 0})\otimes(V/V^\times)$ of $\mathbf{Q}\otimes(V^{\mathrm{gp}}/V^\times)$ under the dual map $\mathrm{Hom}_{\mathbf{Q}}(A, \mathbf{Q}) \to \mathbf{Q}\otimes(V^{\mathrm{gp}}/V^\times)$ of the map $\mathbf{Q}\otimes A' \to A$. Finally, Z is as follows. Take a point $y \in X^{\log}$ lying over x, a family $(f_j)_{1\leq j\leq m}$ ($m = \mathrm{rank}_{\mathbf{Z}}\, V^\times$) of elements of $(\tilde{V})^\times$ (where $\tilde{V} \subset M^{\mathrm{gp}}_{X,x}$ is the inverse image of V) whose image in $V^\times$ is a $\mathbf{Z}$-basis, and fix a branch of $\log(f_j)$ in $\mathcal{O}^{\log}_{X,y}$ and a branch of $\log(h(f_j))$ in $\mathbf{C}$ for each j. Then $Z = \{F(s) \mid s \in \mathrm{sp}(y),\ s(\log(f_j)) = \log(h(f_j))$ for any $1 \leq j \leq m\}$. Here $\log(f_j)$ and $\log(h(f_j))$ are the fixed branches. Then Z is an $\exp(A_{\mathbf{C}})$-orbit. If we change the branches, then Z changes by the action of $\Gamma(\sigma(x))^{\mathrm{gp}}$, where $\sigma(x)$ is the cone in $\mathfrak{g}_{\mathbf{R}}$ generated by $\log\circ\rho(\pi_1^+(x^{\log}))$. Hence $((A, V', Z) \bmod \Gamma)$ is well-defined.

The second map in (1) sends a point $((x, V, h), (x, h')) \in X^{\log}_{\mathrm{val}} \subset X_{\mathrm{val}} \times_X X^{\log}$ to the following point $((A, V', Z) \bmod \Gamma)$ of $\Gamma\backslash D^\sharp_{\mathrm{val}}$ (5.3.3). The pair (A, V') comes from (x, V, h) in the same way as above. Take $(f_j)_{1\leq j\leq m}$ as above and take a family $(f_j)_{m+1\leq j\leq m+n}$ ($n = \mathrm{rank}_{\mathbf{Z}}(V^{\mathrm{gp}}/V^\times)$) of elements of $M^{\mathrm{gp}}_{X,x}$ whose image in $V^{\mathrm{gp}}/V^\times$ is a $\mathbf{Z}$-basis, and fix branches of $\log(f_j)$ for $1 \leq j \leq m+n$ in $\mathcal{O}^{\log}_{X,y}$, and fix branches of $\log h(f_j)$ for $1 \leq j \leq m$ and branches of $\log(h'(f_j))$ ($m+1 \leq j \leq m+n$) in $\mathbf{C}$. Then

$$Z = \left\{ F(s) \;\middle|\; \begin{array}{l} s \in \mathrm{sp}(y),\ s(\log(f_j)) = \log(h(f_j)) \text{ for } 1 \leq j \leq m, \\ s(\log(f_j)) - \log(h'(f_j)) \in \mathbf{R} \text{ for } m+1 \leq j \leq m+n \end{array} \right\}.$$

Hence Z is an $\exp(iA_{\mathbf{R}})$-orbit. If we change the branches, Z changes by the action of $\Gamma(\sigma(x))^{\mathrm{gp}}$. Hence $((A, V', Z) \bmod \Gamma)$ is well defined.

The third map in (1) is the composition

$$X^{\log}_{\mathrm{val}} \to \Gamma\backslash D^\sharp_{\mathrm{val}} \xrightarrow{\psi} \Gamma\backslash D_{\mathrm{SL}(2)},$$

where ψ is the CKS map (5.4.3).

Proposition 8.4.2 *Let the notation be as in* 8.4.1. *Then the period map* $X^{\log}_{\mathrm{val}} \to \Gamma\backslash D_{\mathrm{SL}(2)}$ *is continuous.*

Proof of Theorem 0.5.29 *assuming* 8.4.2. By the nilpotent orbit theorem of Schmid interpreted as Theorem 2.5.14, H extends to a PLH on X. The continuous period map $X^{\log}_{\mathrm{val}} \to \Gamma\backslash D_{\mathrm{SL}(2)}$ (8.4.2) extends the period map $U \to \Gamma\backslash D$ of H. □

To prove 8.4.2, we need the following preparatory lemma.

Lemma 8.4.3 *Let X be an object of $\mathcal{B}(\log)$, let Γ be a neat subgroup of $G_{\mathbf{Z}}$ and let Σ be a fan in $\mathfrak{g}_{\mathbf{Q}}$ which is strongly compatible with Γ. Then there are canonical homeomorphisms*

$$(\Gamma\backslash D_\Sigma)_{\mathrm{val}} \simeq \Gamma\backslash D_{\Sigma,\mathrm{val}}, \quad (\Gamma\backslash D_\Sigma)^{\log}_{\mathrm{val}} \simeq \Gamma\backslash D^\sharp_{\Sigma,\mathrm{val}}.$$

Here $\Gamma\backslash D_{\Sigma,\mathrm{val}}$ means $\Gamma\backslash(D_{\Sigma,\mathrm{val}})$ and $\Gamma\backslash D^\sharp_{\Sigma,\mathrm{val}}$ means $\Gamma\backslash(D^\sharp_{\Sigma,\mathrm{val}})$ as before.

Proof. The first and the second period maps in 8.4.1 (1) for $X = \Gamma\backslash D_\Sigma$ have the form

$$(\Gamma\backslash D_\Sigma)_{\mathrm{val}} \to \Gamma\backslash D_{\mathrm{val}}, \quad (\Gamma\backslash D_\Sigma)^{\log}_{\mathrm{val}} \to \Gamma\backslash D^\sharp_{\mathrm{val}}, \tag{1}$$

respectively. We show that these maps induce the homeomorphisms in 8.4.3. The proofs for these two maps are similar, and so we describe here only the proof for the first map. We have cartesian diagrams of sets with surjective lower rows

$$\begin{array}{ccc} \bigsqcup_{\sigma\in\Sigma} E_{\sigma,\mathrm{val}} & \longrightarrow & (\Gamma\backslash D_\Sigma)_{\mathrm{val}} \\ \downarrow & & \downarrow \\ \bigsqcup_{\sigma\in\Sigma} E_\sigma & \longrightarrow & \Gamma\backslash D_\Sigma, \end{array} \qquad \begin{array}{ccc} \bigsqcup_{\sigma\in\Sigma} E_{\sigma,\mathrm{val}} & \longrightarrow & \Gamma\backslash D_{\Sigma,\mathrm{val}} \\ \downarrow & & \downarrow \\ \bigsqcup_{\sigma\in\Sigma} E_\sigma & \longrightarrow & \Gamma\backslash D_\Sigma, \end{array}$$

and this shows that the first map in (1) induces a bijection $(\Gamma\backslash D_\Sigma)_{\mathrm{val}} \to \Gamma\backslash D_{\Sigma,\mathrm{val}}$. Since the topology of $\Gamma\backslash D_{\Sigma,\mathrm{val}}$ is the quotient of the topology of $\bigsqcup_{\sigma\in\Sigma} E_{\sigma,\mathrm{val}}$, the map $\Gamma\backslash D_{\Sigma,\mathrm{val}} \to (\Gamma\backslash D_\Sigma)_{\mathrm{val}}$ is continuous. This is a homeomorphism since both $\Gamma\backslash D_{\Sigma,\mathrm{val}}$ and $(\Gamma\backslash D_\Sigma)_{\mathrm{val}}$ are proper over $\Gamma\backslash D_\Sigma$. □

8.4.4

Proof of 8.4.2. We may work locally on X and hence, by Proposition 4.3.8, we may assume that there are a subgroup Γ' of Γ such that the Γ-level structure of H comes from a Γ'-level structure of H, a fan Σ in $\mathfrak{g}_\mathbf{Q}$ which is compatible with Γ', and a logarithmic modification $X' \to X$ such that the inverse image of H on X' is of type Φ for this (Γ', Σ). We have the period morphism $X' \to \Gamma'\backslash D_\Sigma$ in $\mathcal{B}(\log)$. This induces a continuous map $X^{\log}_{\mathrm{val}} = (X')^{\log}_{\mathrm{val}} \to (\Gamma'\backslash D_\Sigma)^{\log}_{\mathrm{val}} \simeq \Gamma'\backslash D^\sharp_{\Sigma,\mathrm{val}}$ where the last homeomorphism is due to 8.4.3. The map $X^{\log}_{\mathrm{val}} \to \Gamma\backslash D_{\mathrm{SL}(2)}$ is the composition of this continuous map and the continuous map $\Gamma'\backslash D^\sharp_{\Sigma,\mathrm{val}} \to \Gamma\backslash D_{\mathrm{SL}(2)}$ (Theorem 5.4.4), and hence is continuous. □

Chapter Nine

$\flat$-Spaces

In this chapter, we will consider the relationship between our theory and the theory of Cattani and Kaplan [CK1] and discuss related subjects.

9.1 DEFINITIONS AND MAIN PROPERTIES

Definition 9.1.1

(i) *We define*

$$\mathcal{X}^\flat_{\mathrm{BS}} := \mathcal{X}_{\mathrm{BS}}/\sim, \quad D^\flat_{\mathrm{BS}} := D_{\mathrm{BS}}/\sim, \quad D^\flat_{\mathrm{BS,val}} := D_{\mathrm{BS,val}}/\sim,$$

$$\textit{where } x \sim y \iff \begin{cases} \textit{the parabolic subgroups associated with } x \\ \textit{and } y \textit{ coincide, say } P, \textit{ and } y \in P_u x. \end{cases}$$

(ii) *We define*

$$D^\flat_{\mathrm{SL}(2)} := D_{\mathrm{SL}(2)}/\sim, \quad \textit{where, for } x \in D_{\mathrm{SL}(2),m},\ y \in D_{\mathrm{SL}(2),n},$$

$$x \sim y \iff \begin{cases} m = n, \textit{ the family of weight filtrations of } x \textit{ and} \\ \textit{that of } y \textit{ coinside, say } W, \textit{ and } y \in G_{W,\mathbf{R},u}x. \end{cases}$$

The space $\mathcal{X}^\flat_{\mathrm{BS}}$ was studied by Zucker in [Z1], [Z4]. This space is called the "reductive Borel-Serre space" by him.

In [KU2, 3.15, 3.16], we announced that we will discuss in this book extended period maps into $D^\flat_{\mathrm{SL}(2)}$. However, we later realized that $\Gamma\backslash D^\flat_{\mathrm{SL}(2)}$ for a subgroup Γ of $G_{\mathbf{Z}}$ of finite index is not Hausdorff in general and now we are not sure whether this space is the right object to discuss.

Let $D_{\mathrm{SL}(2),\leq 1} = \cup_{n\leq 1} D_{\mathrm{SL}(2),n}$ (5.2.6). In this book, we consider the image $D^\flat_{\mathrm{SL}(2),\leq 1}$ of $D_{\mathrm{SL}(2),\leq 1}$ in $D^\flat_{\mathrm{SL}(2)}$ (9.1.5 below). This part of $D^\flat_{\mathrm{SL}(2)}$ is certainly a nice object. A special case of $D^\flat_{\mathrm{SL}(2),\leq 1}$ is essentially the space D^* of Cattani and Kaplan, studied in [CK1] (9.4.3).

9.1.2

For a subgroup Γ of $G_{\mathbf{Z}}$ of finite index, we define the topology of $\Gamma\backslash\mathcal{X}^\flat_{\mathrm{BS}}$, $\Gamma\backslash D^\flat_{\mathrm{BS}}$, $\Gamma\backslash D^\flat_{\mathrm{BS,val}}$, and $\Gamma\backslash D^\flat_{\mathrm{SL}(2),\leq 1}$, as quotient spaces of $\mathcal{X}_{\mathrm{BS}}$, D_{BS}, $D_{\mathrm{BS,val}}$, and $D_{\mathrm{SL}(2),\leq 1}$, respectively.

In this book, we consider the topology on $\Gamma\backslash D_\Sigma$ but not the one on D_Σ. Similarly, we do not consider the topologies of $\mathcal{X}^\flat_{\mathrm{BS}}$, $D^\flat_{\mathrm{BS}}$, $D^\flat_{\mathrm{BS,val}}$, and $D^\flat_{\mathrm{SL}(2),\le 1}$ on themselves.

9.1.3

Main results. For a **Q**-parabolic subgroup P of $G_{\mathbf{R}}$, we define $\mathcal{X}^\flat_{\mathrm{BS}}(P) \subset \mathcal{X}^\flat_{\mathrm{BS}}$, $D^\flat_{\mathrm{BS}}(P) \subset D^\flat_{\mathrm{BS}}$, $D^\flat_{\mathrm{BS,val}}(P) \subset D^\flat_{\mathrm{BS,val}}$ to be the images of $\mathcal{X}_{\mathrm{BS}}(P)$, $D_{\mathrm{BS}}(P)$, $D_{\mathrm{BS,val}}(P)$, under the natural projections, respectively.

For an increasing **Q**-rational filtration W on $H_{0,\mathbf{R}}$ such that $\{v \in H_{0,\mathbf{R}} \mid \langle v, W_k\rangle_0 = 0\} = W_{-k-1}$ for all $k \in \mathbf{Z}$, we define $D^\flat_{\mathrm{SL}(2),\le 1}(W) \subset D^\flat_{\mathrm{SL}(2),\le 1}$ to be the image of $D_{\mathrm{SL}(2),\le 1} \cap D_{\mathrm{SL}(2)}(W)$ under the natural projection.

THEOREM 9.1.4 *Let Γ be a subgroup of $G_{\mathbf{Z}}$ of finite index.*

(i) *All the following maps are proper.*

$$\Gamma\backslash\mathcal{X}_{\mathrm{BS}} \to \Gamma\backslash\mathcal{X}^\flat_{\mathrm{BS}}, \quad \Gamma\backslash D_{\mathrm{BS}} \to \Gamma\backslash D^\flat_{\mathrm{BS}},$$
$$\Gamma\backslash D_{\mathrm{BS,val}} \to \Gamma\backslash D^\flat_{\mathrm{BS,val}}, \quad \Gamma\backslash D_{\mathrm{SL}(2),\le 1} \to \Gamma\backslash D^\flat_{\mathrm{SL}(2),\le 1}.$$

Hence $\Gamma\backslash\mathcal{X}^\flat_{\mathrm{BS}}$, $\Gamma\backslash D^\flat_{\mathrm{BS}}$, $\Gamma\backslash D^\flat_{\mathrm{BS,val}}$ are compact, and $\Gamma\backslash D^\flat_{\mathrm{SL}(2),\le 1}$ is Hausdorff.

(ii) *Assume Γ is neat. Then all the following maps are local homeomorphisms:*

$$(\Gamma \cap P_u)\backslash\mathcal{X}^\flat_{\mathrm{BS}}(P) \to \Gamma\backslash\mathcal{X}^\flat_{\mathrm{BS}}, \quad (\Gamma \cap P_u)\backslash D^\flat_{\mathrm{BS}}(P) \to \Gamma\backslash D^\flat_{\mathrm{BS}},$$
$$(\Gamma \cap P_u)\backslash D^\flat_{\mathrm{BS,val}}(P) \to \Gamma\backslash D^\flat_{\mathrm{BS,val}}, \quad (\Gamma \cap G_{W,\mathbf{R},u})\backslash D^\flat_{\mathrm{SL}(2),\le 1}(W) \to \Gamma\backslash D^\flat_{\mathrm{SL}(2),\le 1}.$$

This theorem will be proved in Sections 9.2 and 9.3.

9.1.5

We have the following description of $D^\flat_{\mathrm{SL}(2),1} = D^\flat_{\mathrm{SL}(2)\le 1} - D$.

$$D^\flat_{\mathrm{SL}(2),1} = \left\{ (W, N, F) \;\middle|\; \begin{array}{l} W \text{ is a } \mathbf{Q}\text{-rational increasing filtration on } H_{0,\mathbf{R}}, \\ N \text{ is an } \mathbf{R}\text{-linear endomorphism of } \mathrm{gr}^W_\bullet \text{ of degree } -2, \\ F \text{ is a set of decreasing filtrations } F_{(j)} \text{ on } \mathrm{gr}^W_{j,\mathbf{C}} \ (\forall j), \\ \text{which satisfy the following conditions (1)–(6)} \end{array} \right\} \Big/ \sim$$

(1) $\{v \in H_{0,\mathbf{R}} \mid \langle v, W_k\rangle_0 = 0\} = W_{-k-1} \quad (\forall\, k)$.

(2) $\langle N(x), y\rangle_0 + \langle x, N(y)\rangle_0 = 0 \ (\forall\, k,\ \forall\, x \in \mathrm{gr}^W_k,\ \forall\, y \in \mathrm{gr}^W_{2-k})$. Here the first (resp. the second) $\langle\ ,\ \rangle_0$ denotes the pairing $\mathrm{gr}^W_{k-2} \times \mathrm{gr}^W_{2-k} \to \mathbf{R}$ (resp. $\mathrm{gr}^W_k \times \mathrm{gr}^W_{-k} \to \mathbf{R}$) induced by $\langle\ ,\ \rangle_0$.

(3) $N^j : \mathrm{gr}^W_j \xrightarrow{\sim} \mathrm{gr}^W_{-j} \quad (\forall\, j \ge 0)$.

(4) $N(F^p_{(j)}) \subset F^{p-1}_{(j-2)} \quad (\forall\, j, \forall\, p), \quad N^j(F^p_{(j)}) = F^{p-j}_{(-j)} \quad (\forall\, j \ge 0, \forall\, p)$.

(5) $h^{p,w-p} = \sum_{j\in\mathbf{Z}} \dim(F^p_{(j)}/F^{p+1}_{(j)}) \ (\forall p)$.

(6) (Primitive part of gr^W_j, $\langle \bullet, N^j \bullet\rangle_0$, $F_{(j)}$) is a polarized Hodge structure of weight $w + j$ $(\forall\, j \ge 0)$.

Here

$$(W, N, F) \sim (W', N', F') \Longleftrightarrow W = W', F = F', N = aN' \text{ for some } a \in \mathbf{R}_{>0}.$$

This description is obtained as follows.

For the class of $p = [\rho, \varphi] \in D_{\mathrm{SL}(2),1}$ in $D^{\flat}_{\mathrm{SL}(2),1}$, we have the corresponding triple (W, N, F), where W is the weight filtration associated to p, N is the nilpotent endmorphism on $\mathrm{gr}^W_{\bullet}$ induced by N of (ρ, φ) and F is the set of Hodge filtrations on $\mathrm{gr}^W_{\bullet,\mathbf{C}}$ induced by $F(\varphi(0))$.

The inverse map is given as follows. If a triple (W, N, F) is given, we take any set of $\mathbf{R}$-subspaces S_j $(j \in \mathbf{Z})$ of $H_{0,\mathbf{R}}$ such that $W_k = \sum_{j \le k} S_j$ for all k and $\langle S_j, S_k \rangle_0 = 0$ unless $j + k = 0$. By identifying gr^W_k with S_k, we identify $\mathrm{gr}^W_{\bullet}$ with $H_{0,\mathbf{R}}$. Then F is regarded as a filtration on $H_{0,\mathbf{C}}$ which is $\mathbf{R}$-split with respect to W, and N is regarded as an element of $\mathfrak{g}_{\mathbf{R}}$. By [CKS, 4.66], (N, F) generates a nilpotent orbit. Since F is $\mathbf{R}$-split, (N, F) corresponds to an SL(2)-orbit (ρ, φ) in one variable of rank 1 by 6.1.2 (11). The class of (ρ, φ) in $D^{\flat}_{\mathrm{SL}(2),1}$ does not depend on the choice of $(S_j)_j$.

9.2 PROOFS OF THEOREM 9.1.4 FOR $\Gamma\backslash\mathcal{X}^{\flat}_{\mathrm{BS}}$, $\Gamma\backslash D^{\flat}_{\mathrm{BS}}$, AND $\Gamma\backslash D^{\flat}_{\mathrm{BS,val}}$

Since the proofs for these three spaces are similar, we give the proof here only for $\Gamma\backslash D^{\flat}_{\mathrm{BS}}$ (cf. also [Z4]).

Lemma 9.2.1 *Let P be a $\mathbf{Q}$-parabolic subgroup of $G_{\mathbf{R}}$.*

(i) *There exists a subspace X of $D_{\mathrm{BS}}(P)$ such that*

$$P_u \times X \xrightarrow{\sim} D_{\mathrm{BS}}(P), \quad (g, x) \mapsto gx,$$

as topological spaces.

(ii) *$P_u\backslash D_{\mathrm{BS}}(P)$ is Hausdorff.*

Proof.

(i) follows from $D_{\mathrm{BS}}(P) = D \times^{A_P} \overline{A}_P$ (A_P acts on D here by the Borel-Serre action) and from the relationship with the Iwasawa decomposition 5.1.15.

(ii) follows from (i). □

Lemma 9.2.2 *Let P be as in Lemma 9.2.1. Then $(\Gamma \cap P_u)\backslash D_{\mathrm{BS}}(P) \to P_u\backslash D_{\mathrm{BS}}(P)$ is proper.*

Proof. Let X be as in Lemma 9.2.1. Then, we have a commutative diagram

$$\begin{array}{ccc} (\Gamma \cap P_u)\backslash D_{\mathrm{BS}}(P) & \longrightarrow & P_u\backslash D_{\mathrm{BS}}(P) \\ \uparrow \wr & & \uparrow \wr \\ (\Gamma \cap P_u)\backslash P_u \times X & \longrightarrow & X. \end{array}$$

Since $(\Gamma \cap P_u)\backslash P_u$ is compact, the bottom horizontal arrow is proper. □

LEMMA 9.2.3 *Let P be as in Lemma* 9.2.1. *Then, the action of* $(\Gamma \cap P_u)\backslash(\Gamma \cap P)$ *on* $P_u\backslash D_{\mathrm{BS}}(P)$ *is proper.*

Proof. Since the action of Γ on D_{BS} is proper (Theorem 5.1.14 (iii)), so is the action of $\Gamma \cap P$ on $D_{\mathrm{BS}}(P)$. It follows that the action of $(\Gamma \cap P_u)\backslash(\Gamma \cap P)$ on $(\Gamma \cap P_u)\backslash D_{\mathrm{BS}}(P)$ is proper. That is, in the following commutative diagram in which the horizontal arrows are $(g, x) \mapsto (x, gx)$, the upper horizontal arrow is proper:

$$\begin{array}{ccc}
(\Gamma \cap P_u)\backslash(\Gamma \cap P) \times (\Gamma \cap P_u)\backslash D_{\mathrm{BS}}(P) & \longrightarrow & (\Gamma \cap P_u)\backslash D_{\mathrm{BS}}(P) \times (\Gamma \cap P_u)\backslash D_{\mathrm{BS}}(P) \\
\downarrow & & \downarrow \\
(\Gamma \cap P_u)\backslash(\Gamma \cap P) \times P_u\backslash D_{\mathrm{BS}}(P) & \longrightarrow & P_u\backslash D_{\mathrm{BS}}(P) \times P_u\backslash D_{\mathrm{BS}}(P)
\end{array}$$

Since the right vertical arrow is proper by Lemma 9.2.2 and since the left vertical arrow is surjective, the bottom horizontal arrow is proper. □

LEMMA 9.2.4 *Let P be as in Lemma* 9.2.1. *Then,* $(\Gamma \cap P_u)\backslash D^{\flat}_{\mathrm{BS}}(P)$ *is Hausdorff.*

Proof. Let $x, y \in D_{\mathrm{BS}}(P)$ and $\overline{x}, \overline{y} \in (\Gamma \cap P_u)\backslash D^{\flat}_{\mathrm{BS}}(P)$ be their images, respectively. Assuming that $\overline{x}$ and $\overline{y}$ are not separated by open sets, we show $\overline{x} = \overline{y}$. The map $D_{\mathrm{BS}}(P) \to P_u\backslash D_{\mathrm{BS}}(P)$ factors through $(\Gamma \cap P_u)\backslash D^{\flat}_{\mathrm{BS}}(P)$. Since $P_u\backslash D_{\mathrm{BS}}(P)$ is Hausdorff by 9.2.1 (ii), the images of $\overline{x}, \overline{y}$ in $P_u\backslash D_{\mathrm{BS}}(P)$ coincide. The **Q**-parabolic subgroup $Q \supset P$ of $G_{\mathbf{R}}$ associated to x coincides with the one associated to y. In fact, the **Q**-parabolic subgroup associated to $y = gx$ $(g \in P_u)$ is $gQg^{-1} = Q$. Hence

$$\overline{x}, \overline{y} \in (\Gamma \cap P_u)\backslash D^{\flat}_{\mathrm{BS}}(Q).$$

Applying Lemma 9.2.3 for Q, we see that the action of $(\Gamma \cap Q_u)\backslash(\Gamma \cap Q)$ on $Q_u\backslash D_{\mathrm{BS}}(Q)$ is proper. Hence, using $Q_u \subset P_u \subset P \subset Q$, we see that the action of $(\Gamma \cap Q_u)\backslash(\Gamma \cap P_u)$ on $Q_u\backslash D_{\mathrm{BS}}(Q)$ is proper. Therefore, the quotient space $(\Gamma \cap P_u)\backslash(Q_u\backslash D_{\mathrm{BS}}(Q))$ is Hausdorff (7.2.4). It follows that the images of $\overline{x}$, $\overline{y}$ in $(\Gamma \cap P_u)\backslash(Q_u\backslash D_{\mathrm{BS}}(Q))$ coincide. This shows $\overline{x} = \overline{y}$ since the **Q**-parabolic subgroups of $G_{\mathbf{R}}$ associated with x and y are Q. □

LEMMA 9.2.5 *Let P be as in* 9.2.1. *Then,* $(\Gamma \cap P_u)\backslash D_{\mathrm{BS}}(P) \to (\Gamma \cap P_u)\backslash D^{\flat}_{\mathrm{BS}}(P)$ *is proper.*

Proof. This follows from the fact that the composite map

$$(\Gamma \cap P_u)\backslash D_{\mathrm{BS}}(P) \to (\Gamma \cap P_u)\backslash D^{\flat}_{\mathrm{BS}}(P) \to P_u\backslash D_{\mathrm{BS}}(P)$$

is proper (9.2.2) and that $(\Gamma \cap P_u)\backslash D^{\flat}_{\mathrm{BS}}(P)$ is Hausdorff (9.2.4). □

LEMMA 9.2.6 *Assume Γ is neat. Let P be as in Lemma* 9.2.1. *Then,* $(\Gamma \cap P_u)\backslash D^{\flat}_{\mathrm{BS}}(P) \to \Gamma\backslash D^{\flat}_{\mathrm{BS}}$ *is a local homeomorphism.*

Proof. We apply 7.4.7 by taking $X = D_{\mathrm{BS}}$, $Y = D^{\flat}_{\mathrm{BS}}$, $\Gamma = \Gamma$, $\Gamma' = \Gamma \cap P_u$, $V = (\Gamma \cap P_u)\backslash D^{\flat}_{\mathrm{BS}}(P)$. Then U in 7.4.7 is $(\Gamma \cap P_u)\backslash D_{\mathrm{BS}}(P)$. By 7.4.7, it is sufficient to show that the conditions (1)–(3) in 7.4.7 are satisfied. 7.4.7 (1) is satisfied by Theorem 5.1.14 (iii). 7.4.7 (2) follows from 9.2.5. 7.4.7 (3) is proved as follows.

Let $x \in D_{\mathrm{BS}}$, let $\overline{x}$ be the image of x in $D_{\mathrm{BS}}^{\flat}$, let $\gamma \in \Gamma$ and assume $\gamma\overline{x} = \overline{x}$. It is sufficient to prove that the parabolic subgroup P associated to x satisfies $\gamma \in P_u$. Since $\gamma P \gamma^{-1}$ is the parabolic subgroup associated to $\gamma\overline{x} = \overline{x}$, we have $\gamma P \gamma^{-1} = P$. Since the normalizer of P in $(G^\circ)_{\mathbf{R}}$ is P, there is $n \geq 1$ such that $\gamma^n \in P$. By $\gamma^n\overline{x} = \overline{x}$, there exists $g \in P_u$ such that $\gamma^n x = gx$, i.e., $g^{-1}\gamma^n \in K'_x$. This shows that in P/P_u, γ^n is contained in a discrete subgroup and in a compact subgroup, and hence the image of γ^n in P/P_u is of finite order. Hence there exists $m \geq 1$ such that $\gamma^{mn} \in P_u$. By the neat property of Γ, this shows $\gamma \in P_u$. □

Lemma 9.2.7 *$\Gamma\backslash D_{\mathrm{BS}} \to \Gamma\backslash D_{\mathrm{BS}}^{\flat}$ is proper.*

Proof. Replacing Γ by a subgroup of finite index, we may assume that Γ is neat. Then 9.2.7 follows from 9.2.5 and 9.2.6. □

9.3 PROOF OF THEOREM 9.1.4 FOR $\Gamma\backslash D_{\mathrm{SL}(2),\leq 1}^{\flat}$

9.3.1

Note that

$$D_{\mathrm{SL}(2),\leq 1} = D_{\mathrm{SL}(2),\mathrm{val},\leq 1} \subset D_{\mathrm{BS},\mathrm{val}}. \tag{1}$$

For $p \in D_{\mathrm{SL}(2),1}$, if W denotes the weight filtration associated with p and P denotes the $\mathbf{Q}$-parabolic subgroup of $G_{\mathbf{R}}$ associated with the image of p in D_{BS}, then $P = (G^\circ)_{W,\mathbf{R}}$ and hence $P_u = G_{W,\mathbf{R},u}$. From this, we see that the above inclusion (1) induces an injection $D_{\mathrm{SL}(2),\leq 1}^{\flat} \hookrightarrow D_{\mathrm{BS},\mathrm{val}}^{\flat}$.

Lemma 9.3.2 *$\Gamma\backslash D_{\mathrm{SL}(2),\leq 1} \to \Gamma\backslash D_{\mathrm{SL}(2),\leq 1}^{\flat}$ is proper.*

Proof. Let $(\Gamma\backslash D_{\mathrm{SL}(2),\leq 1}^{\flat})'$ be the set $\Gamma\backslash D_{\mathrm{SL}(2),\leq 1}^{\flat}$ endowed with the topology as a subspace of $\Gamma\backslash D_{\mathrm{BS},\mathrm{val}}^{\flat}$. Then we have a cartesian diagram of topological spaces

$$\begin{array}{ccc}
\Gamma\backslash D_{\mathrm{SL}(2),\leq 1} & \longrightarrow & \Gamma\backslash D_{\mathrm{BS},\mathrm{val}} \\
\downarrow & & \downarrow \\
(\Gamma\backslash D_{\mathrm{SL}(2),\leq 1}^{\flat})' & \longrightarrow & \Gamma\backslash D_{\mathrm{BS},\mathrm{val}}^{\flat}.
\end{array}$$

Since the right vertical arrow is proper and surjective, so is the left vertical arrow. Hence the topology of $(\Gamma\backslash D_{\mathrm{SL}(2),\leq 1}^{\flat})'$ coincides with the quotient of the topology of $\Gamma\backslash D_{\mathrm{SL}(2),\leq 1}$, i.e., $(\Gamma\backslash D_{\mathrm{SL}(2),\leq 1}^{\flat})' = \Gamma\backslash D_{\mathrm{SL}(2),\leq 1}^{\flat}$ as topological spaces. □

Lemma 9.3.3 *$(\Gamma \cap G_{W,\mathbf{R},u})\backslash D_{\mathrm{SL}(2),\leq 1}(W) \to (\Gamma \cap G_{W,\mathbf{R},u})\backslash D_{\mathrm{SL}(2),\leq 1}^{\flat}(W)$ is proper.*

Proof. This follows from 9.3.2 and the following cartesian diagram:

$$\begin{array}{ccc}
(\Gamma \cap G_{W,\mathbf{R},u})\backslash D_{\mathrm{SL}(2),\leq 1}(W) & \longrightarrow & \Gamma\backslash D_{\mathrm{SL}(2),\leq 1} \\
\downarrow & & \downarrow \\
(\Gamma \cap G_{W,\mathbf{R},u})\backslash D_{\mathrm{SL}(2),\leq 1}^{\flat}(W) & \longrightarrow & \Gamma\backslash D_{\mathrm{SL}(2),\leq 1}^{\flat}.
\end{array}$$

□

Lemma 9.3.4 *Let $x \in D^{\flat}_{\mathrm{SL}(2),1}$ and let W be the weight filtration on $H_{0,\mathbf{R}}$ associated to x. Let Γ be a neat subgroup of $G_{\mathbf{Z}}$, let $\gamma \in \Gamma$, and assume $\gamma x = x$. Then $\gamma \in G_{W,\mathbf{R},u}$.*

Proof. Since γW is the weight filtration associated with $\gamma x = x$, we have $\gamma W = W$ and hence $\gamma \in G_{W,\mathbf{R}}$. Let (ρ, φ) be a representative of x. Then, since $F_{\varphi(0)}(\mathrm{gr}^W_{\bullet})$ is polarized, $\mathrm{gr}^W_{\bullet}(\gamma)$ is contained in a discrete subgroup and in a compact subgroup of $\mathrm{Aut}(\mathrm{gr}^W_{\bullet,\mathbf{R}})$ and hence is of finite order. Since Γ is neat, we have $\mathrm{gr}^W_{\bullet}(\gamma) = 1$. Hence $\gamma \in G_{W,\mathbf{R},u}$. □

Lemma 9.3.5 *Assume Γ is neat. Then, $(\Gamma \cap G_{W,\mathbf{R},u})\backslash D^{\flat}_{\mathrm{SL}(2),\le 1}(W) \to \Gamma\backslash D^{\flat}_{\mathrm{SL}(2),\le 1}$ is a local homeomorphism.*

Proof. In 7.4,7, take $X = D_{\mathrm{SL}(2),\le 1}$, $Y = D^{\flat}_{\mathrm{SL}(2),\le 1}$, $\Gamma = \Gamma$, $\Gamma' = \Gamma \cap G_{W,\mathbf{R},u}$, $V = (\Gamma \cap G_{W,\mathbf{R},u})\backslash D^{\flat}_{\mathrm{SL}(2),\le 1}(W)$. Then U in 7.4.7 is $(\Gamma \cap G_{W,\mathbf{R},u})\backslash D_{\mathrm{SL}(2),\le 1}(W)$. By 7.4.7, it is sufficient to prove that the conditions (1)–(3) in 7.4.7 are satisfied. 7.4.7 (1) (resp. (2), (3)) is satisfied by 5.2.15 (iv) (resp. 9.3.3, 9.3.4). □

9.4 EXTENDED PERIOD MAPS

9.4.1

We define a canonical map

$$D_{\Xi} \to D^{\flat}_{\mathrm{SL}(2),\le 1}. \tag{1}$$

Consider the composite map

$$D^{\sharp}_{\Xi} = D^{\sharp}_{\Xi,\mathrm{val}} \xrightarrow{\psi} D_{\mathrm{SL}(2),\le 1} \to D^{\flat}_{\mathrm{SL}(2),\le 1}.$$

We show that this map factors through the projection $D^{\sharp}_{\Xi} \to D_{\Xi}$. In fact, if $p, p' \in D^{\sharp}_{\Xi} - D$ have the same image in D_{Ξ}, then we can write $p = (\sigma, Z)$ and $p' = (\sigma, Z')$, $\sigma = \mathbf{R}_{\ge 0}N$ for some $N \ne 0$, and $Z' = \exp(aN)Z$ for some $a \in \mathbf{R}$. Hence $\psi(p') = \exp(aN)\psi(p)$. Since $\exp(aN) \in G_{W(N),\mathbf{R},u}$, the images of $\psi(p)$ and $\psi(p')$ in $D^{\flat}_{\mathrm{SL}(2),\le 1}$ coincide.

Theorem 9.4.2 *Let X be a connected smooth analytic space endowed with a smooth divisor Y, and let $U = X - Y$. Let H be a variation of polarized Hodge structure on U with unipotent local monodromy along Y. Fix a base point $u \in U$ and let $(H_0, \langle\ ,\ \rangle_0) = (H_{\mathbf{Z},u}, \langle\ ,\ \rangle_u)$. Let Γ be a subgroup of $G_{\mathbf{Z}}$ which contains the global monodromy group* $\mathrm{Image}(\pi_1(U,u) \to G_{\mathbf{Z}})$*, and assume Γ is neat. Then the period map $U \to \Gamma\backslash D$ extends uniquely to a continuous map $X \to \Gamma\backslash D^{\flat}_{\mathrm{SL}(2),\le 1}$.*

Proof. Endow X with the logarithmic structure associated to Y. By the nilpotent orbit theorem of Schmid interpreted as Theorem 2.5.14, H extends to a PLH H' on X. Let Σ be the set of local monodromy cones of H' in $\mathfrak{g}_{\mathbf{R}}$ (2.5.11). By Theorem 4.3.1 (i), the period map $U \to \Gamma\backslash D$ extends to a morphism of logarithmic manifolds

$X \to \Gamma\backslash D_\Sigma$. Since $\Sigma \subset \Xi$, the map $D_\Xi \to D^\flat_{\mathrm{SL}(2),\le 1}$ in 9.4.1 induces a map $\Gamma\backslash D_\Sigma \to \Gamma\backslash D^\flat_{\mathrm{SL}(2),\le 1}$. The following diagram is commutative:

$$\begin{array}{ccccc} X^{\log} & \longrightarrow & \Gamma\backslash D^\sharp_\Sigma & \longrightarrow & \Gamma\backslash D_{\mathrm{SL}(2),\le 1} \\ \downarrow & & \downarrow & & \downarrow \\ X & \longrightarrow & \Gamma\backslash D_\Sigma & \longrightarrow & \Gamma\backslash D^\flat_{\mathrm{SL}(2),\le 1} \end{array} \qquad (1)$$

Since the topology of $\Gamma\backslash D_\Sigma$ is the quotient of that of $\Gamma\backslash D^\sharp_\Sigma$ by the properness of $\Gamma\backslash D^\sharp_\Sigma = (\Gamma\backslash D_\Sigma)^{\log} \to \Gamma\backslash D_\Sigma$ (Theorem A, (vi)), the map $\Gamma\backslash D_\Sigma \to \Gamma\backslash D^\flat_{\mathrm{SL}(2),\le 1}$ is continuous. Hence the composition of the lower row $X \to \Gamma\backslash D^\flat_{\mathrm{SL}(2),\le 1}$ is continuous. Since $\Gamma\backslash D^\flat_{\mathrm{SL}(2),\le 1}$ is Hausdorff (9.1.4), the continuous extension of the period map is unique. □

The composition of the upper row of the diagram (1) coincides with the period map $X^{\log} = X^{\log}_{\mathrm{val}} \to \Gamma\backslash D_{\mathrm{SL}(2)}$ in Section 8.4.

9.4.3

Theorem 9.4.2 reproduces the extended period map of Cattani and Kaplan [CK1] (they considered the case where $h^{p,q} = 0$ unless $(p, q) = (2, 0), (1, 1), (0, 2)$).

Chapter Ten

Local Structures of $D_{\mathrm{SL}(2)}$ and $\Gamma\backslash D^{\flat}_{\mathrm{SL}(2),\leq 1}$

In Section 10.1, for each $p = [\rho, \varphi] \in D_{\mathrm{SL}(2)}$ of rank n, we give a homeomorphism from an open neighborhood of p to a certain subspace of $(\mathbf{R}^n_{\geq 0}) \times \mathfrak{g}_{\mathbf{R}} \times (K_{\mathbf{r}} \cdot \mathbf{r})$, where $\mathbf{r} = \varphi(\mathbf{i})$ and $K_{\mathbf{r}} \cdot \mathbf{r} = \{k \cdot \mathbf{r} \mid k \in K_{\mathbf{r}}\} \subset D$, which sends p to $(0, 0, \mathbf{r})$ (Theorem 10.1.3). Though $D_{\mathrm{SL}(2)}$ is not a real analytic manifold, this homeomorphism is something like a "real analytic local coordinate" of $D_{\mathrm{SL}(2)}$. From this theorem, we can obtain Theorem 10.1.6 which is a criterion for the local compactness of $D_{\mathrm{SL}(2)}$. We give the proof of Theorem 10.1.3 in Section 10.3 after we consider a special open neighborhood $U(p)$ of p in Section 10.2.

In Section 10.4, in the case $n = 1$, we give a description of a neighborhood of the image of p in $\Gamma\backslash D^{\flat}_{\mathrm{SL}(2),\leq 1}$ for a neat subgroup Γ of $G_{\mathbf{Z}}$ of finite index.

We denote by $\theta = \theta_{K_{\mathbf{r}}}$ the Cartan involution on $\mathfrak{g}_{\mathbf{R}}$ associated to $K_{\mathbf{r}}$ all through this chapter.

10.1 LOCAL STRUCTURES OF $D_{\mathrm{SL}(2)}$

10.1.1

We fix our notation.

Let $p \in D_{\mathrm{SL}(2)}$, let n be the rank of p and let (ρ, φ) be an SL(2)-orbit in n variables which represents p. Let $\mathbf{r} = \varphi(\mathbf{i})$ and let $W = (W^{(j)})_{1\leq j\leq n}$ be the family of weight filtrations associated with p. For a subset $J = \{s_1, \ldots, s_m\}$ $(1 \leq s_1 < \cdots < s_m \leq n)$ of $\{1, \ldots, n\}$, let $W^{(J)} := (W^{(s_k)})_{1\leq k\leq m}$ be the corresponding subfamily of weight filtrations.

Let $X = X(\mathbf{G}^n_{\mathbf{m},\mathbf{R}})$ be the character group of the torus $\mathbf{G}^n_{\mathbf{m},\mathbf{R}}$ and let

$$\mathfrak{g}_{\mathbf{R}} = \bigoplus_{\chi \in X} \mathfrak{g}_{\mathbf{R}}(\chi)$$

be the direct sum decomposition into χ-components with respect to the action of $\mathbf{G}^n_{\mathbf{m},\mathbf{R}}$ on $\mathfrak{g}_{\mathbf{R}}$ via $\mathrm{Ad} \circ \tilde{\rho}$ (for $\tilde{\rho}$, see 5.2.2).

For a subset J of $\{1, \ldots, n\}$, let X_J be the submonoid of X consisting of elements χ such that the j-th component of the image of χ in $\mathbf{Z}^n$, under the canonical isomorphism $X \simeq \mathbf{Z}^n$, is ≥ 0 for any $j \in J$. For example, $X_{\emptyset} = X$, and $X_{\{1,\ldots,n\}}$ coincides with the inverse image $X_+ \subset X$ of $\mathbf{N}^n \subset \mathbf{Z}^n$.

We have

$$\mathrm{Lie}(G_{W^{(J)},\mathbf{R}}) = \bigoplus_{\chi \in X_J} \mathfrak{g}_{\mathbf{R}}(\chi^{-1}). \tag{1}$$

We define topological spaces Y_j ($j = 0, 1, 2, 3$) as follows. Put

$$Y_0 := (\mathbf{R}^n_{>0}) \times \mathfrak{g}_\mathbf{R} \times (K_\mathbf{r} \cdot \mathbf{r}), \quad Y_3 := (\mathbf{R}^n_{\geq 0})_{\mathrm{val}} \times \mathfrak{g}_\mathbf{R} \times (K_\mathbf{r} \cdot \mathbf{r}).$$

Let Y_1 (resp. Y_2) be the subspace of $(\mathbf{R}^n_{\geq 0}) \times \mathfrak{g}_\mathbf{R} \times (K_\mathbf{r} \cdot \mathbf{r})$ (resp. Y_3) consisting of all elements (t, a, b) such that

$$a \in \bigoplus_{\chi \in X_J \cup X_J^{-1}} \mathfrak{g}_\mathbf{R}(\chi), \quad b \in (G_{W(J),\mathbf{R}} \cap K_\mathbf{r}) \cdot \mathbf{r}, \quad \text{for } J := \{j \mid t_j = 0\} \subset \{1, \ldots, n\},$$

where $t_j \in \mathbf{R}$ is the j-th component of t (resp. of the image of t in $\mathbf{R}^n_{\geq 0}$). Thus, $Y_2 \subset Y_3$ and we have the canonical projection $Y_2 \to Y_1$ which is proper and surjective. We regard Y_0 as a dense open subset of Y_j for $j = 1, 2, 3$.

We will consider the continuous map

$$\eta_0 : Y_0 \to D, \quad (t, a, b) \mapsto \tilde{\rho}(t) \exp\left(\sum_{\chi \in X} \frac{1}{\chi(t) + \chi(t)^{-1}} a(\chi) \right) b, \tag{2}$$

where $a(\chi)$ denotes the $\mathfrak{g}_\mathbf{R}(\chi)$-component of a.

10.1.2

Fix an $\mathbf{R}$-subspace L of $\mathfrak{g}_\mathbf{R}$ satisfying the following conditions (1) and (2).

(1) $\mathfrak{g}_\mathbf{R} = L \oplus \mathrm{Lie}(\tilde{\rho}(\mathbf{R}^n_{>0})) \oplus \mathrm{Lie}(K_\mathbf{r})$.

(2) L is the sum of $L \cap (\mathfrak{g}_\mathbf{R}(\chi) + \mathfrak{g}_\mathbf{R}(\chi^{-1}))$ where χ ranges over all elements of X.

Such L exists. In fact, since

$$\mathrm{Lie}(K_\mathbf{r}) = \mathfrak{g}^+_\mathbf{R} = \{v \in \mathfrak{g}_\mathbf{R} \mid \theta(v) = v\},$$
$$\mathrm{Lie}(\tilde{\rho}(\mathbf{R}^n_{>0})) \subset \mathfrak{g}^-_\mathbf{R} = \{v \in \mathfrak{g}_\mathbf{R} \mid \theta(v) = -v\},$$
$$\theta(\mathfrak{g}_\mathbf{R}(\chi)) = \mathfrak{g}_\mathbf{R}(\chi^{-1}),$$

where $\theta = \theta_{K_\mathbf{r}}$ is the Cartan involution on $\mathfrak{g}_\mathbf{R}$ associated to $K_\mathbf{r}$, for any $\mathbf{R}$-subspace L' of $\mathfrak{g}_\mathbf{R}(1)^-$ ($\mathfrak{g}_\mathbf{R}(1)$ means $\mathfrak{g}_\mathbf{R}(\chi)$ with $\chi = 1$) such that $\mathfrak{g}_\mathbf{R}(1)^- = L' \oplus \mathrm{Lie}(\tilde{\rho}(\mathbf{R}^n_{>0}))$, the $\mathbf{R}$-subspace $L := L' \oplus \sum_{\chi \neq 1} (\mathfrak{g}_\mathbf{R}(\chi) \oplus \mathfrak{g}_\mathbf{R}(\chi^{-1}))^-$ of $\mathfrak{g}_\mathbf{R}$ satisfies the conditions (1) and (2).

For $j = 0, 1, 2, 3$, let $Y_{j,L}$ be the subspace of Y_j consisting of all elements (t, a, b) satisfying $a \in L$.

THEOREM 10.1.3 *Let p and L be as above.*

(i) *The map η_0 in* 10.1.1 (2) *extends uniquely to the continuous maps η_j ($j = 1, 2, 3$) in the following commutative diagram:*

$$\begin{array}{ccccc} Y_1 & \longleftarrow & Y_2 & \longrightarrow & Y_3 \\ {\scriptstyle \eta_1}\downarrow & & {\scriptstyle \eta_2}\downarrow & & {\scriptstyle \eta_3}\downarrow \\ D_{\mathrm{SL}(2)} & \longleftarrow & D_{\mathrm{SL}(2),\mathrm{val}} & \longrightarrow & D_{\mathrm{BS},\mathrm{val}}. \end{array}$$

The map η_1 sends $(0, 0, \mathbf{r}) \in Y_{1,L} \subset Y_1$ to $p \in D_{\mathrm{SL}(2)}$.

(ii) *There exist open sets U_j of $Y_{j,L}$ for $j = 1, 2, 3$ such that $(0, 0, \mathbf{r}) \in U_1$, that U_2 is the inverse image of U_1 in $Y_{2,L}$, that $U_2 \subset U_3$ and that the restrictions of η_j to U_j are injective open maps for all $j = 1, 2, 3$.*

Thus η_1 induces a homeomorphism from some open neighborhood of $(0, 0, \mathbf{r})$ in $Y_{1,L}$ onto some open neighborhood of p in $D_{\mathrm{SL}(2)}$.

The proof of Theorem 10.1.3 will be given later in Section 10.3.

10.1.4

Example. Consider the case of the standard SL(2)-orbit of the upper half plane. Then $G = \mathrm{SL}(2)$, D is identified with the upper half plane $\mathfrak{h}$, $\check{D} = \mathbf{P}^1(\mathbf{C})$, ρ and φ are the identity maps, $\mathbf{r} = i$ and $D_{\mathrm{SL}(2)}(W) = D_{\mathrm{SL}(2),\mathrm{val}}(W) = D_{\mathrm{BS}}(G_{W,\mathbf{R}}) = \{x + iy \mid x, y \in \mathbf{R}, 0 < y \le \infty\}$. Let

$$L = \left\{ \begin{pmatrix} 0 & x \\ 0 & 0 \end{pmatrix} \;\middle|\; x \in \mathbf{R} \right\} \subset \mathfrak{g}_{\mathbf{R}} = \mathfrak{sl}(2, \mathbf{R}).$$

Then $Y_{1,L} = Y_{2,L} = Y_{3,L} = (\mathbf{R}_{\ge 0}) \times L \times \{i\}$ and $\eta_1 = \eta_2 = \eta_3$ induces a homeomorphism

$$Y_{1,L} \xrightarrow{\sim} D_{\mathrm{SL}(2)}(W), \quad \left(t, \begin{pmatrix} 0 & x \\ 0 & 0 \end{pmatrix}, i\right) \mapsto \frac{x}{1+t^4} + \frac{i}{t^2}.$$

10.1.5

Remark.

(i) Theorem 10.1.3 gives a new proof of Theorem 5.2.15. In fact, $Y_{2,L} \to Y_{1,L}$ is proper and surjective, and hence Theorem 10.1.3 shows that $D_{\mathrm{SL}(2),\mathrm{val}} \to D_{\mathrm{SL}(2)}$ is proper and surjective. Since $D_{\mathrm{SL}(2),\mathrm{val}}$ is Hausdorff, this proves that $D_{\mathrm{SL}(2)}$ is Hausdorff.

(ii) Theorem 10.1.3 also gives a new proof of Proposition 5.2.16.

We can deduce, from Theorem 10.1.3, the following criterion for the local compactness of $D_{\mathrm{SL}(2)}$.

THEOREM 10.1.6 *Let $p \in D_{\mathrm{SL}(2)}$, n, W and $\mathbf{r}$ be as above. Then the following conditions (1)–(4) are equivalent.*

(1) *There exists a compact neighborhood of p in $D_{\mathrm{SL}(2)}$.*
(2) *$G_{W,\mathbf{R}} \cdot \mathbf{r}$ is a neighborhood of $\mathbf{r}$ in D.*

Note that this (2) is equivalent to

(2 bis) $\mathfrak{g}_{\mathbf{R}} = \mathrm{Lie}(G_{W,\mathbf{R}}) + \mathrm{Lie}(K'_{\mathbf{r}})$.

(3) *The following (3.1) and (3.2) are satisfied.*
   (3.1) *$(G^\circ)_{W,\mathbf{R}}$ is parabolic.*
   Here G° denotes the connected component of $1 \in G$ in the Zariski topology (§0.7).
   (3.2) *$(G_{W,\mathbf{R}} \cap K_{\mathbf{r}}) \cdot \mathbf{r}$ is a neighborhood of $\mathbf{r}$ in $K_{\mathbf{r}} \cdot \mathbf{r}$.*

Note that this (3.2) *is equivalent to*

(3.2 bis) $\mathrm{Lie}(K_{\mathbf{r}}) \subset \mathrm{Lie}(G_{W,\mathbf{R}}) + \mathrm{Lie}(K'_{\mathbf{r}})$.

(4) (3.2) *and the following* (4.1) *are satisfied.*
(4.1) *The set of* $\mathbf{R}$*-subspaces* $\{W_k^{(j)} \mid 1 \le j \le n,\ k \in \mathbf{Z}\}$ *of* $H_{0,\mathbf{R}}$ *is linearly ordered by inclusion.*

The proof of this theorem, assuming Theorem 10.1.3, will be given in 10.1.8 below.

10.1.7

Example. Assume $n = 1$. Let

$$d := \dim_{\mathbf{R}} D, \quad e := \dim_{\mathbf{R}}(G_{W,\mathbf{R}} \cdot \mathbf{r}).$$

Then, $d \ge e \ge 1$. Theorem 10.1.3 shows that there is a homeomorphism between some open neighborhood of p in $D_{\mathrm{SL}(2)}$ and some open neighborhood of (0, 0) in the space

$$((\mathbf{R}_{>0}) \times \mathbf{R}^{d-1}) \cup (\{0\}^{1+d-e} \times \mathbf{R}^{e-1}) \tag{1}$$

which is a topological subspace of $\mathbf{R}_{\ge 0} \times \mathbf{R}^{d-1}$. The space (1) is locally compact if and only if $d = e$.

10.1.8

Proof of Theorem 10.1.6, *assuming Theorem* 10.1.3. We first prove that 10.1.6 (1) and 10.1.6 (4) are equivalent. Let L be an $\mathbf{R}$-subspace of $\mathfrak{g}_{\mathbf{R}}$ satisfying 10.1.2 (1) and 10.1.2 (2). By Theorem 10.1.3, there exists a compact neighborhood of p in $D_{\mathrm{SL}(2)}$ if and only if $L \cap (\bigoplus_{\chi \in X_J \cup X_J^{-1}} \mathfrak{g}_{\mathbf{R}}(\chi))$ is independent of J and $\dim((G_{W(J),\mathbf{R}} \cap K_{\mathbf{r}}) \cdot \mathbf{r})$ is independent of J. This condition is satisfied if and only if both (3.2) in 10.1.6 and the following (1) are satisfied.

(1) $L \subset \bigoplus_{\chi \in X_+ \cup X_+^{-1}} \mathfrak{g}_{\mathbf{R}}(\chi)$.

We show that (1) is equivalent to

(2) If $\chi \in X$ and if $\chi, \chi^{-1} \notin X_+$, then $\mathfrak{g}_{\mathbf{R}}(\chi) = 0$.

In fact, it is clear that (2) implies (1). We prove that (1) implies (2). Assuming (1), let χ be an element of X such that $\chi, \chi^{-1} \notin X_+$ and let $v \in \mathfrak{g}_{\mathbf{R}}(\chi)$. We show $v = 0$. Write $v = x + y + z$ with $x \in L$, $y \in \mathrm{Lie}(\tilde{\rho}(\mathbf{R}^n_{>0}))$, $z \in \mathrm{Lie}(K_{\mathbf{r}})$. Then, by (1), we have $x(\chi) = x(\chi^{-1}) = 0$. Since $\mathrm{Ad}(\tilde{\rho}(\mathbf{R}^n_{>0}))$ acts on y trivially, we have $y(\chi) = y(\chi^{-1}) = 0$. Thus $z(\chi^{-1}) = v(\chi^{-1}) = 0$. Since $\theta(z) = z$, we have $z(\chi) = \theta(z(\chi^{-1})) = 0$. This shows $v = v(\chi) = z(\chi) = 0$.

By [KU2, 6.3], the above condition (2) is equivalent to 10.1.6 (4.1). Thus we have proved that 10.1.6 (1) and 10.1.6 (4) are equivalent.

Next we prove implications (4) $\Rightarrow$ (3) $\Rightarrow$ (2) $\Rightarrow$ (1) among the conditions in 10.1.6.

10.1.6 (4) implies 10.1.6 (3), since $(G^\circ)_{W,\mathbf{R}}$ is parabolic if 10.1.6 (4.1) is satisfied.

We prove that 10.1.6 (3) implies 10.1.6 (2 bis). Assume 10.1.6 (3). Since $(G^\circ)_{W,\mathbf{R}}$ is parabolic, we have $\mathfrak{g}_{\mathbf{R}} = \mathrm{Lie}(G_{W,\mathbf{R}}) + \mathrm{Lie}(K_{\mathbf{r}})$. This, together with 10.1.6 (3.2 bis), implies 10.1.6 (2 bis).

We prove that 10.1.6 (2 bis) implies 10.1.6 (1). Assume 10.1.6 (2 bis). Then, we can take an $\mathbf{R}$-subspace L of $\mathfrak{g}_{\mathbf{R}}$ satisfying the conditions 10.1.2 (1) and 10.1.2 (2) inside $\mathrm{Lie}(G_{W,\mathbf{R}})$. Since $\mathrm{Lie}(G_{W,\mathbf{R}}) = \bigoplus_{\chi \in X_+} \mathfrak{g}_{\mathbf{R}}(\chi^{-1})$ (10.1.1 (1)), this L satisfies the above condition (1). Since the condition 10.1.6 (3.2) is clearly satisfied, this shows that 10.1.6 (1) is satisfied. □

COROLLARY 10.1.9 *If $D_{\mathrm{SL}(2)}$ is locally compact, the canonical injection $D_{\mathrm{SL}(2),\mathrm{val}} \to D_{\mathrm{BS},\mathrm{val}}$ is an open map.*

Proof. Assume $D_{\mathrm{SL}(2)}$ is locally compact. Then, by Theorem 10.1.6 (3.2) and 10.1.8 (1), $Y_{2,L} = Y_{3,L}$ is satisfied by any $p \in D_{\mathrm{SL}(2)}$ and any $\mathbf{R}$-subspace L of $\mathfrak{g}_{\mathbf{R}}$ satisfying 10.1.2 (1) and 10.1.2 (2). By Theorem 10.1.3, this shows that any point of $D_{\mathrm{SL}(2),\mathrm{val}}$ has an open neighborhood U in $D_{\mathrm{SL}(2),\mathrm{val}}$ such that U is open in $D_{\mathrm{BS},\mathrm{val}}$. This proves 10.1.9. □

10.2 A SPECIAL OPEN NEIGHBORHOOD $U(p)$

A key for the proof of Theorem 10.1.3 is to define a special open neighborhood $U(p)$ of each $p \in D_{\mathrm{SL}(2)}$.

10.2.1

Let the notation be as in 10.1.1. The definition of $U(p)$ is as follows. For a subset $J = \{s_1, \dots, s_m\}$ $(1 \le s_1 < \cdots < s_m \le n)$ of $\{1, \dots, n\}$, let (ρ_J, φ_J) be the SL(2)-orbit in m variables characterized by the following two properties (cf. 5.2.10).

(1) The family of weight filtrations associated to (ρ_J, φ_J) is $W^{(J)}$.
(2) $\varphi_J(\mathbf{i}_m) = \mathbf{r}$.

Explicitly, (ρ_J, φ_J) is defined by

$$\rho_J(g_1, \dots, g_m) := \rho(h_1, \dots, h_n), \quad \varphi_J(z_1, \dots, z_m) := \varphi(w_1, \dots, w_n),$$

where h_j and w_j $(1 \le j \le n)$ are as follows. If $j \le s_k$ for some k, define $h_j := g_k$ and $w_j := z_k$ for the smallest integer k with $j \le s_k$. Otherwise, $h_j := 1$ and $w_j := i$.

For $(t_1, \dots, t_m) \in \mathbf{R}^m_{>0}$, we have

$$\tilde{\rho}_J(t_1, \dots, t_m) = \tilde{\rho}(t'_1, \dots, t'_n),$$

where $t'_j := t_k$ if $j = s_k$ and $t'_j := 1$ if $j \in \{1, \dots, n\} - J$. That is, $\tilde{\rho}_J$ is the restriction of $\tilde{\rho}$ (5.2.2 (1)) to the J-component of $\mathbf{R}^n_{>0}$.

We denote

$$p_J := [\rho_J, \varphi_J] \in D_{\mathrm{SL}(2)}(W^{(J)}).$$

Using the action of $G_{W^{(J)},\mathbf{R}}$ on $D_{\mathrm{SL}(2)}(W^{(J)})$, we define

$$U(p) := \cup_{J\subset\{1,\dots,n\}} G_{W^{(J)},\mathbf{R}} \cdot p_J \subset D_{\mathrm{SL}(2)}(W). \tag{3}$$

Then $U(p)$ is independent of the choice of the representative (ρ, φ) of p.

As is easily seen,

$$U(p') \subset U(p) \quad \text{for} \quad p' \in U(p).$$

Theorem 10.2.2 *$U(p)$ is open in $D_{\mathrm{SL}(2)}$.*

We deduce Theorem 10.2.2 from Propositions 10.2.4 and 10.2.5 below.

10.2.3

Let $H_{0,\mathbf{C}} = \bigoplus_{\chi\in X} H_{0,\mathbf{C}}(\chi)$ be the decomposition with respect to the action of $\mathbf{G}^n_{\mathbf{m},\mathbf{R}}$ via $\tilde{\rho}$, where $X = X(\mathbf{G}^n_{\mathbf{m},\mathbf{R}})$. Since $\tilde{\rho}(t)\varphi(\mathbf{0}) = \varphi(\mathbf{0})$, we have

$$F(\varphi(\mathbf{0})) = \bigoplus_{\chi\in X} F(\varphi(\mathbf{0}))(H_{0,\mathbf{C}}(\chi)),$$

where $F(\varphi(\mathbf{0}))$ (resp. $F(\varphi(\mathbf{0}))(H_{0,\mathbf{C}}(\chi))$) is the corresponding (resp. induced) Hodge filtration on $H_{0,\mathbf{C}}$ (resp. $H_{0,\mathbf{C}}(\chi)$). If the image of χ under $X \simeq \mathbf{Z}^n$ is $(k_1, \dots, k_n)$, $F(\varphi(\mathbf{0}))(H_{0,\mathbf{C}}(\chi))$ is an $\mathbf{R}$-Hodge structure of weight k_n.

For $\chi \in X$, we call the Hodge numbers of $F(\varphi(\mathbf{0}))(H_{0,\mathbf{C}}(\chi))$ the *χ-Hodge numbers of p*.

Proposition 10.2.4 *Let $p' \in D_{\mathrm{SL}(2)}$ and assume that the family of weight filtrations associated to p' is W. Then $p' = gp$ for some $g \in G_{W,\mathbf{R}}$ if and only if, for each $\chi \in X$, the χ-Hodge numbers of p coincides with the χ-Hodge numbers of p'.*

Here the χ-Hodge numbers of $p' = [\rho', \varphi']$ are defined with respect to $\tilde{\rho}'$.

Proposition 10.2.5 *Let J be a subset of $\{1, \dots, n\}$. Assume that $p_\lambda = [\rho_\lambda, \varphi_\lambda] \in D_{\mathrm{SL}(2)}$ converges to p and that the family of weight filtrations associated to p_λ is $W^{(J)}$ for any λ. Then, if λ is sufficiently large, the χ-Hodge numbers of p_λ coincide with the χ-Hodge numbers of p_J for all $\chi \in X(\mathbf{G}^J_{\mathbf{m},\mathbf{R}})$.*

The proof of Proposition 10.2.4 (resp. 10.2.5) will be given in 10.2.7 (resp. 10.2.16) below.

10.2.6

Reduction of Theorem 10.2.2 *to Propositions* 10.2.4, 10.2.5. It is enough to show the following (1).

(1) If $p' \in U(p)$ and if $p'_\lambda \in D_{\mathrm{SL}(2)}(W)$ converges to p', then $p'_\lambda \in U(p)$ for any sufficiently large λ.

Since $U(p') \subset U(p)$, by replacing p' by p, we can reduce (1) to the following:

(2) If $p_\lambda \in D_{\mathrm{SL}(2)}(W)$ converges to p, then $p_\lambda \in U(p)$ for any sufficiently large λ.

Dividing the sequence $\{p_\lambda\}_\lambda$ into subsequences, we may assume that, for a fixed $J \subset \{1, \ldots, n\}$, the family of weight filtrations associated to p_λ is $W^{(J)}$ for any λ. Then, by Proposition 10.2.5, if λ is sufficiently large, the χ-Hodge numbers of p_λ coincide with those of p_J for all $\chi \in X(\mathbf{G}^J_{\mathbf{m},\mathbf{R}})$. Hence, if λ is sufficiently large, by applying Proposition 10.2.4 taking $(p_J, W^{(J)}, p_\lambda)$ here as (p, W, p') of 10.2.4, we get $g_\lambda \in G_{W^{(J)},\mathbf{R}}$ such that $p_\lambda = g_\lambda p_J \in U(p)$. This shows (2) and hence proves 10.2.2. □

10.2.7

Proof of Proposition 10.2.4. The "only if" part is easy. We prove the "if" part.

For $1 \leq j \leq n$, let N_j^+ be the image of $\left(\begin{smallmatrix} 0 & 0 \\ 1 & 0 \end{smallmatrix}\right)$ of the jth factor of $\mathfrak{sl}(2, \mathbf{R})^{\oplus n}$ under $\rho_* : \mathfrak{sl}(2, \mathbf{R})^{\oplus n} \to \mathfrak{g}_{\mathbf{R}}$. By the theory of representations of the semi-simple Lie algebra $\mathfrak{sl}(2, \mathbf{R})^{\oplus n}$, we have the following. For $\chi \in X$, let

$$A_\chi := \{v \in H_{0,\mathbf{R}}(\chi) \mid N_j^+(v) = 0 \ \ (1 \leq j \leq n)\}$$

and let B_χ be the $\mathbf{R}$-subspace of $H_{0,\mathbf{R}}$ generated by $\rho_*(s)v$ $(s \in \mathfrak{sl}(2, \mathbf{R})^{\oplus n},\ v \in A_\chi)$. Then

$$H_{0,\mathbf{R}} = \bigoplus_{\chi \in X} B_\chi.$$

Let

$$d : X \simeq \mathbf{Z}^n$$

be the isomorphism characterized by the following property: For any $\chi \in X$, the action of $\rho(\Delta(t_1, \ldots, t_n))$ on $H_{0,\mathbf{R}}(\chi)$ coincides with the multiplication by $\prod_{j=1}^n t_j^{m(j)}$ where $m = d(\chi)$. (See 5.2.2 for the notation.) That is, if $m = d(\chi)$ and if $m' \in \mathbf{Z}^n$ is the image of χ under our usual identification $X = \mathbf{Z}^n$ (given by the action of $\tilde{\rho}(t_1, \ldots, t_n)$ on $H_{0,\mathbf{R}}(\chi)$), then $m(j) = m'(j) - m'(j-1)$ ($m'(0)$ means 0). We have $A_\chi = 0$ unless $d(\chi) \geq 0$. Furthermore,

$$B_\chi = \bigoplus_{l \in \mathbf{N}^n,\, l \leq d(\chi)} A_{\chi,l}, \quad \text{where}$$

$$A_{\chi,l} := \left(\prod_{1 \leq j \leq n} N_j^{l(j)}\right) A_\chi, \quad \prod_{1 \leq j \leq n} N_j^{l(j)} : A_\chi \xrightarrow{\sim} A_{\chi,l}.$$

Note that $A_{\chi,l}$ is contained in $H_{0,\mathbf{R}}(\chi')$ where χ' is the element of X satisfying $d(\chi') = d(\chi) - 2l$.

Concerning the filtration $F_{\varphi(\mathbf{0})}$ on $H_{0,\mathbf{C}}$, for all $p \in \mathbf{Z}$, we have that

$$F(\varphi(\mathbf{0}))^p = \bigoplus_{\chi,l} F(\varphi(\mathbf{0}))^p(A_{\chi,l,\mathbf{C}}),$$

where χ ranges over X and, for each χ, l ranges over elements of $\mathbf{N}^n$ such that $l \leq d(\chi)$, and that, for each $\chi \in X$ with $m := d(\chi)$ in $\mathbf{N}^n$, $F(\varphi(\mathbf{0}))(A_{\chi,\mathbf{C}})$ with the intersection form

$$A_\chi \times A_\chi \to \mathbf{R}, \quad (x, y) \mapsto \left\langle x, \left(\prod_{1\leq j\leq n} N_j^{m(j)}\right)(y)\right\rangle_0,$$

is a polarized $\mathbf{R}$-Hodge structure, and

$$\left(\prod_{1\leq j\leq n} N_j^{l(j)}\right) F(\varphi(\mathbf{0}))^p(A_{\chi,\mathbf{C}}) = F(\varphi(\mathbf{0}))^{p-|l|}(A_{\chi,l,\mathbf{C}})$$

for $l \in \mathbf{N}^n$ with $l \leq m$, where $|l| = \sum_{1\leq j\leq n} l(j)$.

Assume that the χ-Hodge numbers of p' coincides with those of p for any $\chi \in X$. Let (ρ', φ') be a representative of p' and define A'_χ, B'_χ, $A'_{\chi,l}$ for (ρ', φ') in the same way as A_χ, B_χ, $A_{\chi,l}$ for (ρ, φ). Then, from the coincidence of the χ-Hodge numbers, we obtain $\dim(A_\chi) = \dim(A'_\chi)$. It follows that the Hodge numbers of $F(\varphi(\mathbf{0}))$ $(A_{\chi,\mathbf{C}})$ and those of $F(\varphi'(\mathbf{0}))(A'_{\chi,\mathbf{C}})$ coincide. Hence there exists an isomorphism of $\mathbf{R}$-vector spaces $g_\chi : A_\chi \xrightarrow{\sim} A'_\chi$ such that $g_\chi(F(\varphi(\mathbf{0}))(A_{\chi,\mathbf{C}})) = F(\varphi'(\mathbf{0}))(A'_{\chi,\mathbf{C}})$ and such that $\langle g_\chi(x), (\prod_{1\leq j\leq n} N_j'^{m(j)})(g_\chi(y))\rangle_0 = \langle x, (\prod_{1\leq j\leq n} N_j^{m(j)})(y)\rangle_0$ for any $x, y \in A_\chi$. Define the isomorphism of $\mathbf{R}$-vector spaces $g_{\chi,l} : A_{\chi,l} \xrightarrow{\sim} A'_{\chi,l}$ $(l \in \mathbf{N}^n,\ l \leq m)$ by $g_{\chi,l}((\prod_{1\leq j\leq n} N_j^{l(j)})(x)) = (\prod_{1\leq j\leq n} N_j'^{l(j)})(g_\chi(x))$ $(x \in A_\chi)$. Define $g : H_{0,\mathbf{R}} \xrightarrow{\sim} H_{0,\mathbf{R}}$ to be the direct sum of $g_{\chi,l}$. Then, as is checked easily, $\mathrm{Int}(g) \circ \tilde{\rho} = \tilde{\rho}'$, $\mathrm{Ad}(g)(N_j) = N'_j$, $g(\varphi(\mathbf{0})) = \varphi'(\mathbf{0})$ and g preserves $\langle\ ,\ \rangle_0$. Hence $g \in G_{W,\mathbf{R}}$ and $gp = p'$. □

We will prove Proposition 10.2.5 in 10.2.16 after preliminaries in 10.2.8–10.2.15.

Proposition 10.2.8 *Let P be a $\mathbf{Q}$-parabolic subgroup of $G_\mathbf{R}$ and let K be a maximal compact subgroup of $G_\mathbf{R}$.*

(i) *The map*

$$P \times A_P \to P, \quad (g, a) \mapsto \mathrm{Int}(a_K)^{-1}(g),$$

extends uniquely to a continuous map

$$\mu : P \times (\overline{A}_P)_{\mathrm{val}} \to P.$$

Here a_K denotes the Borel-Serre lifting of a at K and $(\overline{A}_P)_{\mathrm{val}}$ is as in 5.1.12.

(ii) *The map μ has the following properties.*

(1) $\mu(gs_K, t) = s_K\mu(g, st) \quad (g \in P,\ s \in A_P,\ t \in (\overline{A}_P)_{\mathrm{val}})$.
(2) $\mu(gk, t) = \mu(g, t)k \quad (g \in G_\mathbf{R},\ k \in P \cap K,\ t \in (\overline{A}_P)_{\mathrm{val}})$.

Proof. We prove (i). Let $T \subset G_\mathbf{R}$ be the Borel-Serre lifting of S_P at K, and identify $X(T)$ and $X(S_P)$ via the isomorphism $T \xrightarrow{\sim} S_P$. Let $H_{0,\mathbf{R}} = \bigoplus_{\chi\in X(S_P)} H_{0,\mathbf{R}}(\chi)$ be the decomposition according to the action of T on $H_{0,\mathbf{R}}$. Then, for any $g \in P$,

g preserves $\bigoplus_{\chi \in X(S_P)_+} H_{0,\mathbf{R}}(\alpha\chi^{-1})$ for any $\alpha \in X(S_P)$ (cf. 5.1.6 (3)), where $X(S_P)_+$ is as in 5.1.10. Recall that $(\overline{A}_P)_{\mathrm{val}} = \mathrm{Hom}(X(S_P)_+, \mathbf{R}^{\mathrm{mult}}_{\geq 0})_{\mathrm{val}}$ (5.1.12). To prove (i), since the target space P is a regular space, it is sufficient to prove by Lemma 6.4.7 that, if $g_\lambda \in P$ converges to $g \in P$ and $t_\lambda \in A_P$ converges to $t = (V, h) \in (\overline{A}_P)_{\mathrm{val}}$, then $\mathrm{Int}(t_{\lambda,K})^{-1}(g_\lambda)$ converges in P. To see it, it is sufficient to prove that, for $\alpha \in X(S_P)$ and $v \in H_{0,\mathbf{R}}(\alpha)$, $\mathrm{Int}(t_{\lambda,K})^{-1}(g_\lambda)(v)$ converges. We have $g_\lambda(v) = \sum_{\chi \in X(S_P)_+} (g_\lambda(v))(\alpha\chi^{-1})$ and each $(g_\lambda(v))(\alpha\chi^{-1})$ converges to $(g(v))(\alpha\chi^{-1})$. We have

$$\mathrm{Int}(t_{\lambda,K})^{-1}(g_\lambda)(v) = \sum_{\chi \in X(S_P)_+} \chi(t_\lambda) \cdot (g_\lambda(v))(\alpha\chi^{-1}).$$

Since $\chi \in X(S_P)_+ \subset V$, $\chi(t_\lambda)$ converges (5.1.11 (3)). Hence $\mathrm{Int}(t_{\lambda,K})^{-1}(g_\lambda)(v)$ converges.

Next we prove (ii). When $x \in A_P$ converges to t, $x_K^{-1} g s_K x_K$ converges to $\mu(gs_K, t)$. On the other hand, $x_K^{-1} g s_K x_K = s_K (sx)_K^{-1} g (sx)_K$ converges to $s_K \mu(g, st)$. This proves (1).

(2) is proved in a similar way by using the following Lemma 10.2.9. □

Lemma 10.2.9 *Let P be a $\mathbf{Q}$-parabolic subgroup of $G_{\mathbf{R}}$ and let K be a maximal compact subgroup of $G_{\mathbf{R}}$. If $a \in P \cap K$ and $b \in S_P$, then $ab_K = b_K a$.*

Proof. We have $ab_K a^{-1} \equiv b_K \bmod P_u$, since b belongs to the center of P/P_u. Furthermore, we have $\theta_K(ab_K a^{-1}) = ab_K^{-1} a^{-1} = (ab_K a^{-1})^{-1}$. Hence, by the characterization of the Borel-Serre lifting (5.1.3), we have $ab_K a^{-1} = b_K$. □

10.2.10

Let Ψ be the set of all $\mathbf{Q}$-parabolic subgroups P of $G_{\mathbf{R}}$ satisfying the following conditions (1) and (2).

(1) $(G^\circ)_{W,\mathbf{R}} \subset P$.
(2) The image of $\tilde{\rho}(\mathbf{G}^n_{\mathbf{m},\mathbf{R}})$ in P/P_u is contained in S_P.

Let $P \in \Psi$. Then, the homomorphism $\mathbf{G}^n_{\mathbf{m},\mathbf{R}} \to S_P$ induced by $\tilde{\rho}$ defines a homomorphism $X(S_P) \to X = X(\mathbf{G}^n_{\mathbf{m},\mathbf{R}})$. Define $(\mathbf{R}^n_{\geq 0})_{\mathrm{val}}(P)$ to be the open subset of $(\mathbf{R}^n_{\geq 0})_{\mathrm{val}} = \mathrm{Hom}(X_+, \mathbf{R}^{\mathrm{mult}}_{\geq 0})_{\mathrm{val}}$ consisting of all elements (V, h) such that V contains the image of $X(S_P)_+$ under $X(S_P) \to X = X(\mathbf{G}^n_{\mathbf{m},\mathbf{R}})$.

Let $(\mathbf{R}^n_{\geq 0})_{\mathrm{val}}(P) \to (\overline{A}_P)_{\mathrm{val}}$ be the injective map $(V, h) \mapsto (V', h')$, where V' is the inverse image of $V \subset X$ in $X(S_P)$ and h' is the composite map $V'^{\times} \to V^{\times} \xrightarrow{h} \mathbf{R}_{>0}$. Then, this map is continuous and induces an injective continuous map

$$D \times^{\mathbf{R}^n_{>0}} (\mathbf{R}^n_{\geq 0})_{\mathrm{val}}(P) \hookrightarrow D \times^{A_P} (\overline{A}_P)_{\mathrm{val}} = D_{\mathrm{BS,val}}(P).$$

Here $t \in \mathbf{R}^n_{>0}$ acts on D by the Borel-Serre action of the image of $\tilde{\rho}(t)$ in S_P with respect to P.

We have

(3) For $x = ([\rho', \varphi'], Z, V) \in D_{\mathrm{SL}(2),\mathrm{val}}(W) \cap D_{\mathrm{BS},\mathrm{val}}(P)$, the image of x under the identification $D_{\mathrm{BS},\mathrm{val}}(P) \simeq D \times^{A_P} (\overline{A}_P)_{\mathrm{val}}$ (5.1.12) is $(F, (V', h)) \in D \times^{\mathbf{R}^n_{>0}} (\mathbf{R}^n_{\geq 0})_{\mathrm{val}}(P)$, where F is any element of Z, V' is the inverse image of $V \subset X(\mathbf{G}^J_{\mathbf{m},\mathbf{R}})$ under $X \to X(\mathbf{G}^J_{\mathbf{m},\mathbf{R}})$ (here J is the subset of $\{1, \ldots, n\}$ such that the family of weight filtrations associated to (ρ', φ') is $W^{(J)}$), and h is the homomorphism $V' \to \mathbf{R}^{\mathrm{mult}}_{\geq 0}$ which sends $V'^{\times}$ to 1 and $V' - V'^{\times}$ to 0.

(4) $D_{\mathrm{SL}(2),\mathrm{val}}(W) \subset \bigcup_{P \in \Psi} D_{\mathrm{BS},\mathrm{val}}(P)$.

This assertion (4) follows from 10.2.11 below.

10.2.11

([KU2, 4.11 and the proof of 4.13]). For a valuative submonoid V of X containing X_+, let P_V be the subgroup of $(G^\circ)_{\mathbf{R}}$ consisting of all elements which preserve the $\mathbf{R}$-subspaces $\bigoplus_{\chi \in V} H_{0,\mathbf{R}}(\alpha\chi^{-1})$ of $H_{0,\mathbf{R}}$ for any $\alpha \in X$ (cf. 5.1.6 (3)). Then we have the following:

(1) P_V is a $\mathbf{Q}$-parabolic subgroup of $G_{\mathbf{R}}$.
(2) There exists a $\mathbf{Q}$-parabolic subgroup P of $G_{\mathbf{R}}$ such that $P \subset P_V$ and $P \in \Psi$.
(3) For an element $x = ([\rho', \varphi'], Z, V) \in D_{\mathrm{SL}(2),\mathrm{val}}(W)$, the $\mathbf{Q}$-parabolic subgroup of $G_{\mathbf{R}}$ associated to the image of x under $D_{\mathrm{SL}(2),\mathrm{val}} \to D_{\mathrm{BS},\mathrm{val}} \to D_{\mathrm{BS}}$ is $P_{V'}$ where V' is as in 10.2.10 (3).

By [KU2, 4.12], there is a continuous map $\beta : D \to \mathbf{R}^n_{>0}$ satisfying $\beta(\tilde{\rho}(t)x) = t\beta(x)$ for any $t \in \mathbf{R}^n_{>0}$ and any $x \in D$. Fix such β.

Lemma 10.2.12 *Let $P \in \Psi$. Then, there exists a continuous map $\kappa_P : D \times^{\mathbf{R}^n_{>0}} (\mathbf{R}^n_{\geq 0})_{\mathrm{val}}(P) \to D$ characterized by the property*

$$\kappa_P(gb, t) = \tilde{\rho}(\beta(\mu(g, t)b))^{-1}\mu(g, t)b$$
$$(\forall g \in P,\ \forall b \in K_{\mathbf{r}} \cdot \mathbf{r},\ \forall t \in (\mathbf{R}^n_{\geq 0})_{\mathrm{val}}(P) \subset (\overline{A}_P)_{\mathrm{val}}).$$

Proof. To prove that the map κ_P is well defined, it is sufficient to show the following (1) and (2) for $g \in P, s \in A_P, t \in (\mathbf{R}^n_{\geq 0})_{\mathrm{val}}(P), b \in K_{\mathbf{r}} \cdot \mathbf{r}, k \in P \cap K_{\mathbf{r}}$.

$$\tilde{\rho}(\beta(\mu(gs_{K_{\mathbf{r}}}, t)b))^{-1}\mu(gs_{K_{\mathbf{r}}}, t)b = \tilde{\rho}(\beta(\mu(g, st)b))^{-1}\mu(g, st)b. \tag{1}$$
$$\tilde{\rho}(\beta(\mu(gk, t)b))^{-1}\mu(gk, t)b = \tilde{\rho}(\beta(\mu(g, t)kb))^{-1}\mu(g, t)kb. \tag{2}$$

(1) (resp. (2)) follows from Lemma 10.2.8 (ii) (1) (resp. (2)). The continuity of κ_P follows easily from that of μ (10.2.8 (i)). □

10.2.13

We define a map

$$\kappa : D_{\mathrm{SL}(2)}(W) \to D,$$

which extends $D \to D, x \mapsto \tilde{\rho}(\beta(x))^{-1}x$, as follows. Let $x \in D_{\mathrm{SL}(2)}(W)$ and let J be the subset of $\{1, \ldots, n\}$ such that the family of weight filtrations associated to x is $W^{(J)}$. Put $m = \sharp(J)$. Let (ρ', φ') be an SL(2)-orbit in m variables which represents x. Since both $\tilde{\rho}_J$ and $\tilde{\rho}'$ split $W^{(J)}$, there exists a unique element $u \in G_{W^{(J)},\mathbf{R},u}$ such that $\tilde{\rho}_J = \mathrm{Int}(u) \circ \tilde{\rho}'$. Define

$$\kappa(x) := \tilde{\rho}(\beta(u\varphi'(\mathbf{i}_m)))^{-1}u\varphi'(\mathbf{i}_m),$$

where $\mathbf{i}_m = (i, \ldots, i) \in \mathfrak{h}^m$. As is easily seen, this does not depend on the choice of the representative (ρ', φ').

Lemma 10.2.14

(i) *κ is continuous.*

(ii) *Let $P \in \Psi$. Then, restriction of κ_P in* 10.2.12 *to $D_{\mathrm{SL}(2),\mathrm{val}}(W) \cap D_{\mathrm{BS},\mathrm{val}}(P)$* (10.2.10 (3)) *coincides with the restriction there of the composite map $D_{\mathrm{SL}(2),\mathrm{val}}(W) \to D_{\mathrm{SL}(2)}(W) \xrightarrow{\kappa} D$.*

(iii) *$\kappa(p_J) = \kappa(p)$ for any subset J of $\{1, \ldots, n\}$.*

Proof. (iii) follows from the definition of κ, and (i) follows from (ii) and 10.2.10 (4).

We prove (ii). Let $x = ([\rho', \varphi'], Z, V) \in D_{\mathrm{SL}(2),\mathrm{val}}(W) \cap D_{\mathrm{BS},\mathrm{val}}(P)$. We prove that $\kappa_P(x)$ coincides with $\kappa(\overline{x})$ where $\overline{x}$ denotes the image of x in $D_{\mathrm{SL}(2)}(W)$. Let J, m, u be as in 10.2.13. Let V' be the inverse image of $V \subset X(\mathbf{G}^J_{\mathbf{m},\mathbf{R}})$ in X under $X \to X(\mathbf{G}^J_{\mathbf{m},\mathbf{R}})$. We have $P \subset P_{V'}$ and hence $P_{V',u} \subset P_u$. Write

$$\varphi'(\mathbf{i}_m) = gb \quad (g \in P,\ b \in K_{\mathbf{r}} \cdot \mathbf{r}).$$

We show

(3) $g \in G_{W^{(J)},\mathbf{R}}$.

In fact, since $\tilde{\rho}' \equiv \tilde{\rho}_J \bmod G_{W^{(J)},\mathbf{R},u}$ and $G_{W^{(J)},\mathbf{R},u} \subset P_{V',u} \subset P_u$, we have $\tilde{\rho}' \equiv \tilde{\rho}_J \bmod P_u$. Since $\tilde{\rho}'$ (resp. $\tilde{\rho}_J$) is the Borel-Serre lifting of $\tilde{\rho}' \bmod P_u$ at $K_{g\mathbf{r}} = \mathrm{Int}(g)K_{\mathbf{r}}$ (resp. $\tilde{\rho}_J \bmod P_u$ at $K_{\mathbf{r}}$), we have $\tilde{\rho}' = \mathrm{Int}(g) \circ \tilde{\rho}_J$. On the other hand, $\tilde{\rho}' = \mathrm{Int}(u)^{-1} \circ \tilde{\rho}_J$ and hence ug commutes with $\tilde{\rho}_J(t)$ for any t. This shows $ug \in G_{W^{(J)},\mathbf{R}}$ and hence proves (3).

By 10.2.10 (3), in $D \times^{\mathbf{R}^n_{>0}} (\mathbf{R}^n_{\geq 0})_{\mathrm{val}}(P)$, we have $x = (gb, (V', h))$ where h is the homomorphism $V' \to \mathbf{R}^{\mathrm{mult}}_{\geq 0}$ which sends $V'^{\times}$ to 1 and $V' - V'^{\times}$ to 0. Hence, when $t \in \mathbf{R}^n_{>0}$ converges to (V', h), $\kappa_P(g\tilde{\rho}(t)b)$ converges to $\kappa_P(x)$. On the other hand, $\kappa_P(g\tilde{\rho}(t)b) = \tilde{\rho}(\beta(\mathrm{Int}(\tilde{\rho}(t))^{-1}(g)b))^{-1}\,\mathrm{Int}(\tilde{\rho}(t))^{-1}(g)b$ and $\kappa(\overline{x}) = \tilde{\rho}(\beta(ugb))^{-1}ugb$. Hence, to prove that $\kappa_P(x) = \kappa(\overline{x})$, it is sufficient to show

$$\mathrm{Int}(\tilde{\rho}(t))(g) \to ug. \tag{4}$$

Let $P_{V',0}$ is the subgroup of $(G^\circ)_{\mathbf{R}}$ consisting of all elements which preserve $\bigoplus_{\chi \in V'^{\times}} H_{0,\mathbf{R}}(\alpha\chi)$ for any $\alpha \in X$ and let $(G^\circ)_{W^{(J)},\mathbf{R},0}$ be the subgroup of $(G^\circ)_{W^{(J)},\mathbf{R}}$ consisting of all elements which commute with $\tilde{\rho}_J(t)$ for any t. Then $(G^\circ)_{W^{(J)},\mathbf{R},0} \subset P_{V',0}$. We have a bijection

$$P_{V',u} \times P_{V',0} \to P_{V'}, \quad (a, b) \mapsto ab,$$

which induces a bijection

$$G_{W^{(J)},\mathbf{R},u} \times (G^\circ)_{W^{(J)},\mathbf{R},0} \to (G^\circ)_{W^{(J)},\mathbf{R}}, \quad (a,b) \mapsto ab. \tag{5}$$

Since $\mathrm{Lie}(P_{V',u}) = \bigoplus_{\chi \in V'-V'^\times} \mathfrak{g}_{\mathbf{R}}(\chi^{-1})$, we have

(6) The operator $\mathrm{Ad}(\tilde{\rho}(t_\lambda))^{-1}$ on $\mathrm{Lie}(P_{V',u})$ converges to 0.

Since $\mathrm{Lie}(P_{V',0}) = \bigoplus_{\chi \in V'^\times} \mathfrak{g}_{\mathbf{R}}(\chi)$, we have:

(7) The operator $\mathrm{Ad}(\tilde{\rho}(t_\lambda))^{-1}$ on $\mathrm{Lie}(P_{V',0})$ converges to the identity map.

Write $g = g_u g_0$ ($g_u \in P_{V',u}$, $g_0 \in P_{V',0}$) according to (5). Then, $g_u = u^{-1}$. By (6), $\mathrm{Int}(\tilde{\rho}(t))^{-1}(g_u)$ converges to 1 and, by (7), $\mathrm{Int}(\tilde{\rho}(t))^{-1}(g_0)$ converges to g_0. Hence $\mathrm{Int}(\tilde{\rho}(t))^{-1}(g)$ converges to $g_0 = ug$, i.e., (4) is proved. □

Lemma 10.2.15 *Let $\mathcal{M}$ be the set of all subsets M of $X = X(\mathbf{G}^n_{\mathbf{m},\mathbf{R}})$ satisfying the following condition* (1).

(1) *There exists a finite subset I of X such that $M = \{\alpha\chi^{-1} \mid \alpha \in I, \chi \in X_+\}$.*

For $M \in \mathcal{M}$, let

$$H(M) := \bigoplus_{\chi \in M} H_{0,\mathbf{R}}(\chi) = \sum_{\chi \in M} \bigcap_{1 \le j \le n} W^{(j)}_{j(\chi)},$$

where the χ-component $H_{0,\mathbf{R}}(\chi)$ of $H_{0,\mathbf{R}}$ is defined with respect to $\tilde{\rho}$ and where $j(\chi)$ is the j-th component of the image of χ in $\mathbf{Z}^n$.

Let $\check{D}^W$ be the subset of $\check{D}$ consisting of all elements F satisfying the following condition (2).

(2) *If $M, M' \in \mathcal{M}$ with $M' \subset M$ and if $M - M'$ consists of one element whose image in $\mathbf{Z}^n$ is $k = (k_1, \dots, k_n)$, then $F(H(M)_{\mathbf{C}}/H(M')_{\mathbf{C}})$ is an $\mathbf{R}$-Hodge structure of weight $k_n + w$.*

We endow $\check{D}^W$ with the topology as a subspace of $\check{D}$. Then

(i) $\kappa(p) \in \check{D}^W$ (10.2.13).

(ii) *For any $M, M' \in \mathcal{M}$ such that $M' \subset M$ and $M - M'$ consists of one element $\chi \in X(\mathbf{G}_{\mathbf{m},\mathbf{R}})$, the Hodge numbers of $F(\kappa(p))(H(M)_{\mathbf{C}}/H(M')_{\mathbf{C}})$ coincides with the χ-Hodge numbers of p. Here $F(\kappa(p))$ denotes the Hodge filtration of $\kappa(p)$.*

(iii) *Assume $F_\lambda \in \check{D}^W$ converges to $F \in \check{D}^W$. Then, for any $M \in \mathcal{M}$, the filtration $F_\lambda(H(M)_{\mathbf{C}})$ of $H(M)_{\mathbf{C}}$ converges to $F(H(M)_{\mathbf{C}})$.*

(iv) *Assume $F_\lambda \in \check{D}^W$ converges to $F \in \check{D}^W$. Then, for any $M, M' \in \mathcal{M}$ such that $M' \subset M$, the filtration $F_\lambda(H(M)_{\mathbf{C}}/H(M')_{\mathbf{C}})$ converges to $F(H(M)_{\mathbf{C}}/H(M')_{\mathbf{C}})$.*

Proof. By the definition of κ, we have

$$F(\kappa(p))(H(M)_{\mathbf{C}}/H(M')_{\mathbf{C}}) = gF(\mathbf{r})(H(M)_{\mathbf{C}}/H(M')_{\mathbf{C}})$$

for some $g \in G_{W,\mathbf{R}}$. Furthermore, since $\mathbf{r} = h\varphi(\mathbf{0})$ for some $h \in G_{W,\mathbf{C},u}$, we have

$$F(\mathbf{r})(H(M)_{\mathbf{C}}/H(M')_{\mathbf{C}}) = F(\varphi(\mathbf{0}))(H(M)_{\mathbf{C}}/H(M')_{\mathbf{C}}).$$

This proves (i) and (ii).

(iii) and (iv) are proved in the same way as Lemma 6.1.11. □

10.2.16

Proof of Proposition 10.2.5. Let $\kappa : D_{\mathrm{SL}(2)}(W) \to D$ be as above in 10.2.13. Then $\kappa(p_\lambda)$ converges to $\kappa(p)$ by 10.2.14 (i), and $\kappa(p) = \kappa(p_J)$ by 10.2.14 (iii). Define a set $\mathcal{M}(J)$ of subsets of $X(\mathbf{G}^J_{\mathbf{m},\mathbf{R}})$ in the same way as $\mathcal{M}$ in 10.2.15 by replacing W by $W^{(J)}$. Let $\chi \in X(\mathbf{G}^J_{\mathbf{m},\mathbf{R}})$ and take $M, M' \in \mathcal{M}(J)$ such that $M \supset M'$ and $M' = M - \{\chi\}$. Then $\kappa(p_\lambda), \kappa(p_J) \in \check{D}^{W^{(J)}}$ by 10.2.15 (i), applied by replacing W by $W^{(J)}$. Hence

$$F(\kappa(p_\lambda))(H(M)_{\mathbf{C}}/H(M')_{\mathbf{C}}) \text{ converges to } F(\kappa(p_J))(H(M)_{\mathbf{C}}/H(M')_{\mathbf{C}})$$

by 10.2.15 (iv), applied by replacing W by $W^{(J)}$. Hence, if λ is sufficiently large, the Hodge numbers of $F(\kappa(p_\lambda))(H(M)_{\mathbf{C}}/H(M')_{\mathbf{C}})$ coincide with the Hodge numbers of $F(\kappa(p_J))(H(M)_{\mathbf{C}}/H(M')_{\mathbf{C}})$. By 10.2.15 (ii), applied by replacing W by $W^{(J)}$, the former (resp. the latter) are the χ-Hodge numbers of p_λ (resp. p_J). □

Thus we have completed the proof of Theorem 10.2.2.

10.3 PROOF OF THEOREM 10.1.3

10.3.1

We extend the map $\eta_0 : Y_0 \to D$ to a map

$$\eta_3 : Y_3 = (\mathbf{R}^n_{\geq 0})_{\mathrm{val}} \times \mathfrak{g}_{\mathbf{R}} \times (K_{\mathbf{r}} \cdot \mathbf{r}) \to D_{\mathrm{BS,val}}$$

as follows. We will later show that η_3 is continuous.

We identify $(\mathbf{R}^n_{\geq 0})_{\mathrm{val}}$ with $\mathrm{Hom}(X_+, \mathbf{R}_{\geq 0})_{\mathrm{val}}$ by identifying X_+ with $\mathbf{N}^n$ in the canonical way. Let $(t, a, b) \in Y_3$, and write $t = (V, h)$ where V is a valuative submonoid of X containing X_+ and $h : V \to \mathbf{R}^{\mathrm{mult}}_{\geq 0}$ is a homomorphism such that $h^{-1}(\mathbf{R}_{>0}) = V^\times$. For $\chi \in X$ with $\chi \notin V$, define $h(\chi)$ to be $h(\chi^{-1})^{-1} = 0^{-1} = \infty$. Define $A_1, A_2, A_3 \in \mathfrak{g}_{\mathbf{R}}$ by

$$A_1 = \sum_{\chi \in X} \frac{1}{1 + h(\chi)^{-2}} a(\chi),$$

$$A_2 = \sum_{\chi \in X} \frac{1}{h(\chi)^2 + 1} a(\chi),$$

$$A_3 = \sum_{\chi \in V^\times} \frac{1}{h(\chi)^2 + 1} a(\chi).$$

Here $(1+0^{-2})^{-1}=0$, $(1+\infty^{-2})^{-1}=1$, $(\infty^2+1)^{-1}=0$. Note that

$$A_1 \in \sum_{\chi\in V^{-1}} \mathfrak{g}_{\mathbf{R}}(\chi) = \mathrm{Lie}(P_V),$$
$$A_2 \in \sum_{\chi\in V} \mathfrak{g}_{\mathbf{R}}(\chi) = \theta(\mathrm{Lie}(P_V)),$$
$$A_3 \in \mathrm{Lie}(P_V)\cap\theta(\mathrm{Lie}(P_V)),$$

by the definition of P_V (10.2.11) and by $\theta(\mathfrak{g}_{\mathbf{R}}(\chi)) = \mathfrak{g}_{\mathbf{R}}(\chi^{-1})$. Define

$$\eta_3(t,a,b) := s\tilde{\rho}(t^\circ)x \in D_{\mathrm{BS,val}}(P_V),$$

where $s\in P_V$, $t^\circ \in \mathbf{R}^n_{>0}$ and $x\in D_{\mathrm{BS,val}}(P_V)$ are as follows:

$$s := \exp(A_1)\exp(\theta(A_2))^{-1}\exp(\theta(A_3)) \in P_V,$$
$$t^\circ \in \mathbf{R}^n_{>0} : \text{any element such that } \chi(t^\circ) = h(\chi)\ (\forall \chi\in V^\times),$$
$$x := (T, Z, V'),$$

where $T := \tilde{\rho}(T')$ with T' the subtorus of $\mathbf{G}^n_{\mathbf{m},\mathbf{R}}$ defined to be the annihilator of $V^\times$, Z is the T-orbit $T\cdot b\subset D$ and V' is the inverse image of $V/V^\times \subset X(T')$ in $X(T)$ via the isomorphism $T' \xrightarrow{\sim} T$ induced by $\tilde{\rho}$.

Then, as is easily seen, $\eta_3(t,a,b)$ is independent of the choice of t°. Take $P\in\Psi$ such that $P\subset P_V$ (10.2.11). Then, we have

$$\eta_3(t,a,b) = (sb,t) \in D\times^{\mathbf{R}^n_{>0}} (\mathbf{R}^n_{\geq 0})_{\mathrm{val}}(P) \subset D_{\mathrm{BS,val}}(P) \quad (10.2.10). \tag{1}$$

We prove that η_3 is continuous. Since the target space $D_{\mathrm{BS,val}}$ is a regular space, it is sufficient to show that, if $(t,a,b)\in Y_3$ and if $(t_\lambda,a_\lambda,b_\lambda)\in Y_0$ converges to (t,a,b), then $\eta_0(t_\lambda,a_\lambda,b_\lambda)$ converges in $D_{\mathrm{BS,val}}$ (6.4.7).

Let $a'_\lambda := \sum_{\chi\in X}(\chi(t_\lambda)+\chi(t_\lambda)^{-1})^{-1}a_\lambda(\chi)$. Since t_λ converges to $t=(V,h)$, we have, by 5.1.11 (3),

$$\chi(t_\lambda)\to h(\chi) \quad (\forall\,\chi\in X). \tag{2}$$

Hence a'_λ converges to $A_4 := \sum_{\chi\in V^\times}(h(\chi)+h(\chi)^{-1})^{-1}a(\chi) \in \mathrm{Lie}(P_V)$. We can write

$$\exp(a'_\lambda) = g_\lambda k_\lambda \quad (g_\lambda\in P_V,\ k_\lambda\in K_{\mathbf{r}}), \tag{3}$$

where g_λ converges to $\exp(A_4)$ in P_V and k_λ converges to 1 in $K_{\mathbf{r}}$. Take $P\in\Psi$ such that $P\subset P_V$ (10.2.11). Then

$$\begin{aligned}\eta_0(t_\lambda,a_\lambda,b_\lambda) &= \tilde{\rho}(t_\lambda)\exp(a'_\lambda)b_\lambda = \mathrm{Int}(\tilde{\rho}(t_\lambda))(g_\lambda)\tilde{\rho}(t_\lambda)k_\lambda b_\lambda\\ &= (\mathrm{Int}(\tilde{\rho}(t_\lambda))(g_\lambda)k_\lambda b_\lambda,\ t_\lambda)\\ &\quad \text{in}\quad D\times^{\mathbf{R}^n_{>0}}(\mathbf{R}^n_{\geq 0})_{\mathrm{val}}(P)\subset D_{\mathrm{BS,val}}(P) \quad (10.2.10),\end{aligned} \tag{4}$$

where t_λ (resp. k_λ, b_λ) converges to t in $(\overline{A}_P)_{\mathrm{val}}$ (resp. 1 in $K_{\mathbf{r}}$, b in $K_{\mathbf{r}} \cdot \mathbf{r}$). By comparing (1) and (4), it is sufficient to prove that $\mathrm{Int}(\rho(t_\lambda))(g_\lambda)$ converges to s in P_V. To prove this, we consider the equality

$$g_\lambda = \exp(a'_\lambda)\exp(\theta(a'_\lambda))^{-1}\theta(g_\lambda) \tag{5}$$

obtained from (3), and apply $\mathrm{Int}(\tilde{\rho}(t_\lambda))$ to the right-hand side of (5). We have, by (2),

$$\begin{aligned}\mathrm{Ad}(\tilde{\rho}(t_\lambda))(a'_\lambda) &= \sum_{\chi\in X}\frac{1}{1+\chi(t_\lambda)^{-2}}a_\lambda(\chi) \to A_1,\\ \mathrm{Ad}(\tilde{\rho}(t_\lambda))\theta(a'_\lambda) &= \theta(\mathrm{Ad}(\tilde{\rho}(t_\lambda))^{-1}(a'_\lambda))\\ &= \sum_{\chi\in X}\frac{1}{\chi(t_\lambda)^2+1}\theta(a_\lambda(\chi)) \to \theta(A_2),\end{aligned}$$

where we used the fact $\theta(v(\chi)) = \theta(v)(\chi^{-1})$ for $v \in \mathfrak{g}_{\mathbf{R}}$. Furthermore, since g_λ converges to $\exp(A_4)$ and since the eigenvalues of $\mathrm{Ad}(\tilde{\rho}(t_\lambda))^{-1}$ on $\mathrm{Lie}(P_V)$ converge, we have

$$\mathrm{Int}(\tilde{\rho}(t_\lambda))(\theta(g_\lambda)) = \theta(\mathrm{Int}(\tilde{\rho}(t_\lambda))^{-1}(g_\lambda)) \to \theta(\exp(A_3)) = \exp(\theta(A_3)).$$

Hence

$$\mathrm{Int}(\tilde{\rho}(t_\lambda))(g_\lambda) \to \exp(A_1)\exp(\theta(A_2))^{-1}\exp(\theta(A_3)) = s \quad \text{in } P_V,$$

as desired.

10.3.2

Let $\eta_2 : Y_2 \to D_{\mathrm{BS,val}}$ be the restriction of η_3 to Y_2. We prove

$$\eta_2(Y_2) \subset U(p)_{\mathrm{val}}, \tag{1}$$

where $U(p)_{\mathrm{val}}$ is the inverse image of $U(p) \subset D_{\mathrm{SL}(2)}$ (10.2.1 (3)) in $D_{\mathrm{SL}(2),\mathrm{val}}$.

Let $(t, a, b) \in Y_2$, let $J := \{j \mid t_j = 0\} \subset \{1, \dots, n\}$ where $t_j \in \mathbf{R}$ is the j-th component of the image of t in $\mathbf{R}^n_{\geq 0}$ and let $A_1, A_2, A_3 \in \mathfrak{g}_{\mathbf{R}}$ be as in 10.3.1. We prove first

$$A_1, \theta(A_2), \theta(A_3) \in \mathrm{Lie}(G_{W(J),\mathbf{R}}). \tag{2}$$

By the definitions of A_1, A_2, A_3, the χ-components of A_1, $\theta(A_2)$, $\theta(A_3)$ are zero unless $\chi^{-1} \in V$. Since $(t, a, b) \in Y_2$ (10.1.1), these χ components are zero unless $\chi \in X_J \cup X_J^{-1}$. Hence by (1) in 10.1.1, it is sufficient to prove $(X_J \cup X_J^{-1}) \cap V \subset X_J$. But this follows from

$$X_J \subset V, \quad X_J \cap V^\times = X_J \cap X_J^{-1}. \tag{3}$$

Next we prove that

(4) Each element of $G_{W(J),\mathbf{R}} \cap K_{\mathbf{r}}$ commutes with each element of $\tilde{\rho}(\mathbf{R}^J_{>0})$.

To prove (4), take a $\mathbf{Q}$-parabolic subgroup P of $G_{\mathbf{R}}$ such that the $(G^\circ)_{W(J),\mathbf{R}} \subset P$ and such that the image of $\tilde{\rho}(\mathbf{G}^J_{\mathbf{m},\mathbf{R}})$ in P/P_u is contained in S_P (cf. 10.2.10). Then (4) is reduced to Lemma 10.2.9.

Now we prove (1). Write $b = k\mathbf{r}$, $k \in G_{W(J),\mathbf{R}} \cap K_{\mathbf{r}}$ (10.1.1). We have $Z = T \cdot b = Tk \cdot \mathbf{r} = kT \cdot \mathbf{r}$ by (4), and hence

$$\eta_2(t, a, b) = s\tilde{\rho}(t^\circ)(T, Z, V') = s\tilde{\rho}(t^\circ)k(T, T \cdot \mathbf{r}, V')$$

with s, t° and $x = (T, Z, V')$ as in 10.3.1. We have $s\tilde{\rho}(t^\circ)k \in G_{W(J),\mathbf{R}}$ by (2), and $(T, T \cdot \mathbf{r}, V') = (p_J, T \cdot \mathbf{r}, V/V^\times) \in U(p)_{\mathrm{val}}$. Hence we have (1).

The map $\eta_2 : Y_2 \to U(p)_{\mathrm{val}}$ is continuous since the topology of $U(p)_{\mathrm{val}}$ (as a subspace of $D_{\mathrm{SL}(2),\mathrm{val}}$) coincides with the topology as a subspace of $D_{\mathrm{BS},\mathrm{val}}$ by 10.2.1 (3) and 5.2.13.

10.3.3

Since the image of $\eta_2(t, a, b)$ ($(t, a, b) \in Y_2$) under $U(p)_{\mathrm{val}} \to U(p)$ coincides with $s\tilde{\rho}(t^\circ)kp_J$ and this depends only on the image of (t, a, b) in Y_1, the composite map $Y_2 \xrightarrow{\eta_2} U(p)_{\mathrm{val}} \to U(p)$ factors through the canonical projection $Y_2 \to Y_1$. Thus η_2 induces a continuous map $\eta_1 : Y_1 \to U(p)$.

Lemma 10.3.4 *Let L be as in 10.1.2. Then, there exist an open neighborhood O of 0 in $\mathfrak{g}_{\mathbf{R}}$ and a real analytic function $f = (f_j)_{1\le j\le 3} : O \to \mathrm{Lie}(\tilde{\rho}(\mathbf{R}^n_{>0})) \times L \times \mathrm{Lie}(K_{\mathbf{r}})$ having the following properties* (1)–(4).

(1) *For any $x \in O$,* $\exp(x) = \exp(f_1(x))\exp(f_2(x))\exp(f_3(x))$.
(2) $f(0) = 0$.
(3) $\exp : O \to G_{\mathbf{R}}$ *is an injective open map.*
(4) *For $k = 2, 3$, f_k has the form of absolutely convergent series $f_k = \sum_{m=0}^{\infty} f_{k,m}$, where $f_{k,m}$ is the part of degree m in the Taylor expansion of f_k at 0, such that $f_{k,m}(x) = h_{k,m}(x \otimes \cdots \otimes x)$ for some linear map $h_{k,m} : \mathfrak{g}_{\mathbf{R}}^{\otimes m} \to \mathfrak{g}_{\mathbf{R}}$ having the following property: If $x_j \in \mathfrak{g}_{\mathbf{R}}(\chi_j)$ for $1 \le j \le m$ and for $\chi_j \in X$, then $h_{k,m}(x_1 \otimes \cdots \otimes x_m) \in \sum_{\psi} \mathfrak{g}_{\mathbf{R}}(\psi)$ where ψ ranges over all elements of X of the form $\prod_{1\le j\le m} \chi_j^{e_j}$ with $e_j = 1$ or -1 for each j.*

Proof. We apply the following. For $v \in \mathfrak{g}_{\mathbf{R}}$, write $v = v_1 + v_2 + v_3$ ($v_1 \in \mathrm{Lie}(\tilde{\rho}(\mathbf{R}^n_{>0}))$, $v_2 \in L$, $v_3 \in \mathrm{Lie}(K_{\mathbf{r}})$). If $\chi \in X$ and $v \in \mathfrak{g}_{\mathbf{R}}(\chi)$, then $v_2, v_3 \in \mathfrak{g}_{\mathbf{R}}(\chi) + \mathfrak{g}_{\mathbf{R}}(\chi^{-1})$ by the property 10.1.2 (2) of L. Furthermore

$$\exp(v) = \exp(v_1 + v_2 + v_3) = \exp(v_1)\exp(v_2')\exp(v_3)$$

with

$$v_2' = v_2 + \frac{1}{2}([v_2, v_1] + [v_3, v_1] + [v_3, v_2]) + \cdots .$$

By repeating this, replacing v_2 with v_2', we have 10.3.4. □

10.3.5

For $P \in \Psi$, define an open subset U_P of $D \times^{\mathbf{R}^n_{>0}} (\mathbf{R}^n_{\geq 0})_{\mathrm{val}}(P)$ in 10.2.10 by

$$U_P = \{(gb, t) \mid g \in P,\ b \in K_{\mathbf{r}} \cdot \mathbf{r},\ t \in (\mathbf{R}^n_{\geq 0})_{\mathrm{val}}(P),\ \mu(g, t) \in \exp(O)\} \quad (10.2.8).$$

For $j = 1, 2, 3$, define an open set U_j of $Y_{j,L}$ as follows. There exists an open neighborhood O' of 0 in L such that $\exp(O') \subset \exp(\mathrm{Lie}(P) \cap O) \exp(K_{\mathbf{r}})$ for any $P \in \Psi$. Fix such O'. Define U_j as the set of all $(t, a, b) \in Y_{j,L}$ such that $a \in O'$. Then $\eta_3(U_3) \subset \bigcup_{P \in \Psi} U_P$. More precisely, if $x = ((V, h), a, b) \in U_3$ and $P \in \Psi$ with $P \subset P_V$, then $\eta_3(x) \in U_P$.

To complete the proof of Theorem 10.1.3, it is sufficient to prove that the restriction of η_j to U_j is an injective open map for $j = 1, 2, 3$. To do it, a key is to define partial inverses η'_P of η_3 for $P \in \Psi$ as in 10.3.7 below.

Lemma 10.3.6 *Let $P \in \Psi$. Then there exists a continuous map*

$$\nu : \{(g, t) \in P \times (\mathbf{R}^n_{\geq 0})_{\mathrm{val}}(P) \mid \mu(g, t) \in \exp(O)\} \to L,$$
$$\textit{such that } \ \nu(g, t) = h\big(f_2(\log(\mu(g, t))), t\big) \ \textit{ if } \ t \in \mathbf{R}^n_{>0},$$
$$\textit{where } \ h(v, t) = \sum_{\chi \in X} (\chi(t) + \chi(t)^{-1}) v(\chi) \quad (v \in \mathfrak{g}_{\mathbf{R}},\ t \in \mathbf{R}^n_{>0}).$$

Here log : $\exp(O) \to O$ *denotes the inverse of* exp.

Proof. Since the target space L is a regular space, it is sufficient to prove that, if $g \in P$, $t \in (\mathbf{R}^n_{\geq 0})_{\mathrm{val}}(P)$, with $\mu(g, t) \in \exp(O)$, and if $(g_\lambda, t_\lambda)_\lambda$ converges to (g, t) satisfying $g_\lambda \in P$, $t_\lambda \in \mathbf{R}^n_{>0}$, with $\mu(g_\lambda, t_\lambda) \in \exp(O)$, then $h\big(f_2(\log(\mu(g_\lambda, t_\lambda))), t_\lambda\big)$ converges.

Let B be the set of $(\chi, \psi) \in X \times X$ such that $\psi \in \chi X(S_P)_+ \cap \chi^{-1} X(S_P)_+$. (Note that $\chi X(S_P)_+ \cap \chi^{-1} X(S_P)_+ \subset X(S_P)_+$. In fact, if $\alpha, \beta \in X(S_P)_+$ and $\psi = \chi\alpha = \chi^{-1}\beta$, then $\psi^2 = \chi\alpha\chi^{-1}\beta = \alpha\beta \in X(S_P)_+$ and hence $\psi \in X(S_P)_+$.)

By the property (4) of $f_{2,m}$ in 10.3.4, for some open neighborhood U of 0 in $\mathrm{Lie}(P)$ and for some open neighborhood U' of t in $(\mathbf{R}^n_{\geq 0})_{\mathrm{val}}(P)$,

$$f_2(\log(\mu(g \exp(v), t'))) = \sum_{(\chi, \psi) \in B} c_{\chi, \psi} \psi(t') f_\chi(v) \quad (v \in U,\ t' \in U' \cap \mathbf{R}^n_{>0}),$$

where $f_\chi : U \to \mathfrak{g}_{\mathbf{R}}(\chi)$ is a real analytic function satisfying $f_\chi(0) = 0$, $c_{\chi, \psi} \in \mathbf{C}$ and $\sum$ converges absolutely. Write $g_\lambda = g \exp(v_\lambda)$ with $v_\lambda \to 0$. Then

$$h\big(f_2(\log(\mu(g \exp(v_\lambda), t_\lambda))), t_\lambda\big) = \sum_{(\chi, \psi)} c_{\chi, \psi} (\chi(t_\lambda)\psi(t_\lambda) + \chi(t_\lambda)^{-1}\psi(t_\lambda)) f_\chi(v_\lambda)$$

converges, since $\chi(t_\lambda)\psi(t_\lambda)$ and $\chi^{-1}(t_\lambda)\psi(t_\lambda)$ converge because $\chi\psi, \chi^{-1}\psi \in X(S_P)_+ \subset V$. $\square$

10.3.7

Define a continuous map $\eta'_P : U_P \to Y_3$ by

$$\eta'_P(gb,t) := \big(\exp(f_1(\log(\mu(g,t)))),\ \nu(g,t),\ \exp(f_3(\log(\mu(g,t))))b\big)$$
$$(g \in P,\ b \in K_{\mathbf{r}} \cdot \mathbf{r},\ t \in (\mathbf{R}^n_{\geq 0})_{\mathrm{val}}(P)),\ \text{ with } \mu(g,t) \in \exp(O)).$$

This is well defined by 10.2.8 (ii).

LEMMA 10.3.8 *$\eta_3\eta'_P(x) = x$ for all $x \in U_P$.*

This is clear.

LEMMA 10.3.9 *If $x \in U_3$ and $\eta_3(x) \in U_P$, then $\eta'_P\eta_3(x) = x$.*

Proof. This is true when $x \in Y_0$, as is easily seen. The general case follows from this by density. □

LEMMA 10.3.10 *The restriction of η_3 to U_3 is injective.*

Proof. Assume $x, y \in U_3$ and $\eta_3(x) = \eta_3(y)$. Then, by the definition of η_3, the valuative submonoid V of X associated to x coincides with that of y. Take $P \in \Psi$ such that $P \subset P_V$. Then, by 10.3.9,

$$x = \eta'_P\eta_3(x) = \eta'_P\eta_3(y) = y.$$ □

LEMMA 10.3.11 *The restriction of η_3 to U_3 is an open map.*

Proof. Let U be an open set of U_3. It is sufficient to prove that if $x \in U$, then $\eta_3(U)$ is a neighborhood of $\eta_3(x)$. Take V, P for x as in the proof of 10.3.10. Let $U' = \eta'^{-1}_P(U) \subset U_P$. Then, U' is open, $\eta_3(x) \in U'$ and $U' = \eta_3\eta'_P(U') \subset \eta_3(U)$. □

LEMMA 10.3.12 *The restriction of η_2 to U_2 is injective.*

Proof. This is clear by 10.3.10. □

LEMMA 10.3.13 *$\eta'_P(U(p)_{\mathrm{val}} \cap U_P) \subset Y_2$.*

Proof. Let $x \in U(p)_{\mathrm{val}} \cap U_P$ and take $P \in \Psi$ such that $P \subset P_V$ where V is the valuative submonoid of X associated to x. Let J be the subset of $\{1, \ldots, n\}$ such that the family of weight filtrations associated with the image of x in $D_{\mathrm{SL}(2)}$ is $W^{(J)}$. Then x is written in the form (gb, t) with $g \in G_{W^{(J)},\mathbf{R}} \cap P$, $b \in (G_{W^{(J)},\mathbf{R}} \cap K_{\mathbf{r}}) \cdot \mathbf{r}$, $t \in (\mathbf{R}^n_{\geq 0})_{\mathrm{val}}$. Let $k = 2$ or 3. Let C be the subset of $X \times X$ consisting of all $(\chi, \psi) \in X \times X$ such that $\psi\chi, \psi\chi^{-1} \in X_J$. By the property (4) of $f_{k,m}$ in 10.3.4, for some open neighborhood U of t in $(\mathbf{R}^n_{\geq 0})_{\mathrm{val}}(P)$,

$$f_k(\log(\mu(g,t'))) = \sum_{(\chi,\psi) \in C} \psi(t')c_{k,\chi,\psi} \quad (t' \in U \cap \mathbf{R}^n_{>0}),$$

where $c_{k,\chi,\psi} \in \mathfrak{g}_{\mathbf{R}}(\chi)$ and $\sum$ converges absolutely.

We prove that $\eta'_P(x) \in Y_2$. First we prove that the third component (i.e., $(K_{\mathbf{r}} \cdot \mathbf{r})$-component) of $\eta'_P(x)$ is contained in $(G_{W^{(J)},\mathbf{R}} \cap K_{\mathbf{r}}) \cdot \mathbf{r}$. For this, it is sufficient to prove that $f_3(\log(\mu(g,t))) \in \mathrm{Lie}(G_{W^{(J)},\mathbf{R}} \cap K_{\mathbf{r}})$. It is the limit of $\sum_{(\chi,\psi)\in C} \chi(t')c_{3,\chi,\psi}$ where t' converges to t. Let $(\chi, \psi) \in C$. Then $\psi \in X_J \subset V$ (10.3.2 (3)). If $\psi \notin V^\times$, then $\psi(t')$ converges to 0. If $\psi \in V^\times$, then $\psi(t')$ converges, and $\psi \in X_J \cap V^\times = X_J \cap X_J^{-1}$ (10.3.2 (3)) and hence $\chi \in X_J$. This proves $f_3(\log(\mu(g,t))) \in \mathrm{Lie}(G_{W(J),\mathbf{R}} \cap K_{\mathbf{r}})$.

It remains to prove that the second component (i.e., L-component) $\nu(g,t)$ of $\eta'_p(x)$ is contained in $\sum_{\chi \in X_J \cup X_J^{-1}} \mathfrak{g}_{\mathbf{R}}(\chi)$. When t' converges to t, $\nu(g,t') = \sum_{(\chi,\psi)\in C}(\chi(t')+\chi(t')^{-1})\psi(t')c_{2,\chi,\psi}$ converges to $\nu(g,t)$. Let $(\chi,\psi) \in C$. Then $\chi\psi, \chi^{-1}\psi \in X_J \subset V$ and hence $\chi(t')\psi(t')$ and $\chi(t')^{-1}\psi(t')$ converge. These converge to 0 if $\chi\psi \notin V^\times$ and $\chi^{-1}\psi \notin V^\times$.

If either one of $\chi\psi, \chi^{-1}\psi$ belongs to $V^\times$, then either one of $\chi\psi, \chi^{-1}\psi$ belongs to $V^\times \cap X_J = X_J \cap X_J^{-1}$. Since $\psi \in X_J$, we see that either $\chi \in \psi^{-1}\cdot(X_J \cap X_J^{-1}) \subset X_J^{-1}$ or $\chi \in \psi \cdot (X_J \cap X_J^{-1}) \subset X_J$. This proves $\nu(g,t) \in \sum_{\chi \in X_J \cup X_J^{-1}} \mathfrak{g}_{\mathbf{R}}(\chi)$. □

Lemma 10.3.14 *The restriction of η_2 to U_2 is an open map.*

Proof. By virtue of Lemma 10.3.13, this can be proved in the same way as Lemma 10.3.11. □

Lemma 10.3.15 *The restriction of η_1 to U_1 is an injective open map.*

Proof. By the description of the restriction of η_2 to U_2 given in 10.3.3, we see that the following diagram is set-theoretically cartesian.

$$\begin{array}{ccc} U_2 & \xrightarrow{\eta_2} & U(p)_{\mathrm{val}} \\ \downarrow & & \downarrow \\ U_1 & \xrightarrow{\eta_1} & U(p). \end{array} \qquad (1)$$

We prove that the restriction of η_1 to U_1 is injective. If $x, y \in U_1$ and $\eta_1(x) = \eta_1(y)$, then there exists $z \in U(p)_{\mathrm{val}}$ whose image in $U(p)$ is $\eta_1(x) = \eta_1(y)$. Since (1) is set-theoretically cartesian, there exist $u, v \in U_2$ whose images in U_1 are x, y, respectively, and $z = \eta_2(u) = \eta_2(v)$. By 10.3.12, u and v coincide, and hence so do x and y.

We prove that the restriction of η_1 to U_1 is an open map. Let U be an open set of U_1. Since the topology of $U(p)$ is the quotient of that of $U(p)_{\mathrm{val}}$, to see that $\eta_1(U)$ is open in $U(p)$, it is sufficient to prove that the inverse image of $\eta_1(U)$ in $U(p)_{\mathrm{val}}$ is open in $U(p)_{\mathrm{val}}$. Since (1) is set-theoretically cartesian, this inverse image coincides with the image under η_2 of the inverse image of U in U_2, and hence it is open by 10.3.14. □

10.4 LOCAL STRUCTURES OF $D_{\mathrm{SL}(2),\le 1}$ AND $\Gamma\backslash D^\flat_{\mathrm{SL}(2),\le 1}$

In this section, we describe local structures of $D_{\mathrm{SL}(2)}$ around a point of rank 1 and local structures of $\Gamma\backslash D^\flat_{\mathrm{SL}(2),\le 1}$ for a subgroup Γ of $G_{\mathbf{Z}}$ of finite index.

Proposition 10.4.1 *Let $p \in D_{\mathrm{SL}(2)}$ be a point of rank 1. Let (ρ,φ) be an* SL(2)*-orbit in one variable representing p, let $\mathbf{r} = \varphi(i)$ and let W be the weight filtration associated with p. Fix a closed subgroup S' of $S := \{g \in G_{\mathbf{R}} \mid g\tilde\rho(t) = \tilde\rho(t)g \ (\forall t \in \mathbf{R}^\times)\}$ such that S is the direct product of $\tilde\rho(\mathbf{R}_{>0})$ and S'. Then we have a* *homeomorphism*

$$U(p) \xrightarrow{\sim} \{(t, u, a) \mid \textit{if } t = 0, \textit{ then } a \in S' \cdot \mathbf{r}\} \subset (\mathbf{R}_{\geq 0}) \times G_{W,\mathbf{R},u} \times (S'K_{\mathbf{r}} \cdot \mathbf{r}),$$

$$\begin{cases} u\tilde{\rho}(t)sk\mathbf{r} \mapsto (t, u, sk\mathbf{r}), \\ usp \mapsto (0, u, s\mathbf{r}), \end{cases} \qquad (t \in \mathbf{R}_{>0},\ u \in G_{W,\mathbf{R},u},\ s \in S',\ k \in K_{\mathbf{r}}).$$

Proof. This follows from Theorem 10.1.3 and the definition of η_1 given in 10.3.3. □

10.4.2

For a topological space X, let $C(X)$ be the quotient of $\mathbf{R}_{\geq 0} \times X$ obtained by collapsing $\{0\} \times X$ into one point. We denote the image of $\{0\} \times X$ in $C(X)$ simply by 0.

Proposition 10.4.3 *Let $p \in D_{\mathrm{SL}(2)}$ be a point of rank* 1 *and let* (ρ, φ), $\mathbf{r}$, W, S' *be as in Proposition* 10.4.1. *Let Γ be a neat subgroup of $G_{\mathbf{Z}}$ of finite index.*

(i) *Let $\overline{U}(p)$ be the image of $U(p)$ in $D^{\flat}_{\mathrm{SL}(2),\leq 1}$. Endow $(\Gamma \cap G_{W,\mathbf{R},u})\backslash \overline{U}(p)$ with the quotient topology via $U(p) \to (\Gamma \cap G_{W,\mathbf{R},u})\backslash \overline{U}(p)$. Then*

$$(\Gamma \cap G_{W,\mathbf{R},u})\backslash \overline{U}(p) \to \Gamma \backslash D^{\flat}_{\mathrm{SL}(2),\leq 1}$$

is a local homeomorphism. Furthermore, we have a homeomorphism

$$(\Gamma \cap G_{W,\mathbf{R},u})\backslash \overline{U}(p) \xrightarrow{\sim} \{(x, a) \mid \textit{if } x = 0, \quad \textit{then } a \in S' \cdot \mathbf{r}\}$$
$$\subset C\big((\Gamma \cap G_{W,\mathbf{R},u})\backslash G_{W,\mathbf{R},u}\big) \times S'K_{\mathbf{r}} \cdot \mathbf{r}$$

induced by the homeomorphism in Proposition 10.4.1.

(ii) *Let* $l = \dim_{\mathbf{R}}(K_{\mathbf{r}} \cdot \mathbf{r}) - \dim_{\mathbf{R}}((K_{\mathbf{r}} \cap G_{W,\mathbf{R}}) \cdot \mathbf{r})$, $m = \dim_{\mathbf{R}} G_{W,\mathbf{R},u}$, $n = \dim_{\mathbf{R}} D - m - 1$. *Then $n \geq l$. There exists a homeomorphism from an open neighborhood of the image of p in $\Gamma\backslash D^{\flat}_{\mathrm{SL}(2),\leq 1}$ onto an open neighborhood of* $(0, \mathbf{0})$ *in the space*

$$\{(x, (a_j)_{1\leq j\leq n}) \in C((\Gamma \cap G_{W,\mathbf{R},u})\backslash G_{W,\mathbf{R},u})$$
$$\times \mathbf{R}^n \mid \textit{if } x = 0, \quad \textit{then } a_j = 0 \textit{ for } 1 \leq j \leq l\}.$$

Proof. The first statement in (i) follows from Theorem 9.1.4 (ii), and the second statement follows from the first statement and from Proposition 10.4.1. (ii) follows from (i). □

10.4.4

For example, in the classical case of the upper half plane, if p denotes the point $i\infty$, then Proposition 10.4.1 shows that

$$U(p) \xrightarrow{\sim} \mathbf{R}_{\geq 0} \times \mathbf{R}, \quad x + iy \mapsto (\sqrt{y}^{-1}, x) \quad (x \in \mathbf{R},\ 0 < y \leq \infty),$$

and Proposition 10.4.3 shows that we have a homeomorphism from the image of $U(p)$ in $\Gamma\backslash D^{\flat}_{\mathrm{SL}(2)}$ onto $C(\mathbf{S}^1) \simeq \mathbf{C}$ which sends the image of $x + iy$ ($x \in \mathbf{R}, 0 < y \leq \infty$) to $\exp(2\pi i a(x + iy)) \in \mathbf{C}$, where a is the positive integer such that $\Gamma \cap G_{W,\mathbf{R},u}$ is generated by $z \mapsto z + a$.

Chapter Eleven

Moduli of PLH with Coefficients

In this chapter, we generalize Theorems A and B to the case of polarized logarithmic Hodge structures with coefficients. The key observation is Lemma 11.1.3. Then the generalized results Theorems 11.1.17 and 11.3.1 follow from Theorems A and B, respectively.

11.1 SPACE $\Gamma\backslash D_{\Sigma}^{A}$

11.1.1

In §11, let A be a semi-simple $\mathbf{Q}$-subalgebra of $\mathrm{End}_{\mathbf{Q}}(H_{0,\mathbf{Q}})$ such that, for any $a \in A$, the dual mapping ${}^t a : H_{0,\mathbf{Q}} \to H_{0,\mathbf{Q}}$ with respect to $\langle\ ,\ \rangle_0$ belongs to A. Here ${}^t a$ is characterized by $\langle ax, y\rangle_0 = \langle x, {}^t ay\rangle_0$ for any $x, y \in H_{0,\mathbf{Q}}$.

11.1.2

Let $\check{D}^A \subset \check{D}$ be the subset of $\check{D}$ consisting of all $F \in \check{D}$ such that the F^p are A-submodules of $H_{0,\mathbf{C}}$ for all $p \in \mathbf{Z}$. Let

$$D^A := D \cap \check{D}^A \quad \text{in } \check{D}.$$

For $R = \mathbf{Z}, \mathbf{Q}, \mathbf{R}, \mathbf{C}$, let

$$G_R^A := \{g \in G_R \mid ga = ag \ \ (\forall a \in A)\}.$$

For $R = \mathbf{Q}, \mathbf{R}, \mathbf{C}$, let

$$\mathfrak{g}_R^A := \{X \in \mathfrak{g}_R \mid Xa = aX \ \ (\forall a \in A)\}.$$

Lemma 11.1.3 *As a topological subspace of $\check{D}$, $\check{D}^A$ is a disjoint union of a finite number of $G_{\mathbf{C}}^A$-orbits.*

Proof. Write $A_{\mathbf{C}} = \prod_{j\in J} B_j$, where J is a finite index set and the B_j are simple algebras over $\mathbf{C}$. Each B_j is isomorphic over $\mathbf{C}$ to the algebra $M(n_j, \mathbf{C})$ of (n_j, n_j)-matrices with entries in $\mathbf{C}$ for some $n_j \geq 1$. We have a direct decomposition $H_{0,\mathbf{C}} = \bigoplus_{j\in J} V_j$, where V_j is the part of $H_{0,\mathbf{C}}$ on which $A_{\mathbf{C}}$ acts through B_j. For $F \in \check{D}^A$, we have $F = \bigoplus_{j\in J} F(V_j)$. It is sufficient to show that, if $F, F' \in \check{D}^A$ and $\dim F^p(V_j) = \dim F'^p(V_j)$ for any p, j, then there exists $g \in G_{\mathbf{C}}^A$ such that $F' = gF$. We define $g := \bigoplus_{j\in J} g_j$, where $g_j : V_j \to V_j$ are constructed as follows.

For $j \in J$, as a subspace of $A_{\mathbf{C}}$, we have ${}^tB_j = B_k$ for some $k \in J$. For $l \in J$, we have

$$\langle V_j, V_l\rangle_0 = \langle B_j V_j, V_l\rangle_0 = \langle V_j, B_k V_l\rangle_0.$$

Hence $\langle V_j, V_l\rangle_0 = 0$ for $l \neq k$ and the restriction of $\langle\ ,\ \rangle_0$ to $V_j \times V_k$ is nondegenerate. Let E_j be a simple B_j-module which is unique up to isomorphisms. For each $j \in J$, since B_j is isomorphic to $M(n_j, \mathbf{C})$, two functors

$$M \mapsto \theta_j(M) := \mathrm{Hom}_{B_j}(E_j, M),$$
$$M \mapsto \theta'_j(M) := E_k \otimes_{B_j} M$$

are both equivalences from the category of B_j-modules to the category of $\mathbf{C}$-modules. Here, in the definition of θ'_j, E_k is regarded as a right B_j-module on which $b \in B_j$ acts by ${}^tb$. Furthermore, there is an isomorphism $\theta_j \simeq \theta'_j$ of functors. We fix such an isomorphism. Then $\langle\ ,\ \rangle_0 : V_j \times V_k \to \mathbf{C}$ induces a pairing

$$\begin{aligned}(\ ,\ ) : \theta_j(V_j) \otimes \theta_k(V_k) &\simeq \theta'_j(V_j) \otimes \theta_k(V_k)\\ &= E_k \otimes_{B_j} V_j \otimes \mathrm{Hom}_{B_k}(E_k, V_k) \xrightarrow{\nu} V_j \otimes V_k \to \mathbf{C}\end{aligned}$$

where $\nu(a \otimes b \otimes h) = b \otimes h(a)$. Since the annihilator of F^p with respect to $\langle\ ,\ \rangle_0$ is F^{w+1-p}, the annihilator of $F^p(V_j)$ in V_k with respect to $\langle\ ,\ \rangle_0$ is $F^{w+1-p}(V_k)$. It follows that the annihilator of $\theta_j(F^p(V_j))$ in $\theta_k(V_k)$ with respect to $(\ ,\ )$ is $\theta_k(F^{w+1-p}(V_k))$. The same holds for F'.

Let $J_1 := \{j \in J \mid {}^tB_j = B_j\}$ and take subsets J_2 and J_3 of J such that J is the disjoint union of J_1, J_2, J_3 and that, if $j \in J_2$ (resp. J_3), then ${}^tB_j = B_k$ for some $k \in J_3$ (resp. J_2). We define g_j $(j \in J)$.

First assume $j \in J_1$. Then, the annihilator of $\theta_j(F^p(V_j))$ in $\theta_j(V_j)$ with respect to $(\ ,\ )$ is $\theta_j(F^{w+1-p}(V_j))$ for any p, the same holds to F', and $\dim \theta_j(F^p(V_j)) = \dim \theta_j(F'^p(V_j))$ for all p. Hence, by the same reason as the transitivity of the action of $G_{\mathbf{C}}$ on $\check{D}$, there is an automorphism g_j of $(\theta_j(V_j), (\ ,\ ))$ such that $\theta_j(F'(V_j)) = g_j(\theta_j(F(V_j)))$. Denote the B_j-automorphism of V_j corresponding to g_j in the categorical equivalence, by the same letter g_j. This is the definition of g_j. Then, $F'(V_j) = g_j(F(V_j))$ and g_j preserves the restriction of $\langle\ ,\ \rangle_0$ to V_j.

Next assume $j \in J_2$, ${}^tB_j = B_k$ $(k \in J_3)$. We define g_j and g_k as follows. Since $\dim \theta_j(F^p(V_j)) = \dim \theta_j(F'^p(V_j))$ for all p, there exists an automorphism g_j of $\theta_j(V_j)$ such that $\theta_j(F'(V_j)) = g_j(\theta_j(F(V_j)))$. Denote the B_j-automorphism of V_j corresponding to g_j by the same letter g_j, and denote the inverse of ${}^tg_j : V_k \to V_k$ by g_k. These are the definitions of g_j and g_k. Then, $F'(V_j) = g_j(F(V_j))$, $F'(V_k) = g_k(F(V_k))$, and $g_j \oplus g_k$ preserves the restriction of $\langle\ ,\ \rangle_0$ to $V_j \oplus V_k$. □

By 11.1.3, $\check{D}^A$ is an analytic manifold and D^A is an open submanifold of $\check{D}^A$.

11.1.4

In the classical situation (8.2.7), for a subgroup Γ of $G^A_{\mathbf{Z}}$ of finite index, $\Gamma\backslash D^A$ is a Shimura variety [D3].

11.1.5

By an *A-nilpotent cone*, we mean a nilpotent cone that is contained in $\mathfrak{g}_{\mathbf{R}}^A$. By an *$A$-fan in $\mathfrak{g}_{\mathbf{Q}}$*, we mean a fan in $\mathfrak{g}_{\mathbf{Q}}$ consisting of A-nilpotent cones. For an A-fan Σ in $\mathfrak{g}_{\mathbf{Q}}$, we define the subset D_Σ^A of D_Σ by

$$D_\Sigma^A := \{(\sigma, Z) \in D_\Sigma \mid Z \subset \check{D}^A\}.$$

We have

$$D_\Sigma^A = \bigcup_{\sigma \in \Sigma} D_\sigma^A \quad \text{where} \quad D_\sigma^A = D_{\{\text{face of } \sigma\}}^A.$$

We define $D_\Sigma^{\sharp, A} \subset D_\Sigma^\sharp$ similarly.

11.1.6

Let Σ be an A-fan, and let Γ be a subgroup of $G_{\mathbf{Z}}^A$ which is strongly compatible with Σ (as a subgroup of $G_{\mathbf{Z}}$).

For $\sigma \in \Sigma$, let

$$\check{E}_\sigma^A := \text{toric}_\sigma \times \check{D}^A \subset \check{E}_\sigma, \quad E_\sigma^A := E_\sigma \cap \check{E}_\sigma^A, \quad E_\sigma^{\sharp, A} = E_\sigma^\sharp \cap \check{E}_\sigma^A \ \text{ in } \ \check{E}_\sigma.$$

We endow E_σ^A with the following structure of a logarithmic local ringed space over $\mathbf{C}$. The topology of E_σ^A is the strong topology in $\check{E}_\sigma^A$ and the $\mathcal{O}$ (resp. M) of E_σ^A is the inverse image of $\mathcal{O}$ (resp. M) of $\check{E}^A$. We endow $E_\sigma^{\sharp, A}$ with the topology as a subspace of E_σ^A. Then the topology and the logarithmic structure of E_σ^A are the inverse images of those of E_σ, and the topology of $E_\sigma^{\sharp, A}$ is the inverse image of that of $E_\sigma^\sharp$.

The surjection $E_\sigma \to \Gamma(\sigma)^{\text{gp}} \backslash D_\sigma$ induces a surjection $E_\sigma^A \to \Gamma(\sigma)^{\text{gp}} \backslash D_\sigma^A$.

We endow $\Gamma \backslash D_\Sigma^A$ with the following structure of a logarithmic local ringed space over $\mathbf{C}$. A subset U of $\Gamma \backslash D_\Sigma^A$ is open if and only if, for any $\sigma \in \Sigma$, the inverse image U_σ of U in E_σ^A is open. For an open set U of $\Gamma \backslash D_\Sigma^A$, $\mathcal{O}(U)$ (resp. $M(U)$) is the set of all $\mathbf{C}$-valued functions on U whose pullback on U_σ belongs to $\mathcal{O}(U_\sigma)$ (resp. $M(U_\sigma)$) for any $\sigma \in \Sigma$.

We endow $D_\Sigma^{\sharp, A}$ with the strongest topology for which $E_\sigma^{\sharp, A} \to D_\Sigma^{\sharp, A}$ are continuous for all $\sigma \in \Sigma$.

Theorem 11.1.7 *Let Σ be an A-fan in $\mathfrak{g}_{\mathbf{Q}}$ and let Γ be a neat subgroup of $G_{\mathbf{Z}}^A$ which is strongly compatible with Σ. Then we have*

(i) *E_σ^A $(\sigma \in \Sigma)$ and $\Gamma \backslash D_\Sigma^A$ are logarithmic manifolds.*

(ii) *For $\sigma \in \Sigma$, E_σ^A is a $\sigma_{\mathbf{C}}$-torsor over $\Gamma(\sigma)^{\text{gp}} \backslash D_\sigma^A$ in the category of logarithmic manifolds.*

(iii) *For $\sigma \in \Sigma$, the map*

$$\Gamma(\sigma)^{\text{gp}} \backslash D_\sigma^A \to \Gamma \backslash D_\Sigma^A$$

is open and locally an isomorphism of logarithmic manifolds.

(iv) *There exists a canonical homeomorphism*

$$(\Gamma\backslash D_\Sigma^A)^{\log} \simeq \Gamma\backslash D_\Sigma^{\sharp,A}.$$

The rest of this section is devoted to the proof of Theorem 11.1.7.

11.1.8

First, we prove that E_σ^A is a logarithmic manifold.

Locally on $\check{E}_\sigma$, $\tilde{E}_\sigma$ is the set of common "zeros" of a finite set of logarithmic differential forms on $\check{E}_\sigma$ in the sense of 3.5.7 and 0.4.17. Hence locally on $\check{E}_\sigma^A$, $\tilde{E}_\sigma^A := \tilde{E}_\sigma \cap E_\sigma^A$ is the set of common "zeros" (in the sense of 3.5.7 and 0.4.17) of a finite set of logarithmic differential forms on the logarithmically smooth fs logarithmic analytic space $\check{E}_\sigma^A$. Since E_σ is open in $\tilde{E}_\sigma$ in the strong topology, E_σ^A is open in $\tilde{E}_\sigma^A$ in the strong topology.

11.1.9

We prove that $\Gamma\backslash D_\Sigma^A$ is a logarithmic manifold, and prove 11.1.7 (ii) and 11.1.7 (iii).

Let $(\sigma, Z) \in D_\Sigma^A$, let $F_{(0)} \in Z$ and let p be the image of (σ, Z) in $\Gamma\backslash D_\Sigma$. Let $Y \subset \check{D}$ and $S \subset \check{E}_\sigma$ be as in 7.3.11, and take an open neighborhood U_1 of $(0, F_{(0)})$ in S and an open neighborhood U_2 of p in $\Gamma\backslash D_\Sigma$ such that $U_1 \subset E_\sigma$, that the map $E_\sigma \to \Gamma\backslash D_\Sigma$ induces an isomorphism $U_2 \xrightarrow{\sim} U_1$ and that $\sigma_{\mathbf{C}} \times U_1 \to E_\sigma$ is an open immersion. Let $U_1^A := U_1 \cap \check{E}_\sigma^A$, $U_2^A := U_2 \cap \Gamma\backslash D_\Sigma^A$. Then U_1^A is a logarithmic manifold and $U_1^A \xrightarrow{\sim} U_2^A$ follows from $U_1 \xrightarrow{\sim} U_2$. This proves that $\Gamma\backslash D_\Sigma$ is a logarithmic manifold and also proves 11.1.7 (iii). Furthermore, $\sigma_{\mathbf{C}} \times U_1^A \to E_\sigma^A$ is an open immersion. Hence 11.1.7 (ii) follows.

11.1.10

We prove 11.1.7 (iv).

By 11.1.7 (ii), 11.1.7 (iii), the logarithmic structure of $\Gamma\backslash D_\Sigma^A$ is the inverse image of the logarithmic structure of $\Gamma\backslash D_\Sigma$. Hence $(\Gamma\backslash D_\Sigma^A)^{\log}$ is the fiber product of $(\Gamma\backslash D_\Sigma)^{\log}$ and $\Gamma\backslash D_\Sigma^A$ over $\Gamma\backslash D_\Sigma$. By Theorem A (vi), it is the fiber product of $\Gamma\backslash D_\Sigma^\sharp$ and $\Gamma\backslash D_\Sigma^A$ over $\Gamma\backslash D_\Sigma$, which is homeomorphic to $\Gamma\backslash D^{\sharp,A}$.

This completes the proof of Theorem 11.1.7.

11.2 PLH WITH COEFFICIENTS

11.2.1

Let Σ be an A-fan in $\mathfrak{g}_{\mathbf{Q}}$ and let Γ be a subgroup of $G_{\mathbf{Z}}^A$ which is strongly compatible with Σ. As in 2.5.8, we denote by

$$\Phi := (w, (h^{p,q})_{p,q\in\mathbf{Z}}, H_0, \langle\ ,\ \rangle_0, \Gamma, \Sigma). \qquad (1)$$

11.2.2

Let X be an object of $\overline{\mathcal{A}}_2(\log)$ (3.2.4). By a *polarized logarithmic Hodge structure of type Φ with coefficients in A (A-PLH of type Φ*, for short) on X, we mean a PLH $(H_{\mathbf{Z}}, \langle\ ,\ \rangle, F, \mu)$ on X of type Φ endowed with an A-module structure on $H_{\mathbf{Q}}$ satisfying the following conditions (1) and (2).

(1) The F^p are A-submodules of $\mathcal{O}_X^{\log} \otimes H_{\mathbf{Z}}$ for all $p \in \mathbf{Z}$.
(2) The image of μ under $\Gamma\backslash \underline{\mathrm{Isom}}((H_{\mathbf{Z}}, \langle\ ,\ \rangle), (H_0, \langle\ ,\ \rangle_0)) \to \Gamma\backslash \underline{\mathrm{Isom}}(H_{\mathbf{Q}}, H_{0,\mathbf{Q}})$ belongs to $\Gamma\backslash \underline{\mathrm{Isom}}_A(H_{\mathbf{Q}}, H_{0,\mathbf{Q}})$.

LEMMA 11.2.3 *Let X be an object of $\overline{\mathcal{A}}_2(\log)$.*

(i) *A PLH $(H_{\mathbf{Z}}, \langle\ ,\ \rangle, F, \mu)$ on X of type Φ can be endowed with a structure of an A-PLH of type Φ if and only if the following condition* (1) *is satisfied.*

(1) *For any local representative $\tilde{\mu} : H_{\mathbf{Z}} \to H_0$ of μ, the $\tilde{\mu}(F^p)$ is A-submodules of $\mathcal{O}_X^{\log} \otimes H_0$.*

If (1) *is satisfied, the action of $a \in A$ on $H_{\mathbf{Q}}$, in the unique structure of A-PLH of type Φ, is given by $a \mapsto \tilde{\mu}^{-1} \circ a \circ \tilde{\mu}$.*

(ii) *The forgetful functor from the category of A-PLH on X of type Φ to the category of PLH on X of type Φ is fully faithful.*

Proof. Easy. □

11.2.4

By a *polarized logarithmic Hodge structure with coefficients in A (A-PLH)* on X, we mean a PLH $(H_{\mathbf{Z}}, \langle\ ,\ \rangle, F)$ on X endowed with an A-module structure on $H_{\mathbf{Q}}$ satisfying the conditions (1) in 11.2.2 and the condition

$$\langle ax, y\rangle = \langle x, {}^t a y\rangle \quad (\forall a \in A). \tag{1}$$

An A-PLH of type Φ is an A-PLH. In fact, the last condition (1) is satisfied since, for a local representative $\tilde{\mu}$ of μ, we have

$$\begin{aligned}\langle ax, y\rangle &= \langle \tilde{\mu}(ax), \tilde{\mu}(y)\rangle_0 = \langle a\tilde{\mu}(x), \tilde{\mu}(y)\rangle_0 \\ &= \langle \tilde{\mu}(x), {}^t a\tilde{\mu}(y)\rangle_0 = \langle \tilde{\mu}(x), \tilde{\mu}({}^t ay)\rangle_0 = \langle x, {}^t ay\rangle.\end{aligned}$$

11.3 MODULI

THEOREM 11.3.1 *Let Σ be an A-fan in $\mathfrak{g}_{\mathbf{Q}}$ and let Γ be a neat subgroup of $G_{\mathbf{Z}}^A$ which is strongly compatible with Σ. Define a contravariant functor $\underline{\mathrm{PLH}}_\Phi^A$ from the category $\overline{\mathcal{A}}_2(\log)$ (3.2.4) to the category of sets by*

$$\underline{\mathrm{PLH}}_\Phi^A(X) := (\textit{isomorphism classes of } A\textit{-PLH on } X \textit{ of type } \Phi).$$

Then, $\underline{\mathrm{PLH}}_\Phi^A$ is represented by $\Gamma\backslash D_\Sigma^A$, i.e., there exists an isomorphism of functors

$$\underline{\mathrm{PLH}}_\Phi^A \simeq \mathrm{Mor}(\ \ , \Gamma\backslash D_\Sigma^A).$$

Proof. By Lemma 11.2.3 (ii), we have the injectivity of $\underline{\mathrm{PLH}}_{\Phi}^{A} \to \underline{\mathrm{PLH}}_{\Phi}$.

Let X be an object of $\overline{\mathcal{A}}_2(\log)$. We identify $\underline{\mathrm{PLH}}_{\Phi}(X)$ with $\mathrm{Mor}(X, \Gamma\backslash D_{\Sigma})$ by Theorem B. We prove that $\underline{\mathrm{PLH}}_{\Phi}^{A}(X)$ and $\mathrm{Mor}(X, \Gamma\backslash D_{\Sigma}^{A})$ coincide as subsets of $\underline{\mathrm{PLH}}_{\Phi}(X) = \mathrm{Mor}(X, \Gamma\backslash D_{\Sigma})$.

Let $h \in \underline{\mathrm{PLH}}_{\Phi}(X) = \mathrm{Mor}(X, \Gamma\backslash D_{\Sigma})$. Locally on X, h comes from $\tilde{h} \in \mathrm{Mor}(X, E_{\sigma})$ for some $\sigma \in \Sigma$. If h comes from $\tilde{h} \in \mathrm{Mor}(X, E_{\sigma})$, then, as is easily seen, h belongs to $\underline{\mathrm{PLH}}_{\Phi}^{A}(X)$ if and only if $\tilde{h} \in \mathrm{Mor}(X, E_{\sigma}^{A})$.

It remains to prove that, if h comes from $\tilde{h} \in \mathrm{Mor}(X, E_{\sigma})$, then $h \in \mathrm{Mor}(X, \Gamma\backslash D_{\Sigma}^{A})$ if and only if $\tilde{h} \in \mathrm{Mor}(X, E_{\sigma}^{A})$. The "if" part is clear. The "only if" part follows from Theorem 11.1.7. □

Chapter Twelve

Examples and Problems

In this chapter, we give some examples and open problems.

12.1 SIEGEL UPPER HALF SPACES

In this section, we consider Siegel upper half spaces. We describe some points at infinity of Siegel upper half spaces, and consider their behaviors in the map $\psi : D^{\#}_{\Sigma,\mathrm{val}} \to D_{\mathrm{SL}(2)}$ (Section 5.4, Chapter 6).

12.1.1

In Section 12.1, we assume $w = 1$, $h^{1,0} = h^{0,1} = g$, $h^{p,q} = 0$ if $(p, q) \neq (1, 0)$, $(0, 1)$, and there is a $\mathbf{Z}$-basis $(e_j)_{1\le j\le 2g}$ of H_0 such that

$$((\langle e_j, e_k\rangle_0)_{j,k} = \begin{pmatrix} 0 & -1_g \\ 1_g & 0 \end{pmatrix}.$$

By using this basis, we identify $H_{0,R} = R^{2g}$ for $R = \mathbf{Z}, \mathbf{Q}, \mathbf{R}, \mathbf{C}$. The algebraic group G is identified with

$$\mathrm{Sp}(g) = \left\{ \begin{pmatrix} A & B \\ C & D \end{pmatrix} \,\middle|\, {}^tAC = {}^tCA,\ {}^tBD = {}^tDB,\ {}^tAD - {}^tCB = 1_g \right\}.$$

12.1.2

By [KU2, 6.7], in this case, the fundamental diagram (5.0.1) becomes

$$\begin{array}{ccccccc}
 & & D_{\mathrm{SL}(2),\mathrm{val}} & = & D_{\mathrm{BS},\mathrm{val}} & & \\
 & & \downarrow & & \downarrow & & \\
 & D^{\sharp}_{\Sigma,\mathrm{val}} & \to & D_{\mathrm{SL}(2)} & = & D_{\mathrm{BS}} & = \mathcal{X}_{\mathrm{BS}} \\
 & \downarrow & & & & & \\
\Gamma\backslash D_{\Sigma} \leftarrow & D^{\sharp}_{\Sigma} & & & & &
\end{array}$$

That is, $D_{\mathrm{SL}(2),\mathrm{val}} = D_{\mathrm{BS},\mathrm{val}}$, the canonical map $D_{\mathrm{BS}} \to \mathcal{X}_{\mathrm{BS}}$ is a homeomorphism, and there exists a unique homeomorphism $D_{\mathrm{SL}(2)} \simeq D_{\mathrm{BS}}$ which makes the square in the above diagram commutative. We will identify $D_{\mathrm{SL}(2)}$ with D_{BS} via this canonical homeomorphism.

12.1.3

Description of $D \simeq \mathfrak{h}_g$. We first recall the well-known description of D. Let $M(g, \mathbf{C})$ be the set of all $g \times g$ complex matrices and let $M(g, \mathbf{C})_{\mathrm{sym}}$ be the set of all $g \times g$ symmetric complex matrices. For $\tau \in M(g, \mathbf{C})_{\mathrm{sym}}$, let $F(\tau)$ be the corresponding Hodge filtration defined by

$$F^0(\tau) := H_{0,\mathbf{C}}, \ F^1(\tau) := \begin{pmatrix} \text{subspace of } H_{0,\mathbf{C}} \text{ spanned} \\ \text{by the column vectors of } \begin{pmatrix} \tau \\ 1_g \end{pmatrix} \end{pmatrix}, \ F^2(\tau) := \{0\}. \tag{1}$$

Let

$$\mathfrak{h}_g := \{\tau \in M(g, \mathbf{C})_{\mathrm{sym}} \mid \mathrm{Im}(\tau) \text{ is positive definite}\}.$$

Then, we have the following identifications:

$$\begin{array}{ccc} \mathfrak{h}_g & \xrightarrow{\sim} & D \\ \cap\big\downarrow & & \cap\big\downarrow \\ M(g, \mathbf{C})_{\mathrm{sym}} & \xrightarrow{\subset} & \check{D}, \end{array} \tag{2}$$

where the top horizontal arrow is an isomorphism and the bottom horizontal arrow is an open injection both of which send $\tau \mapsto F(\tau)$, and the vertical arrows are open inclusions. From now on, we identify $\mathfrak{h}_g = D$ and $M(g, \mathbf{C})_{\mathrm{sym}} \subset \check{D}$ by (2).

The action of $\left(\begin{smallmatrix} A & B \\ C & D \end{smallmatrix}\right) \in G_{\mathbf{R}} = \mathrm{Sp}(g, \mathbf{R})$ on $D = \mathfrak{h}_g$ is given by

$$\tau \mapsto (A\tau + B)(C\tau + D)^{-1}.$$

For example, for $\tau, \tau' \in M(g, \mathbf{C})_{\mathrm{sym}}$, we have

$$\exp\begin{pmatrix} 0 & \tau \\ 0 & 0 \end{pmatrix} F(\tau') = F(\tau + \tau') \quad \text{where } 0 = 0_g, \tag{3}$$

in particular,

$$F(\tau) = \exp\begin{pmatrix} 0 & \tau \\ 0 & 0 \end{pmatrix} F(0).$$

12.1.4

In the rest of Section 12.1, we consider some special points at infinity for D, which can be described simply, and their behaviors in the map $\psi : D^\sharp_{\mathrm{val}} \to D_{\mathrm{SL}(2)}$ (5.4.3).

For $1 \le j \le g$, let $N_j \in \mathfrak{g}_{\mathbf{Q}}$ be the element defined by

$$N_j(e_{j+g}) = e_j, \quad N_j(e_k) = 0 \ \text{ for } k \ne j + g.$$

We have

$$\exp\left(\sum_{j=1}^{g} z_j N_j\right) F(\tau) = F(\tau + \mathrm{diag}(z_1, \ldots, z_g)) \quad (\tau \in M(g, \mathbf{C})_{\mathrm{sym}}, \ \ z_j \in \mathbf{C})$$

where $\mathrm{diag}(z_1, \ldots, z_g)$ denotes the diagonal matrix. Define a rational nilpotent cone σ by

$$\sigma = (\mathbf{R}_{\geq 0})N_1 + \cdots + (\mathbf{R}_{\geq 0})N_g.$$

For $\tau \in M(g, \mathbf{C})_{\mathrm{sym}}$, $\exp(i\sigma_{\mathbf{R}})F(\tau)$ is a σ-nilpotent i-orbit. Let

$$p_\tau = (\sigma, \exp(i\sigma_{\mathbf{R}})F(\tau)) \in D_\sigma^\sharp.$$

We have

(1) In $D_\sigma^\sharp$, $F(z) \in D$ $(z = (z_{jk}) \in \mathfrak{h}_g)$ converges to p_τ if and only if $z_{jk} \to \tau_{jk}$ for any (j, k) such that $j \neq k$, $\mathrm{Re}(z_{jj}) \to \mathrm{Re}(\tau_{jj})$ for any j, and $\mathrm{Im}(z_{jj}) \to \infty$ for any j.

For $\tau \in M(g, \mathbf{C})_{\mathrm{sym}}$, let $\tilde{p}_\tau \in D^\sharp_{\sigma,\mathrm{val}}$ be the element (A, V, Z) where $A = \sum_{j=1}^g \mathbf{Q}N_j$, $Z = \exp(i\sigma_{\mathbf{R}})F(\tau)$, and V is the valuative cone of $A^* = \mathrm{Hom}_{\mathbf{Q}}(A, \mathbf{Q})$ consisting of all elements $\sum_{j=1}^n x_j N_j^*$ ($(N_j^*)_j$ is the basis of A^* dual to the basis $(N_j)_j$ of A) such that $(x_j)_j$ is ≥ 0 for the lexicographic order of $\mathbf{Q}^g$. We have

(2) In $D^\sharp_{\sigma,\mathrm{val}}$, $F(z) \in D$ $(z \in \mathfrak{h}_g)$ converges to $\tilde{p}_\tau$ if and only if $z_{jk} \to \tau_{jk}$ for any (j, k) such that $j \neq k$, $\mathrm{Re}(z_{jj}) \to \mathrm{Re}(\tau_{jj})$ for any j, $\mathrm{Im}(z_{jj}) \to \infty$ for any j, and $\mathrm{Im}(z_{jj})/\mathrm{Im}(z_{j+1,j+1}) \to \infty$ for $1 \leq j < g$.

12.1.5

Let $\psi : D^\sharp_{\sigma,\mathrm{val}} \to D_{\mathrm{SL}(2)} = D_{\mathrm{BS}}$ be the map defined in 5.4.3. Then we have

$$\psi(\tilde{p}_0) = [\rho, \varphi] \in D_{\mathrm{SL}(2)},$$

where (ρ, φ) is the SL(2)-orbit of rank g defined as follows. For $a = (a_j)_{1 \leq j \leq g} \in \mathrm{SL}(2, \mathbf{C})^g$, $\rho(a) = \bigoplus_{j=1}^g \rho_j(a_j)$ where $\rho_j(a_j)$ is the action of $a_j \in \mathrm{SL}(2, \mathbf{C})$ on the 2-dimensional vector space $\mathbf{C}e_j \oplus \mathbf{C}e_{j+g}$ with respect to the basis (e_j, e_{j+g}) and we regard $H_{0,\mathbf{C}}$ as the direct sum of these two-dimensional subspaces. For $(z_j)_{1 \leq j \leq g} \in \mathbf{C}^g$, $\varphi(z) = F(\mathrm{diag}(z_1, \ldots, z_g))$. The family of weight filtrations $W = (W^{(j)})_{1 \leq j \leq g}$ associated to (ρ, φ) is given by $W_{-2}^{(j)} = 0$, $W_{-1}^{(j)} = \sum_{k=1}^j \mathbf{R}e_k$, $W_0^{(j)} = \sum_{k=1}^g \mathbf{R}e_k + \sum_{k=j+1+g}^{2g} \mathbf{R}e_k$, $W_1^{(j)} = H_{0,\mathbf{R}}$. Let P be the rational parabolic subgroup of $G_{\mathbf{R}}$ consisting of all elements a such that $aW^{(j)} = W^{(j)}$ for any j. Then

$$D_{\mathrm{SL}(2)}(W) = D_{\mathrm{BS}}(P).$$

For $\tau \in M(g, \mathbf{C})_{\mathrm{sym}}$, we have

$$\psi(\tilde{p}_\tau) = [\mathrm{Int}(b) \circ \rho,\ b\varphi] \in D_{\mathrm{SL}(2)}(W) \quad \text{with } b = \exp\begin{pmatrix} 0 & \mathrm{Re}(\tau) \\ 0 & 0 \end{pmatrix} \in G_W.$$

Thus $\psi(\tilde{p}_\tau)$ depends only on the real part $\mathrm{Re}(\tau)$ of τ. As an element of D_{BS}, we have

$$\psi(\tilde{p}_\tau) = (P, F(\mathrm{Re}(\tau) + \mathrm{diag}((\mathbf{R}_{>0})i, \ldots, (\mathbf{R}_{>0})i))) \in D_{\mathrm{BS}}(P).$$

12.1.6

The continuity of $\psi : D^{\sharp}_{\sigma,\mathrm{val}} \to D_{\mathrm{SL}(2)} = D_{\mathrm{BS}}$ (5.4.4) shows that when $F \in D$ converges to $\tilde{p}_\tau$ in $D^{\sharp}_{\sigma,\mathrm{val}}$, then F converges to the above point $\psi(\tilde{p}_\tau)$ of $D_{\mathrm{BS}}(P)$. Here we prove this convergence directly and see to what kind of (down to earth) problem on matrices the continuity of ψ at these points $\tilde{p}_\tau$ corresponds.

We have a homeomorphism

$$P_u \times \mathbf{R}^g_{\geq 0} \simeq D_{\mathrm{BS}}(P)$$

which sends (u, a) $(u \in P_u,\ a = (a_j)_j \in \mathbf{R}^g_{>0})$ to $uF(i \cdot d(a))$ where $d(a)$ is the diagonal matrix whose jth diagonal component is $\prod_{k=j}^{g} a_j^{-1}$. We have

$$P_u = \left\{ \begin{pmatrix} A & B \\ 0 & {}^tA^{-1} \end{pmatrix} \;\middle|\; \begin{matrix} A \in \mathrm{GL}(g, \mathbf{R}),\ A \text{ is strictly lower} \\ \text{trianglar, } B \in M(g, \mathbf{R})_{\mathrm{sym}} \end{matrix} \right\}.$$

(Strictly lower triangular matrices are lower triangular matrices whose diagonal components are 1.) If $F(z)$ $(z \in \mathfrak{h}_g)$ is the image of $(u, a) \in P_u \times \mathbf{R}^g_{>0}$ and if we represent u as the matrix in the above form, we have

$$F(z) = \begin{pmatrix} A & B \\ 0 & {}^tA^{-1} \end{pmatrix} F(i \cdot d(a))$$

and this is rewritten as

$$z = i \cdot A\, d(a)\, {}^tA + B\, {}^tA.$$

It is sufficient to prove that when $F(z) \to \tilde{p}_\tau$, $A \to 1$ and $a_j \to 0$ for $1 \leq j \leq n$. By (2) in 12.1.4, we are reduced to the following statement on matrices which we apply by taking the imaginary part $A\, d(a)\, {}^tA$ of z as S and taking $d(a)$ and ${}^tA$ as Λ and U, respectively.

Proposition 12.1.7 *For a positive definite real symmetric matrix $S \in M(n, \mathbf{R})$, consider the unique factorization*

$$S = {}^tU \Lambda U,$$

where Λ is a diagonal matrix $\mathrm{diag}(\lambda_1, \ldots, \lambda_n)$ *with $\lambda_j \in \mathbf{R}_{>0}$ and $U \in \mathrm{GL}(n, \mathbf{R})$ is a strictly upper triangular matrix. Assume $S = (s_{jk})$ varies satisfying the following conditions* (1) *and* (2).

(1) *If $j \neq k$, s_{jk} converges.*
(2) *$s_{jj} \to \infty$ for any j.*

Then we have the following.

(i) *U converges to* 1.
(ii) *$\lambda_j - s_{jj}$ converges to* 0 *for any j.*

Proof. Let u_{jk} be the (j, k)-component of U and let c_{jk} be the limit of s_{jk}. We prove (ii) and the following stronger version (i)′ of (i) by induction on j.

(i)′ If $1 \leq j < k \leq n$, $\lambda_j u_{jk}$ converges to c_{jk}, and u_{jk} converges to 0.

If $1 \leq j \leq k \leq n$, we have $s_{jk} = \lambda_j u_{jk} + \sum_{\ell=1}^{j-1} \lambda_\ell u_{\ell j} u_{\ell k}$. By induction on j, $\lambda_\ell u_{\ell j}$ converges and $u_{\ell k} \to 0$, and hence $\lambda_\ell u_{\ell j} u_{\ell k} \to 0$. Hence $s_{jk} - \lambda_j u_{jk} \to 0$.

Taking $k = j$, we obtain (ii). Taking $k > j$, we obtain $\lambda_j u_{jk} \to c_{jk}$. Since $\lambda_j \to \infty$, we have $u_{jk} \to 0$ proving (i)$'$. □

12.2 CASE $G_{\mathbf{R}} \simeq O(1, n-1, \mathbf{R})$

In Section 12.2, we assume that $G_{\mathbf{R}}$ is isomorphic to $O(1, n-1, \mathbf{R})$ for some $n \geq 1$. We assume also that D is not empty.

We give explicit descriptions of various enlargements of D and the canonical maps among them in this case. We show (12.2.13) that, if $n \geq 3$ and if we adopt the weak toplogy on E_σ ($\sigma \in \Xi$), instead of the strong topology, in the definition of the topology of $D_\Xi^\sharp$, then the map $\psi : D_\Xi^\sharp \to D_{\mathrm{SL}(2)}$ would not be continuous.

In the present case, w is an even integer. By twist, we assume $w = 0$ without loss of generality. We have $h^{0,0} = 1$, and $h^{j,-j} = 0$ for any non-zero even integer j, as is seen from the signature condition. So n is odd. We have

$$\dim_{\mathbf{C}} D = \left(\sum_{j \geq 1} h^{j,-j} \right)^2 - \frac{1}{2} \sum_{j \geq 1} h^{j,-j}(h^{j,-j} - 1).$$

Lemma 12.2.1 *Assume that the following condition* (1) *is not satisfied.*

(1) $h^{1,-1} \neq 0$ *and there exists an element* $x \neq 0$ *of* $H_{0,\mathbf{Q}}$ *such that* $\langle x, x \rangle_0 = 0$.

Then $D_\sigma = D$ *for any rational nilpotent cone* σ, *and* $D_{\mathrm{SL}(2)} = D$.

Proof. If $D_\sigma \neq D$, then the theory of the associated SL(2)-orbit (6.1.1) tells $D_{\mathrm{SL}(2)} \neq D$. It is sufficient to prove that, if $D_{\mathrm{SL}(2)} \neq D$, then the condition (1) is satisfied. Assume there is an element of $D_{\mathrm{SL}(2)}$ of rank n with $n \geq 1$, and let $(W^{(j)})_{1 \leq j \leq n}$ be the associated family of weight filtrations. Consider $W^{(n)}$. We have $W^{(n)}_{-1} \neq 0$ and $\langle W^{(n)}_{-1}, W^{(n)}_{-1} \rangle_0 = 0$. Since $G_{\mathbf{R}} \simeq O(1, n-1, \mathbf{R})$, this shows that $\dim W^{(n)}_{-1} = 1$. Hence there exists a unique $k \leq -1$ such that $\mathrm{gr}^{W^{(n)}}_k \neq 0$. By the definition of the weight filtration, we have $k = -1$ or $k = -2$. Since $\mathrm{gr}^{W^{(n)}}_k$ is one-dimensional with weight k, k must be even. Hence k must be -2. Hence the Hodge type of $\mathrm{gr}^{W^{(n)}}_{-2}$ is $(-1, -1)$. This shows $h^{1,-1} \neq 0$. Since the filtration $W^{(n)}$ is rational, we have a nonzero element x of $H_0 \cap W^{(n)}_{-1}$ which satisfies $\langle x, x \rangle_0 = 0$. □

12.2.2

In the rest of Section 12.2, we assume that the condition 12.2.1 (1) is satisfied.

Because of the existence of x in the condition 12.2.1 (1), there exist a finitely generated $\mathbf{Z}$-submodule H'_0 of H_0 and elements e_1, e_2 of $H_{0,\mathbf{Q}}$ satisfying the following conditions:

$$\mathrm{rank}\, H'_0 = \mathrm{rank}\, H_0 - 2, \quad H_{0,\mathbf{Q}} = H'_{0,\mathbf{Q}} \oplus \mathbf{Q}e_1 \oplus \mathbf{Q}e_2,$$
$$\langle e_1, e_2 \rangle_0 = 1, \quad \langle e_j, e_j \rangle_0 = 0 \ \text{ and } \ \langle e_j, H'_0 \rangle_0 = 0 \ \text{ for } j = 1, 2.$$

We fix such H'_0 and e_1, e_2.

Let W be the $\mathbf{Q}$-rational increasing filtration on $H_{0,\mathbf{R}}$ defined by $W_k = H_{0,\mathbf{R}}$ if $k \geq 2$, $W_k = H'_{0,\mathbf{R}} + \mathbf{R}e_1$ if $k = 1, 0$, $W_k = \mathbf{R}e_1$ if $k = -1, -2$, and $W_k = 0$ if $k \leq -3$. Let $P = (G^\circ)_{W,\mathbf{R}}$ be the $\mathbf{Q}$-parabolic subgroup of $G_\mathbf{R}$ associated to W.

For $a \in H'_{0,\mathbf{C}}$, let $N_a \in \mathfrak{g}_\mathbf{C}$ be the nilpotent operator defined by

$$\begin{cases} N_a(e_2) = a, \\ N_a(b) = -\langle a, b\rangle_0 e_1 \quad (b \in H'_{0,\mathbf{C}}), \\ N_a(e_1) = 0. \end{cases} \tag{1}$$

For any non-zero element $a \in H'_{0,\mathbf{R}}$, the weight filtration on $H_{0,\mathbf{R}}$ defined by N_a coincides with the above filtration W.

It is easily seen that we have

$$N_a N_b = N_b N_a \quad \text{for } a, b \in H'_{0,\mathbf{C}}.$$

PROPOSITION 12.2.3

(i) *We have*

$$D_\Xi = \bigcup_{g \in G_\mathbf{Q},\, v \in H'_{0,\mathbf{Q}}} g(D_{(\mathbf{R}_{\geq 0})N_v}),$$

$$D^\sharp_\Xi = \bigcup_{g \in G_\mathbf{Q},\, v \in H'_{0,\mathbf{Q}}} g(D^\sharp_{(\mathbf{R}_{\geq 0})N_v}),$$

$$D_{\mathrm{SL}(2)} = \bigcup_{g \in G_\mathbf{Q}} g(D_{\mathrm{SL}(2)}(W)),$$

$$D_{\mathrm{BS}} = \bigcup_{g \in G_\mathbf{Q}} g(D_{\mathrm{BS}}(P)).$$

(ii) *If σ is a rational nilpotent cone and there is a σ-nilpotent orbit, then $\sigma \in \Xi$.*

(iii) $D_{\mathrm{SL}(2)} = D_{\mathrm{SL}(2),\leq 1}$, *hence* $D_{\mathrm{SL}(2),\mathrm{val}} \xrightarrow{\sim} D_{\mathrm{SL}(2)}$. *Furthermore, we have* $D_{\mathrm{BS},\mathrm{val}} \xrightarrow{\sim} D_{\mathrm{BS}}$.

(iv) *Regard $D_{\mathrm{SL}(2)}$ as a subset of D_{BS} via $D_{\mathrm{SL}(2)} \xleftarrow{\sim} D_{\mathrm{SL}(2),\mathrm{val}} \to D_{\mathrm{BS},\mathrm{val}} \xrightarrow{\sim} D_{\mathrm{BS}}$. Then, the topology of $D_{\mathrm{SL}(2)}$ coincides with the topology as a subspace of D_{BS}. We have*

$$D_{\mathrm{SL}(2)} \cap D_{\mathrm{BS}}(P) = D_{\mathrm{SL}(2)}(W).$$

The proof of Proposition 12.2.3 will be given in 12.2.8 below.

Next we describe the set D_σ and the map ψ in Section 5.4 (12.2.5 below).

12.2.4

Let $\langle\ ,\ \rangle'_0$ be the restriction of $\langle\ ,\ \rangle_0$ to $H'_{0,\mathbf{C}}$. Then $\langle\ ,\ \rangle'_0$ is negative definite. Let

$$h'^{p,q} = \begin{cases} h^{p,q} & \text{if } (p,q) \neq (1,-1), (-1,1), \\ h^{p,q} - 1 & \text{if } (p,q) = (1,-1), (-1,1), \end{cases}$$

and let $\check{D}'$ be the set of all decreasing filtrations F on $H'_{0,\mathbf{C}}$ satisfying $\dim \mathrm{gr}^p_F = h'^{p,-p}$ for all $p \in \mathbf{Z}$ and $\langle F^p, F^q\rangle'_0 = 0$ for any $p, q \in \mathbf{Z}$ with $p+q > 0$. This space $\check{D}'$ is the "$\check{D}$ for $(H'_0, \langle\ ,\ \rangle'_0, (h'^{p,q}))$" and is compact. We have

(1) (H'_0, F) for $F \in \check{D}'$ is a Hodge structure of weight 0.

To show this, it is sufficient to prove that if $p+q=1$, $H'_{0,\mathbf{C}}$ is the direct sum of F^p and $\bar{F}^q$. For $a+ib \in F^p \cap \overline{F}^q$ with $a, b \in H'_{0,\mathbf{R}}$, we have

$$0 = \langle a+ib,\ a-ib\rangle'_0 = \langle a, a\rangle'_0 + \langle b, b\rangle'_0.$$

Since $\langle\ ,\ \rangle_0$ is negative definite, this shows $a=b=0$. Since $\dim F^p + \dim \overline{F}^q = \operatorname{rank} H'_0$ by the definition of $h'^{\bullet,\bullet}$, this shows that $H'_{0,\mathbf{C}}$ is the direct sum of F^p and $\overline{F}^q$ and this proves (1).

In particular, for $F \in \check{D}'$, we have a direct sum decomposition

$$H'_{0,\mathbf{C}} = H'_{0,\mathbf{R}} \oplus i \cdot (H'_{0,\mathbf{R}} \cap F^0) \oplus F^1$$

and $H'_{0,\mathbf{R}} \cap F^0$ is one-dimensional over $\mathbf{R}$. For $a \in H'_{0,\mathbf{C}}$, define $\operatorname{Re}_F(a) \in H'_{0,\mathbf{R}}$ and $\operatorname{Im}_F(a) \in H'_{0,\mathbf{R}} \cap F^0$ by

$$a \equiv \operatorname{Re}_F(a) + i \cdot \operatorname{Im}_F(a) \mod F^1. \tag{2}$$

Define a map

$$s : \check{D}' \to \check{D} \quad \text{by} \quad s(F)^p = \begin{cases} F^p & \text{if } p > 1, \\ F^p \oplus \mathbf{C}e_2 & \text{if } p = 0, 1, \\ F^p \oplus \mathbf{C}e_2 \oplus \mathbf{C}e_1 & \text{if } p \le -1. \end{cases} \tag{3}$$

Then $s(F)$ is an $\mathbf{R}$-split mixed Hodge structure with respect to W.

Proposition 12.2.5 *Let $v \in H'_{0,\mathbf{Q}} - \{0\}$ and $\sigma = (\mathbf{R}_{\ge 0})N_v$.*

(i) *$D_\sigma = D \cup \{(\sigma, \exp(\sigma_{\mathbf{C}})\exp(N_a)s(F)) \mid a \in H'_{0,\mathbf{C}},\ F \in \check{D}'$ with $F^0 \ni v\} \ne D$.*

(ii) *For $a \in H'_{0,\mathbf{C}}$ and $F \in \check{D}'$ with $F^0 \ni v$, the* SL(2)*-orbit (ρ, φ) in one variable associated to $(N_v, \exp(N_a)s(F))$ is characterized by $\rho_*\left(\begin{smallmatrix} 0 & 1 \\ 0 & 0 \end{smallmatrix}\right) = N_v$ and $\varphi(0) = \exp(N_{\operatorname{Re}_F(a)})s(F)$* (12.2.4).

The proof of 12.2.5 will be given in 12.2.7 below.

Lemma 12.2.6

(i) *For $a, b \in H'_{0,\mathbf{C}}$ and for $F, F' \in \check{D}'$, $\exp(N_a)s(F) = \exp(N_b)s(F')$ if and only if $F' = F$ and $a \equiv b \bmod F^1$.*

(ii) *For $a \in H'_{0,\mathbf{C}}$ and $F \in \check{D}'$, the following conditions* (1) *and* (2) *are equivalent.*

(1) *$\exp(N_a)s(F) \in D$.*
(2) *$\operatorname{Im}_F(a) \ne 0$.*

(iii) *For $v \in H'_{0,\mathbf{R}} - \{0\}$, $a \in H'_{0,\mathbf{C}}$ and $F \in \check{D}'$, the following conditions* (3)–(5) *are equivalent.*

(3) *$\exp(\mathbf{C}N_v)\exp(N_a)s(F)$ is an $((\mathbf{R}_{\ge 0})N_v)$-nilpotent orbit.*
(4) *$N_v \exp(N_a)s(F)^p \subset \exp(N_a)s(F)^{p-1}$ (i.e., $N_v s(F)^p \subset s(F)^{p-1}$) for all $p \in \mathbf{Z}$.*
(5) *$v \in F^0$.*

Proof. (i) is proved easily.

We prove (ii). First assume $\mathrm{Im}_F(a) = 0$ and let $b = \mathrm{Re}_F(a)$. Then $\exp(N_a) s(F) = \exp(N_b)s(F)$ by (i). Since $\exp(N_b) \in G_{\mathbf{R}}$, if $\exp(N_a)s(F) \in D$, then we should have $s(F) = \exp(-N_b)\exp(N_a)s(F) \in D$. But $e_2 \in s(F)^1 \cap \overline{s(F)}^1$ shows that $s(F)$ does not belong to D.

Next assume $\mathrm{Im}_F(a) \neq 0$. To show (1), we may assume $a = ib$ for a non-zero element b of $H'_{0,\mathbf{R}} \cap F^0$. We assume $a = ib$. We can see easily $H_{0,\mathbf{C}} = \bigoplus_{j\in\mathbf{Z}} H^{j,-j}_{\exp(iN_b)s(F)}$ with

$$H^{j,-j}_{\exp(iN_b)s(F)} = \exp(iN_b)s(F)^j \cap (\text{complex conjugate of } \exp(iN_b)s(F)^{-j})$$
$$= \begin{cases} H_F^{j,-j} & \text{for } j \geq 2, \\ H_F^{1,-1} \oplus \mathbf{C}\left(e_2 + ib + \frac{1}{2}\langle b, b\rangle_0 e_1\right) & \text{for } j = 1, \\ \mathbf{C}\left(e_2 - \frac{1}{2}\langle b, b\rangle_0 e_1\right) & \text{for } j = 0, \end{cases}$$

and $H^{-j,j}_{\exp(iN_b)s(F)} = (\text{complex conjugation of } H^{j,-j}_{\exp(iN_b)s(F)})$. We have the desired positivity as

$$-\left\langle e_2 + ib + \frac{1}{2}\langle b, b\rangle_0 e_1,\ e_2 - ib + \frac{1}{2}\langle b, b\rangle_0 e_1\right\rangle_0 = -2\langle b, b\rangle_0 > 0,$$
$$\left\langle e_2 - \frac{1}{2}\langle b, b\rangle_0 e_1,\ e_2 - \frac{1}{2}\langle b, b\rangle_0 e_1\right\rangle_0 = -\langle b, b\rangle_0 > 0.$$

Thus, we have (1).

We prove (iii). First we prove that (4) and (5) are equivalent. (4) is equivalent to the following two conditions.

(6) $$N_v(e_2) \in F^0,$$

(7) $$N_v(F^1) = 0.$$

But (6) is equivalent to $v \in F^0$. If $v \in F^0$, then $N_v(F^1) = -\langle v, F^1\rangle_0 e_1 = 0$ because $\langle F^0, F^1\rangle_0 = 0$. Hence (6) implies (7).

Clearly (3) implies (4). Finally, if (4) and (5) are satisfied, then we have the small Griffiths transversality by (4), and we have, by (ii) (note $\mathrm{Im}(zv) = \mathrm{Im}(z)v$ by (5)), the positivity

$$\exp(zN_v)\exp(N_a)s(F) \in D \quad \text{for } z \in \mathbf{C} \text{ with } \mathrm{Im}(z) \gg 0$$ □

12.2.7

Proof of Proposition 12.2.5. We prove 12.2.5 (i). For the first equality, the inclusion $\supset$ follows from Lemma 12.2.6 (iii). We show the converse. Assume $(N_v, \tilde{F})$ ($\tilde{F} \in \check{D}$) generates a nilpotent orbit. Let $F = \tilde{F}(\mathrm{gr}_0^W) \in \check{D}'$. Since $(W, \tilde{F})$ is a mixed Hodge structure, $\tilde{F}(\mathrm{gr}_1^W)$ has the Hodge type $(1, 1)$. Hence there exist $b \in H'_{0,\mathbf{C}}$ and $c \in \mathbf{C}$ such that $e_2 + b + ce_1 \in \tilde{F}^1$. By $\langle e_2 + b + ce_1, e_2 + b + ce_1\rangle_0 = 0$, we have

$c = -\frac{1}{2}\langle b, b\rangle_0$. Hence $e_2 + b + ce_1 = \exp(N_b)(e_2)$. On the other hand, if $x \in \tilde{F}^p \cap W_0$ with $p \geq 0$, $x = x' + c'e_1$ for some $x' \in s(F)^p$ and some $c' \in \mathbf{C}$. By $\langle e_2 + b + ce_1, x' + c'e_1\rangle_0 = 0$, we have $c' = -\langle b, x'\rangle_0$. Hence $x = \exp(N_b)(x')$. These prove $\tilde{F} = \exp(N_b)s(F)$ by the definition of $s(F)$ in 12.2.4 (3).

As is easily seen, there exists $F \in \check{D}'$ such that $v \in F^0$. Since $(\sigma, \exp(\sigma_{\mathbf{C}})s(F)) \in D_\sigma$, we have $D_\sigma \neq D$.

Next we prove 12.2.5 (ii). Write $\mathrm{Im}_F(a) = yv$ with $y \in \mathbf{R}$. Since $\exp(N_a)s(F) = \exp(iyN_v) \cdot \exp(N_{\mathrm{Re}_F(a)})s(F)$, (ρ, φ) coincides with the SL(2)-orbit in one variable associated to $(N_v, \exp(N_{\mathrm{Re}_F(a)})s(F))$. Since $\exp(N_{\mathrm{Re}_F(a)})s(F)$ is $\mathbf{R}$-split with respect to W, we have $\varphi(0) = (W, \exp(N_{\mathrm{Re}_F(a)})s(F))^\wedge = \exp(N_{\mathrm{Re}_F(a)})s(F)$ (6.1.2 (11)). □

12.2.8

Proof of Proposition 12.2.3. First we prove 12.2.3 (i) for D_{BS}. It is sufficient to prove that any $\mathbf{Q}$-parabolic subgroup Q of $G_{\mathbf{R}}$ such that $Q \neq (G^\circ)_{\mathbf{R}}$ is conjugate to P under some element of $G_{\mathbf{Q}}$. By [KU2, 6.4], there exist $\mathbf{Q}$-subspaces I_j ($0 \leq j \leq m$, $m > 1$) of $H_{0,\mathbf{Q}}$ such that $0 = I_0 \subsetneq I_1 \subsetneq \cdots \subsetneq I_m = H_{0,\mathbf{Q}}$, the annihilator of I_j under $\langle\ ,\ \rangle_0$ is I_{m-j} for any j, and that $Q = \{g \in (G^\circ)_{\mathbf{R}} \mid gI_{j,\mathbf{R}} = I_{j,\mathbf{R}}\ (\forall\, j)\}$. Let $r \geq 1$ be the largest integer such that $2r \leq m$. Since $\langle I_r, I_r\rangle_0 = 0$, the assumption $G_{\mathbf{R}} \simeq O(1, n-1, \mathbf{R})$ shows $r = 1$ and $\dim I_1 = 1$. Take any $\mathbf{Q}$-subspace J of $H_{0,\mathbf{Q}}$ such that I_{m-1} is the direct sum of I_1 and J. Take any basis e of I_1 and take any element f of $H_{0,\mathbf{Q}}$ such that $\langle e, f\rangle_0 = 1$. Replacing f by $f - \frac{1}{2}\langle f, f\rangle_0 e$, we may assume $\langle f, f\rangle_0 = 0$. As a $\mathbf{Q}$-vector space with a nondegenerate symmetric bilinear form $\langle\ ,\ \rangle_0$, $H_{0,\mathbf{Q}}$ is decomposed into the direct sum of an anisotropic part J and a hyperbolic part $\mathbf{Q}e + \mathbf{Q}f$. Hence, by the uniqueness of such decomposition up to $G_{\mathbf{Q}}$-conjugacy, there exists an element g of $G_{\mathbf{Q}}$ such that $g(H'_{0,\mathbf{Q}}) = J$, $g(e_1) = e$, $g(e_2) = f$. We have $Q = gPg^{-1}$, as desired.

12.2.3 (iii) for D_{BS} follows from 12.2.3 (i) for D_{BS} and from rank $S_P = 1$.

Next we prove 12.2.3 (i) for D_Ξ and $D_\Xi^\sharp$. Let N be a non-zero nilpotent element of $\mathfrak{g}_{\mathbf{Q}}$. It is sufficient to show that N is conjugate to N_v for some $v \in H'_{0,\mathbf{Q}}$. Let $W' := W(N)$. Then, by the argument as above, there exists a unique integer $k \leq -1$ such that $\dim \mathrm{gr}_k^{W'} = 1$, and this k is -1 or -2. But, if $\mathrm{gr}_{-1}^{W'} \neq 0$, it has a Hodge structure of weight -1 and hence cannot be one dimensional. Thus $\mathrm{gr}_k^{W'} = 0$ for $k \neq 2, 0, -2$ and $\dim \mathrm{gr}_{-2}^{W'} = 1$. Hence, by the argument as above, we have $W' = gW$ for some $g \in G_{\mathbf{Q}}$. Replacing N by $\mathrm{Ad}(g)^{-1}N$, we may assume $W' = W$. Write $N(e_2) = v + ce_1$ with $v \in H'_{0,\mathbf{Q}}$ and $c \in \mathbf{Q}$. Then, since $H_{0,\mathbf{Q}} \times H_{0,\mathbf{Q}} \to \mathbf{Q}$, $(x, y) \mapsto \langle x, N(y)\rangle_0$ is antisymmetric, $c = \langle e_2, N(e_2)\rangle_0 = 0$. Hence $N(e_2) = N_v(e_2)$. We have $N(e_1) = 0 = N_v(e_1)$ since $W' = W$. For $a \in H'_{0,\mathbf{Q}}$, $\langle e_2, N(a)\rangle_0 = -\langle N(e_2), a\rangle_0 = -\langle N_v(e_2), a\rangle_0 = \langle e_2, N_v(a)\rangle_0$ and hence $N(a) = N_v(a)$. Thus we obtained $N = N_v$.

Next, we prove 12.2.3 (ii). Let (σ, Z) be a nilpotent orbit with σ rational. Assume $\sigma \neq 0$, and take rational elements N, N' of the interior (0.7.7) of σ. It is sufficient to prove $N' \in \mathbf{Q}N$. By the above argument, we may assume $N = N_v$ for some

$v \in H'_{0,\mathbf{Q}}$. Since N and N' have a common weight filtration (5.2.4), the above argument shows $N' = N_{v'}$ for some $v' \in H'_{0,\mathbf{Q}}$. By Lemma 12.2.5 (i), there exist $a \in H'_{0,\mathbf{C}}$ and $F \in \check{D}'$ such that $F^0 \ni v$ and $\exp(N_a)s(F) \in Z$. Since $(N_{v'}, \exp(N_a)s(F))$ also generates a nilpotent orbit, we have $v' \in F^0$ (12.2.6 (iii)). Thus v, v' belong to the part of Hodge type (0, 0) for the Hodge structure (H'_0, F) which is one dimensional. Hence $v' = cv$ for some $c \in \mathbf{C}$ and we have $c \in \mathbf{Q}$ since both N and N' are rational.

Next, we prove 12.2.3 (i) for $D_{\mathrm{SL}(2)}$ and 12.2.3 (iii) for $D_{\mathrm{SL}(2)}$. Let (ρ, φ) be an SL(2)-orbit in n variables and of rank n with $n \geq 1$ whose associated family of weight filtrations is rational. Let $(N_j)_{1\leq j\leq n}$ be the associated family of nilpotent elements, and let $(W^{(j)})_{1\leq j\leq n}$ be the associated family of weight filtrations. The argument in the proof of 12.2.3 (ii) shows that there is $g \in G_{\mathbf{Q}}$ such that $W^{(n)} = gW$. By the similar argument as in the proof of 12.2.3 (ii), for any $c_1, \ldots, c_n \in \mathbf{R}_{>0}$, we have that $\sum_{1\leq j\leq n} c_j N_j$ is a multiple of $N_1 + \cdots + N_n$. Hence $n = 1$ and we also have $[\rho, \varphi] \in g(D_{\mathrm{SL}(2)}(W))$, as desired.

Finally, we prove 12.2.3 (iv). The statement concerning the topology follows from the fact $D_{\mathrm{SL}(2)} \cap D_{\mathrm{BS}}(P) = D_{\mathrm{SL}(2)}(W)$. The last fact follows from 12.2.3 (i) and from the fact that, if $g \in G_{\mathbf{Q}}$ and if $gPg^{-1} = P$, then $gW = W$ since W consists of all P-stable $\mathbf{R}$-subspaces of $H_{0,\mathbf{R}}$. □

Lemma 12.2.9 *For $F \in \check{D}'$ and a nonzero element b of the one dimensional $\mathbf{R}$-space $H'_{0,\mathbf{R}} \cap F^0$, consider the maximal compact subgroup $K_{\mathbf{r}}$ for $\mathbf{r} = \exp(iN_b)s(F)$.*

(i) *$K_{\mathbf{r}}$ and $K_{\mathbf{r}} \cdot \mathbf{r} \subset D$ depends only on $\langle b, b\rangle_0$. They are independent of F.*
(ii) *The Borel-Serre lifting*

$$\nu : \mathbf{G}_{\mathbf{m},\mathbf{R}} \to G_{W,\mathbf{R}}$$

of the weight map $\mathbf{G}_{\mathbf{m},\mathbf{R}} \to G_{W,\mathbf{R}}/G_{W,\mathbf{R},u}$ at $K_{\mathbf{r}}$ does not depend on the choice of (F, b) as above, and is given by

$$\nu(t)e_1 = t^{-2}e_1, \quad \nu(t)e_2 = t^2 e_2, \quad \nu(t)a = a \quad (\forall a \in H'_{0,\mathbf{R}}).$$

Proof. The Cartan involution associated with $K_{\mathbf{r}}$ coincides with $\mathrm{Int}(C_{\mathbf{r}})$ where $C_{\mathbf{r}} \in G_{\mathbf{R}}$ is the Weil operator at $\mathbf{r}$ which acts on $H_{\mathbf{r}}^{p,q}$ $(p+q=0)$ as the multiplication by i^{p-q}. From the description of $H_{\mathbf{r}}^{p,q}$ in the proof of Lemma 12.2.6 (ii), we can deduce that $C_{\mathbf{r}}$ acts on $H'_{0,\mathbf{R}}$ as the multiplication by -1, sends e_2 to $-\frac{1}{2}\langle b, b\rangle_0 e_1$, and sends e_1 to $-2\langle b, b\rangle_0^{-1} e_2$. Lemma 12.2.9 follows from this. □

The local structures of $D_{\mathrm{SL}(2)}$ and D_{BS} are described as follows.

Proposition 12.2.10

(i) *Fix $v \in H'_{0,\mathbf{Q}} - \{0\}$ and $F \in \check{D}'$ with $v \in F^0$, let (ρ, φ) be the* SL(2)*-orbit in one variable associated to $(N_v, s(F))$ and let $\mathbf{r} = \varphi(i) = \exp(iN_v)s(F)$. Define a topological subspace Y of $(\mathbf{R}_{\geq 0}) \times H'_{0,\mathbf{R}} \times (K_{\mathbf{r}} \cdot \mathbf{r})$ by*

$$Y = \{(r, x, y) \in (\mathbf{R}_{\geq 0}) \times H'_{0,\mathbf{R}} \times (K_{\mathbf{r}} \cdot \mathbf{r}) \mid \textit{if } r = 0, \textit{ then } y \in (G_{W,\mathbf{R}} \cap K_{\mathbf{r}}) \cdot \mathbf{r}\}.$$

Then we have a commutative diagram

$$\begin{array}{ccc}
Y & \xrightarrow[\beta_W]{\sim} & D_{\mathrm{SL}(2)}(W) \\
\cap\downarrow & & \downarrow \\
(\mathbf{R}_{\geq 0}) \times H'_{0,\mathbf{R}} \times (K_{\mathbf{r}} \cdot \mathbf{r}) & \xrightarrow[\beta_P]{\sim} & D_{\mathrm{BS}}(P) \\
\downarrow & & \downarrow \\
(\mathbf{R}_{\geq 0}) \times H'_{0,\mathbf{R}} & \xrightarrow[\overline{\beta}_P]{\sim} & X_{\mathrm{BS}}(P),
\end{array}$$

where the vertical arrows are the canonical ones and the horizontal arrow β_W (resp. $\beta_P, \overline{\beta}_P$) is the unique homeomorphism which is compatible with the homeomorphism

$$\beta : (\mathbf{R}_{>0}) \times H'_{0,\mathbf{R}} \times (K_{\mathbf{r}} \cdot \mathbf{r}) \xrightarrow{\sim} D, \quad (t, a, F) \mapsto \exp(N_a)\nu(\sqrt{t})F,$$

with ν as in 12.2.9 (ii). *The group $G_{W,\mathbf{R}} \cap K_{\mathbf{r}}$ coincides with the subgroup of $G_{\mathbf{R}}$ consisting of all elements which preserve $H'_{0,\mathbf{R}}$ and send (e_1, e_2) to (e_1, e_2) or to $(-e_1, -e_2)$.*

(ii) *$D_{\mathrm{SL}(2)}$ is locally compact if and only if $h^{j,-j} = 0$ for any $j \geq 2$. If this condition is satisfied, then $D_{\mathrm{SL}(2)} \to D_{\mathrm{BS}}$ is bijective.*

This follows from results in Chapter 10.

Proposition 12.2.11 *We use the notation in* 12.2.10.

(i) *Let $K := \{g \in G_{W,\mathbf{R}} \cap K_{\mathbf{r}} \mid gv = v\} \subset J := \{g \in G_{W,\mathbf{R}} \cap K_{\mathbf{r}} \mid gv \in H'_{0,\mathbf{Q}}\}$. Then, the image of $D^{\sharp}_{\sigma}$ in $D_{\mathrm{BS}}(P)$ coincides with the union of D and $\beta_P(\{0\} \times H'_{0,\mathbf{R}} \times (K \cdot \mathbf{r}))$. On the other hand, the intersection of $D_{\mathrm{BS}}(P)$ and the image of $D^{\sharp}_{\Xi}$ in D_{BS} coincides with the union of D and $\beta_P(\{0\} \times H'_{0,\mathbf{R}} \times (J \cdot \mathbf{r}))$.*

(ii) *The map $\psi : D^{\sharp}_{\Xi} \to D_{\mathrm{SL}(2)}$* (5.4.4) *is injective and the image is dense.*

(iii) *If $n = 3$, the maps $\psi : D^{\sharp}_{\Xi} \to D_{\mathrm{SL}(2)}$ and $D_{\mathrm{SL}(2)} \to D_{\mathrm{BS}}$ are homeomorphisms. If $n \geq 5$, the map $\psi : D^{\sharp}_{\Xi} \to D_{\mathrm{SL}(2)}$ is not bijective, and the topology of $D^{\sharp}_{\Xi}$ does not coincide with the topology as a subspace of $D_{\mathrm{SL}(2)}$.*

(iv) *$\Gamma\backslash D_{\Xi}$ (resp. $D^{\sharp}_{\Xi}$) is locally compact if and only if $n = 3$.*

Proof. (i) follows by 12.2.3 and 12.2.5.

We prove (ii). The injectivity of ψ follows from Proposition 12.2.5 (ii) which shows that $p \in D^{\sharp}_{\Xi}$ is recovered from $\psi(p) = [\rho, \varphi]$ as $((\mathbf{R}_{\geq 0})N, \varphi(i\mathbf{R}))$. The density of the image follows from (i).

We prove (iii). The case $n = 3$ is a "classical situation" (0.4.14) and hence we have the result. Nonbijectivity in the case $n \geq 5$ follows from (i). The difference of the topology in the case $n \geq 5$ is shown as follows. Take $v \in H'_{0,\mathbf{Q}} - \{0\}$ and let $\sigma = (\mathbf{R}_{\geq 0})N_v$. Take a directed family of elements $g_\lambda \neq 1$ of $\mathrm{Aut}(H'_{0,\mathbf{Q}}, \langle\ ,\ \rangle'_0)$ which converges to 1 (in $\mathrm{Aut}(H'_{0,\mathbf{R}})$), and extend g_λ to an element of $G_{\mathbf{Q}}$ by $g_\lambda(e_j) = e_j$ for $j = 1, 2$. We can take such g_λ so that $\mathrm{Ad}(g_\lambda)(\sigma) \neq \sigma$ for every λ. Take an element F of $\check{D}'$ with $F^0 \ni v$. Then $(\mathrm{Ad}(g_\lambda)(\sigma),\ g_\lambda \exp(i\sigma_{\mathbf{R}})s(F)) \in D^{\sharp}_{\Xi}$

does not converge to $(\sigma, \exp(i\sigma_{\mathbf{R}})s(F))$ in $D^{\sharp}_{\Xi}$, because $\mathrm{Ad}(g_\lambda)(\sigma) \neq \sigma$ and $D^{\sharp}_{\sigma}$ is open in $D^{\sharp}_{\Xi}$. On the other hand, we see $W(\mathrm{Ad}(g_\lambda)(\sigma)) = W(\sigma) = W$, and $\psi(\mathrm{Ad}(g_\lambda)(\sigma), g_\lambda \exp(i\sigma_{\mathbf{R}})s(F)) = \beta_W(0, 0, g_\lambda \exp(iN_v)s(F))$ converges to $\psi(\sigma, \exp(i\sigma_{\mathbf{R}})s(F)) = \beta_W(0, 0, \exp(iN_v)s(F))$, since $g_\lambda \exp(iN_v)s(F)$ converges to $\exp(iN_v)s(F)$.

(iv) follows from the general criterion 7.4.12. □

Lemma 12.2.12 *Let the notation be as in* 12.2.4 *and Proposition* 12.2.10. *Then for* $a \in H'_{0,\mathbf{C}}$ *and* $F \in \check{D}'$ *with* $\mathrm{Im}_F(a) \neq 0$, *we have*

$$\exp(N_a)s(F) = \beta_P(t, \mathrm{Re}_F(a), \exp(iN_{t\,\mathrm{Im}_F(a)})s(F))$$

where t *is the positive square root of* $\langle v, v\rangle_0 \langle \mathrm{Im}_F(a), \mathrm{Im}_F(a)\rangle_0^{-1}$.

Proof. We have

$$\begin{aligned}\exp(N_a)s(F) &= \exp(N_{\mathrm{Re}_F(a)}) \exp(iN_{\mathrm{Im}_F(a)})s(F)\\ &= \exp(N_{\mathrm{Re}_F(a)})\nu(\sqrt{t}) \exp(iN_{t\,\mathrm{Im}_F(a)})s(F).\end{aligned}$$

Since $\langle t\,\mathrm{Im}_F(a), t\,\mathrm{Im}_F(a)\rangle'_0 = \langle v, v\rangle'_0$, we have $\exp(iN_{t\,\mathrm{Im}_F(a)})s(F) \in K_{\mathbf{r}} \cdot \mathbf{r}$ by Lemma 12.2.9 (i). □

12.2.13

In this book, we use the strong topology of E_σ in $\check{E}_\sigma$, not the weak topology (= the topology as a subspace of $\check{E}_\sigma$) (Section 3.1). Here and in 12.3.6 in the next section, we show how bad things can happen if we use the weak topology.

Assume $n \geq 5$ and let $v \in H'_{0,\mathbf{Q}} - \{0\}$, $\sigma = (\mathbf{R}_{\geq 0})N_v$. We show that, if we endow $E^{\sharp}_{\sigma}$ with the topology as a subspace of $\check{E}^{\sharp}_{\sigma}$, then the composite map $E^{\sharp}_{\sigma} \to D^{\sharp}_{\sigma} \xrightarrow{\psi} D_{\mathrm{SL}(2)}$ is not continuous.

In fact, there is a holomorphic map $\Delta \to \check{D}'$, $z \mapsto F(z)$, such that $F^0(0) \ni v$ and $F^0(z) \not\ni v$ for any $z \neq 0$. For $z \neq 0$, since $iv \notin F^0(z)$, we have $\mathrm{Re}_{F(z)}(iv) \neq 0$. For $z \neq 0$, take $y_z > 0$ such that $y_z \mathrm{Re}_{F(z)}(iv)$ does not converge to 0 when $z \neq 0$ converges to 0. Since $\mathrm{Im}_{F(0)}(iv) = v$, $\mathrm{Im}_{F(z)}(iv) \neq 0$ if z is near to 0. Hence, if z is near to 0, $\exp(iy_z N_v)s(F(z)) \in D$ by 12.2.6 (ii) and hence $(\mathbf{e}(iy_z N_v), s(F(z))) \in E^{\sharp}_{\sigma}$ (3.3.4). When $z \to 0$, $(\mathbf{e}(iy_z N_v), s(F(z)))$ converges to $(0, s(F(0)))$ in $E^{\sharp}_{\sigma}$ in the weak topology. However, since $\mathrm{Re}_{F(z)}(iy_z v)$ $(z \neq 0, z \to 0)$ diverges, the image

$$\exp(iy_z N_v)s(F(z)) = \beta_W(t, \mathrm{Re}_{F(z)}(iy_z v), \exp(iN_{t\,\mathrm{Im}_{F(z)}(iy_z v)})s(F(z))) \in D$$

of $(\mathbf{e}(iy_z N_v), s(F(z)))$ does not converge to the image

$$\psi(\sigma, \exp(i\sigma_{\mathbf{R}})s(F(0))) = \beta_W(0, 0, \exp(iN_v)s(F(0))) \in D_{\mathrm{SL}(2)}$$

of $(0, s(F(0)))$, as is seen from Lemma 12.2.12.

12.2.14

Let Γ be a neat subgroup of $G_{\mathbf{Z}}$ of finite index, let $v \in H'_{0,\mathbf{Q}} - \{0\}$, and let $\sigma = (\mathbf{R}_{\geq 0})N_v$. We assume $h^{j,-j} = 0$ for $j \geq 2$. Under this assumption, we give in 12.2.15

explicit descriptions of open neighborhoods of boundary points of $\Gamma(\sigma)^{\mathrm{gp}}\backslash D_\sigma$ and of $D_\sigma^\sharp$.

Let

$$S := \{(q, F) \in \mathbf{C} \times \check{D}' \mid \text{if } q = 0, \text{then } F^0 \ni v\},$$

and regard S as a logarithmic manifold in the natural way.

Let

$$S^\sharp = \{(q, F) \in S \mid q \in \mathbf{R}_{\geq 0}\} \subset S,$$

and endow $S^\sharp$ with the topology as a subspace of S. We have a continuous map

$$S^\sharp \times H'_{0,\mathbf{R}} \to E_\sigma^\sharp\ ,\ (q, F, a) \mapsto (q, \exp(N_a)s(F)),$$

where we identify $|\mathrm{toric}|_\sigma$ with $\mathbf{R}_{\geq 0}$, via the unique isomorphism $\Gamma(\sigma) \simeq \mathbf{N}$, and $\check{E}_\sigma^\sharp$ with $\mathbf{R}_{\geq 0} \times \check{D}$. By composing with $E_\sigma^\sharp \to D_\sigma^\sharp$, we have a continuous map

$$\alpha_\sigma^\sharp : S^\sharp \times H'_{0,\mathbf{R}} \to D_\sigma^\sharp.$$

Take a $\mathbf{C}$-subspace B of $H'_{0,\mathbf{C}}$ such that $2\dim B + 1 = \dim H'_{0,\mathbf{C}}$ and $B \not\ni v$. Let X_B be the open set of the logarithmic manifold $S \times B$ consisting of all elements (q, F, a) $((q, F) \in S,\ a \in B)$ satisfying the following conditions (1) and (2).

(1) If $q \neq 0$, then $\mathrm{Im}_F(a) \neq \frac{1}{2\pi}\log(|q|)\,\mathrm{Im}_F(iv)$.
(2) $H'_{0,\mathbf{C}}$ is the direct sum of B, $\mathbf{C}v$ and F^1.

We have a morphism of logarithmic manifolds

$$X_B \to E_\sigma\ ,\ (q, F, a) \mapsto (q, \exp(N_a)s(F)),$$

where we identify toric_σ with $\mathbf{C}$, via the unique isomorphism $\Gamma(\sigma) \simeq \mathbf{N}$, and $\check{E}_\sigma$ with $\mathbf{C} \times \check{D}$. By composing with $E_\sigma \to \Gamma(\sigma)^{\mathrm{gp}}\backslash D_\sigma$, we have a morphism of logarithmic manifolds

$$\alpha_{\sigma,B} : X_B \to \Gamma(\sigma)^{\mathrm{gp}}\backslash D_\sigma.$$

Proposition 12.2.15 *Assume $h^{j,-j} = 0$ for all $j \geq 2$. Let the notation be as in* 12.2.14.

(i) *The continuous map*

$$\alpha_\sigma^\sharp : S^\sharp \times H'_{0,\mathbf{R}} \to D_\sigma^\sharp$$

is an injective open map whose image contains $D_\sigma^\sharp - D$.

(ii) *For B as in* 12.2.14, *the morphism $\alpha_{\sigma,B} : X_B \hookrightarrow \Gamma(\sigma)^{\mathrm{gp}}\backslash D_\sigma$ is an open immersion of logarithmic manifolds. When B varies, the union of images of $\alpha_{\sigma,B}$ contain $\Gamma(\sigma)^{\mathrm{gp}}\backslash(D_\sigma - D)$.*

(iii) *We have a commutative diagram*

$$\begin{array}{ccc} S^\sharp \times H'_{0,\mathbf{R}} & \xrightarrow{\alpha_\sigma^\sharp} & D_\sigma^\sharp \\ \downarrow & & \downarrow \psi \\ (\mathbf{R}_{\geq 0}) \times H'_{0,\mathbf{R}} \times (K_{\mathbf{r}} \cdot \mathbf{r}) & \xrightarrow[\beta_P]{\sim} & D_{\mathrm{BS}}(P). \end{array}$$

Here the left vertical arrow is the unique continuous map which sends $(q, F, a) \in S^\sharp \times H'_{0,\mathbf{R}}$ *with* $q \neq 0$ *to* $\big(t, a + \mathrm{Re}_F(u), \exp(iN_{t\,\mathrm{Im}_F(u)})s(F)\big)$, *where* $u = -(\frac{\log(q)}{2\pi})iv$ *and* t *the positive square root of* $\langle v, v\rangle_0 \langle \mathrm{Im}_F(u), \mathrm{Im}_F(u)\rangle_0^{-1}$.

Proof. Let U be the open set of $\check{D}$ consisting of all $\tilde{F} \in \check{D}$ such that $\tilde{F}^1 \to \mathrm{gr}^W_{1,\mathbf{C}}$ is surjective. We prove

$$U = \cup_{b \in H'_{0,\mathbf{C}}} \exp(N_b)s(\check{D}'). \tag{1}$$

In fact, if $\tilde{F} \in U$, there exist $b \in H'_{0,\mathbf{C}}$ and $c \in \mathbf{C}$ such that $e_2 + b + ce_1 \in \tilde{F}^1$. By the argument in 12.2.7, $c = -\langle b, b\rangle_0$. From the assumption $h^{j,-j} = 0$ for all $j \geq 2$, we see that $F := \tilde{F}(\mathrm{gr}^W_{0,\mathbf{C}})$ belongs to $\check{D}'$, and we have $\tilde{F} = \exp(N_b)s(F)$. By (1) and 12.2.5, we can deduce (i) and (ii) from results in §7.

(iii) follows from 12.2.12. □

12.3 EXAMPLE OF WEIGHT 3 (A)

In this section, we examine an example for which if we use the weak topology of E_σ (i.e., the topology as a subspace of $\check{E}_\sigma$), in stead of the strong topology, then the space $\Gamma\backslash D_\Xi$ becomes not Hausdorff for any congruence subgroup Γ of $G_{\mathbf{Z}}$ (see 12.3.6). We also give an example such that E_σ is not open in the weak topology of $\tilde{E}_\sigma$ in $\check{E}_\sigma$ (12.3.10), though E_σ is open in the strong topology of $\tilde{E}_\sigma$ in $\check{E}_\sigma$ by Theorem A (i).

Let H_0, $\langle\ ,\ \rangle_0$, $(e_j)_{1\leq j\leq 4}$ be as in the case of 12.1.1 for $g = 2$. Let $w = 3$ and let

$$h^{p,q} = 1 \ \ (p+q = 3,\ p, q \geq 0), \quad h^{p,q} = 0 \ \ \text{otherwise}.$$

12.3.1

For $\tau = (\tau_{ij}) \in M(2, \mathbf{C})_{\mathrm{sym}}$ and $\lambda \in \mathbf{C}$, define $F(\tau, \lambda) \in \check{D}$ as follows. Let

$$v_1 = \tau_{11}e_1 + \tau_{21}e_2 + e_3, \qquad v_2 = \tau_{12}e_1 + \tau_{22}e_2 + e_4.$$

Define

$$\begin{aligned}
&F^0(\tau, \lambda) = H_{0,\mathbf{C}}, \quad F^4(\tau, \lambda) = 0,\\
&F^2(\tau, \lambda) = (\text{the } \mathbf{C}\text{-subspace of } H_{0,\mathbf{C}} \text{ spanned by } v_1 \text{ and } v_2),\\
&F^3(\tau, \lambda) = (\text{the } \mathbf{C}\text{-subspace of } H_{0,\mathbf{C}} \text{ spanned by the vector } \lambda v_1 + v_2),\\
&F^1(\tau, \lambda) = F^3(\tau, \lambda)^\perp.
\end{aligned}$$

Then we have an open immersion:

$$M(2, \mathbf{C})_{\mathrm{sym}} \times \mathbf{C} \hookrightarrow \check{D}, \quad (\tau, \lambda) \mapsto F(\tau, \lambda).$$

PROPOSITION 12.3.2 *We have*

$$F(\tau, \lambda) \in D \Longleftrightarrow \begin{cases} \lambda\bar{\lambda}\,\mathrm{Im}(\tau_{11}) + (\lambda + \bar{\lambda})\,\mathrm{Im}(\tau_{12}) + \mathrm{Im}(\tau_{22}) < 0,\\ \det(\mathrm{Im}(\tau)) < 0. \end{cases}$$

12.3.3

The action of $G_{\mathbf{C}} = \mathrm{Sp}(2, \mathbf{C})$ on these $F(\tau, \lambda) \in \check{D}$ is given by

$$F(\tau, \lambda) \mapsto F((A\tau + B)(C\tau + D)^{-1}, \lambda') \quad \left(\begin{pmatrix} A & B \\ C & D \end{pmatrix} \in G_{\mathbf{C}}\right),$$

$$\text{where } \lambda' = (p\lambda + q)(r\lambda + s)^{-1} \text{ if } C\tau + D = \begin{pmatrix} p & q \\ r & s \end{pmatrix}.$$

Here we assume that $C\tau + D$ is invertible and $r\lambda + s \neq 0$.

12.3.4

Define the two nilpotent elements N_α and N_β of $\mathfrak{g}_{\mathbf{Q}}$ by

$$N_\alpha(e_3) = e_1, \quad N_\alpha(e_j) = 0 \ (j \neq 3);$$
$$N_\beta(e_4) = e_3, \quad N_\beta(e_3) = -e_1, \quad N_\beta(e_1) = -e_2, \quad N_\beta(e_2) = 0.$$

Let S be the set of all square free positive integers. Define also nilpotent elements N_m $(m \in S)$ of $\mathfrak{g}_{\mathbf{Q}}$ by

$$N_m(e_1) = e_3, \quad N_m(e_4) = -me_2, \quad N_m(e_j) = 0 \text{ for } j = 2, 3.$$

Then

$$N_\alpha^2 = N_m^2 = 0, \quad N_\beta^4 = 0, \ N_\beta^3 \neq 0.$$

Put

$$\sigma_\alpha := (\mathbf{R}_{\geq 0})N_\alpha, \quad \sigma_\beta := (\mathbf{R}_{\geq 0})N_\beta, \quad \sigma_m := (\mathbf{R}_{\geq 0})N_m \quad (m \in S).$$

We have

$$\exp(zN_\alpha)F(\tau, \lambda) = F\left(\tau + \begin{pmatrix} z & 0 \\ 0 & 0 \end{pmatrix}, \lambda\right),$$

$$\exp(zN_\beta)F\left(\begin{pmatrix} 0 & 0 \\ 0 & w \end{pmatrix}, 0\right) = F\left(\begin{pmatrix} -z & \frac{1}{2}z^2 \\ \frac{1}{2}z^2 & w - \frac{1}{3}z^3 \end{pmatrix}, z\right),$$

$$\exp(zN_m)F\left(\begin{pmatrix} 0 & b \\ b & w \end{pmatrix}, \lambda\right) = F\left(\begin{pmatrix} 0 & b \\ b & w \end{pmatrix}, \lambda + bz\right) \quad \text{if } b = \pm i\sqrt{m},$$

for $z, \lambda, w \in \mathbf{C}$.

PROPOSITION 12.3.5

(i) *If σ is a rational nilpotent cone and if there is a σ-nilpotent orbit, then $\sigma \in \Xi$.*

(ii) *D_Ξ is the union of $g(D_{\sigma_\alpha})$ $(g \in G_{\mathbf{Q}})$, $g(D_{\sigma_\beta})$ $(g \in G_{\mathbf{Q}})$ and $g(D_{\sigma_m})$ $(g \in G_{\mathbf{Q}}, m \in S)$.*

(iii) *Let $\sigma = \sigma_\alpha$ (resp. σ_β, σ_m with $m \in S$), and let*

$$U = \{(w, z) \in \mathbf{C}^2 \mid \mathrm{Im}(z) < 0\} \quad (\textit{resp. } \mathbf{C}, \ \mathbf{C} \times \{\pm 1\}).$$

Then, we have a bijection $U \to D_\sigma - D$ given by

$$(w, z) \mapsto \left(\sigma, \exp(\sigma_{\mathbf{C}})F\left(\begin{pmatrix} 0 & w \\ w & z \end{pmatrix}, 0\right)\right)$$

$$\left(\textit{resp. } z \mapsto \left(\sigma,\ \exp(\sigma_{\mathbf{C}})F\left(\begin{pmatrix} 0 & 0 \\ 0 & z \end{pmatrix}, 0\right)\right),\right.$$

$$\left.(z, \varepsilon) \mapsto \left(\sigma,\ \exp(\sigma_{\mathbf{C}})F\left(\begin{pmatrix} 0 & \varepsilon i\sqrt{m} \\ \varepsilon i\sqrt{m} & z \end{pmatrix}, 0\right)\right)\right).$$

This map induces a homeomorphism from U onto the topological subspace $\Gamma(\sigma)^{\mathrm{gp}}\backslash D_\sigma - \Gamma(\sigma)^{\mathrm{gp}}\backslash D$ of $\Gamma(\sigma)^{\mathrm{gp}}\backslash D_\sigma$.

Proof. We first prove (ii). Let N be a non-zero nilpotent element of $\mathfrak{g}_{\mathbf{Q}}$. Then there are four cases.

(1) $N^2 = 0$ and $\dim(\mathrm{Image}(N)) = 1$.
(2) $N^2 = 0$ and $\dim(\mathrm{Image}(N)) = 2$.
(3) $N^3 = 0$ and $N^2 \neq 0$.
(4) $N^4 = 0$ and $N^3 \neq 0$.

Assume (N, F) generates a nilpotent orbit for some $F \in \check{D}$. We show that (3) does not happen. In the case (3), $F\left(\mathrm{gr}^{W(N)}_{2,\mathbf{C}}\right)$ must be a Hodge structure of dimension one of odd weight $3+2 = 5$, but it is impossible.

In the case (1), since $(W(N)[-3], F)$ is a mixed Hodge structure, $F\left(\mathrm{gr}^{W(N)}_{1,\mathbf{C}}\right)$ is a one-dimensional polarized Hodge structure of weight $3+1 = 4$ with respect to the pairing $(x, y) \mapsto \langle x, N(y)\rangle_0$. Hence its Hodge type must be $(2, 2)$ and this pairing on $\mathrm{gr}^{W(N)}_1$ must be positive definite. Since $\langle e_3, N_\alpha(e_3)\rangle_0 = \langle e_3, e_1\rangle_0 = 1 > 0$, there exists an element $g \in G_{\mathbf{Q}}$ such that $N = c\cdot \mathrm{Ad}(g)(N_\alpha)$ for some $c \in \mathbf{Q}_{>0}$.

In the case (2), since $(W(N)[-3], F)$ is a mixed Hodge structure, $F\left(\mathrm{gr}^{W(N)}_{1,\mathbf{C}}\right)$ is a two-dimensional polarized Hodge structure of weight $3+1 = 4$ with respect to the pairing $(x, y) \mapsto \langle x, N(y)\rangle_0$. Hence its Hodge type must be $(3, 1)+(1, 3)$ (all other possibilities do not fit the assumption on $(h^{p,q})$), and this pairing on $\mathrm{gr}^{W(N)}_1$ must be negative definite. The quadratic form associated to this pairing must be isomorphic to $-c(x^2+my^2)$ over $\mathbf{Q}$ for some $m \in S$ and $c \in \mathbf{Q}_{>0}$. Since $\langle xe_1 + ye_4, N_m(xe_1+ye_4)\rangle_0 = \langle xe_1+ye_4, xe_3 - mye_2\rangle_0 = -(x^2+my^2)$, there exists an element $g \in G_{\mathbf{Q}}$ such that $N = c\cdot \mathrm{Ad}(g)(N_m)$ for some $c \in \mathbf{Q}_{>0}$.

In the case (4), since $(W(N)[-3], F)$ is a mixed Hodge structure, $F\left(\mathrm{gr}^{W(N)}_{3,\mathbf{C}}\right)$ is a one-dimensional polarized Hodge structure of weight $3+3 = 6$ with respect to the pairing $(x, y) \mapsto \langle x, N^3(y)\rangle_0$. Hence its Hodge type must be $(3, 3)$ and this pairing on $\mathrm{gr}^{W(N)}_3$ must be positive definite. Since $\langle e_4,\ N_\beta^3(e_4)\rangle_0 = \langle e_4,\ e_2\rangle_0 = 1 > 0$, there exists an element $g \in G_{\mathbf{Q}}$ such that $N = c\cdot \mathrm{Ad}(g)(N_\beta)$ for some $c \in \mathbf{Q}_{>0}$.

Next we prove (iii). The proofs for σ_a, σ_b, σ_m are similar, and so we give here only the proof for $\sigma = \sigma_m$. Put $W := W(N_m)$. If $(\sigma, Z) \in D_\sigma - D$, then, for $F \in Z$, $(W[-3], F)$ is a mixed Hodge structure, $F\left(\mathrm{gr}^W_{1,\mathbf{C}}\right)$ must be of Hodge type $(3, 1)+(1, 3)$ and $F\left(\mathrm{gr}^W_{-1,\mathbf{C}}\right)$ must be of Hodge type $(2, 0)+(0, 2)$. Hence there exist $p, q, r, b \in \mathbf{C}$ such that $pe_1+qe_2+re_3+e_4 \in F^3$ and $be_2+e_3 \in F^2$. From $\langle F^2, F^3\rangle_0 = 0$, we have $p = b$. From $N_m(F^3) \subset F^2$, we have $m = -pb$. Hence $b^2 = -m$. Replacing F by $\exp(-b^{-1}rN_m)F$, we have that $be_1+ze_2+e_4 \in F^3$ and $be_2+e_3 \in F^2$ for some $b, z \in \mathbf{C}$ such that $b^2 = -m$. Thus $F = F\left(\left(\begin{smallmatrix} 0 & b \\ b & z \end{smallmatrix}\right), 0\right)$.

Finally, we prove (i). Let (σ, Z) be a nilpotent orbit with σ rational. Take rational elements N, N' of the interior (0.7.7) of σ. It is sufficient to prove $N' \in \mathbf{Q}N$.

By (ii), there exists $g \in G_{\mathbf{Q}}$ and $c \in \mathbf{Q}_{>0}$ such that $c \cdot \mathrm{Ad}(g)(N)$ is equal either one of N_α, N_β, N_m $(m \in S)$. Hence we may assume N is either one of N_α, N_β, N_m $(m \in S)$. Note that N' has the same weight filtration as N by (5.2.4). Let $F \in Z$. Then $((\mathbf{R}_{\geq 0})N, \exp(\mathbf{C}N)F)$ belongs to $D_{(\mathbf{R}_{\geq 0})N}$. By using (iii), we can check easily that, if an element N' of $\mathfrak{g}_{\mathbf{Q}}$ satisfies $NN' = N'N$ and $N'(W(N)_k) \subset W(N)_{k-2}$ for all $k \in \mathbf{Z}$ and $N'(F^p) \subset F^{p-1}$ for all $p \in \mathbf{Z}$, then $N' = cN$ for some $c \in \mathbf{Q}$. □

The rest of Section 12.3 is devoted to the proof of

Proposition 12.3.6 *Let Γ be a subgroup of $G_{\mathbf{Z}} = \mathrm{Sp}(2, \mathbf{Z})$ which contains*

$$\mathrm{Ker}(\mathrm{Sp}(2, \mathbf{Z}) \to \mathrm{Sp}(2, \mathbf{Z}/n\mathbf{Z}))$$

for some $n \geq 1$. Let $(\Gamma \backslash D_\Xi)_{\mathrm{weak}}$ be the set $\Gamma \backslash D_\Xi$ endowed with the strongest topology in which, for any $\sigma \in \Xi$, $E_\sigma \to \Gamma \backslash D_\Xi$ is continuous in the topology of E_σ as a subspace of $\check{E}_\sigma$. Then, $(\Gamma \backslash D_\Xi)_{\mathrm{weak}}$ is not Hausdorff. Moreover, there exists a point of $(\Gamma \backslash D_\Xi)_{\mathrm{weak}}$ which has no Hausdorff neighborhoods.

12.3.7

In the rest of Section 12.3, let $N = N_\alpha$, $\sigma = \sigma_\alpha$.

For $t \in \mathbf{R}_{>0}$, put

$$p(t) := \left(0,\ F\left(\begin{pmatrix} 0 & 0 \\ 0 & -it \end{pmatrix}, 0\right)\right) \in E_\sigma,$$

$$\overline{p}(t) \in \Gamma \backslash D_\Xi : \text{the image of } p(t).$$

Here 0 is the origin of toric_σ. For the proof of Proposition 12.3.6, it is enough to prove the following (1) and (2).

(1) For any $t, t' \in \mathbf{R}_{>0}$, any pair of neighborhoods of $\overline{p}(t)$ and $\overline{p}(t')$ in $(\Gamma \backslash D_\Xi)_{\mathrm{weak}}$ intersect each other.

(2) If $t \in \mathbf{R}$ and if $t' \neq t$ is close to t, then $\overline{p}(t') \neq \overline{p}(t)$ in $\Gamma \backslash D_\Xi$

Note that, if $t' \to t$ then $p(t') \to p(t)$ in the weak topology of E_σ, and hence $\overline{p}(t')$ converges to $\overline{p}(t)$ in $(\Gamma \backslash D_\Xi)_{\mathrm{weak}}$. Hence (1) and (2) show that $\overline{p}(t)$ has no Hausdorff neighborhoods in $(\Gamma \backslash D_\Xi)_{\mathrm{weak}}$.

12.3.8

Proof of 12.3.7 (1). It is enough to show that there exists a sequence of points x_m $(m \geq 1)$ in $\Gamma \backslash D$ such that

$$x_m \to \overline{p}(t) \ \text{ and } \ x_m \to \overline{p}(t') \ \text{ in } (\Gamma \backslash D_\Xi)_{\mathrm{weak}}. \tag{1}$$

Take

$$A_m = \begin{pmatrix} a_m & b_m \\ c_m & d_m \end{pmatrix} \in \mathrm{Ker}(\mathrm{SL}(2, \mathbf{Z}) \to \mathrm{SL}(2, \mathbf{Z}/n\mathbf{Z})) \quad \text{such that}$$

$$a_m, b_m, c_m, d_m > 0, \ \frac{b_m}{c_m} \to \infty, \ \frac{d_m}{a_m} \to \frac{t'}{t}, \ \frac{c_m}{a_m} \to 0. \tag{2}$$

This is possible. In fact, take $f_m, g_m > 0$ so that $f_m, g_m \to \infty$, $\frac{f_m}{g_m} \to \frac{t'}{t}$, and put

$$A_m = \begin{pmatrix} a_m & b_m \\ c_m & d_m \end{pmatrix} := \begin{pmatrix} n^2 g_m + 1 & n(n^2 f_m g_m + f_m + g_m) \\ n & n^2 f_m + 1 \end{pmatrix}$$

Then this satisfies all the conditions for A_m in (2).

Put

$$\gamma_m := \begin{pmatrix} A_m & 0 \\ 0 & {}^t A_m^{-1} \end{pmatrix} \in \Gamma,$$

$$p_m := \left(\mathbf{e}\left(i\frac{tb_m d_m}{a_m c_m} N\right),\ F\left(\begin{pmatrix} 0 & 0 \\ 0 & -it \end{pmatrix}, 0\right)\right) \in E_\sigma .$$

Then p_m clearly converges to $p(t)$ in the weak topology of E_σ.

The image of p_m in $\Gamma\backslash D$ is the image of the element $q_m \in D$ where

$$q_m := \exp\left(i\frac{tb_m d_m}{a_m c_m} N\right) F\left(\begin{pmatrix} 0 & 0 \\ 0 & -it \end{pmatrix}, 0\right).$$

By 12.3.3, we have

$$\begin{aligned} \gamma_m q_m &= \gamma_m F\left(\begin{pmatrix} i\dfrac{tb_m d_m}{a_m c_m} & 0 \\ 0 & -it \end{pmatrix}, 0\right) \\ &= F\left(\begin{pmatrix} i\dfrac{tb_m}{c_m} & 0 \\ 0 & -i\dfrac{td_m}{a_m} \end{pmatrix}, -\dfrac{c_m}{a_m}\right) \\ &= \exp\left(i\frac{tb_m}{c_m} N\right) F\left(\begin{pmatrix} 0 & 0 \\ 0 & -i\dfrac{td_m}{a_m} \end{pmatrix}, -\frac{c_m}{a_m}\right). \end{aligned}$$

Let

$$p'_m := \left(\mathbf{e}\left(i\frac{tb_m}{c_m} N\right),\ F\left(\begin{pmatrix} 0 & 0 \\ 0 & -i\frac{td_m}{a_m} \end{pmatrix}, -\frac{c_m}{a_m}\right)\right) \in E_\sigma .$$

Then, by (2), p'_m clearly converges to $p(t')$ in the weak topology of E_σ, and the image of p'_m in $\Gamma\backslash D$ coincides with the image of $\gamma_m q_m$. Hence the images of p_m and p'_m in $\Gamma\backslash D$ coincide. Let $x_m \in \Gamma\backslash D$ be this image. Since p_m converges to $p(t)$ and p'_m converges to $p(t')$ in the weak topology, x_m converges to both $\overline{p}(t)$ and $\overline{p}(t')$ in $(\Gamma\backslash D_\Xi)_{\mathrm{weak}}$. □

12.3.9

Proof of 12.3.7(2). Let $t, t' \in \mathbf{R}_{>0}$ and assume $\overline{p}(t') = \overline{p}(t)$. We will show that if t' is close to t, then $t' = t$.

There exists $\gamma \in \Gamma$ such that

$$\gamma\left(\sigma,\ \exp(\sigma_{\mathbf{C}})F\left(\begin{pmatrix} 0 & 0 \\ 0 & -it \end{pmatrix}, 0\right)\right) = \left(\sigma,\ \exp(\sigma_{\mathbf{C}})F\left(\begin{pmatrix} 0 & 0 \\ 0 & -it' \end{pmatrix}, 0\right)\right) \quad \text{in } D_\Xi .$$

Hence, $\mathrm{Ad}(\gamma)(\sigma) = \sigma$ and so $\mathrm{Ad}(\gamma)(N) = cN$ for some $c \in \mathbf{Q}_{>0}$. It follows that γ preserves the N-filtration $W := W(N)$ and also the induced Hodge filtration on $\mathrm{gr}^W_{0,\mathbf{C}}$, i.e.,

$$\gamma F = F, \quad \text{where} \quad F := F\left(\begin{pmatrix} 0 & 0 \\ 0 & -it' \end{pmatrix}, 0\right)(\mathrm{gr}^W_0). \tag{1}$$

Under the identification $\mathrm{gr}^W_0 \simeq \mathbf{R}e_2 \oplus \mathbf{R}e_4$, the induced Hodge filtration F becomes

$$F^0 = \mathrm{gr}^W_{0,\mathbf{C}}, \; F^4 = \{0\}, \; F^1 = F^2 = F^3 = \mathbf{C}(-ite_2 + e_4).$$

and the action of γ on gr^W_0 belongs to $\mathrm{SL}(2, \mathbf{Z})$ with respect to the basis e_2, e_4. Denote this element of $\mathrm{SL}(2, \mathbf{Z})$ by $\left(\begin{smallmatrix} a & b \\ c & d \end{smallmatrix}\right)$. Then (1) becomes $\frac{ita-b}{-itc+d} = it'$. This means that there is an element of $\mathrm{SL}(2, \mathbf{Z})$ which sends $it \in \mathfrak{h}$ to $it' \in \mathfrak{h}$. But this is impossible if $t' \neq t$ and t' is sufficiently close to t. □

12.3.10

We prove that E_σ ($\sigma = \sigma_\alpha$) is not open in the weak topology of $\tilde{E}_\sigma$ in $\check{E}_\sigma$. Let

$$p = \left(0, F\left(\begin{pmatrix} 0 & 0 \\ 0 & -i \end{pmatrix}, 0\right)\right) \in E_\sigma.$$

Take a real number $c > 2$. For $y \in \mathbf{R}$ with $y > 0$, let

$$p(y) = \left(\mathbf{e}(iy^c N), F\left(\begin{pmatrix} 0 & 0 \\ 0 & -i \end{pmatrix}, \frac{1}{y}\right)\right) \in \tilde{E}_\sigma.$$

Then, $p(y) \to p$ when $y \to \infty$ in the weak topology of $\tilde{E}_\sigma$. However, when $y > 1$, $p(y) \notin E_\sigma$, because, by 12.3.2,

$$\exp(iy^c N) F\left(\begin{pmatrix} 0 & 0 \\ 0 & -i \end{pmatrix}, \frac{1}{y}\right) = F\left(\begin{pmatrix} iy^c & 0 \\ 0 & -i \end{pmatrix}, \frac{1}{y}\right) \notin D.$$

12.3.11

Remark. The example, we treated in this section, is the Hodge structures of weight 3 of the mirrors of quintic hypersurfaces in $\mathbf{P}^4$ (cf., for example, [Mo]).

12.4 EXAMPLE OF WEIGHT 3 (B)

In Sections 12.2 and 12.3, where we considered nonclassical examples, we had only boundary points of rank 1. Here we consider boundary points of rank 2 in a nonclassical situation. We observe how some boundary points of rank 2 behave under the maps

$$D^\sharp_{\sigma,\mathrm{val}} \to D_{\mathrm{SL}(2)} \leftarrow D_{\mathrm{SL}(2),\mathrm{val}} \to D_{\mathrm{BS}}.$$

Let H_0, $\langle\ ,\ \rangle_0$ and $(e_j)_{1\le j\le 6}$ be as in Section 12.1 for $g=3$. Let $w=3$ and let

$$h^{p,q}=2 \quad \text{if } (p,q)=(2,1),(1,2), \quad h^{p,q}=1 \quad \text{if } (p,q)=(3,0),(0,3),$$
$$h^{p,q}=0 \quad \text{otherwise.}$$

12.4.1

Let

$$N_1 := \begin{pmatrix} 0&0&0&0&0&-1\\ 0&0&0&0&-1&0\\ 0&0&0&-1&0&0\\ 0&0&0&0&0&0\\ 0&0&0&0&0&0\\ 0&0&0&0&0&0 \end{pmatrix}, \quad N_2 := \begin{pmatrix} 0&0&0&0&0&0\\ 1&0&0&0&0&0\\ 0&-1&0&0&0&0\\ 0&0&0&0&-1&0\\ 0&0&0&0&0&1\\ 0&0&0&0&0&0 \end{pmatrix},$$

$$\sigma := (\mathbf{R}_{\ge 0})N_1 + (\mathbf{R}_{\ge 0})N_2 \subset \mathfrak{g}_{\mathbf{R}}.$$

For $a,b,c\in\mathbf{R}^\times$, let

$$d(a,b,c) = \begin{pmatrix} a^{-1}&0&0&0&0&0\\ 0&b^{-1}&0&0&0&0\\ 0&0&c^{-1}&0&0&0\\ 0&0&0&a&0&0\\ 0&0&0&0&b&0\\ 0&0&0&0&0&c \end{pmatrix} \in G_{\mathbf{R}}.$$

Define $F_{(0)}\in\check{D}$ by

$$F^4_{(0)}=0, \quad F^3_{(0)}=\mathbf{C}e_6, \quad F^2_{(0)}=\mathbf{C}e_1+\mathbf{C}e_5+\mathbf{C}e_6,$$
$$F^1_{(0)}=\mathbf{C}e_1+\mathbf{C}e_2+\mathbf{C}e_4+\mathbf{C}e_5+\mathbf{C}e_6, \quad F^0_{(0)}=H_{0,\mathbf{C}}.$$

Then $(\sigma, F_{(0)})$ generates a nilpotent orbit.

There is a unique continuous map

$$\alpha : |\Delta|^2 \to D^\sharp_\sigma$$

satisfying

$$\alpha(r_1,r_2)=\exp(iy_1N_1+iy_2N_2)F_{(0)} \quad \text{for } (r_1,r_2)\in|\Delta^*|^2,$$

where $y_j=-\frac{1}{2\pi}\log(r_j)>0$ and $|\Delta^*|:=|\Delta|-\{0\}$. This map satisfies $\alpha(0,0)=(\sigma,\exp(i\sigma_{\mathbf{R}})F_{(0)})$. There is a unique continuous map

$$\alpha_{\mathrm{val}} : (|\Delta|^2)_{\mathrm{val}} \to D^\sharp_{\sigma,\mathrm{val}}$$

which is compatible with α.

12.4.2

There exists a unique SL(2)-orbit in two variables (ρ,φ) whose N_j are the above N_j $(j=1,2)$ and $\varphi(0,0)=F_{(0)}$. We have $\tilde\rho(t_1,t_2)=d(t_1t_2^{-1},t_1t_2,t_1t_2^3)$. Let

$$\mathbf{r}=\varphi(i,i)=\exp(iN_1+iN_2)F_{(0)}.$$

Then there exists a unique continuous map

$$\beta : \mathbf{R}^2_{\geq 0} \to D_{\mathrm{SL}(2)}$$

such that

$$\beta(t_1, t_2) = \tilde{\rho}(t_1, t_2) \cdot \mathbf{r} \quad \text{for} \quad (t_1, t_2) \in \mathbf{R}^2_{>0}.$$

This map satisfies $\beta(0, 0) = [\rho, \varphi]$. There exists a unique continuous map

$$\beta_{\mathrm{val}} : (\mathbf{R}^2_{\geq 0})_{\mathrm{val}} \to D_{\mathrm{SL}(2),\mathrm{val}},$$

which is compatible with β.

12.4.3

Let P (resp. Q) be the $\mathbf{Q}$-parabolic subgroup of $G_{\mathbf{R}}$ consisting of all elements which preserve the $\mathbf{R}$-subspaces

$$\mathbf{R}e_3, \quad \mathbf{R}e_2 + \mathbf{R}e_3, \quad \mathbf{R}e_1 + \mathbf{R}e_2 + \mathbf{R}e_3 + \mathbf{R}e_4, \quad \mathbf{R}e_1 + \mathbf{R}e_2 + \mathbf{R}e_3 + \mathbf{R}e_4 + \mathbf{R}e_5$$

and the $\mathbf{R}$-subspace $\mathbf{R}e_1 + \mathbf{R}e_2 + \mathbf{R}e_3$ (resp. $\mathbf{R}e_2 + \mathbf{R}e_3 + \mathbf{R}e_4$) of $H_{0,\mathbf{R}}$. Then there exists a unique continuous map

$$\beta_P : \mathbf{R}^3_{\geq 0} \to D_{\mathrm{BS}}(P),$$

which satisfies

$$\beta_P(t_1, t_2, t_3) = d(t_1, t_1t_2, t_1t_2t_3) \cdot \mathbf{r} \quad \text{for} \quad (t_1, t_2, t_3) \in \mathbf{R}^3_{>0},$$

and there exists a unique continuous map

$$\beta_Q : \mathbf{R}^3_{\geq 0} \to D_{\mathrm{BS}}(Q),$$

which satisfies

$$\beta_Q(t_1, t_2, t_3) = d(t_1^{-1}, t_1t_2, t_1t_2t_3) \cdot \mathbf{r} \quad \text{for} \quad (t_1, t_2, t_3) \in \mathbf{R}^3_{>0}.$$

We have

$$\beta_P(0, 0, 0) = (P,\ d(\mathbf{R}^3_{>0}) \cdot \mathbf{r}), \quad \beta_Q(0, 0, 0) = (Q,\ d(\mathbf{R}^3_{>0}) \cdot \mathbf{r}) \ \text{ in } \ D_{\mathrm{BS}}.$$

Proposition 12.4.4 *Let V be the open set of $(|\Delta|^2)_{\mathrm{val}}$ consisting of (r_1, r_2) $(r_1, r_2 \in |\Delta|,\ r_2 \neq 0)$ and all elements lying over $(0, 0) \in |\Delta|^2$ which are not equal to $(0, 0)_\infty$. (See* 0.5.21 *for the notation for elements of $(|\Delta|^2)_{\mathrm{val}}$ lying over $(0, 0) \in |\Delta|^2$.) Then we have a commutative diagram*

$$\begin{array}{ccc} (|\Delta|^2)_{\mathrm{val}} \supset V & \xrightarrow{\alpha_{\mathrm{val}}} & D^\sharp_{\sigma,\mathrm{val}} \\ \downarrow & & \downarrow \psi \\ \mathbf{R}^2_{\geq 0} & \xrightarrow{\beta} & D_{\mathrm{SL}(2)}. \end{array}$$

Here the left vertical arrow sends $(r_1, r_2) \in V$ $(r_1, r_2 \in |\Delta|, r_2 \neq 0)$ *to*

$$\left(\frac{\log(r_2)}{\log(r_1)}, \left(-\frac{2\pi}{\log(r_2)} \right)^{1/2} \right),$$

and sends $(0, 0)_s$ *and* $(0, 0)_{s,z}$ *to* $(s, 0)$.

Proof. Easy. □

12.4.5

Note that $(\mathbf{R}^2_{\geq 0})_{\mathrm{val}}$ is covered by the two open subsets $(\mathbf{R}^2_{\geq 0})_{\mathrm{val}}(P)$ and $(\mathbf{R}^2_{\geq 0})_{\mathrm{val}}(Q)$ defined as follows. The former consists of all (r_1, r_2) $(r_1, r_2 \in \mathbf{R}_{\geq 0}, r_2 \neq 0)$, points $(0, 0)_s$ and $(0, 0)_{s,z}$ with $s < 1$, and points $(0, 0)_{1,z}$ with $0 \leq z < \infty$. The latter consists of all (r_1, r_2) $(r_1, r_2 \in \mathbf{R}_{\geq 0}, r_1 \neq 0)$, points $(0, 0)_s$ and $(0, 0)_{s,z}$ with $s > 1$, and points $(0, 0)_{1,z}$ with $0 < z \leq \infty$.

Proposition 12.4.6 *We have commutative diagrams*

$$\begin{array}{ccc} (\mathbf{R}^2_{\geq 0})_{\mathrm{val}}(P) & \xrightarrow{\beta_{\mathrm{val}}} & D_{\mathrm{SL}(2),\mathrm{val}} \\ {\scriptstyle\delta}\downarrow & & \downarrow \\ \mathbf{R}^3_{\geq 0} & \xrightarrow{\beta_P} & D_{\mathrm{BS}}, \end{array} \qquad \begin{array}{ccc} (\mathbf{R}^2_{\geq 0})_{\mathrm{val}}(Q) & \xrightarrow{\beta_{\mathrm{val}}} & D_{\mathrm{SL}(2),\mathrm{val}} \\ {\scriptstyle\delta'}\downarrow & & \downarrow \\ \mathbf{R}^3_{\geq 0} & \xrightarrow{\beta_Q} & D_{\mathrm{BS}}. \end{array}$$

Here the map δ *sends* (r_1, r_2) *with* $r_2 \neq 0$ *to*

$$\left(\frac{r_1}{r_2}, r_2^2, r_2^2 \right),$$

$(0, 0)_s$ *and* $(0, 0)_{s,z}$ *with* $s < 1$ *to* $(0, 0, 0)$, *and* $(0, 0)_{1,z}$ *to* $(z, 0, 0)$. *The map* δ' *sends* (r_1, r_2) *with* $r_1 \neq 0$ *to*

$$\left(\frac{r_2}{r_1}, r_1^2, r_2^2 \right),$$

$(0, 0)_s$ *and* $(0, 0)_{s,z}$ *with* $s > 1$ *to* $(0, 0, 0)$, *and* $(0, 0)_{1,z}$ *to* $(z^{-1}, 0, 0)$.

Proof. Easy. □

12.4.7

Contrary to the case of §12.1, there is no continuous map $D^\sharp_{\sigma,\mathrm{val}} \to D_{\mathrm{BS}}$ and there is no continuous map $D_{\mathrm{SL}(2)} \to D_{\mathrm{BS}}$ which extend the identity map of D. In fact, by 12.4.2, 12.4.3, when $t \in \mathbf{R}_{>0}$ converges to 0, $\beta(t^2, t) \in D$ converges to $\beta_P(0, 0, 0) = (P, d(\mathbf{R}^3_{>0}) \cdot \mathbf{r})$ in D_{BS} and $\beta(t, t^2) \in D$ converges to $\beta_Q(0, 0, 0) = (Q, d(\mathbf{R}^3_{>0}) \cdot \mathbf{r}) \neq (P, d(\mathbf{R}^3_{>0}) \cdot \mathbf{r})$ in D_{BS}. On the other hand, both $\beta(t^2, t) = \exp(it^{-6}N_1 + it^{-2}N_2)F_{(0)}$ and $\beta(t, t^2) = \exp(it^{-6}N_1 + it^{-4}N_2)F_{(0)}$ converge to $\alpha_{\mathrm{val}}((0, 0)_0)$ in $D^\sharp_{\sigma,\mathrm{val}}$ and also both converge to $\beta(0, 0)$ in $D_{\mathrm{SL}(2)}$.

12.5 RELATIONSHIP WITH [U2]

We describe the relationship of this book with the work [U2], which is the prototype of the present work.

12.5.1

Let Ξ' be the subset of Ξ consisting of $\{0\}$ and all elements of the form $(\mathbf{R}_{\geq 0})N$ where N is any element in $\mathfrak{g}_{\mathbf{Q}}$ satisfying the following condition (1).

$$N^2 = 0, \quad \dim(\mathrm{Image}(N)) = \begin{cases} 1 & \text{if } w \text{ is odd,} \\ 2 & \text{if } w \text{ is even.} \end{cases} \tag{1}$$

In [U2], the space $D_{\Xi'}$, extended period maps into $\Gamma\backslash D_{\Xi'}$ ($\Gamma = G_{\mathbf{Z}}$ is considered there), and their differentials are constructed. The notation and the topology of $\Gamma\backslash D_{\Xi'}$ in [U2] are different from those of this book. We describe the relationship with the formulation in this book, together with additional comments.

Note that N, as in (1) with $w = 2$, arises, for example, as the logarithm of the local monodromy of degenerations of surfaces with a simple elliptic singularity. The motivation of [U2] was the hope to start to extend Griffiths' theory of period maps and their differentials in such situations.

12.5.2

Let I be the set of rational increasing filtrations W on $H_{0,\mathbf{R}}$ satisfying the following (1) and (2).

$$W_1 = H_{0,\mathbf{R}}, \quad W_{-2} = 0, \quad \langle W_0, W_{-1}\rangle_0 = 0. \tag{1}$$

$$\dim(\mathrm{gr}^W_1) = \begin{cases} 1 & \text{if } w \text{ is odd,} \\ 2 & \text{if } w \text{ is even.} \end{cases} \tag{2}$$

We will consider the open set $\cup_{W\in I} D_{\mathrm{SL}(2)}(W)$ of $D_{\mathrm{SL}(2)}$.

PROPOSITION 12.5.3

(i) *The continuous map*

$$D^\sharp_{\Xi'} = D^\sharp_{\Xi',\mathrm{val}} \xrightarrow{\psi} \cup_{W\in I} D_{\mathrm{SL}(2)}(W)$$

is bijective.

(ii) *The canonical map*

$$\bigcup_{W\in I} D_{\mathrm{SL}(2)}(W) = \bigcup_{W\in I} D_{\mathrm{SL}(2),\mathrm{val}}(W) \to D_{\mathrm{BS}}$$

is injective, and the topology of $\bigcup_{W\in I} D_{\mathrm{SL}(2)}(W)$ *coincides with the topology induced from the topology of* D_{BS}.

Proof. (i) is proved in [U2, (3.12) (ii)]. We reprove (i) here using terminologies in the this book. Let $W \in I$, and define a rational $\mathbf{R}$-subspace L_W of $\mathfrak{g}_{\mathbf{R}}$ by

$$L_W = \{X \in \mathfrak{g}_{\mathbf{R}} \mid X(W_0) = 0\}. \tag{1}$$

Then if w is odd, L_W is identified with the space of all $\mathbf{R}$-linear maps from the one-dimensional $\mathbf{R}$-space gr_1^W to the one-dimensional $\mathbf{R}$-space gr_{-1}^W. In the case w is even, the map from L_W to the set of all antisymmetric bilinear forms $\mathrm{gr}_1^W \times \mathrm{gr}_1^W \to \mathbf{R}$ which sends $X \in L_W$ to the bilinear form $(x, y) \mapsto \langle X(x), y\rangle_0$, is bijective. Hence, in any case, we have

$$\dim_{\mathbf{R}} L_W = 1. \tag{2}$$

We define a map of the converse direction

$$f : \bigcup_{W \in I} D_{\mathrm{SL}(2)}(W) \to D^{\sharp}_{\Xi'} \tag{3}$$

as follows. Let $W \in I$ and $p \in D_{\mathrm{SL}(2)}(W) - D$, let (ρ, φ) be an SL(2)-orbit in one variable representing p, and let $N = \rho_* \left(\begin{smallmatrix} 0 & 1 \\ 0 & 0 \end{smallmatrix}\right)$. Let $\sigma = (\mathbf{R}_{\geq 0})N$. Then, since $N \in L_W$, we have by (2) that σ is rational. Define $f(p) := (\sigma, \varphi(i\mathbf{R})) \in D^{\sharp}_{\Xi'}$. Then $f(p)$ is independent of the choice of the representative (ρ, φ) of p. We show that $\psi \circ f$ and $f \circ \psi$ are the identity maps. This is clear for $\psi \circ f$ by 6.1.1 (8). We consider $f \circ \psi$. Let $((\mathbf{R}_{\geq 0})N, Z) \in D^{\sharp}_{\Xi'} - D$, and let $W = W(\sigma) \in I$ for $\sigma := (\mathbf{R}_{\geq 0})N$. Let $F \in Z$, let (ρ, φ) be the SL(2)-orbit in one variable associated to (N, F) and let $\varepsilon = \varepsilon(W[-w], F) \in \mathfrak{g}_{\mathbf{C}}$. Since $\varepsilon(W_{k,\mathbf{C}}) \subset W_{k-2,\mathbf{C}}$ for any $k \in \mathbf{Z}$, we have $\varepsilon \in L_{W,\mathbf{C}}$. By (2), we have $\varepsilon = (a + ib)N$ for some $a, b \in \mathbf{R}$, and $F = \exp((a + ib)N)\varphi(0)$. Hence $(\sigma, Z) = (\sigma, \exp(aN)\exp(i\mathbf{R}N)\varphi(0))$ and hence

$$[\rho, \varphi] = \psi(\sigma, Z) = \exp(aN)\psi(\sigma, \exp(i\mathbf{R}N)\varphi(0)) = \exp(aN)[\rho, \varphi].$$

This shows that $\exp(aN)\varphi(0) = \varphi(0)$. But $aN \in L^{-1,-1}(W[-w], \varphi(0))$ and hence $a = 0$ by Lemma 6.1.8 (iii). This shows that

$$(\sigma, Z) = (\sigma, \exp(i\mathbf{R}N)\varphi(0)) = (\sigma, \varphi(i\mathbf{R})) = f([\rho, \varphi]) = f(\psi(\sigma, Z)).$$

The proof of (ii) is easy. For $W \in I$, $D_{\mathrm{SL}(2),\mathrm{val}}(W)$ is the intersection of $D_{\mathrm{SL}(2),\mathrm{val}}$ with the open set $D_{\mathrm{BS},\mathrm{val}}(P)$ with $P = (G^{\circ})_{W,\mathbf{R}}$. By the definition of the topology of $D_{\mathrm{SL}(2),\mathrm{val}}$, this shows that $\bigcup_{W \in I} D_{\mathrm{SL}(2),\mathrm{val}}(W)$ is a topological subspace of $D_{\mathrm{BS},\mathrm{val}}$. Furthermore, for $W \in I$ and $P = (G^{\circ})_{W,\mathbf{R}}$, the map $D_{\mathrm{BS},\mathrm{val}}(P) \to D_{\mathrm{BS}}(P)$ is a homeomorphism since $\dim S_P = 1$. This proves (ii). □

12.5.4

The continuous bijection $\psi : D^{\sharp}_{\Xi'} \to \bigcup_{W \in I} D_{\mathrm{SL}(2)}(W)$ (12.5.3 (i)) is not necessarily a homeomorphism in general. In [U2], the topology of $\Gamma\backslash D_{\Xi'}$ for $\Gamma = G_{\mathbf{Z}}$ is not the one defined in this book, but is defined as the quotient of the topology of $\bigcup_{W \in I} D_{\mathrm{SL}(2)}(W)$. (In [U2], the topology of $\cup_{W \in I} D_{\mathrm{SL}(2)}(W)$ is defined in terms of Siegel sets, but this definition is equivalent to the one as a subset of D_{BS} in the present case.)

12.5.5

The relationship of notation. $D_{\Xi'}$ in this book is the union of D and $B(W(N)[-w], p, N)$ of [U2], where N ranges over all elements of $\mathfrak{g}_{\mathbf{Q}}$ satisfying 12.5.1 (1), and p ranges over all possible choices of Hodge numbers on $\mathrm{gr}^{W(N)}$.

For N as above and for $\Gamma = G_{\mathbf{Z}}$, $\Gamma(\sigma)^{\mathrm{gp}}\backslash D_\sigma$ for $\sigma = (\mathbf{R}_{\geq 0})N$ of this book is $\tilde{D}_{W,N}$ with $W = W(N)[-w]$ in [U2].

$\Gamma\backslash D_{\Xi'}$ of this book is denoted as $\overline{D/\Gamma}$ in [U2].

Proposition 12.5.6 *Let $p \in D^\sharp_{\Xi'} - D$, let (ρ, φ) be an* SL(2)*-orbit in one variable whose class coincides with $\psi(p) \in D_{\mathrm{SL}(2)}$, let $\mathbf{r} = \varphi(i)$ and let W be the weight filtration associated to (ρ, φ). Let Γ be a subgroup of $G_{\mathbf{Z}}$ of finite index. Then the following conditions* (1)–(4) *are equivalent.*

(1) *The image of p in $\Gamma\backslash D_{\Xi'}$ has a compact neighborhood.*
(2) *$\psi(p) \in D_{\mathrm{SL}(2)}$ has a compact neighborhood in $D_{\mathrm{SL}(2)}$.*
(3) *$G_{W,\mathbf{R}} \cdot \mathbf{r}$ is a neighborhood of $\mathbf{r}$ in D.*
(4) $\dim K_{\mathbf{r}} = \dim K'_{\mathbf{r}}$.

Proof. Write $p = ((\mathbf{R}_{\geq 0})N, \exp(i\mathbf{R}N)F)$ where N is an element of $\mathfrak{g}_{\mathbf{Q}}$ that satisfies 12.5.1 (1). Let (ρ, φ) be the SL(2)-orbit in one variable associated with (N, F), let $\mathbf{r} = \varphi(i)$, and let $W = W(N)$. By Theorem 10.1.6, (2) is equivalent to (3).

Assume (3). We prove (1). By the proof of Proposition 12.5.3 (i), we have $\mathbf{r} \in Z$. By Theorem 7.4.12, it is sufficient to prove that, if $F' \in D$ is sufficiently near to $\mathbf{r}$, then (N, F') satisfies the small Griffiths transversality. By (3), $F' = g \cdot \mathbf{r}$ for some $g \in G_{W,\mathbf{R}}$. Since $\mathrm{Ad}(g)(N) \in L_W = \mathbf{R}N$ with L_W in 12.5.3 (1), we have $\mathrm{Ad}(g)(N) = aN$ for some $a \in \mathbf{R}^\times$. Hence

$$NF'^p = NgF^p = a^{-1}gNF^p \subset a^{-1}gF^{p-1} = F'^{p-1}.$$

Hence we have (1).

Assume (1). We prove (3). If $F' \in D$ is near to $\mathbf{r}$, then, by Theorem 7.4.12, (N, F') generates a nilpotent orbit. If F' converges to $\mathbf{r}$, the class of the SL(2)-orbit (ρ', φ') associated to (N, F') converges to $\psi(p)$ in $D_{\mathrm{SL}(2)}$. By Theorem 10.2.2, $\varphi'(i) = g \cdot \mathbf{r}$ for some $g \in G_{W,\mathbf{R}}$. Furthermore, since $\varepsilon(W[-w], F') \in i\mathbf{R}N$ by the proof of 12.5.3 (i) and since it converges to $\varepsilon(W[-w], \mathbf{r}) = iN$, $F' = \exp(iaN)\varphi'(0)$ for some $a \in \mathbf{R}_{>0}$. Hence $F' = \tilde{\rho}(\sqrt{a})^{-1}\varphi'(i) = \tilde{\rho}(\sqrt{a})^{-1}g \cdot \mathbf{r}$ with $\tilde{\rho}(\sqrt{a})^{-1}g \in G_{W,\mathbf{R}}$.

Finally by [U2, (3.12) (iii)], (3) and (4) are equivalent. □

12.6 COMPLETE FANS

As is seen from the examples in Sections 12.2 and 12.3, it is often impossible to construct a compact $\Gamma\backslash D_\Sigma$ (it is often even impossible to construct a locally compact $\Gamma\backslash D_\Sigma$ which has a point outside $\Gamma\backslash D$). We think that the right generalization of the compactness of toroidal compactifications to the general case is the "completeness" in the following sense.

Definition 12.6.1 *A fan Σ in $\mathfrak{g}_{\mathbf{Q}}$ is* complete *if $D_{\Sigma,\mathrm{val}} = D_{\mathrm{val}}$.*

As is easily seen, Σ is complete if and only if $D^\sharp_{\Sigma,\mathrm{val}} = D^\sharp_{\mathrm{val}}$.

12.6.2

In the examples in Sections 12.2 and 12.3, the fan Ξ is complete; whereas, in Section 12.1 (the case $D = \mathfrak{h}_g$), Ξ is complete if $g = 1$, but Ξ is not complete if $g \geq 2$.

CONJECTURE 12.6.3 *There exists a complete fan which is strongly compatible with $G_{\mathbf{Z}}$ (and hence with any subgroup of $G_{\mathbf{Z}}$ of finite index).**

PROPOSITION 12.6.4 *Let Γ be a neat subgroup of $G_{\mathbf{Z}}$, let Σ be a fan which is strongly compatible with Γ, and assume that $\Gamma\backslash D_\Sigma$ is compact. Then Σ is complete.*

Proof. Let $p \in D^\sharp_{\mathrm{val}}$. We prove $p \in D^\sharp_{\Sigma,\mathrm{val}}$. Since $\Gamma\backslash D^\sharp_\Sigma \to \Gamma\backslash D_\Sigma$ and $\Gamma\backslash D^\sharp_{\Sigma,\mathrm{val}} \to \Gamma\backslash D^\sharp_\Sigma$ are proper, $\Gamma\backslash D^\sharp_{\Sigma,\mathrm{val}}$ is compact.

Take a rational nilpotent cone τ such that $p \in D^\sharp_{\tau,\mathrm{val}}$. Since D is dense in $D^\sharp_{\tau,\mathrm{val}}$, there exists a directed family $(x_\lambda)_\lambda$ of points of D which converges to p in $D^\sharp_{\tau,\mathrm{val}}$. Since $\Gamma\backslash D^\sharp_{\Sigma,\mathrm{val}}$ is compact, replacing $(x_\lambda)_\lambda$ by a cofinal subfamily if necessary, we may assume that the image of $(x_\lambda)_\lambda$ in $\Gamma\backslash D$ converges to an element (α mod Γ) in $\Gamma\backslash D^\sharp_{\Sigma,\mathrm{val}}$ for some $\alpha \in D^\sharp_{\Sigma,\mathrm{val}}$. Since $\Gamma\backslash D_{\mathrm{SL}(2)}$ is Hausdorff, the images of p and α in $\Gamma\backslash D_{\mathrm{SL}(2)}$ coincide. Replacing α by a suitable translation of α by Γ, we may assume that the images of p and α in $D_{\mathrm{SL}(2)}$ coincide.

Since $D^\sharp_{\Sigma,\mathrm{val}} \to \Gamma\backslash D^\sharp_{\Sigma,\mathrm{val}}$ is a local homeomorphism, there exist $\gamma_\lambda \in \Gamma$ such that $(\gamma_\lambda x_\lambda)_\lambda$ converges to α in $D^\sharp_{\Sigma,\mathrm{val}}$. Then, in $D_{\mathrm{SL}(2)}$, both $(x_\lambda)_\lambda$ and $(\gamma_\lambda x_\lambda)_\lambda$ converge to $\psi(p)$. Since $D_{\mathrm{SL}(2)} \to \Gamma\backslash D_{\mathrm{SL}(2)}$ is a local homeomorphism, we see that $\gamma_\lambda = 1$ if λ is sufficiently large.

Thus $(x_\lambda)_\lambda$ converges to p in $D^\sharp_{\tau,\mathrm{val}}$ and also converges to α in $D^\sharp_{\Sigma,\mathrm{val}}$. Take $\sigma \in \Sigma$ such that $\alpha \in D^\sharp_{\sigma,\mathrm{val}}$. Then $(x_\lambda)_\lambda$ converges to α in $D^\sharp_{\sigma,\mathrm{val}}$ by 7.3.2 (ii). Hence $p = \alpha$ in $D^\sharp_{\mathrm{val}}$ by 7.3.10 (1). Thus $p \in D^\sharp_{\Sigma,\mathrm{val}}$, as desired. □

12.6.5

In the classical situation (0.4.14), [AMRT] constructed a fan Σ in $\mathfrak{g}_{\mathbf{Q}}$ which is strongly compatible with $G_{\mathbf{Z}}$ (and hence strongly compatible with any subgroup Γ of $G_{\mathbf{Z}}$ of finite index) and for which $G_{\mathbf{Z}}\backslash D_\Sigma$ (and hence $\Gamma\backslash D_\Sigma$ for such Γ) is compact. For a subgroup Γ of $G_{\mathbf{Z}}$ of finite index, $\Gamma\backslash D_\Sigma$ is the toroidal compactification of $\Gamma\backslash D$ by Mumford and others associated to Σ. By 12.6.4, this Σ is complete.

THEOREM 12.6.6 *Let X be a connected, logarithmically smooth, fs logarithmic analytic space and let $U = X_{\mathrm{triv}}$ be the open set of X consisting of all points at which the logarithmic structure is trivial. Let H be a variation of polarized Hodge structure on U with unipotent local mondromy along $X - U$. Let $u \in U$ and let $(H_0, \langle\ ,\ \rangle_0) = (H_{\mathbf{Z},u}, \langle\ ,\ \rangle_u)$. Let Γ be a neat subgroup of $G_{\mathbf{Z}}$ of finite index, let Σ be*

*See the end of section 12.7.

a fan in $\mathfrak{g}_{\mathbf{Q}}$, *and assume that* Γ *contains the global monodromy* $\mathrm{Image}(\pi_1(U, u) \to G_{\mathbf{Z}})$, *that* Σ *is strongly compatible with* Γ, *and that* Σ *is complete. Then* $X(\Sigma)$ *in* 4.3.5 *is a logarithmic modification of* X *and the period map* $U \to \Gamma\backslash D$ *associated to* H *extends to a morphism* $X(\Sigma) \to \Gamma\backslash D_{\Sigma}$ *of logarithmic manifolds.*

By the nilpotent orbit theorem of Schmid interpreted as Theorem 2.5.14, H extends to a PLH on X. Hence Theorem 12.6.6 is reduced to

PROPOSITION 12.6.7 *Let* Γ *be a neat subgroup of* $G_{\mathbf{Z}}$ *and let* Σ *be a fan in* $\mathfrak{g}_{\mathbf{Q}}$ *which is strongly compatible with* Γ. *Assume* Σ *is complete. Let* X *be an object of* $\mathcal{B}(\log)$ *and let* H *be a polarized logarithmic Hodge structure of weight* w *and of Hodge type* $(h^{p,q})$ *endowed with a* Γ*-level structure. Then*:

(i) *The object* $X(\Sigma)$ *of* $\mathcal{B}(\log)$ *over* X (4.3.5) *is a logarithmic modification of* X (3.6.12).

(ii) *The inverse image of* H *on* $X(\Sigma)$ *is of type* $(w, (h^{p,q}), H_0, \langle\ ,\ \rangle_0, \Gamma, \Sigma)$ *and hence we have the associated period morphism* $X(\Sigma) \to \Gamma\backslash D_{\Sigma}$.

Proof. To prove (i), consider the maps

$$X_{\mathrm{val}} \to \Gamma\backslash D_{\mathrm{val}} = \Gamma\backslash D_{\Sigma,\mathrm{val}} \to \Gamma\backslash D_{\Sigma},$$

where the first arrow is as in 8.4.1.

Let $(x, V, h) \in X_{\mathrm{val}}$. Let y be a point of $X^{\log}$ lying over x, and take a representative $\tilde{\mu}_y : (H_{\mathbf{Z},y}, \langle\ ,\ \rangle_y) \xrightarrow{\sim} (H_0, \langle\ ,\ \rangle_0)$ of the germ μ_y of the Γ-level structure μ of H. The image of $(x, V, h) \in X_{\mathrm{val}}$ in $\Gamma\backslash D_{\mathrm{val}}$ has the form $((A, V', Z) \bmod \Gamma)$ where $(A, V', Z) \in D_{\mathrm{val}}$ is defined as in 8.4.1 with respect to the above choice of $\tilde{\mu}_y$. The image of (A, V', Z) under $D_{\mathrm{val}} = D_{\Sigma,\mathrm{val}} \to D_{\Sigma}$ is the pair (σ, Z'), where σ is the smallest element of Σ such that $\{h \in \mathrm{Hom}_{\mathbf{Q}}(A, \mathbf{Q}) \mid h(\sigma \cap A) \subset \mathbf{Q}_{\geq 0}\} \subset V'$ (5.3.2 (3)), and $Z' = \exp(\sigma_{\mathbf{C}})Z$.

Via $\tilde{\mu}_y$, the local monodromy action of $\pi_1(x^{\log})$ on $H_{\mathbf{Z},y}$ induces a homomorphism $\pi_1(x^{\log}) \to G_{\mathbf{Z}}$ and its logarithm $s_x : \pi_1(x^{\log}) \to \mathfrak{g}_{\mathbf{Q}}$. Let $\sigma_1 = \pi_1^{+}(x^{\log}) \cap s_x^{-1}(\sigma)$ and, as in 3.6.14, let $P(\sigma_1) = \{a \in (M_X^{\mathrm{gp}}/\mathcal{O}_X^{\times})_x \mid h_\gamma(a) \geq 0\ (\forall\, \gamma \in \sigma_1)\}$, where h_γ is the homomorphism $(M_X^{\mathrm{gp}}/\mathcal{O}_X^{\times})_x \to \mathbf{Z}$ induced by γ. (Here we use the duality of $\pi_1(x^{\log})$ and $(M_X^{\mathrm{gp}}/\mathcal{O}_X^{\times})_x$.) Then it is easy to prove $(x, \sigma_1) \in Q_1'(X)$ (3.6.14) and $P(\sigma_1) \subset V$. As in 4.3.5, $\tilde{\mu}_y$ defines a lifting to $(M_X^{\mathrm{gp}}/\mathcal{O}_X^{\times})_x \otimes \mathfrak{g}_{\mathbf{Q}}$ of the germ of $\mathcal{N} \in \Gamma(X, \Gamma\backslash((M_X^{\mathrm{gp}}/\mathcal{O}_X^{\times}) \otimes \mathfrak{g}_{\mathbf{Q}}))$ at x, and this lifting coincides with s_x via the identification $(M_X^{\mathrm{gp}}/\mathcal{O}_X^{\times})_x \otimes \mathfrak{g}_{\mathbf{Q}} = \mathrm{Hom}(\pi_1(x^{\log}), \mathfrak{g}_{\mathbf{Q}})$. Hence the inclusion $P(\sigma_1) \subset V$ shows that the condition (6) of 3.6.28 is satisfied. By Theorem 3.6.28, $X(\Sigma)$ is a logarithmic modification of X and (i) is proved. Furthermore, the image of (x, V, h) under $X_{\mathrm{val}} = X(\Sigma)_{\mathrm{val}} \to X(\Sigma)$ coincides with $(x, \sigma_1, h_1) \in X(\Sigma) \subset Q'(X)$, where h_1 is the restriction of h.

To prove (ii), let $x_1 = (x, \sigma_1, h_1)$ be any element of $X(\Sigma) \subset Q'(X)$. Take a point (x, V, h) of X_{val} lying over x_1. Fix $\tilde{\mu}_y$ as above, and define $(A, V', Z) \in D_{\mathrm{val}}$ and $(\sigma, Z') \in D_{\Sigma}$ from (x, V, h) as above with respect to $\tilde{\mu}_y$. Also let $(\sigma_1', Z_1') \in \check{D}_{\mathrm{orb}}$ be the representative of the image of x_1 under the period map $X(\Sigma) \to \Gamma\backslash\check{D}_{\mathrm{orb}}$ associated to the pullback of H to $X(\Sigma)$, defined with respect to $\tilde{\mu}_y$. Then σ_1' is generated by $s_x(\sigma_1)$ as a cone. It is sufficient to prove the following (1) and (2).

(1) $\sigma_1' \subset \sigma$. Furthermore, σ is the smallest element of Σ which contains σ_1'.
(2) $\exp(\sigma_{\mathbf{C}})Z_1'$ is a σ-nilpotent orbit.

We prove (1). Since $\sigma_1 = \pi_1^+(x^{\log}) \cap s_x^{-1}(\sigma)$, we have $\sigma_1' \subset \sigma$. If $\tau \in \Sigma$ contains σ_1', then $\sigma_1' \subset \sigma \cap \tau$ and hence $(\sigma \cap \tau \cap A)^\vee \subset (\sigma_1' \cap A)^\vee \subset V'$. Since σ is the smallest element of Σ satisfying $(\sigma \cap A)^\vee \subset V'$, we have $\sigma \cap \tau = \sigma$, that is, $\sigma \subset \tau$.

We prove (2). We have

$$Z' = \exp(\sigma_{\mathbf{C}})Z, \quad Z_1' = \exp(\sigma_{1,\mathbf{C}}')Z.$$

Hence $\exp(\sigma_{\mathbf{C}})Z_1'$ coincides with the σ-nilpotent orbit Z'. □

Proposition 12.6.8 *Assume we are in the classical situation* 0.4.14. *Let Σ be a fan in $\mathfrak{g}_{\mathbf{Q}}$, let Γ be a subgroup of $G_{\mathbf{Z}}$ of finite index, and assume that Γ is strongly compatible with Σ. Then, $\Gamma \backslash D_\Sigma$ is compact if and only if Σ is complete.*

Proof. The "only if" part is shown in 12.6.4. We prove the "if" part. Replacing Γ by a neat subgroup of Γ of finite index, we may assume that Γ is neat. By the existence of a toroidal compactification proved in [AMRT], there is a fan Σ' which is strongly compatible with Γ such that $\Gamma \backslash D_{\Sigma'}$ is compact. Write $\Gamma \backslash D_{\Sigma'}$ by X. By 12.6.6, the universal PLH on X defines the period map $X(\Sigma) \to \Gamma \backslash D_\Sigma$. The image of this map is dense since it contains $\Gamma \backslash D$. Furthermore, $X(\Sigma)$ is compact since $X(\Sigma)$ is a logarithmic modification of X and X is compact. Hence $X(\Sigma) \to \Gamma \backslash D_\Sigma$ is surjective, and hence $\Gamma \backslash D_\Sigma$ is compact. □

12.7 PROBLEMS

12.7.1

Construct in general a complete fan (12.6.1) that is strongly compatible with $G_{\mathbf{Z}}$ (Conjecture 12.6.3).*

12.7.2

Relate our theory to the theory of Hodge modules of Morihiko Saito.

12.7.3

Study p-adic analogues of our theory.

12.7.4

Give structures of ringed spaces on D_{BS} and on $D_{\mathrm{SL}(2)}$ and extend the standard Hodge filtration over these spaces. What kind of functors would D_{BS} and $D_{\mathrm{SL}(2)}$ with

*See the end of section 12.7.

these structures represent? What kind of morphism of functors would $D^{\sharp}_{\mathrm{val}} \to D_{\mathrm{SL}(2)}$ represent?

12.7.5

Let $f : Y \to X$ be a logarithmically smooth morphism of logarithmically smooth, fs logarithmic analytic spaces (2.1.11), whose underlying morphism of analytic spaces is projective, and assume that $f^{-1}(X_{\mathrm{triv}}) = Y_{\mathrm{triv}}$ and that the cokernel of $(M_X^{\mathrm{gp}}/\mathcal{O}_X^{\times})_{f(y)} \to (M_Y^{\mathrm{gp}}/\mathcal{O}_Y^{\times})_y$ is torsion free for any $y \in Y$. Assume that there exists the corresponding period map $\varphi : X \to \Gamma\backslash D_{\Sigma}$. Then, φ gives a global invariant of degenerations of the fibers of f. We have a decomposition $\Gamma\backslash D_{\Sigma} = \bigsqcup_j Z_j$, according to the Γ-equivalence class of the cone σ of a $(\sigma, Z) \in D_{\Sigma}$. For each j, investigate the common property among the fibers $f^{-1}(x)$ for $x \in \varphi^{-1}(Z_j)$. For each $z \in Z_j$, describe geometric relation among the fibers $f^{-1}(x)$ for $x \in \varphi^{-1}(z)$.

12.7.6

As we have seen in this book, our moduli space $\Gamma\backslash D_{\Sigma}$ is not necessarily an analytic space and not necessarily locally compact. However, can we expect that, if X is an algebraic variety over $\mathbf{C}$, the closure $\overline{\varphi(X)}$ in $\Gamma\backslash D_{\Sigma}$ of the image of a period map $\varphi : X \to \Gamma\backslash D$ is always an algebraic variety (and hence is locally compact)? Is $\overline{\varphi(X)}$ compact when Σ is complete?

Assume, furthermore, that X is a moduli space of algebraic varieties, and φ is the corresponding period map. Assume Σ is complete and φ is an immersion. Then, can we expect that $\overline{\varphi(X)}$ is a natural compactification of the the moduli space X?

12.7.7

Generalize the present results to the case when D is the classifying space of mixed Hodge structures with polarized graded quotients.

Added in the Proof

Recently, Kenta Watanabe proved that Conjecture 12.6.3 was false. The authors expect that the following modified version of that conjecture is correct. The idea of the following new definition is due to Chikara Nakayama.

Let N be the set of all rational nilpotent cones σ in $\mathfrak{g}_{\mathbf{R}}$ such that (σ, Z) is a nilpotent orbit for some subset Z of $\check{D}$.

For a fan Σ in $\mathfrak{g}_{\mathbf{Q}}$, we say Σ is *complete in the new sense* (resp. *in the weak sense*) if

$$\bigcup_{\sigma \in N} \sigma = \bigcup_{\sigma \in \Sigma} \sigma \quad (\text{resp. } \bigcup_{\sigma \in N} \sigma \subset \bigcup_{\sigma \in \Sigma} \sigma).$$

Then, Σ is complete in the weak sense if Σ is either complete (in the sense of 12.6.1) or complete in the new sense.

CONJECTURE (modified version) There is a fan Σ in $\mathfrak{g}_{\mathbf{Q}}$ which is complete in the new sense and strongly compatible with $G_{\mathbf{Z}}$.

Problem 12.7.1 should be rewritten as follows.

PROBLEM (1) For each given $\Phi_0 = \big(w, (h^{p,q})_{p,q}, H_0, \langle\ ,\ \rangle_0\big)$, determine whether there exists or not a complete fan in $\mathfrak{g}_{\mathbf{Q}}$ which is strongly compatible with $G_{\mathbf{Z}}$.

(2) Construct a fan in $\mathfrak{g}_{\mathbf{Q}}$ which is complete in the new sense and strongly compatible with $G_{\mathbf{Z}}$.

Appendix

A1 POSITIVE DIRECTION OF LOCAL MONODROMY

A1.1

The points at infinity of the classifying space of polarized logarithmic Hodge structures appear in various directions of degeneration of polarized Hodge structures. To consider the direction of the degeneration, it is important to define the monodromy cone in the group of local monodromy consisting of local monodromy of positive direction. For this, we have to fix which direction in the local monodromy group is positive. In particular, we have to fix one of the two isomorphisms $\pi_1(\Delta^*) \simeq \mathbf{Z}$ and call the element of $\pi_1(\Delta^*)$ corresponds to 1 the positive generator of $\pi_1(\Delta^*)$, and define the monodromy cone in $\pi_1(\Delta^*)$ as the submonoid generated by this element, that is, as the image of $\mathbf{N} \subset \mathbf{Z}$ under this isomorphism. This is in fact a rather delicate point. The authors think that the following is the best definition which is compatible with the works of many people (Schmid [Sc], etc.). In our definition:

(i) When we regard the upper half plane $\mathfrak{h}$ as the universal covering of Δ^* via the projection $\mathfrak{h} \to \Delta^*, \tau \mapsto e^{2\pi i\tau}$, the positive generator of $\pi_1(\Delta^*)$ is the automorphism $\tau \mapsto \tau + 1$ of $\mathfrak{h}$.

(ii) The positive generator of $\pi_1(\Delta^*)$ is the class of a route in Δ^* in the counter-clockwise directed circle $[0, 1] \to \Delta, t \mapsto re^{2\pi it}$ for any fixed r $(0 < r < 1)$.

(iii) Let q be the coordinate function of Δ. Then the logarithms of q form a local system on Δ^* and hence the group $\pi_1(\Delta^*)$ acts on stalks of this sheaf. If γ is the positive generator of $\pi_1(\Delta^*)$, γ acts on the germs l of logarithms of q by $l \mapsto l - 2\pi i$, not $l \mapsto l + 2\pi i$.

The reader may feel that (iii) is strange. In what follows, we explain that (i)–(iii) are compatible.

A1.2

Let X be a topological space and fix $a \in X$. The fundamental group $\pi_1(X, a)$ of X with base point a is the group of all equivalence classes of continuous maps $\gamma : [0, 1] \to X$ such that $\gamma(0) = \gamma(1) = a$. (We omit the definition of the equivalence (see [Sp]).) For such $\gamma, \delta : [0, 1] \to X$, the product class$(\gamma)$class$(\delta)$ is the class of the route $\gamma\delta : [0, 1] \to X$ which is "go through γ first, and then go through δ".

A1.3

The universal covering $\tilde{X}$ of X for base point a is defined to be the set of all equivalence classes of continuous maps $f : [0, 1] \to X$ such that $f(0) = a$. (We omit the definition of the equivalence (see [Sp]).) We have the canonical projection $p : \tilde{X} \to X$, $f \mapsto f(1)$. The fundamental group $\pi_1(X, a)$ is identified with the group of all automorphisms of $\tilde{X}$ over X. In this identification, the class of $\gamma : [0, 1] \to X$ ($\gamma(0) = \gamma(1) = a$) in $\pi_1(X, a)$ sends the class of $f : [0, 1] \to X$ ($f(0) = a$) to the class of the route $\gamma f : [0, 1] \to X$ which is "go through γ first and then go through f".

A1.4

Example 1. Regard the upper half plane $\mathfrak{h}$ as the universal covering of Δ^* via the projection $\tau \mapsto e^{2\pi i\tau}$. Fix any element $a \in \Delta^*$. Then the class in $\pi_1(\Delta^*, a)$ of $[0, 1] \to \Delta$, $t \mapsto ae^{2\pi it}$, is identified (by A1.3) with the automorphism $\mathfrak{h} \to \mathfrak{h}$, $\tau \mapsto \tau + 1$.

Example 2. The class of $[0, 1] \to \mathbf{R}/\mathbf{Z}$, $t \mapsto (t \bmod \mathbf{Z})$, in $\pi_1(\mathbf{R}/\mathbf{Z}, 0)$ is identified with the automorphism $x \mapsto x + 1$ of the universal covering $\mathbf{R}$ of $\mathbf{R}/\mathbf{Z}$.

A1.5

Let X and a be as in A1.2, and let L be a locally constant sheaf on X. Then $\pi_1(X, a)$ acts on the stalk L_a of L at a. For the class of $\gamma : [0, 1] \to X$ ($\gamma(0) = \gamma(1) = a$) in $\pi_1(X, a)$, the action of γ is the composition, from the left to the right, of all isomorphisms or their inverses in

$$L_a \xrightarrow{\sim} \gamma^{-1}(L)_1 \xleftarrow{\sim} \Gamma([0, 1], \gamma^{-1}(L)) \xrightarrow{\sim} \gamma^{-1}(L)_0 \xleftarrow{\sim} L_a.$$

This action coincides with the composition, from the left to the right, of all isomorphisms or their inverses in

$$L_a \xrightarrow{\sim} p^{-1}(L)_\gamma \xleftarrow{\sim} \Gamma(\tilde{X}, p^{-1}(L)) \xrightarrow{\sim} p^{-1}(L)_{\tilde{a}} \xleftarrow{\sim} L_a.$$

Here we regard γ as an element of $\tilde{X}$, and $\tilde{a}$ denotes the point of $\tilde{X}$ which is the class of the constant map $[0, 1] \to X$ with value a.

Thus we have a group homomorphism:

$$\pi_1(X, a) = \mathrm{Aut}_X(\tilde{X}) \to \mathrm{Aut}(L_a), \quad \gamma \mapsto \text{(rewinding along } \gamma).$$

Be careful that if we define the action of $\gamma \in \mathrm{Aut}_X(\tilde{X})$ on L_a as the composition, from the left to the right, of all isomorphisms or their inverses in

$$L_a \xrightarrow{\sim} p^{-1}(L)_{\tilde{a}} \xleftarrow{\sim} \Gamma(\tilde{X}, p^{-1}(L)) \xrightarrow{\sim} p^{-1}(L)_\gamma \xleftarrow{\sim} L_a,$$

the group structure of $\mathrm{Aut}_X(\tilde{X})$ could not be preserved (the order of the product should become the converse).

A1.6

Let x be an fs logarithmic point. Let $a = (x, h_0) \in x^{\log}$.

We have an isomorphism

$$\pi_1(x^{\log}, a) \simeq \mathrm{Hom}(M_x^{\mathrm{gp}}/\mathcal{O}_x^\times, \mathbf{Z}) \tag{1}$$

defined as follows. We have an exact sequence

$$0 \to \mathrm{Hom}(M_x^{\mathrm{gp}}/\mathcal{O}_x^\times, \mathbf{Z}) \to \mathrm{Hom}(M_x^{\mathrm{gp}}/\mathcal{O}_x^\times, \mathbf{R}) \to \mathrm{Hom}(M_x^{\mathrm{gp}}/\mathcal{O}_x^\times, \mathbf{S}^1) \to 0$$

where the third arrow is induced by $\mathbf{R} \to \mathbf{S}^1$, $t \mapsto e^{2\pi i t}$, and we have a homeomorphism $x^{\log} \xrightarrow{\sim} \mathrm{Hom}(M_x^{\mathrm{gp}}/\mathcal{O}_x^\times, \mathbf{S}^1)$, $(x, h) \mapsto hh_0^{-1}$. Via this exact sequence and the last isomorphism, we identify $\mathrm{Hom}(M_x^{\mathrm{gp}}/\mathcal{O}_x^\times, \mathbf{R})$ with the universal covering of $x^{\log}$, and $\mathrm{Hom}(M_x^{\mathrm{gp}}/\mathcal{O}_x^\times, \mathbf{Z})$ with the fundamental group of $x^{\log}$ acting on the universal covering $\mathrm{Hom}(M_x^{\mathrm{gp}}/\mathcal{O}_x^\times, \mathbf{R})$ by translations.

In the isomorphism (1), for $f \in \mathrm{Hom}(M_x^{\mathrm{gp}}/\mathcal{O}_x^\times, \mathbf{Z})$, the corresponding element of $\pi_1(x^{\log}, a)$ coincides with the class of $[0, 1] \to x^{\log}$, $t \mapsto (x, h_t)$ where $h_t(q) = e^{2\pi i t f(q)} h_0(q)$ for $q \in M_{X,x}$.

Example. Consider the logarithmic point $x = 0 \in \Delta$ whose logarithmic structure is the inverse image of M_Δ. Let $a = (x, h_0) \in x^{\log}$ with $h_0 : M_x \to \mathbf{S}^1$, $q \mapsto 1$, where q is the coordinate function of Δ regarded as a section of M_x. Then $x^{\log} \simeq \mathrm{Hom}(M_x^{\mathrm{gp}}/\mathcal{O}_x^\times, \mathbf{S}^1) \simeq \mathbf{S}^1$, $(x, h) \mapsto hh_0^{-1} \mapsto h(q)$. Let f_0 be an isomorphism $M_x^{\mathrm{gp}}/\mathcal{O}_x^\times \xrightarrow{\sim} \mathbf{Z}$, (class of q) $\mapsto 1$. Then, under the above isomorphism (1), the corresponding element of $\pi_1(x^{\log}, a)$ is the class of $\gamma : [0, 1] \to x^{\log} = \mathbf{S}^1$, $t \mapsto e^{2\pi i t}$, since $f_0(\text{class of } q) = 1$ and $h_0(q) = 1$. From (1), it follows that $\pi_1(x^{\log}, a) \simeq \mathbf{Z}$, (class of γ) $\mapsto 1$. This isomorphism sends $1 \in \mathbf{Z}$ to the automorphism $\xi \mapsto \xi + 1$ of $\mathbf{R}$ which is regarded as the universal covering of $x^{\log} = \mathbf{S}^1$ via $\mathbf{R} \to \mathbf{S}^1$, $\xi \mapsto e^{2\pi i \xi}$.

A1.7

Let x be an fs logarithmic point. We define the monodromy cone $\pi_1^+(x^{\log}) \subset \pi_1(x^{\log})$ as the image of $\mathrm{Hom}(M_x/\mathcal{O}_x^\times, \mathbf{N})$ under the isomorphism (1) in A1.6.

For example, in the case $x = 0 \in \Delta$, let $a \in x^{\log}$ be as in Example in A1.6, $b \in \Delta^*$ be any point, and δ be a path in $\Delta^{\log}$ joining from a to b. Then $\pi_1(\Delta^{\log}, a) \simeq \pi_1(\Delta^{\log}, b)$, $\gamma \mapsto \delta^{-1}\gamma\delta$, and via the isomorphisms $\pi_1(x^{\log}, a) \simeq \pi_1(\Delta^{\log}, a) \simeq \pi_1(\Delta^{\log}, b) \simeq \pi_1(\Delta^*, b)$, the generator of $\pi_1^+(x^{\log}, a) \simeq \mathbf{N}$ corresponds to the class in $\pi_1(\Delta^*, b)$ of $[0, 1] \mapsto \Delta^*$, $t \mapsto be^{2\pi i t}$.

A1.8

Example. Let L be the locally constant sheaf on $X = \Delta^*$ defined to be the inverse image of $q^{\mathbf{Z}}$ under $\mathcal{O}_X \to \mathcal{O}_X^\times$, $f \mapsto \exp(2\pi i f)$, where q is the coordinate function of Δ. Then $L = \mathbf{Z}\tau + \mathbf{Z}$ where τ is a local section of $\mathcal{O}_X$ such that $q = \exp(2\pi i \tau)$. Note that L is the local system of the first homology groups H_1 of the standard family of elliptic curves parametrized by Δ^* (see 0.1.4, 0.2.1, 0.2.10). Let $x = 0 \in \Delta$, $a \in x^{\log}$ and $b \in \Delta^*$ be as in A1.6 and A1.7, and let $\gamma \in \pi_1(\Delta^*, b)$ be the element corresponding to the generator of $\pi_1^+(x^{\log}, a)$. Then the action of γ on the stalk L_b sends $1 \in L_b$ to 1 and $\tau \in L_b$ to $\tau - 1$.

Let $L^* = \mathcal{H}om_{\mathbf{Z}}(L, \mathbf{Z})$ which is the local system of the first cohomology groups H^1 of the standard family of elliptic curves. Let (e_1, e_2) be the local $\mathbf{Z}$-basis of L^* which is dual to $(\tau, 1)$. Then the action of γ on the stalk L_b^* sends $e_1 \in L_b^*$ to e_1 and $e_2 \in L_b^*$ to $e_1 + e_2$.

A1.9

Let x be an fs logarithmic point. Then $\mathcal{O}_x^{\log}$ is a locally constant sheaf on $x^{\log}$ and hence $\pi_1(x^{\log})$ acts on the stalk $\mathcal{O}_{x,y}^{\log}$ at $y \in x^{\log}$.

For example, in the case $x = 0 \in \Delta$, the action of the generator of $\pi_1^+(x^{\log})$ on $\mathcal{O}_{x,y}^{\log}$ $(y \in x^{\log})$ sends $(2\pi i)^{-1} \log(q)$ (q is the coordinate function of Δ) to $(2\pi i)^{-1} \log(q) - 1$ (A1.8). Note that the corresponding element of $\pi_1(\Delta^*)$ is the automorphism of $\mathfrak{h}$ which sends $\tau \in \mathfrak{h}$ to $\tau + 1 \in \mathfrak{h}$ (A1.4, Example 1). This is a little confusing since the coordinate function τ of $\mathfrak{h}$ is $(2\pi i)^{-1} \log(q)$. There is duality between the point and the coordinate function.

A1.10

Let x be an fs logarithmic point. For an abelian group A, normalize the isomorphism

$$H^1(x^{\log}, A) \simeq \mathrm{Hom}(\pi_1(x^{\log}), A)$$

as follows. Let $\mathcal{F}$ be an A-torsor on $x^{\log}$. Since $\mathcal{F}$ is a locally constant sheaf, $\pi_1(x^{\log})$ acts on stalks of $\mathcal{F}$. We define the homomorphism $\pi_1(x^{\log}) \to A$ corresponding to the class of $\mathcal{F}$ in $H^1(x^{\log}, A)$ by $\gamma \mapsto a$ $(\gamma \in \pi_1(x^{\log}))$ where a is the element of A such that $f = a + \gamma(f)$ (not $\gamma(f) = a + f$) for any element f of any stalk of $\mathcal{F}$. Here $a+$ denotes the action of a on $\mathcal{F}$.

Consider the exact sequence

$$0 \to \mathbf{Z} \xrightarrow{2\pi i} \mathcal{L}_x \xrightarrow{\exp} \tau^{-1}(M_x^{\mathrm{gp}}) \to 0$$

and let $\delta : M_x^{\mathrm{gp}} \to H^1(x^{\log}, \mathbf{Z})$ be the connecting homomorphism which sends $q \in M_x^{\mathrm{gp}}$ to the $\mathbf{Z}$-torsor of $(2\pi i)^{-1} \log(f)$. Then the composition $M_x^{\mathrm{gp}} \to H^1(x^{\log}, \mathbf{Z}) \xrightarrow{\sim} \mathrm{Hom}(\pi_1(x^{\log}), \mathbf{Z})$, where the second arrow is as in our normalization, sends $q \in M_x^{\mathrm{gp}}$ to the homomorphism $\pi_1(x^{\log}) \to \mathbf{Z}$ induced by q via the isomorphism (1) in Section A1.6. This normalization is useful in Chapter 8.

A2 PROPER BASE CHANGE THEOREM FOR TOPOLOGICAL SPACES

The following proper base change theorem A2.1 is well-known (Deligne [SGA4$\frac{1}{2}$, page 39], Kashiwara and Schapira [KS, Remark 2.5.3]). In [KS, Prop. 2.6.7] and in [V] by Verdier (see also Godement [Go, II Theorem 4.11.1]), this theorem is proved for locally compact spaces. But in this book, we use this theorem for more general topological spaces. The proof of this theorem in the general case is outlined by Kajiwara and Nakayama [KjNc, §2]. For the convenience of the reader, we give here the details of the proof in the general case.

Theorem A2.1 *Let X and Y be topological spaces and let $f : X \to Y$ be a proper (cf. Section 0.7) continuous map. Let Y' be a topological space, let $g : Y' \to Y$ be a continuous map, let $f' : X' = X \times_Y Y' \to Y'$ be the map induced by f, and let $g' : X' \to X$ be the map induced by g.*

(i) *Let $\mathcal{F}$ be a sheaf of abelian groups on X. Then*

$$g^{-1} R^m f_* \mathcal{F} \xrightarrow{\sim} R^m f'_* (g')^{-1} \mathcal{F} \quad \text{for any } m \geq 0.$$

(ii) *Let $\mathcal{F}$ be a sheaf of groups on X. Then*

$$g^{-1} R^1 f_* \mathcal{F} \xrightarrow{\sim} R^1 f'_* (g')^{-1} \mathcal{F}.$$

(iii) *Let $\mathcal{F}$ be a sheaf of sets on X. Then*

$$g^{-1} f_* \mathcal{F} \xrightarrow{\sim} f'_* (g')^{-1} \mathcal{F}.$$

Here in (ii), $R^1 f_* \mathcal{F}$ is defined to be the sheaf on Y associated to the presheaf $U \mapsto H^1(f^{-1}(U), \mathcal{F})$ where H^1 is the set of isomorphism classes of $\mathcal{F}$-torsors.

The authors learned the proof of this theorem, given below, from Chikara Nakayama.

Corollary A2.2 *Let X and Y be topological spaces, let $f : X \to Y$ be a proper continuous map, and let $y \in Y$. Let $i : f^{-1}(y) \hookrightarrow X$ be the inclusion map.*

(i) *Let $\mathcal{F}$ be a sheaf of abelian groups on X. Then*

$$(R^m f_* \mathcal{F})_y \xrightarrow{\sim} H^m(f^{-1}(y), i^{-1}\mathcal{F}) \quad \text{for any } m \geq 0.$$

(ii) *Let $\mathcal{F}$ be a sheaf of groups on X. Then*

$$(R^1 f_* \mathcal{F})_y \xrightarrow{\sim} H^1(f^{-1}(y), i^{-1}\mathcal{F}).$$

(iii) *Let $\mathcal{F}$ be a sheaf of sets on X. Then*

$$(f_* \mathcal{F})_y \xrightarrow{\sim} \Gamma(f^{-1}(y), i^{-1}\mathcal{F}).$$

This corollary is the case $Y' = \{y\}$ of the theorem. Conversely, the theorem follows from this corollary easily. This corollary is deduced from the following proposition.

Proposition A2.3 *Let X be a topological space and let K be a compact subspace of X satisfying the following condition* (H).

(H) *If $x, y \in K$ and $x \neq y$, there are open subsets U, V of X such that $x \in U$, $y \in V$, and $U \cap V = \emptyset$.*

Let $i : K \to X$ be the iclusion map. In the following, $\varinjlim_U$ denotes the inductive limit where U ranges over all open neighborhoods of K in X.

(i) *Let $\mathcal{F}$ be a sheaf of abelian groups on X. Then*

$$\varinjlim_U H^m(U, \mathcal{F}) \xrightarrow{\sim} H^m(K, i^{-1}\mathcal{F}) \quad \text{for any } m \geq 0.$$

(ii) *Let $\mathcal{F}$ be a sheaf of groups on X. Then*

$$\varinjlim_{U} H^1(U, \mathcal{F}) \xrightarrow{\sim} H^1(K, i^{-1}\mathcal{F}).$$

(iii) *Let $\mathcal{F}$ be a sheaf of sets on X. Then*

$$\varinjlim_{U} \Gamma(U, \mathcal{F}) \xrightarrow{\sim} \Gamma(K, i^{-1}\mathcal{F}).$$

A2.4

We can deduce Corollary A2.2 from Proposition A2.3 as follows. Apply A2.3 by taking K to be $f^{-1}(y)$ in A2.2. In this situation, $f^{-1}(V)$ for open neighborhoods V of y form a basis of neighborhoods of $f^{-1}(y)$. (In fact, if U is an open neighborhood of $f^{-1}(y)$, $f(X-U)$ is closed in Y since f is a closed map, and $V := Y - f(X-U)$ is an open neighborhood of y and satisfies $f^{-1}(V) \subset U$.) Hence $(R^m f_*\mathcal{F})_y = \varinjlim_U H^m(U, \mathcal{F})$, where U ranges over all open neighborhoods of $f^{-1}(y)$ (we take $m = 1$ in (ii), and $m = 0$ in (iii)).

In the rest of A2, we prove Proposition A2.3.

A2.5

We first give the proof of (iii) of Proposition A2.3.

Proof of (iii) The injectivity of $\varinjlim_U \Gamma(U, \mathcal{F}) \to \Gamma(K, i^{-1}\mathcal{F})$ is easily seen by looking at stalks at points of K. We prove the surjectivity. Let $s \in \Gamma(K, i^{-1}\mathcal{F})$.

Claim 1 *There are a finite family $(U_k)_{1\le k\le n}$ of open sets of X such that $K \subset \bigcup_{k=1}^n U_k$ and elements $s_k \in \Gamma(U_k, \mathcal{F})$ such that the germ of s_k at any point of $K \cap U_k$ coincides with the germ of s.*

Proof. For each $x \in K$, there is an open neighborhood $U(x)$ of x in X and $s(x) \in \Gamma(U(x), \mathcal{F})$ such that at any point of $K \cap U(x)$, the germ of $s(x)$ and that of s coincide. Since K is compact, a finite subfamily of $(U(x))_{x\in K}$ covers K. □

Claim 2 *Let U and V be open sets of X such that $K \subset U \cup V$. Then there are open sets U', V' of X such that $K - V \subset U' \subset U$ and $K - U \subset V' \subset V$ and that $U' \cap V' = \emptyset$.*

Proof. For each $x \in K - V$ and $y \in K - U$, since $x \neq y$, the condition (H) shows that there are open sets $U(x, y)$ and $V(x, y)$ of X such that $x \in U(x, y)$ and $y \in V(x, y)$ and that $U(x, y) \cap V(x, y) = \emptyset$. For each $x \in K - V$, since $K - U \subset \bigcup_{y\in K-U} V(x, y)$ and $K - U$ is compact, there is a finite family $y_1, \ldots, y_{m(x)}$ of elements of $K - U$ such that $K - U$ is contained in $V(x) := \bigcup_{k=1}^{m(x)} V(x, y_k)$. Let $U(x) = \bigcap_{k=1}^{m(x)} U(x, y_k)$. Then $U(x) \cap V(x) = \emptyset$. Since $K - V$ is compact and $K - V \subset \bigcup_{x\in K-V} U(x)$, there is a finite family $x_1, \ldots, x_r$ of elements of $K - V$ such that $K - V \subset U' := U \cap \bigcup_{k=1}^r U(x_k)$. Let $V' = V \cap \bigcap_{k=1}^r V(x_k)$. Then $K - V \subset U' \subset U$, $K - U \subset V' \subset V$, and $U' \cap V' = \emptyset$. □

Now we can prove (iii) of proposition A2.3. Let $(U_k)_{1 \le k \le n}$ and s_k be as in Claim 1. We prove by induction on n. Assume $n \geq 2$. Let $K' = K - U_n$. Since K' is compact and $K' \subset \bigcup_{k=1}^{n-1} U_k$, by the induction hypothesis, there are an open neighborhood W of $K - U_n$ in X and $t \in \Gamma(W, \mathcal{F})$ such that at any point of K', the germ of t and that of s coincide. By shrinking W, we may assume that the germ of s and that of t coincide at any point of $K \cap W$. Hence we are reduced to the case $n = 2$.

Now assume $n = 2$ (so $K \subset U_1 \cup U_2$). By Claim 2, there are open sets U_1' and U_2' of X such that $K - U_2 \subset U_1' \subset U_1$, $K - U_1 \subset U_2' \subset U_2$, and $U_1' \cap U_2' = \emptyset$. Let U_3' be the open subset of $U_1 \cap U_2$ consisting of all points at which the germ of s_1 and that of s_2 coincide. Let $U = U_1' \cup U_2' \cup U_3'$. Then since

$$K - U_1' - U_2' \subset K \cap U_1 \cap U_2 \subset U_3',$$

we have $K \subset U$. Since $U_1' \cap U_2' = \emptyset$, there is an element $s' \in \Gamma(U, \mathcal{F})$ whose restriction to U_k' coincides with that of s_k for $k = 1, 2$, and whose restriction to U_3' coincides with that of s_1 (= that of s_2). Thus, the germ of s' at any point of K coincides with that of s. □

A2.6

We review flabby sheaves and soft sheaves.

A sheaf $\mathcal{F}$ on a topological space X is said to be *flabby* if, for any open set U of X, the map $\Gamma(X, \mathcal{F}) \to \Gamma(U, \mathcal{F})$ is surjective.

A sheaf $\mathcal{F}$ on a compact topological space X is said to be *soft* if for any closed set C of X, the map $\Gamma(X, \mathcal{F}) \to \Gamma(C, i^{-1}\mathcal{F})$ is surjective, where $i : C \hookrightarrow X$ is the inclusion map.

By Proposition A2.3 (iii), which we just proved, we have

Lemma A2.7 *Let X be a topological space and let K be a compact subspace of X satisfying the condition* (H) *in* A2.3. *Let $i : K \hookrightarrow X$ be the inclusion map. Then if $\mathcal{F}$ is a flabby sheaf on X, $i^{-1}\mathcal{F}$ is a soft sheaf on K.* □

A2.8

We review the cohomology of sheaves. Let X be a topological space.

If $\mathcal{F}$ is a sheaf of abelian groups on X, there is an exact sequence

$$0 \to \mathcal{F} \to \mathcal{Q}^0 \to \mathcal{Q}^1 \to \mathcal{Q}^2 \to \cdots \tag{1}$$

of sheaves of abelian groups with $\mathcal{Q}^k$ $(k \geq 0)$ flabby [Go, II 4.3]. For such sequence, the m-th cohomology group $H^m(X, \mathcal{F})$ is identified with the m-th cohomology group $H^m(\Gamma(X, \mathcal{Q}^\bullet))$ of the complex $\Gamma(X, \mathcal{Q}^\bullet)$ of abelian groups.

In the case X is compact, if we have an exact sequence (1) with $\mathcal{Q}^k$ $(k \geq 0)$ soft, then the m-th cohomology group $H^m(X, \mathcal{F})$ is identified with $H^m(\Gamma(X, \mathcal{Q}^\bullet))$ (see [KS, Exercise II.5]).

If $\mathcal{F}$ is a sheaf of groups on X, there is an injective homomorphism $\mathcal{F} \to \mathcal{Q}$ of sheaves of groups with $\mathcal{Q}$ flabby. For such injective homomorphism, $H^1(X, \mathcal{F})$ is identified with the quotient set $\Gamma(X, \mathcal{F}\backslash\mathcal{Q})/\Gamma(X, \mathcal{Q})$ of $\Gamma(X, \mathcal{F}\backslash\mathcal{Q})$ under the natural right action of $\Gamma(X, \mathcal{Q})$.

In the case X is compact, if we have an injective homomorphism $\mathcal{F} \to \mathcal{Q}$ with $\mathcal{Q}$ soft, then $H^1(X, \mathcal{F})$ is identified with $\Gamma(X, \mathcal{F}\backslash\mathcal{Q})/\Gamma(X, \mathcal{Q})$.

(These assertions for sheaves of groups can be proved in the same way as in the above case of sheaves of abelian groups.)

A2.9

We prove (i) and (ii) of Proposition A2.3.

Proof of (i). Take an exact sequence A2.8 (1) with $\mathcal{Q}^k$ $(k \geq 0)$ flabby. Then we have a commutative diagram

$$\begin{array}{ccc} \varinjlim_U H^m(U, \mathcal{F}) & \xrightarrow{\sim} & \varinjlim_U H^m(\Gamma(U, \mathcal{Q}^\bullet)) \\ \downarrow & & \downarrow \\ H^m(K, i^{-1}\mathcal{F}) & \xrightarrow{\sim} & H^m(\Gamma(K, i^{-1}\mathcal{Q}^\bullet)). \end{array}$$

Here the lower horizontal isomorphism is due to the fact that $i^{-1}\mathcal{Q}^k$ are soft by A2.7 and to A2.8. Since $\varinjlim_U H^m(\Gamma(U, \mathcal{Q}^\bullet)) \xrightarrow{\sim} H^m(\varinjlim_U \Gamma(U, \mathcal{Q}^\bullet))$, the right vertical arrow is an isomorphism by (iii) of A2.3. Hence the left vertical arrow is an isomorphism. □

Proof of (ii). Take an injective homomorphism $\mathcal{F} \to \mathcal{Q}$ with $\mathcal{Q}$ flabby. Then we have a commutative diagram

$$\begin{array}{ccc} \varinjlim_U H^1(U, \mathcal{F}) & \xrightarrow{\sim} & \varinjlim_U \Gamma(U, \mathcal{F}\backslash\mathcal{Q})/\Gamma(U, \mathcal{Q}) \\ \downarrow & & \downarrow \\ H^1(K, i^{-1}\mathcal{F}) & \xrightarrow{\sim} & \Gamma(K, i^{-1}\mathcal{F}\backslash i^{-1}\mathcal{Q})/\Gamma(K, i^{-1}\mathcal{Q}). \end{array}$$

Here the lower horizontal isomorphism is due to the fact that $i^{-1}\mathcal{Q}$ is soft by A2.7 and to A2.8. Since $\varinjlim_U(\Gamma(U, \mathcal{F}\backslash\mathcal{Q})/\Gamma(U, \mathcal{Q})) \xrightarrow{\sim} (\varinjlim_U \Gamma(U, \mathcal{F}\backslash\mathcal{Q}))/(\varinjlim_U \Gamma(U, \mathcal{Q}))$, the right vertical arrow is a bijection by (iii) of A2.3. Hence the left vertical arrow is a bijection. □

References

[Ak] D. Abramovich and K. Karu, Weak semistable reduction in characteristic 0, *Invent. Math.* **139** (2000), 241–273.

[AHV] J. M. Aroca, H. Hironaka, and J. L. Vicente, *Desingularization theorems*, Memorias de Math. del Instituto "Jorge Juan" No. 30, Consejo Superior de Investigaciones Cientificas, Madrid, 1964.

[AMRT] A. Ash, D. Mumford, M. Rapoport, and Y. S. Tai, *Smooth compactification of locally symmetric varieties*, Math. Sci. Press, Brookline, MA, 1975.

[BB] L. Baily and A. Borel, Compactification of arithmetic quotients of bounded symmetric domains, *Ann. Math.* **84** (1966), 442–528.

[B] A. Borel, *Introduction aux groupes arithmétiques*, Hermann, Paris, 1969.

[BJ] A. Borel and L. Ji, Compactifications of locally symmetric sapces, *J. Differential Geom.* **73** (2006), 263–317.

[BS] A. Borel and J.-P. Serre, Corners and arithmetic groups, *Comment. Math. Helv.* **48** (1973), 436–491.

[Bn] N. Bourbaki, *Topologie générale I*, Éléments de Mathématique, Vol. 2179, Hermann, Paris, 1966 (English translation: Hermann and Addison-Wesley, Reading, MA, 1966).

[CCK] J. A. Carlson, E. Cattani, and A. Kaplan, Mixed Hodge structures and compactifications of Siegel's space, in *Journées de géométrie algébrique d'Angers 1979* (A. Beauville, ed.), Sijthoff & Noordhoff, Dordrecht, 1980, pp. 77–105.

[CK1] E. Cattani and A. Kaplan, Extension of period mappings for Hodge structures of weight 2, *Duke Math. J.* **44** (1977), 1–43.

[CK2] ——, Polarized mixed Hodge structures and the local monodromy of a variation of Hodge structure, *Invent. Math.* **67** (1982), 101–115.

[CKS] E. Cattani, A. Kaplan, and W. Schmid, Degeneration of Hodge structures, *Ann. Math.* **123** (1986), 457–535.

[D1] P. Deligne, *Équations differentielles à points singuliers réguliers*, Lecture Notes in Mathematics No. 163, Springer-Verlag, Berlin, 1970.

[D2] ——, *Travaux de Griffiths*, Sém. Bourbaki No. 376 (1969/70), Lecture Notes in Mathematics No.180, Springer-Verlag, Berlin, 1971, pp. 213–238.

[D3] ——, *Travaux de Shimura*, Sém. Bourbaki No. 389 (1970/71), Lecture Notes in Mathematics No.244, Springer-Verlag, Berlin, 1971, pp. 123–165.

[D4] ——, Théorie de Hodge, II, *Publ. Math. IHES* **40** (1972), 5–57.

[D5] ——, La conjecture de Weil, II, *Publ. Math. IHES* **52** (1980), 137–252.

[F] T. Fujisawa, Limits of Hodge structures in several variables, *Compositio Math.* **115** (1999), 129–183.

[Go] R. Godement, *Topologie algébrique et théorie des faiseaux*, Hermann, Paris 1958.

[G1] P. A. Griffiths, Periods of integrals on algebraic manifolds. I. Construction and properties of modular varieties, *Am. J. Math.* **90** (1968), 568–626.

[G2] ——, Periods of integrals on algebraic manifolds. II. Local study of period mappings, *Am. J. Math.* **90** (1968), 805–865.

[G3] ——, Periods of integrals on algebraic manifolds. III. Some global differential-geometric properties of the period mapping, *Publ. Math. IHES* **38** (1970), 125–180.

[G4] ——, On periods of certain rational integrals. I, II, *Ann. Math.* **90** (1969), 460–541.

[G5] ——, Periods of integrals on algebraic manifolds: Summary of main results and discussions of open problems, *Bull. Am. Math. Soc.* **76** (1970), 228–296.

[G.et] P. A. Griffiths et al., *Topics in transcendental algebraic geometry*, Ann. Math. Studies Vol. 106, Princeton University Press, Princeton, NJ, 1984.

[GH] P. A. Griffiths and J. Harris, *Principles of algebraic geometry*, John Wiley & Sons, New York, 1994.

[HZ] M. Harris and S. Zucker, Boundary cohomology of Shimura varieties I. Coherent cohomology on toroidal compactifications, *Ann. Scient. Éc. Norm. Sup.* **27** (1994), 249–344.

[H] H. Hironaka, Resolution of singularities of an algebraic variety over a field of characteristic zero. I, II., *Ann. Math.* **79** (1964), 109–203, 205–326.

[HK] O. Hyodo and K. Kato, Semi-stable reduction and crystalline cohomology with logarithmic poles, *Astérisque* (1994), 221–268.

[I1] L. Illusie, Cohomologie de de Rham et cohomologie étale p-adique (d'après G. Faltings, J.-M. Fontaine et al.), *Astérisque* **189–190** (1990), 325–374.

[I2] ——, Logarithmic spaces (according to K. Kato), in *Barsotti Symposium in Algebraic Geometry* (V. Critstante and W. Messing, eds.), Perspectives in Mathematics No.15, Academic Press, New York, 1994, pp. 183–203.

[IKN] L. Illusie, K. Kato, and C. Nakayama, Quasi-unipotent logarithmic Riemann-Hilbert correspondences, *J. Math. Sci. Univ. Tokyo* **12** (2005), 1–66.

[KKN] T. Kajiwara, K. Kato, and C. Nakayama, Logarithmic abelian varieties, Part I: Complex analytic theory, *J. Math. Sci. Univ. Tokyo.* **15** (2008), 69–193.

[KKN2] ——, *Logarithmic abelian varieties, Part II. Algebraic theory*, Nagoya Math. J. **189** (2008), 63–138.

[KjNc] T. Kajiwara and C. Nakayama, Higher direct images of local systems in log Betti cohomology, preprint.

[K] M. Kashiwara, The asymptotic behavior of a variation of polarized Hodge structure, *Publ. RIMS Kyoto Univ.* **21** (1985), 853–875.

[KS] M. Kashiwara and P. Schapira, *Sheaves on manifolds*, Grund. math. Wissen. No.292, Springer, Berlin, 1990.

[Kf1] F. Kato, Log smooth deformation theory, *Tôhoku Math. J.* **48** (1996), 317–354.

[Kf2] ——, The relative log Poincaré lemma and relative log de Rham theory, *Duke Math. J.* **93** (1998), 179–206.

[Kk1] K. Kato, Logarithmic structures of Fontaine-Illusie, in *Algebraic analysis, geometry, and number theory* (J.-I. Igusa ed.), Perspectives in Mathematics, Johns Hopkins University Press, Baltimore, 1989, pp. 191–224.

[Kk2] ——, Toric singularities, *Am. J. Math.* **116** (1994), 1073–1099.

[KMN] K. Kato, T. Matsubara, and C. Nakayama, Log C^∞-functions and degenerations of Hodge structures, *Adv. Stud. Pure Math.* **36** Algebraic Geometry 2000, Azumino (2002), 269–320.

[KkNc] K. Kato and C. Nakayama, Log Betti cohomology, log étale cohomology, and log de Rham cohomology of log schemes over **C**, *Kodai Math. J.* **22** (1999), 161–186.

[KU1] K. Kato and S. Usui, Logarithmic Hodge structures and classifying spaces (summary), in *CRM Proceedings & Lecture Notes: The Arithmetic and Geometry of Algebraic Cycles*, NATO Advanced Study Institute/CRM Summer School, 1998. Banff Canada Vol. 24 pp. 115–130.

[KU2] ——, Borel-Serre spaces and spaces of SL(2)-orbits, *Adv. Studies Pure Math.* **36**, Algebraic Geometry 2000, Azumino (2002), 321–382.

[KyNy] Y. Kawamata and Y. Namikawa, Logarithmic deformations of normal crossing varieties and smoothing of degenerate Calabi-Yau varieties, *Invent. Math.* **118** (1994), 395–409.

[KKMS] G. Kempf, F. Knudsen, D. Mumford, and B. Saint-Donat, *Toroidal embeddings I*, Lecture Notes in Mathematics Vol. 339, Springer-Verlag, Berlin, 1973.

[Ma1] T. Matsubara, On log Hodge structures of higher direct images, *Kodai Math. J.* **21** (1998), 81–101.

[Ma2] ——, Log Hodge structures of higher direct images in several variables, preprint.

[Mo] D. R. Morrison, Mirror symmetry and rational curves on quintic three-folds: A guide for mathematicians, *J. Am. Math. Soc.* **6** (1993), 223–247.

[Od] T. Oda, *Convex bodies and algebraic geometry*, Ergebnisse Math. No.15, Springer-Verlag, Berlin, 1988.

[Og] A. Ogus, On the logarithmic Riemann-Hilbert correspondences, *Documenta Math.* extra volume: Kazuya Kato's fiftieth birthday (2003), 655–724.

[R] P. Ribenboim, *Théorie des valuationes*, University of Montreal Press, Montreal, 1964.

[Ss] S. Saito, *Infinitesimal logarithmic Torelli problem for degenerating hypersurfaces in* $\mathbf{P}^n$, with Appendix by A. Ikeda, Adv. Stud. Pure Math. **36**, Algebraic Geometry 2000, Azumino (2002), 401–442.

[Sa1] I. Satake, On compactifications of the quotient spaces for arithmetically defined discontinuous groups, *Ann. Math.* **72** (1960), 555–580.

[Sa2] ——, *Algebraic structures of symmetric domains*, Iwanami Shoten, Tokyo Princeton University Press, Princeton, NJ, 1980.

[Sc] W. Schmid, Variation of Hodge structure: The singularities of the period mapping, *Invent. Math.* **22** (1973), 211–319.

[SGA $4\frac{1}{2}$] Sém. géométrie algébrique, by P. Deligne, in *Cohomologie étale*, Lecture Notes in Mathematics No.569, Springer-Verlag, Berlin, 1977.

[Sp] E. H. Spanier, *Algebraic topology*, Springer-Verlag, Berlin, 1966.

[St] J.H.M. Steenbrink, Limits of Hodge structures, *Invent. Math.* **31** (1976), 229–257.

[SZ] J.H.M. Steenbrink and S. Zucker, Variation of mixed Hodge structure I, *Invent. Math.* **80** (1985), 489–542.

[U1] S. Usui, A numerical criterion for admissibility of semi-simple elements, *Tôhoku Math. J.* **45** (1993), 471–484.

[U2] ——, Complex structures on partial compactifications of arithmetic quotients of classifying spaces of Hodge structures, *Tôhoku Math. J.* **47** (1995), 405–429.

[U3] ——, Recovery of vanishing cycles by log geometry, *Tôhoku Math. J.* **53** (2001), 1–36.

[U4] ——, Images of extended period maps, *J. Alg. Geom.* **15** (2006), 603–621.

[Z1] S. Zucker, L_2 cohomology of warped products and arithmetic groups, *Invent. Math.* **70** (1982), 169–218.

[Z2] ——, Satake compactifications, *Comment. Math. Helv.* **58** (1983), 312–343.

[Z3] ——, Variation of mixed Hodge structure II, *Invent. Math.* **80** (1985), 543–565.

[Z4] ——, On the reductive Borel-Serre compactification: L^p-cohomology of arithmetic groups (for large p), *Amr. J. Math.* **123** (2001), 951–984.

List of Symbols

CHAPTER 0

$\Omega_X^\bullet$: de Rham complex of an analytic space X 0.1.7
$(\ ,\ )_F : H_{\mathbf{C}} \times H_{\mathbf{C}} \to \mathbf{C}$, Hermitian form associated with F 0.1.8
$|\Delta| := \{r \in \mathbf{R} \mid 0 \leq r < 1\}$ 0.2.9
$X_{\text{triv}} := \{x \in X \mid M_{X,x} = \mathcal{O}_{X,x}^\times\} \subset X$ 0.4.30
$\mathbf{N} := \mathbf{Z}_{\geq 0}$ 0.7.1
L_R for $\mathbf{Z}$-module L and $R = \mathbf{Q}, \mathbf{R}, \mathbf{C}$ 0.7.3
$\Phi_0 = \big(w, (h^{p,q})_{p,q\in\mathbf{Z}}, H_0, \langle\ ,\ \rangle_0\big)$ 0.7.3
$(h^{p,q})_{p,q\in\mathbf{Z}}$ 0.7.3
H_0 : $\mathbf{Z}$-module 0.7.3
$\langle\ ,\ \rangle_0$: $\mathbf{Q}$-rational non-degenerate $(-1)^w$-symmetric $\mathbf{C}$-bilinear form on $H_{0,\mathbf{C}}$ 0.7.3
$G_R := \text{Aut}(H_{0,R}, \langle\ ,\ \rangle_0)$ for $R = \mathbf{Z}, \mathbf{Q}, \mathbf{R}, \mathbf{C}$ 0.7.3
$\mathfrak{g}_R := \text{Lie}\, G_R$ for $R = \mathbf{Q}, \mathbf{R}, \mathbf{C}$ 0.7.3
G° : connected component of G in Zariski topology containing 1 0.7.3
$\mathcal{A}$: category of analytic spaces 0.7.4
$\mathcal{A}(\log)$: category of fs logarithmic analytic spaces 0.7.4
$f^{-1}(\mathcal{F})$: sheaf-theoretic inverse image 0.7.6
$f^*(\mathcal{F})$: module-theoretic inverse image 0.7.6

CHAPTER 1

$(H_{\mathbf{Z}}, F)$: Hodge structure 1.1.1, 0.1.5
$F = (F^p)_{p\in\mathbf{Z}}$: Hodge filtration 1.1.1, 0.1.5
$\langle\ ,\ \rangle$: bilinear form on $H_{\mathbf{Q}}$ 1.1.2, 0.1.8
$C_F \in \text{Aut}(H_{\mathbf{C}})$: Weil operator associated with F 1.1.2
D : classifying space of polarized Hodge structures of type Φ_0 1.2.1
$\check{D}$: compact dual of D 1.2.2
σ : a nilpotent cone in $\mathfrak{g}_{\mathbf{R}}$ 1.3.5
$\check{D}_{\text{orb}} := \{(\sigma, \exp(\sigma_{\mathbf{C}})F) \mid$ nilpotent cone σ in $\mathfrak{g}_{\mathbf{R}}$, $F \in \check{D}\}$ 1.3.6
$\check{D}_{\text{orb}}^\sharp := \{(\sigma, \exp(i\sigma_{\mathbf{R}})F) \mid$ nilpotent cone σ in $\mathfrak{g}_{\mathbf{R}}$, $F \in \check{D}\}$ 1.3.6
Σ : a fan in $\mathfrak{g}_{\mathbf{Q}}$ 1.3.8
D_Σ : space of nilpotent orbits in directions in Σ 1.3.8
$D_\Sigma^\sharp$: space of nilpotent i-orbits in directions in Σ 1.3.8
$D_\sigma := D_{\{\text{face of } \sigma\}}$ 1.3.9
$D_\sigma^\sharp := D_{\{\text{face of } \sigma\}}^\sharp$ 1.3.9

$D_\Sigma^\sharp \to D_\Sigma$, canonical projection 1.3.9
$D_\Sigma = \bigcup_{\sigma\in\Sigma} D_\sigma$ 1.3.9
$D_\Sigma^\sharp = \bigcup_{\sigma\in\Sigma} D_\sigma^\sharp$ 1.3.9
$\Gamma(\sigma) := \Gamma \cap \exp(\sigma)$ 1.3.10
Ξ : fan in $\mathfrak{g}_{\mathbf{Q}}$ of rank ≤ 1 1.3.11

CHAPTER 2

$\alpha : M \to \mathcal{O}_X$, logarithmic structure 2.1.1, 0.2.4
$f^*(M, \alpha)$: inverse image of logarithmic structure 2.1.3
$\mathcal{S}^\times \subset \mathcal{S}$: group of invertible elements in a monoid $\mathcal{S}$ 2.1.4
$\mathcal{S}^{\mathrm{gp}} \supset \mathcal{S}$: group associated to an integral monoid $\mathcal{S}$ 2.1.4
$(\mathcal{S}_\lambda, \theta_\lambda)_\lambda$: chart of a logarithmic structure 2.1.5
$\mathbf{C}[\mathcal{S}]$: monoid ring of $\mathcal{S}$ over $\mathbf{C}$ 2.1.6
$\omega^q_{X/Y}$: sheaf of logarithmic differential q-forms on X over Y 2.1.7
$(\omega^\bullet_{X/Y}, d)$: logarithmic de Rham complex of X over Y 2.1.7
$X \times_Z Y$: fiber product in the category $\mathcal{A}(\log)$ 2.1.10
$\overline{\mathcal{A}}_1$: category 2.2.1
$\overline{\mathcal{A}}_1(\log)$: category 2.2.1
Z_{an} : analytic space associated to a scheme Z locally of finite type over $\mathbf{C}$ 2.2.1
$X^{\log}$ for $X \in \overline{\mathcal{A}}_1(\log)$ 2.2.3
$\tau : X^{\log} \to X$, canonical projection 2.2.3
$\mathcal{L}$: sheaf of logarithms of M_X^{gp} 2.2.4
$\mathcal{O}_X^{\log}$: sheaf of rings on $X^{\log}$ 2.2.4
$(X^{\log}, \mathcal{O}_X^{\log})$: ringed space associated with $X \in \overline{\mathcal{A}}_1(\log)$ 2.2.4
$\tau : (X^{\log}, \mathcal{O}_X^{\log}) \to (X, \mathcal{O}_X)$, canonical projection 2.2.4
$f^{\log} : (X^{\log}, \mathcal{O}_X^{\log}) \to (Y^{\log}, \mathcal{O}_Y^{\log})$, associated morphism 2.2.4
$\omega^{q,\log}_{X/Y}$: sheaf of logarithmic differential q-forms on $X^{\log}$ over $Y^{\log}$ 2.2.6
$(\omega^{\bullet,\log}_{X/Y}, d)$: logarithmic de Rham complex of $X^{\log}$ over $Y^{\log}$ 2.2.6
$\pi_1^+(x^{\log}) \simeq \mathrm{Hom}(M_x/\mathcal{O}_x^\times, \mathbf{N})$: positive monoid in $\pi_1(x^{\log})$ 2.2.9
$\gamma(\log(f)) = \log(f) - 2\pi[\gamma, f]$ for $\gamma \in \pi_1(x^{\log})$ and $f \in M_x$ 2.2.9, A1
$\nu : \mathcal{O}_X^{\log} \otimes_A L \xrightarrow{\sim} \mathcal{O}_X^{\log} \otimes_A L_0$ 2.3.2
$\xi = \exp(\sum_{j=1}^n (2\pi i)^{-1} \log(q_j) \otimes N_j) : \mathcal{O}_X^{\log} \otimes_A L_0 \xrightarrow{\sim} \mathcal{O}_X^{\log} \otimes_A L_0$ 2.3.2
$\nu : L \xrightarrow{\sim} \xi^{-1}(1 \otimes L_0)$ 2.3.2
$\nu_y(1 \otimes v) = \xi_y^{-1}(1 \otimes v)$ for $v \in L_0$ 2.3.2
$\mathcal{N} : L \to (M_X^{\mathrm{gp}}/\mathcal{O}_X^\times) \otimes L$, logarithm of local monodromy actions 2.3.4
$(H_{\mathbf{Z}}, F)$: prelogarithmic Hodge structure 2.4.1
$(H_{\mathbf{Z}}, \langle\ ,\ \rangle, F)$: prepolarized logarithmic Hodge structure 2.4.3
$\mathcal{H}$: category 2.4.4
$\mathcal{H}'$: category 2.4.4
$\mathcal{P}$: category 2.4.4
$\mathcal{P}'$: category 2.4.4

$\mathrm{sp}(y) := \mathrm{Hom}_{\mathbf{C}\text{-alg}}(\mathcal{O}^{\log}_{x,y}, \mathbf{C})$, space of specializations over $y \in x^{\log}$ 2.4.6
$\tilde{s} : M_x^{\mathrm{gp}} \to \mathbf{C}^\times,\ a \mapsto \exp(s(\log(a)))$ 2.4.6
$F(s)$: specializaion of F at $s \in \mathrm{sp}(y)$ 2.4.6
$(H_{\mathbf{Z}}, \langle\ ,\ \rangle, F)$: polarized logarithmic Hodge structure 2.4.8, 2.4.7
$\check{\varphi} : X \to \Gamma\backslash\check{D}_{\mathrm{orb}}$, period map 2.5.3
$\Phi := (w, (h^{p,q})_{p,q\in\mathbf{Z}}, H_0, \langle\ ,\ \rangle_0, \Gamma, \Sigma)$ 2.5.8
$\varphi : X \to \Gamma\backslash D_\Sigma$, period map (of H with respect to Σ) 2.5.10
$\varphi^{\log} : X^{\log} \to \Gamma\backslash D^\sharp_\Sigma$ 2.5.10
$\mathbf{R}^{\mathrm{add}}_{\geq 0}$: monoid 2.6.2
$(H_{\mathbf{Z}}, W, F)$: logarithmic mixed Hodge structure 2.6.4, 2.6.2

CHAPTER 3

$S_{\mathrm{weak}/X}$ 3.1.1
$S_{\mathrm{str}/X}$ 3.1.1
$U(\varepsilon)$: a neighborhood in the strong topology 3.1.3
$U(s, \varepsilon)$: a neighborhood in the strong topology 3.1.4, 3.1.5
$\overline{\mathcal{D}}$: categorical closure of a category $\mathcal{D}$ 3.2.2
$\overline{\mathcal{A}}$: category 3.2.3
$\overline{\mathcal{A}(\log)}$: category 3.2.3
$\overline{\mathcal{A}(\log)}'$: category 3.2.3
$\overline{\mathcal{A}(\log)}'_{\mathrm{fs}}$: category 3.2.3
$\overline{\mathcal{A}}_2$: category 3.2.4
$\mathcal{B}$: category 3.2.4
$\mathcal{B}^*$: category 3.2.4
$\overline{\mathcal{A}}_1(\log)$: category 3.2.4
$\overline{\mathcal{A}}_2(\log)$: category 3.2.4
$\overline{\mathcal{A}}(\log)$: category 3.2.4
$\mathcal{B}(\log)$: category 3.2.4
$\mathcal{B}^*(\log)$: category 3.2.4
$\mathcal{S}^\vee := \mathrm{Hom}(\mathcal{S}, \mathbf{N})$, dual monoid 3.3.1
$\delta : \{\text{face of } \mathcal{S}\} \xrightarrow{\sim} \{\text{face of } \mathcal{S}^\vee\},\ \mathcal{S}' \mapsto \{h \in \mathcal{S}^\vee \mid h(\mathcal{S}') = \{0\}\}$ 3.3.1
$\{\text{face of } \sigma\} \xrightarrow{\sim} \{\text{face of } \Gamma(\sigma)\},\ \tau \mapsto \Gamma(\tau)$, bijection 3.3.1
$\mathbf{C}^{\mathrm{mult}}$: monoid 3.3.2
$\mathbf{R}^{\mathrm{mult}}_{\geq 0}$: monoid 3.3.2
$\mathrm{toric}_\sigma := \mathrm{Spec}(\mathbf{C}[\Gamma(\sigma)^\vee])_{\mathrm{an}}$, toric variety 3.3.2
$\mathrm{torus}_\sigma := \mathbf{C}^\times \otimes \Gamma(\sigma)^{\mathrm{gp}}$, torus 3.3.2
$\check{E}_\sigma := \mathrm{toric}_\sigma \times \check{D}$ 3.3.2, 3.4.1
$\pi_1(\mathrm{toric}_\sigma^{\log}) \simeq \pi_1(\mathrm{torus}_\sigma) \simeq \Gamma(\sigma)^{\mathrm{gp}}$, canonical isomorphisms 3.3.2
$\pi_1^+(q^{\log}) = \pi_1(q^{\log}) \cap \Gamma(\sigma)$ for $q \in \mathrm{toric}_\sigma$ 3.3.2
$\sigma(q)$: face of σ corresponding to a point $q \in \mathrm{toric}_\sigma$ 3.3.2
$H_\sigma = (H_{\sigma,\mathbf{Z}}, \langle\ ,\ \rangle_\sigma, F_\sigma)$: canonical pre-PLH on $\check{E}_\sigma$ 3.3.3
$(H_{\sigma,\mathbf{Z}}, \langle\ ,\ \rangle_\sigma)$: canonical local system on $\mathrm{toric}_\sigma^{\log}$, and also on $\check{E}_\sigma^{\log}$ 3.3.3

$\nu : \mathcal{O}^{\log}_{\mathrm{toric}_\sigma} \otimes_{\mathbf{Z}} H_{\sigma,\mathbf{Z}} \simeq \mathcal{O}^{\log}_{\mathrm{toric}_\sigma} \otimes_{\mathbf{Z}} H_0$ 3.3.3

$\mathcal{O}^{\log}_{\check{E}_\sigma} \otimes_{\mathbf{Z}} H_{\sigma,\mathbf{Z}} = \mathcal{O}^{\log}_{\check{E}_\sigma} \otimes_{\mathbf{Z}} H_0$, identification via ν 3.3.3

μ_σ : canonical $\Gamma(\sigma)^{\mathrm{gp}}$-level structure of H_σ 3.3.3

$\xi(1 \otimes v) = 1 \otimes \mu_\sigma(v)$ $(v \in H_{\sigma,\mathbf{Z}})$, expression of μ_σ 3.3.3

$\tilde{E}_\sigma \subset \check{E}_\sigma$ 3.3.4

$E_\sigma \subset \tilde{E}_\sigma \subset \check{E}_\sigma$ 3.3.4, 3.4.1

$\mathbf{e} : \sigma_{\mathbf{C}} \to \mathrm{torus}_\sigma$, surjective homomorphism 3.3.5

$|\mathrm{toric}|_\sigma := \mathrm{Hom}(\Gamma(\sigma)^\vee, \mathbf{R}^{\mathrm{mult}}_{\geq 0}) \subset \mathrm{toric}_\sigma$ 3.3.9

$|\mathrm{torus}|_\sigma := \mathbf{R}_{>0} \otimes \Gamma(\sigma)^{\mathrm{gp}} \subset \mathrm{torus}_\sigma$ 3.3.9

$\check{E}^\sharp_\sigma := |\mathrm{toric}|_\sigma \times \check{D}$ 3.3.9

$E^\sharp_\sigma \subset \check{E}^\sharp_\sigma$ 3.3.9, 3.4.3

$\varphi : E_\sigma \to \Gamma(\sigma)^{\mathrm{gp}} \backslash D_\sigma$, period map 3.3.10

$\varphi^\sharp : E^\sharp_\sigma \to D^\sharp_\sigma$, period map 3.3.10

E_σ : topological space 3.4.1

$\mathcal{O}_{E_\sigma}$: sheaf of rings of E_σ 3.4.1

M_{E_σ} : logarithmic structure of E_σ 3.4.1

$\Gamma(\sigma)^{\mathrm{gp}} \backslash D_\sigma$: topological space 3.4.2

$\Gamma \backslash D_\Sigma$: topological space 3.4.2

$\mathcal{O}_{\Gamma \backslash D_\Sigma}$: sheaf of rings of $\Gamma \backslash D_\Sigma$ 3.4.2

$M_{\Gamma \backslash D_\Sigma}$: logarithmic structure of $\Gamma \backslash D_\Sigma$ 3.4.2

$X \times_Z Y$: fiber product in the category $\mathcal{B}(\log)$ 3.5.1

$X \times^{\mathrm{cl}}_Z Y$: fiber product in the category of topological spaces (cl : classical) 3.5.1

ω^1_X : sheaf of logarithmic differential 1-forms on X for $X \in \mathcal{B}(\log)$ 3.5.2

$\omega^\bullet_X$ logarithmic de Rham complex on X for $X \in \mathcal{B}(\log)$ 3.5.2

$\omega^{\bullet,\log}_X$: logarithmic de Rham complex on $X^{\log}$ for $X \in \mathcal{B}(\log)$ 3.5.2

θ_X : sheaf of logarithmic vector fields for logarithmically smooth $X \in \mathcal{B}(\log)$ 3.5.4

T_X : logarithmic tangent bundle for logarithmically smooth $X \in \mathcal{B}(\log)$ 3.5.6

$B_I(X)$: logarithmic blow-up of X with respect to I for $X \in \mathcal{B}(\log)$ 3.6.6

$B^*_I(X)$: variant of $B_I(X)$ for $X \in \mathcal{B}(\log)$ 3.6.6

$(N_{\mathbf{Q}}, s)$ 3.6.7

$s \in \Gamma(X, M^{\mathrm{gp}}_X / \mathcal{O}^\times_X) \otimes N_{\mathbf{Q}}$ for $X \in \mathcal{B}(\log)$ 3.6.7

$X(\sigma)$ for $X \in \mathcal{B}(\log)$ and a finitely generated rational cone σ in $N_{\mathbf{R}}$ 3.6.8

$X(\Sigma)$ for $X \in \mathcal{B}(\log)$ and a rational fan Σ in $N_{\mathbf{R}}$ 3.6.10

$Q_1(X)$: a set of (x, P) for $X \in \mathcal{B}(\log)$ 3.6.14

$Q(X)$: a set of (x, P, h) for $X \in \mathcal{B}(\log)$ 3.6.14

$Q'_1(X)$: a set of (x, σ) for $X \in \mathcal{B}(\log)$ 3.6.14

$Q'(X)$: a set of (x, σ, h) for $X \in \mathcal{B}(\log)$ 3.6.14

$q_Y : Y \to Q(X)$ for $X \in \mathcal{B}(\log)$ and $Y \in \mathcal{B}(\log)$ over X 3.6.14

$q'_Y : Y \to Q'(X)$ for $X \in \mathcal{B}(\log)$ and $Y \in \mathcal{B}(\log)$ over X 3.6.14

$q_{Y,1} : Y \to Q_1(X)$ for $X \in \mathcal{B}(\log)$ and $Y \in \mathcal{B}(\log)$ over X 3.6.14

$q'_{Y,1} : Y \to Q'_1(X)$ for $X \in \mathcal{B}(\log)$ and $Y \in \mathcal{B}(\log)$ over X 3.6.14

X_{val} for $X \in \mathcal{B}(\log)$ 3.6.18

$X^{\log}_{\mathrm{val}}$ for $X \in \mathcal{B}(\log)$ 3.6.26

CHAPTER 4

$\underline{\mathrm{PLH}}_\Phi : \overline{\mathcal{A}}_2(\log) \to (\mathrm{Sets})$, functor 4.2.1
$\underline{\mathrm{PLH}}_\Phi|_{\mathcal{A}(\log)} : \mathcal{A}(\log) \to (\mathrm{Sets})$, functor 4.2.2
$\mathcal{N} : H_{\mathbf{Q}} \to (M_X^{\mathrm{gp}}/\mathcal{O}_X^\times) \otimes H_{\mathbf{Q}}$ for a pre-PLH H on $X \in \mathcal{B}(\log)$ 4.3.5
$\mathcal{N} \in \Gamma(X, \Gamma\backslash((M_X^{\mathrm{gp}}/\mathcal{O}_X^\times) \otimes \mathfrak{g}_{\mathbf{Q}}))$ for H on $X \in \mathcal{B}(\log)$ with a Γ-level structure 4.3.5
$\nabla : \mathcal{M} \to \omega_X^1 \otimes_{\mathcal{O}_X} \mathcal{M}$ for a PLH H on a logarithmically smooth $X \in \mathcal{B}(\log)$ 4.4.2
$\theta_X \to \bigoplus_p \mathcal{H}om_{\mathcal{O}_X}(\mathcal{M}^p, \mathcal{M}/\mathcal{M}^p)$ 4.4.2
$\theta_X \simeq \mathcal{E}nd_{\langle\,,\,\rangle}(\mathcal{M})/F^0\mathcal{E}nd_{\langle\,,\,\rangle}(\mathcal{M})$ 4.4.3
$T_X^h \subset T_X$: horizontal logarithmic tangent bundle 4.4.4
$\theta_X^h \subset \theta_X$: horizontal submodule 4.4.4
$\theta_X^h \simeq \mathrm{gr}_F^{-1} \mathcal{E}nd_{\langle\,,\,\rangle}(\mathcal{M})$ 4.4.4

CHAPTER 5

$\mathcal{X}$: set of all maximal compact subgroups of $G_{\mathbf{R}}$ 5.1.1
K_F : maximal compact subgroup of $G_{\mathbf{R}}$ associated to F 5.1.2
K'_F : isotropy subgroup of $G_{\mathbf{R}}$ at $F \in D$ 5.1.2
P_u : unipotent radical of a parabolic subgroup P 5.1.3
C : center of P/P_u 5.1.3
θ_K : Cartan involuton of $G_{\mathbf{R}}$ at K 5.1.3
a_K : Borel-Serre lifting of $a \in C \subset P/P_u$ at K 5.1.3
$a \circ F := a_{K_F} F$: Borel-Serre action of $a \in C$ on D 5.1.3
$a \circ K := \mathrm{Int}(a_K)$: Borel-Serre action of $a \in C$ on $\mathcal{X}$ 5.1.3
S_P : maximal $\mathbf{Q}$-split torus of the center C of P/P_u 5.1.4
A_P : connected component of $\mathbf{R}$-valued points of S_P containing 1 5.1.4
$D_{\mathrm{BS}} \supset D$: Borel-Serre space 5.1.5
$\mathcal{X}_{\mathrm{BS}} \supset \mathcal{X}$: Borel-Serre space 5.1.5
$V^\times$: set of invertible elements of V 5.1.5
$D_{\mathrm{BS,val}}$: valuative Borel-Serre space 5.1.6
$T_{>0}$ for a torus T 5.1.6
$W_\alpha := \bigoplus_{\chi \in V} H_{0,\mathbf{R}}(\alpha\chi^{-1}) \subset H_{0,\mathbf{R}}$ for $\alpha \in X(T)$ and $V \subset X(T)$: valuative 5.1.6
$P_{T,V}$: parabolic subgroup associated with (T, V) 5.1.6
$D_{\mathrm{BS}}(P) \subset D_{\mathrm{BS}}$ 5.1.8
$\mathcal{X}_{\mathrm{BS}}(P) \subset \mathcal{X}_{\mathrm{BS}}$ 5.1.8
$D_{\mathrm{BS,val}}(P) \subset D_{\mathrm{BS,val}}$ 5.1.8
$\Delta'_P \subset X(S_P)$ 5.1.9
$\Delta_P \subset \Delta'_P \subset X(S_P)$ 5.1.9
$X(S_P)_+ \subset X(S_P)$ 5.1.10
$\overline{A}_P \supset A_P$ 5.1.10
$\overline{A}_P \simeq \mathrm{Map}(\Delta_P, \mathbf{R}_{\geq 0}) \simeq \mathbf{R}^r_{\geq 0}$ 5.1.10

$D \times^{A_P} \overline{A}_P$ 5.1.10
$D_{\mathrm{BS}}(P) \simeq D \times^{A_P} \overline{A}_P$ 5.1.10
$\mathrm{Spec}(\mathbf{C}[\mathcal{S}])_{\mathrm{an,val}} = \mathrm{Hom}(\mathcal{S}, \mathbf{C}^{\mathrm{mult}})_{\mathrm{val}}$, valuative toric variety 5.1.11
$\mathrm{Hom}(\mathcal{S}, \mathbf{R}^{\mathrm{mult}}_{\geq 0})_{\mathrm{val}}$ 5.1.11
$(\overline{A}_P)_{\mathrm{val}} := \mathrm{Hom}(X(S_P)_+, \mathbf{R}^{\mathrm{mult}}_{\geq 0})_{\mathrm{val}}$ 5.1.12
$D_{\mathrm{BS,val}}(P) \simeq D \times^{A_P} (\overline{A}_P)_{\mathrm{val}}$ 5.1.12
$\mathcal{X}_{\mathrm{BS}}(P) \simeq \mathcal{X} \times^{A_P} \overline{A}_P$ 5.1.13
$G_{\mathbf{R}} \simeq P_u \times \mathbf{R}^r_{>0} \times K$, Iwasawa decomposition 5.1.15
$S^{\mathbf{R}}_P$: maximal $\mathbf{R}$-split torus of the center C of P/P_u 5.1.15
$A^{\mathbf{R}}_P$: connected component of $\mathbf{R}$-valued points of $S^{\mathbf{R}}_P$ 5.1.15
$\mathcal{X} \simeq P_u \times \mathbf{R}^r_{>0}$ 5.1.15
$D \simeq P_u \times \mathbf{R}^r_{>0} \times (K_F/K'_F)$ 5.1.15
$\mathcal{X}_{\mathrm{BS}}(Q) \simeq P_u \times (\mathbf{R}^r_{>0} \times^{A_Q} \overline{A}_Q)$ 5.1.15
$D_{\mathrm{BS}}(Q) \simeq P_u \times (\mathbf{R}^r_{>0} \times^{A_Q} \overline{A}_Q) \times (K_F/K'_F)$ 5.1.15
$D_{\mathrm{BS,val}}(Q) \simeq P_u \times (\mathbf{R}^r_{>0} \times^{A_Q} \overline{A}_{Q,\mathrm{val}}) \times (K_F/K'_F)$ 5.1.15
$\mathcal{X}_{\mathrm{BS}}(P) \simeq P_u \times \mathbf{R}^r_{\geq 0}$ 5.1.15
$D_{\mathrm{BS}}(P) \simeq P_u \times \mathbf{R}^r_{\geq 0} \times (K_F/K'_F)$ 5.1.15
$D_{\mathrm{BS,val}}(P) \simeq P_u \times (\mathbf{R}^r_{\geq 0})_{\mathrm{val}} \times (K_F/K'_F)$ 5.1.15
(ρ, φ) : SL(2)-orbit 5.2.1
$\mathbf{i} \in \mathfrak{h}^n$ 5.2.1
$N_j := \rho_{*j}\begin{pmatrix} 0 & 1 \\ 0 & 0 \end{pmatrix} \in \mathfrak{g}_{\mathbf{R}}$ 5.2.2
$\rho_{*j} : \mathfrak{sl}(2, \mathbf{C}) \to \mathfrak{g}_{\mathbf{C}}$ 5.2.2
$\Delta : \mathbf{G}^n_{\mathbf{m},\mathbf{R}} \to \mathrm{SL}(2, \mathbf{R})^n$ 5.2.2
$\tilde{\rho} : \mathbf{G}^n_{\mathbf{m},\mathbf{R}} \to G_{\mathbf{R}}$ 5.2.2
$\tilde{\rho}_j : \mathbf{G}_{\mathbf{m},\mathbf{R}} \to G_{\mathbf{R}}$ 5.2.2
$W(N)$: weight filtration associated with a nilpotent endomorphism N 5.2.4
$W(\sigma)$: weight filtration associated with a nilpotent orbit (σ, Z) 5.2.4
$(W(\sigma_j))_{1 \leq j \leq n}$: family of weight filtrations associated to an SL(2)-orbit 5.2.5
$D_{\mathrm{SL}(2)}$: space of SL(2)-orbits 5.2.6
$D_{\mathrm{SL}(2),n}$: space of SL(2)-orbits of rank n 5.2.6
$[\rho, \varphi] \in D_{\mathrm{SL}(2)}$ 5.2.6
$D_{\mathrm{SL}(2),\mathrm{val},n}$: space of valuative SL(2)-orbits of rank n 5.2.7
$D_{\mathrm{SL}(2),\mathrm{val}}$: space of valuative SL(2)-orbits 5.2.7
$X(\mathbf{G}^n_{\mathbf{m},\mathbf{R}})_+ \subset X(\mathbf{G}^n_{\mathbf{m},\mathbf{R}})$ 5.2.7
$\mathbf{r} = \varphi(\mathbf{i}) \in D$: reference point 5.2.8
$W^{(j)} := W(\sigma_j)$ 5.2.9
$G_{W,\mathbf{R}} \subset G_{\mathbf{R}}$ 5.2.12
$D_{\mathrm{SL}(2)}(W) \subset D_{\mathrm{SL}(2)}$ 5.2.12
$D_{\mathrm{SL}(2),\mathrm{val}}(W) \subset D_{\mathrm{SL}(2),\mathrm{val}}$ 5.2.12
$B(U, U', U'')$: a basis of the filter in D of a point of $D_{\mathrm{SL}(2)}$ 5.2.16
$\mathcal{V}$: a set of pairs of a subspace of $\mathfrak{g}_{\mathbf{Q}}$ and a valuative submonoid in its dual 5.3.1
$\mathfrak{F}(A, V)$: set of rational nilpotent cones in $\mathfrak{g}_{\mathbf{R}}$ associated to $(A, V) \in \mathcal{V}$ 5.3.2
$\check{D}_{\mathrm{val}} \supset D_{\mathrm{val}}$ 5.3.3
$\check{D}^{\sharp}_{\mathrm{val}} \supset D^{\sharp}_{\mathrm{val}}$ 5.3.3

D_{val} : space of valuative nilpotent orbits 5.3.3
$D^{\sharp}_{\text{val}}$: space of valuative nilpotent i-orbits 5.3.3
$D^{\sharp}_{\text{val}} \to D_{\text{val}}$ 5.3.3
$D_{\Sigma,\text{val}} \subset D_{\text{val}}$: space of valuative nilpotent orbits in directions in Σ 5.3.5
$D^{\sharp}_{\Sigma,\text{val}} \subset D^{\sharp}_{\text{val}}$: space of valuative nilpotent i-orbits in directions in Σ 5.3.5
$D_{\sigma,\text{val}} = D_{\{\text{face of }\sigma\},\text{val}}$ 5.3.5
$D^{\sharp}_{\sigma,\text{val}} = D^{\sharp}_{\{\text{face of }\sigma\},\text{val}}$ 5.3.5
$D^{\sharp}_{\Sigma,\text{val}} \to D_{\Sigma,\text{val}}$ 5.3.5
$D_{\Sigma,\text{val}} \to D_{\Sigma}$ 5.3.5
$D^{\sharp}_{\Sigma,\text{val}} \to D^{\sharp}_{\Sigma}$ 5.3.5
$\text{toric}_{\sigma,\text{val}} := \text{Spec}(\mathbf{C}[\Gamma(\sigma)^{\vee}])_{\text{an,val}}$, valuative toric variety 5.3.6
$|\text{toric}|_{\sigma,\text{val}} \subset \text{toric}_{\sigma,\text{val}}$ 5.3.6
$E_{\sigma,\text{val}} := \text{toric}_{\sigma,\text{val}} \times_{\text{toric}_{\sigma}} E_{\sigma}$ 5.3.7
$E^{\sharp}_{\sigma,\text{val}} := |\text{toric}|_{\sigma,\text{val}} \times_{|\text{toric}|_{\sigma}} E^{\sharp}_{\sigma} \subset E_{\sigma,\text{val}}$ 5.3.7
$E_{\sigma,\text{val}} \to E_{\sigma}$ 5.3.7
$E^{\sharp}_{\sigma,\text{val}} \to E^{\sharp}_{\sigma}$ 5.3.7
$\psi : D^{\sharp}_{\text{val}} \to D_{\text{SL}(2)}$, CKS map 5.4.3

CHAPTER 6

$w_N : \mathbf{G}_{\mathbf{m},\mathbf{R}} \to P/P_u$, weight map 6.1.1
$W[l]$: shift of filtration 6.1.2
$\varepsilon = \varepsilon(W, F) \in \mathfrak{g}_{\mathbf{C}}$ 6.1.2
$\hat{F} = (W, F)^{\wedge}$: associated $\mathbf{R}$-split Hodge filtration 6.1.2
$\mathfrak{F}^W$: set of decreasing filtrations F on $H_{\mathbf{C}}$ such that (W, F) is a MHS 6.1.2
$I^{p,q}$: (p, q)-component of Deligne splitting 6.1.2
$L^{-1,-1} = L^{-1,-1}(W, F) \subset \text{End}_{\mathbf{C}}(V_{\mathbf{C}})$ 6.1.2
$\delta \in L^{-1,-1}_{\mathbf{R}}$ 6.1.2
$\tilde{I}^{p,q} := I^{p,q}(W, \exp(-i\delta)F)$, $\mathbf{R}$-splitting of $\mathbf{R}$-split MHS $(W, \exp(-i\delta)F)$ 6.1.2
$\delta_{p,q}$: (p, q)-component of δ with respect to $(\tilde{I}^{p,q})$ 6.1.2
$\zeta \in L^{-1,-1}_{\mathbf{R}}$ 6.1.2
$\zeta_{p,q}$: (p, q)-component of ζ with respect to $(\tilde{I}^{p,q})$ 6.1.2
$\hat{F}_{(j)} := (W(\sigma_j)[-w], \exp(iN_{j+1})\hat{F}_{(j+1)})^{\wedge}$ 6.1.3
$\mathbf{0}_j \in \mathbf{C}^j$ 6.1.3
$\mathbf{i}_k \in \mathfrak{h}^k$ 6.1.3
$\hat{N}_j \in \mathfrak{g}_{\mathbf{R}}$ 6.1.3
$a_h \in \mathfrak{g}_{\mathbf{R}}$ $(h \in \mathbf{N}^n)$ 6.1.5
$b_h \in \mathfrak{g}_{\mathbf{R}}$ $(h \in \mathbf{N}^n)$ 6.1.5
$\mathfrak{g}^{\pm}_{\mathbf{R}} = \mathfrak{g}^{\pm,\mathbf{r}}_{\mathbf{R}} \subset \mathfrak{g}_{\mathbf{R}}$ 6.2.1
$c_h \in \mathfrak{g}^{-}_{\mathbf{R}}$ $(h \in \mathbf{N}^n)$ 6.2.1
$k_h \in \mathfrak{g}^{+}_{\mathbf{R}}$ $(h \in \mathbf{N}^n)$ 6.2.1
$V = V^0 \supsetneq V^1 \supsetneq \cdots \supsetneq V^n = \{0\}$, for a valuative submonoid V of A^* 6.3.1
$\nu_j : V^{j-1}_{\mathbf{Q}}/V^j_{\mathbf{Q}} \hookrightarrow \mathbf{R}$ 6.3.1

$\{0\} = A_0 \subsetneq A_1 \subsetneq \cdots \subsetneq A_n = A$, annihilators of the V^j 6.3.1
$S = \bigsqcup_{1\le j\le n} S_j$: a subdivision of index set S 6.3.3
$(\sum_{s\in S_j} a_s N_s)_{1\le j\le n} : A^* \to \mathbf{R}^n$ 6.3.3
$S_{\le j} := \bigsqcup_{k\le j} S_k \subset S$ 6.3.11
$S_{\ge j} := \bigsqcup_{k\ge j} S_k \subset S$ 6.3.11
$\sigma_{c,\alpha,\beta}$: a nilpotent cone in $\mathfrak{g}_{\mathbf{R}}$ associated with c, α, β 6.3.11
$\check{D}_j$: set of all $F' \in \check{D}$ such that $((N_s)_{s\in S_{\le j}}, F')$ generates a nilpotent orbit 6.4.1
$\tilde{\psi} : E^\sharp_{\sigma,\mathrm{val}} \to D_{\sigma,\mathrm{val}} \xrightarrow{\psi} D_{\mathrm{SL}(2)}$ 6.4.8

CHAPTER 7

$\tilde{E}_{\sigma,\mathrm{val}} := \mathrm{toric}_{\sigma,\mathrm{val}} \times_{\mathrm{toric}_\sigma} \tilde{E}_\sigma$ 7.1.2
$|\ | : \mathrm{toric}_{\sigma,\mathrm{val}} \to |\mathrm{toric}|_{\sigma,\mathrm{val}}$ 7.1.2
$\beta : D \to \mathbf{R}^m_{>0}$, continuous map s. t. $\beta(\tilde{\rho}(t)\cdot x) = t\beta(x)\ (x \in D,\ t \in \mathbf{R}^m_{>0})$ 7.2.10
$U(\tau) \subset \mathrm{toric}_\sigma$, for a face τ of σ 7.3.4
$|U|(\tau) \subset |\mathrm{toric}|_\sigma$, for a face τ of σ 7.3.4

CHAPTER 8

$\underline{\mathrm{LS}}_\Gamma$: sheaf of classes of logarithmic local systems of type $(H_0, \langle\ ,\ \rangle_0, \Gamma)$ 8.1.3
$\underline{\mathrm{LS}}_\Gamma \simeq R^1\tau_*(\Gamma)$ 8.1.3
$R^1\tau_*(\mathbf{Z}) \simeq M^{\mathrm{gp}}_X/\mathcal{O}^\times_X$ 8.1.4
$\underline{\mathrm{LS}}_\Gamma \simeq \Gamma \otimes (M^{\mathrm{gp}}_X/\mathcal{O}^\times_X)$ 8.1.5
$\Psi := (H_0, \langle\ ,\ \rangle_0, \Gamma, \Sigma)$ 8.1.6
$\Psi_\sigma := (H_0, \langle\ ,\ \rangle_0, \Gamma(\sigma)^{\mathrm{gp}}, \{\text{face of } \sigma\})$ 8.1.6
$\tilde{\mu}_y : (H_{\mathbf{Z},y}, \langle\ ,\ \rangle_y) \xrightarrow{\sim} (H_0, \langle\ ,\ \rangle_0)$, a lifting of μ_y 8.1.7
$(H_\sigma, \langle\ ,\ \rangle_\sigma, \mu_\sigma)$: canonical logarithmic local system of type Ψ_σ on toric_σ 8.1.7
$\underline{\mathrm{LS}}_\Psi$: sheaf of classes of logarithmic local systems of type Ψ 8.1.8
$C := \underline{\mathrm{PLH}}_\Phi : \overline{\mathcal{A}}_2(\log) \to (\mathrm{Sets})$, functor 8.2.1
$C_\sigma := \underline{\mathrm{PLH}}_{\Phi_\sigma} : \overline{\mathcal{A}}_2(\log) \to (\mathrm{Sets})$, functor 8.2.1
$\Phi_\sigma := (w, (h^{p,q})_{p,q\in\mathbf{Z}}, H_0, \langle\ ,\ \rangle_0, \Gamma(\sigma)^{\mathrm{gp}}, \{\text{face of } \sigma\})$ 8.2.1
$B_\sigma : \overline{\mathcal{A}}_2(\log) \to (\mathrm{Sets})$, functor 8.2.1
$(H_{\theta,\mathbf{Z}}, \langle\ ,\ \rangle_\theta, \mu_\theta)$: logarithmic local system induced by $\theta : X \to \mathrm{toric}_\sigma$ 8.2.1
$\check{B}_\sigma : \overline{\mathcal{A}}_2(\log) \to (\mathrm{Sets})$, functor 8.2.3
$F^{\log} := \mathcal{O}^{\log}_X \otimes_{\tau^{-1}(\mathcal{O}_X)} \tau^{-1}(F)$ 8.2.3
$\overline{C} := \underline{\mathrm{LS}}_\Psi : \overline{\mathcal{A}}_2(\log) \to (\mathrm{Sets})$, functor 8.2.4
$\overline{C}_\sigma := \underline{\mathrm{LS}}_{\Psi_\sigma} : \overline{\mathcal{A}}_2(\log) \to (\mathrm{Sets})$, functor 8.2.4
$(\bigsqcup_{\sigma\in\Sigma} C_\sigma)/\sim\ \xrightarrow{\sim} C$ 8.2.6
$(\bigsqcup_{\sigma\in\Sigma} \overline{C}_\sigma)/\sim\ \xrightarrow{\sim} \overline{C}$ 8.2.6
$X_{\mathrm{val}} \to \Gamma\backslash D_{\mathrm{val}}$, period map 8.4.1
$X^{\log}_{\mathrm{val}} \to \Gamma\backslash D^\sharp_{\mathrm{val}}$, period map 8.4.1
$X^{\log}_{\mathrm{val}} \to \Gamma\backslash D_{\mathrm{SL}(2)}$, period map 8.4.1

CHAPTER 9

$\mathcal{X}^\flat_{\mathrm{BS}}$: $\flat$-Borel-Serre space 9.1.1
$D^\flat_{\mathrm{BS}}$: $\flat$-Borel-Serre space 9.1.1
$D^\flat_{\mathrm{BS,val}}$: $\flat$-valuative Borel-Serre space 9.1.1
$D^\flat_{\mathrm{SL}(2)}$: $\flat$-SL(2)-orbit space 9.1.1
$D^\flat_{\mathrm{SL}(2),\leq 1} \subset D^\flat_{\mathrm{SL}(2)}$: $\flat$-SL(2)-orbit space of rank ≤ 1 9.1.1
D^* : space of Cattani-Kaplan 9.1.1
$\mathcal{X}^\flat_{\mathrm{BS}}(P) \subset \mathcal{X}^\flat_{\mathrm{BS}}$ 9.1.3
$D^\flat_{\mathrm{BS}}(P) \subset D^\flat_{\mathrm{BS}}$ 9.1.3
$D^\flat_{\mathrm{BS,val}}(P) \subset D^\flat_{\mathrm{BS,val}}$ 9.1.3
$D^\flat_{\mathrm{SL}(2),\leq 1}(W) \subset D^\flat_{\mathrm{SL}(2),\leq 1}$ 9.1.3
$D^\flat_{\mathrm{SL}(2),1} = D^\flat_{\mathrm{SL}(2),\leq 1} - D$ 9.1.5
$D_\Xi \to D^\flat_{\mathrm{SL}(2),\leq 1}$, period map 9.4.1

CHAPTER 10

$W^{(J)}$ for $J \subset \{1,\ldots,n\}$ 10.1.1
$\mathfrak{g}_{\mathbf{R}}(\chi) \subset \mathfrak{g}_{\mathbf{R}}$ 10.1.1
$X_J \subset X = X(\mathbf{G}^n_{\mathbf{m},\mathbf{R}})$ 10.1.1
$X_+ \subset X$ 10.1.1
$Y_0 := \mathbf{R}^n_{>0} \times \mathfrak{g}_{\mathbf{R}} \times K_{\mathbf{r}} \cdot \mathbf{r}$ 10.1.1
$Y_3 := (\mathbf{R}^n_{\geq 0})_{\mathrm{val}} \times \mathfrak{g}_{\mathbf{R}} \times K_{\mathbf{r}} \cdot \mathbf{r}$ 10.1.1
$Y_1 \subset \mathbf{R}^n_{\geq 0} \times \mathfrak{g}_{\mathbf{R}} \times K_{\mathbf{r}} \cdot \mathbf{r}$ 10.1.1
$Y_2 \subset Y_3$ 10.1.1
$\eta_0 : Y_0 \to D$, continuous map 10.1.1
$L \subset \mathfrak{g}_{\mathbf{R}}$: an $\mathbf{R}$-subspace 10.1.2
$Y_{j,L} \subset Y_j$ $(j = 0, 1, 2, 3)$ 10.1.2
$\eta_1 : Y_1 \to D_{\mathrm{SL}(2)}$ 10.1.3, 10.3.3
$\eta_2 : Y_2 \to D_{\mathrm{SL}(2),\mathrm{val}}$ 10.1.3, 10.3.2
$\eta_3 : Y_3 \to D_{\mathrm{BS,val}}$ 10.1.3, 10.3.1
U_j : an open set of $Y_{j,L}$ $(j = 1, 2, 3)$ 10.1.3
$p_J = [\rho_J, \varphi_J] \in D_{\mathrm{SL}(2)}(W^{(J)})$ for $J \subset \{1,\ldots,n\}$ 10.2.1
$U(p) := \cup_{J\subset\{1,\ldots,n\}} G_{W^{(J)},\mathbf{R}} \cdot p_J$, special open neighborhood of $p \in D_{\mathrm{SL}(2)}$ 10.2.1
$\mu : P \times (\bar{A}_P)_{\mathrm{val}} \to P$ 10.2.8
Ψ : a set of $\mathbf{Q}$-parabolic subgroups of $G_{\mathbf{R}}$ 10.2.10
$(\mathbf{R}^n_{\geq 0})_{\mathrm{val}}(P) \subset (\mathbf{R}^n_{\geq 0})_{\mathrm{val}}$: open subset 10.2.10
$(\mathbf{R}^n_{\geq 0})_{\mathrm{val}}(P) \to (\bar{A}_P)_{\mathrm{val}}$ 10.2.10
$D \times^{\mathbf{R}^n_{>0}} (\mathbf{R}^n_{\geq 0})_{\mathrm{val}}(P) \hookrightarrow D \times^{A_P} (\bar{A}_P)_{\mathrm{val}} = D_{\mathrm{BS,val}}(P)$ 10.2.10
$P_V \subset (G^\circ)_{\mathbf{R}}$: parabolic subgroup associated with a valuative submonoid V 10.2.11
$\kappa_P : D \times^{\mathbf{R}^n_{>0}} (\mathbf{R}^n_{\geq 0})_{\mathrm{val}}(P) \to D$ 10.2.12

$\kappa : D_{\mathrm{SL}(2)}(W) \to D$ 10.2.13
$\check{D}^W \subset \check{D}$ 10.2.15
$U(p)_{\mathrm{val}} \subset D_{\mathrm{SL}(2),\mathrm{val}}$: inverse image of $U(p) \subset D_{\mathrm{SL}(2)}$ 10.3.2
$C(X)$: cone over a topological space X 10.4.2
$\overline{U}(p) \subset D^{\flat}_{\mathrm{SL}(2),\leq 1}$: image of $U(p) \subset D_{\mathrm{SL}(2),\leq 1}$ 10.4.3

CHAPTER 11

$A \subset \mathrm{End}_{\mathbf{Q}}(H_{0,\mathbf{Q}})$: a semisimple $\mathbf{Q}$-algebra 11.1.1
$\check{D}^A \subset \check{D}$ 11.1.2
$D^A := D \cap \check{D}^A$ in $\check{D}$ 11.1.2
$G^A_R \subset G_{\mathbf{R}}$ $(R = \mathbf{Z}, \mathbf{Q}, \mathbf{R}, \mathbf{C})$ 11.1.2
$\mathfrak{g}^A_R \subset \mathfrak{g}_{\mathbf{R}}$ $(R = \mathbf{Q}, \mathbf{R}, \mathbf{C})$ 11.1.2
$\Gamma\backslash D^A$ 11.1.4
$D^A_\Sigma \subset D_\Sigma$ 11.1.5
$D^A_\sigma \subset D_\sigma$ 11.1.5
$D^{\sharp,A}_\Sigma \subset D^\sharp_\Sigma$ 11.1.5
$\check{E}^A_\sigma \subset \check{E}_\sigma$ 11.1.6
$E^A_\sigma := E_\sigma \cap \check{E}^A_\sigma$ 11.1.6
$E^{\sharp,A}_\sigma := E^\sharp_\sigma \cap \check{E}^A_\sigma$ in $\check{E}_\sigma$ 11.1.6
$\Gamma\backslash D^A_\Sigma$ 11.1.6
$\underline{\mathrm{PLH}}^A_\Phi : \overline{\mathcal{A}}_2(\log) \to (\mathrm{Sets})$, functor 11.3.1
$\underline{\mathrm{PLH}}^A_\Phi \simeq \mathrm{Mor}(\ \ ,\Gamma\backslash D^A_\Sigma)$ 11.3.1

CHAPTER 12

$\mathrm{Sp}(g)$: symplectic group 12.1.1
$\mathfrak{h}_g$: Siegel upper half space 12.1.3
$|\Delta^*| := |\Delta| - \{0\}$ 12.4.1

Index

www.ingramcontent.com/pod-product-compliance
Ingram Content Group UK Ltd.
Pitfield, Milton Keynes, MK11 3LW, UK
UKHW020238161224
452563UK00006B/218

9 780691 138220